21st International Laser Physics Workshop

AA008885

Journal of Physics: Conference Series Volume 414

Calgary, Canada
23-27 July 2012

ISBN: 978-1-62276-992-6
ISSN: 1742-6588

Printed from e-media with permission by:

Curran Associates, Inc.
57 Morehouse Lane
Red Hook, NY 12571

Some format issues inherent in the e-media version may also appear in this print version.

Printed by Curran Associates, Inc. (2013)

For permission requests, please contact the Institute of Physics
at the address below.

Institute of Physics
Dirac House, Temple Back
Bristol BS1 6BE UK

Phone: 44 1 17 929 7481
Fax: 44 1 17 920 0979

techtracking@iop.org

Additional copies of this publication are available from:

Curran Associates, Inc.
57 Morehouse Lane
Red Hook, NY 12571 USA
Phone: 845-758-0400
Fax: 845-758-2634
Email: curran@proceedings.com
Web: www.proceedings.com

TABLE OF CONTENTS

MODERN TRENDS IN LASER PHYSICS

012001 Coherent Manipulation of Quantum States in a Coupled Cavity-Atom System 1
Y Wang, J Wan, B Zou, J Zhang, Y Zhu

012002 Self-amplified Spontaneous Gamma-ray Emission 10
A A Zadernovsky

012003 Recent Progress on Compact and Portable Terahertz Source Based on Parametric Conversion from Dual-Frequency Solid-State Laser Pulses 19
Y J Ding, P Zhao, D Li

012004 Interplay of Light and Collisions in Photon Echo Formation 64
N N Rubtsova, V G Gol'Dort, V N Ishchenko, E B Khvorostov, S A Kochubei, V A Reshetov, I V Yevseyev

012005 Switching Between Superluminal to Subluminal Velocities and Tunable Slow Light in a Four Level Atomic System 76
S Davuluri, Y V Rostovtsev

012006 Room-temperature Single-photon Sources with Definite Circular and Linear Polarizations Based on Single-emitter Fluorescence in Liquid Crystal Hosts 86
J M Winkler, S G Lukishova, L J Bissell

012007 Generation of Polarization-Entangled Photons in a Standard Polarization-Maintaining Fiber 100
V O Lorenz, B Fang, O Cohen

012008 X-ray Emission from Relativistic Electrons in a Transverse High Intensity Optical Lattice 105
I A Andriyash, E D'Humières, V T Tikhonchuk, Ph Balcou

012009 The Applicability Range of Different Forms of the Radiation Friction Force in the Ultrarelativistic Electron Interaction with Electromagnetic Wave (Exact Solutions) 112
S S Bulanov, T Z Esirkepov, M Kando, J K Koga, S V Bulanov

012010 Photo-proton Emission from Halo Nuclei in the Presence of a Linearly Polarized Laser Field 122
A Dadi, C Müller

012011 Towards High Intensity Few-cycle Pulses Using Plasma Wakefield Self-compression Effect 131
A Pipahl, E A Anashkina, M Toncian, T Toncian, S A Skobelev, A V Bashinov, A A Gonoskov, O Willi, A V Kim

STRONG FIELD & ATTOSECOND PHYSICS

012012 Spin Effects in Strong-field Laser-electron Interactions 139
S Ahrens, T-O Müller, S Villalba-Chávez, H Bauke, C Müller

012013 New Mechanism for the Generation of Electron-positron Pairs in Laser-matter Interaction: Resonantly Enhanced Pair Production in a Molecular System 149
F Fillion-Gourdeau, E Lorin, A D Bandrauk

012014 Tailoring Isolated Attosecond Pulses Using Quantum Path Interferences 159
C Hernández-García, L Plaja

012015 Population Trapping in Xe Atoms 165
A Azarm, S M Sharifi, A Sridharan, S Hosseini, Q Q Wang, A M Popov, O V Tikhonova, E A Volkova, S L Chin

012016 Pair Production in a Plane Wave Due to a Thermal Background 175
B King, A D Piazza

BIOPHOTONICS

012017 Further Development of the Laser Tweezers Technique for Biomedical Applications 184
K Afanasiev, A Korobtsov, S Kotova, N Losevsky, A Mayorova, V Patlan, V Volostnikov

012018 Noninvasive Control of Dental Calculus Removal: Qualification of Two Fluorescence Methods 192
S Gonchukov, A Sukhinina, D Bakhmutov, T Biryukova

012019 Magnetic Nanoparticles As a Seed Layer for Growing ZnO Nanowires for Optical Applications 199
M S Alsalhi, M Atif, A A Ansari, K Khun, Z H Ibupoto, M Willander

012020 Experimental Verification of the Wave Model of the Genetic Coding 212
T Y Moguilnaya, V I Tomilin, A A Agibalov

012021 Effect of CO_2 Pulsed Laser Irradiation on Improving the Biocompatibility of a Polyethersulfone Film 224
S Jelvani, H Pazokian, S M Farisar
012022 Laser Surface Modification of Polyethersulfone Films: Effect of Laser Wavelength on Biocompatibility 233
H Pazokian, M Mollabashi, S Jelvani, J Barzin
012023 Monitoring Toxicants by Stimulated Mandelshtam-brillouin Scattering (SMBS) in a Turbulent Water Flow 243
T Y Moguilnaya, P V Bobkov, V A Tomilin
012024 Potentiometric Urea Biosensor Utilizing Nanobiocomposite of Chitosan-iron Oxide Magnetic Nanoparticles 255
A Ali, M S Alsalhi, M Atif, A A Ansari, M Q Israr, J R Sadaf, E Ahmed, O Nur, M Willander
012025 In Vitro Studies of Photosensitizer Fluorescence Changes on Singlet Oxygen Mediated Photobleaching 266
M Atif

PHYSICS OF LASERS

012026 Semiconductor A3B5 Nanostructures for Infrared Femtosecond Lasers 277
N N Rubtsova, S A Kochubei, A A Kovalyov, V V Preobrazhenskii, M A Putyato, B R Semyagin, T S Shamirzaev, G M Borisov, V E Kisel', N V Kuleshov, S V Kuril'Chik, O V Buganov, S A Tikhomirov
012027 Precise Synchronization of Qcw Pumped Active-passive Mode Locked Picosecond Lasers 288
A A Karnaukhov, V B Morozov, A N Olenin, D V Yakovlev

PHYSICS OF COLD TRAPPED ATOMS

012028 Superfluid to Normal Phase Transition in Strongly Correlated Bosons in Two Dimensions 292
J Carrasquilla, M Rigol
012029 Dynamics of the Mott Insulator to Superfluid Quantum Phase Transition in the Truncated Wigner Approximation 301
M Tylutki, J Dziarmaga, W H Zurek
012030 Mixing Dipolar Condensates: A New Opportunity for Enhancing Superfluid Pairing in a Spin-polarized Fermi Gas 315
B Kain, H Y Ling
012031 Quasicondensation Reexamined 323
P Bienias, K Pawlowski, M Gajda, K Rzazewski
012032 Effective Time-reversal Via Periodic Shaking 333
C Weiss
012033 Expansion After a Geometric Quench of an Atomic Polarized Attractive Fermi Gas in One Dimension 343
C J Bolech, F Heidrich-Meisner, S Langer, I P McCulloch, G Orso, M Rigol
012034 Vortex Phases of Rotating Superfluids 351
M Correggi, F Pinsker, N Rougerie, J Yngvason
012035 Composite Vortices in Two-component Bose-Einstein Condensates 358
J Tempere, S Ceuppens, E Vermeyen
012036 Roton Instability and Phonon Collapse of Two-dimensional Tilted Dipoles 365
A K Fedorov, I L Kurbakov, Y E Lozovik

QUANTUM INFORMATION SCIENCE

012037 Ghost Imaging of Space Objects 372
D V Strekalov, B I Erkmen, N Yu
012038 Ideal Quantum Reading of Optical Memories 390
M Dall'Arno, A Bisio, G M D'Ariano
012039 Quantum Minimax Receiver for Ternary Coherent State Signal in the Presence of Thermal Noise 402
K Kato
012040 Quantum Tomograms and Their Application in Quantum Information Science 408
A K Fedorov, S O Yurchenko

012041 The Digital Randomness of Black-body Radiation 417
S Varró
Author Index

Journal of Physics: Conference Series **414** (2013) 011001 doi:10.1088/1742-6596/414/1/011001

Peer review statement

All papers published in this volume of *Journal of Physics: Conference Series* have been peer reviewed through processes administered by the proceedings Editors. Reviews were conducted by expert referees to the professional and scientific standards expected of a proceedings journal published by IOP Publishing.

Coherent manipulation of quantum states in a coupled cavity-atom system

Yanhua Wang[1,2], Jinyin Wan[1,3], Bichen Zou[1], Jiepeng Zhang[4], and Yifu Zhu[1]

[1]*Department Physics, Florida International University, Miami, FL, USA*
[2]*College of Physics and Electronics, Shanxi University, Taiyuan, 030006, China*
[3]*Shanghai Institute of Optics and Fine Mechanics, Chinese Academy of Sciences, Shanghai, China*
[4]*Physics Division, Los Alamos National Laboratory, Los Alamos, NM, USA*

Abstract

We study atomic coherence and interference in four-level atoms confined in an optical cavity and explores the interplay between cavity QED and electromagnetically induced transparency (EIT). The destructive interference can be induced in the coupled cavity-atom system with a free-space control laser tuned to the normal mode resonance and leads to suppression of the normal mode excitation. Then by adding a pump laser coupled to the four-level atoms from free space, the control-laser induced destructive interference can be reversed and the normal mode excitation is restored. When the free-space control laser is tuned to the atomic resonance and forms a Λ-type EIT configuration with the cavity-atom system, EIT is manifested as a narrow transmission peak of a weak probe laser coupled into the cavity mode. With the free-space pump laser driving the cavity-confined atoms in a four-level configuration, the narrow transmission peak of the cavity EIT can be split into two peaks and the dressed intra-cavity dark states are created analogous to the dressed states in free space. We report experimental studies of such coherently coupled cavity-atom system realized with cold Rb atoms confined in an optical cavity and discuss possible applications in quantum nonlinear optics and quantum information science.

1. Introduction

Cavity QED has been a subject of many recent studies and has a variety of applications in quantum physics and quantum electronics [1]. The basic cavity QED system consists of a single two-level atom coupled to a single cavity mode [2]. The composite atom-cavity system exhibits a double-peaked transmission spectrum representing the two normal modes of the first excited eigenstates. When the system is resonantly coupled, the two normal modes are separated in frequency by 2g ($g = \mu\sqrt{\omega_a / 2\hbar\varepsilon_0 V}$ is the atom-cavity coupling coefficient), commonly referred to as the vacuum Rabi splitting [1-2]. Observation of the two normal modes in the optical wavelength range requires a cavity-atom system with a g value greater or comparable with the decay rates of the cavity and the atomic system, which can be realized in a high finesse cavity with a small mode volume [2]. On the other hand, if N two-level atoms collectively interact with the cavity mode, the coupling coefficient becomes $G = \sqrt{N}g$ and the vacuum Rabi splitting of the normal modes for the collectively coupled atom-cavity system becomes 2G and may then be observed in a cavity with a moderate mode volume and finesse [3-5].

Electromagnetically induced transparency (EIT) can be created in various atomic systems via coherent interactions of radiations fields and atoms [6-7]. EIT has been shown to be important for various applications in quantum optics and nonlinear optics [8-10]. Recent studies of EIT and related phenomena have been extended to coherent coupled atom-cavity systems [11-14]. It has been shown that in a coherently coupled cavity and multi-atom system, the interplay of the collective coupling of the atoms and the cavity mode, and the atomic coherence and interference manifested by EIT may lead to interesting linear and nonlinear optical phenomena [15-20].

Here we present studies of an atom-cavity system consisting of N four-level atoms confined in an optical cavity and coherently coupled from free space by two laser fields: one acts as a coupling laser and forms a Λ-type standard EIT configuration with the cavity mode; another acts as a pump laser and forms a N-type coupled atomic system with the control laser and the cavity mode. By varying the frequency detuning of the control laser, several distinct phenomena are manifested in the coupled cavity-atom system. When the control laser is tuned to the atomic resonance and the pump laser is absent, the cavity-atom system exhibit cavity EIT [11-15]. If the pump laser is turned on and tuned to the atomic resonance, the transmission peak of the cavity EIT can be split into two peaks, indicating generation of two dressed intra-cavity dark states [21-22]. On the other hand, when the control laser is tuned to the resonance of one of the two normal modes and the pump laser is off, the destructive interference is induced for the normal mode excitation, which is manifested as a narrow dip in the transmission spectrum of the probe laser coupled into the cavity mode [23-24]. When the pump laser is on and tuned to the atomic resonance, the destructive interference induced by the control laser is reversed and the light transmission of the probe laser at the resonance of the normal mode is restored. Such manipulation of the coupled cavity-atom system shows a simple way to control the quantum states of the cavity QED by the laser induced coherence and interference and may be useful in a variety of applications in nonlinear optics, and quantum information science.

2. Theoretical analysis

We consider a composite atom-cavity system that consists of a single mode cavity confining N identical four-level atoms driven by a control laser and a pump laser from free space as shown in Fig. 1. The cavity mode couples the atomic transition |1>-|3> and $\Delta_c = \nu_c - \nu_{13}$ is the cavity-atom detuning. The classical control laser drives the atomic transition |2>-|3> with Rabi frequency 2Ω, and the classical pump laser drives the atomic transition |2>-|4> with Rabi frequency $2\Omega_S$. $\Delta = \nu - \nu_{23}$ is the control frequency detuning and $\Delta_s = \nu_s - \nu_{24}$ is the pump laser detuning. We calculate the light transmission of a weak probe laser coupled into the cavity mode as the probe frequency detuning $\Delta_p = \nu_p - \nu_{13}$ is scanned across the atomic transition frequency ν_{13}. The interaction Hamiltonian for the cavity-atom system is

$$H = -\hbar(\sum_{i=1}^{N} \Omega \hat{\sigma}_{32}^{(i)} + \sum_{i=1}^{N} \Omega_s \hat{\sigma}_{42}^{(i)} + \sum_{i=1}^{N} g\hat{a}\hat{\sigma}_{31}^{(i)}) + H.C. , \quad (1)$$

where $\hat{\sigma}_{lm}^{(i)}$ (l, m=1-4) is the atomic operator for the ith atom and $\hat{a}$ is the annihilation operator of the cavity photons. The resulting operator equations of motion for the intra-cavity light field (two-sided cavity, one input) is given by [25-26]

$$\dot{\hat{a}} = -\frac{i}{\hbar}[\hat{a}, H] - \frac{\kappa_1 + \kappa_2}{2}\hat{a} + \sqrt{\kappa_1}\hat{a}_p^{in}, \quad (2)$$

where $\hat{a}_p^{in}$ is the input probe field, and κ_1 and κ_2 are the loss rates of the cavity mirrors. The equation of the motion for the expectation value of the intra-cavity probe field is [20]

$$\dot{a} = -((\kappa_1 + \kappa_2)/2 - i\Delta_c)a + \sum_{i=1}^{N} ig\sigma_{31}^{(i)} + \sqrt{\kappa_1}a_p^{in}, \quad (3)$$

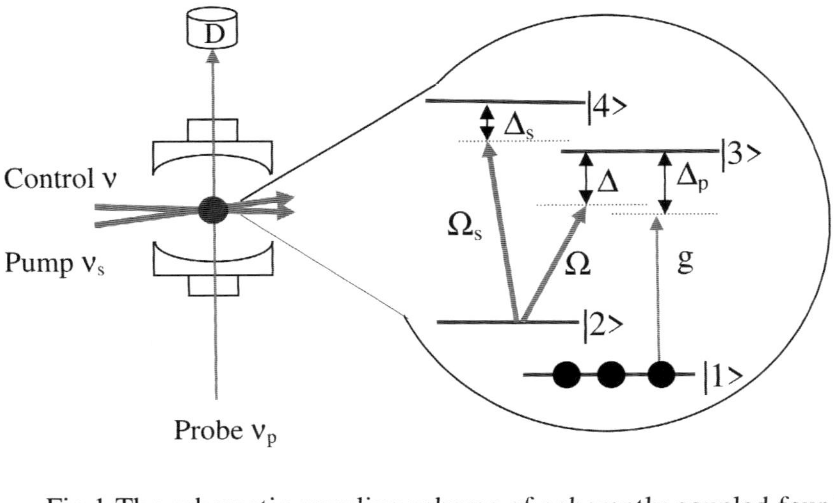

Fig.1 The schematic coupling scheme of coherently coupled four-level atoms in a cavity. A control laser drives |2> - |3> transition with Rabi frequency 2Ω and a pump laser couples |2> - |4> transition with Rabi frequency $2\Omega_S$. Δ (Δ_s) is the control (pump) detuning. The cavity mode is coupled to the atomic transition |1> - |3> with the collective coupling coefficient $\sqrt{N}g$ ($g = \mu\sqrt{\omega_a / 2\hbar\varepsilon_0 V}$) ($\Delta_c$ is the cavity-atom detuning). A weak probe laser is coupled into the cavity and Δ_p is its frequency detuning from the atomic transition.

For a symmetric cavity (as in our experiment), $\kappa_1 = \kappa_2 = \kappa$. Under the condition of g<<Ω, the atomic population is concentrated in |1> and the steady-state solution of the intra-cavity probe field is given by

$$a = \frac{\sqrt{\kappa} a_p^{in}}{\kappa - i\Delta_c - i\chi}, \qquad (4)$$

where χ is the atomic susceptibity given by

$$\chi = \frac{ig^2 N}{\Gamma_3 - i\Delta_p + \dfrac{\Omega^2 (\Gamma_4 + \gamma_{12} - i(\Delta_s + \Delta_p - \Delta))}{(\gamma_{12} - i(\Delta_p - \Delta))(\Gamma_4 + \gamma_{12} - i(\Delta_s + \Delta_p - \Delta)) + \Omega_s^2}}. \qquad (5)$$

Here Γ_3 (Γ_4) is the decay rate of the excited state |3> (|4>) and γ_{12} is the decoherence rate of the ground states |1> and |2>. The transmitted probe field is then given by $a_p^{out} = \sqrt{\kappa} a$.

First, we consider manipulation of cavity EIT in the four-level system. Fig. 2 plot the normalized transmission intensity $\frac{I_t}{I_{in}} = \frac{|a_p^{out}|^2}{|a_p^{in}|^2}$ of the probe laser coupled into the cavity mode

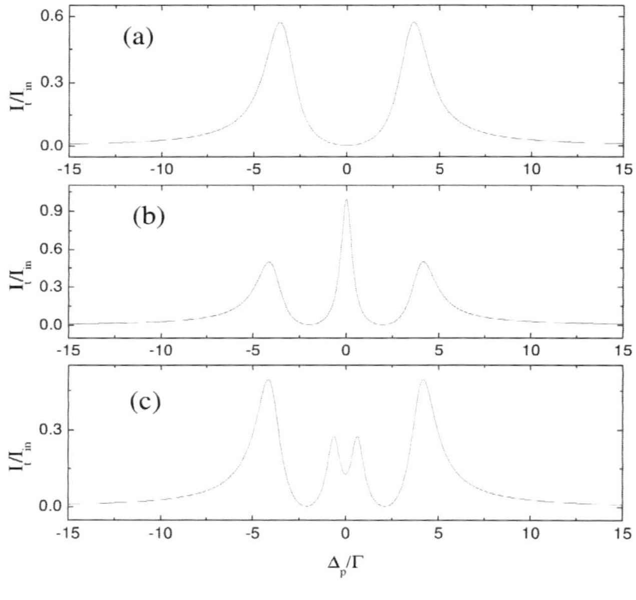

Fig. 2 The transmission intensity of the probe laser coupled into the cavity-atom system versus the probe detuning Δ_p/Γ. (a) Without the control laser (Ω=0) and the pump laser (Ω_s=0). (b) With the control laser at Δ=0 (Ω=2Γ), but without the pump laser (Ω_s=0), (c) With both the control laser (Ω=2Γ) and the pump laser at Δ_s=0 (Ω_s=Γ). The other parameters are $\Gamma_3 = \Gamma_4 = \Gamma$, $g\sqrt{N} = 3.5\Gamma$, $\kappa = 1.5\Gamma$, $\gamma_{12} = 0.001\Gamma$, and Δ_c=0.

versus the probe frequency detuning Δ_p/Γ. Fig. 2(a) depicts the probe transmission spectrum without both the control laser and the pump laser, in which the two transmission peaks represent the two normal modes separated in frequency by the vacuum Rabi frequency $2G = 2\sqrt{N}g$ [3-5]. When the control laser is present and tuned to the atomic resonance at $\Delta=0$ (without the pump laser), cavity EIT is created in the cavity-atom system, and the probe transmission spectrum exhibits three peaks as shown in Fig. 2(b): the central peak at $\Delta p=0$ represents the cavity EIT, or intra-cavity dark state [11-13] and the two sideband peaks represent the normal modes of the coupled cavity-atom system [3-5], which are modified by the free-space control laser [21]. Fig. 2(c) plots the probe transmission spectrum when a resonant pump laser ($\Delta_s=0$) and a Rabi frequency $\Omega_s=\Gamma$ is present. It shows that the central EIT peak is split into two peaks and the peak separation is approximately equal to $2\Omega_s$. This is analogous to the dressed states of a laser coupled two-level system in free space [27]. Next we consider the effect of the control-laser-induced interference on the normal mode excitation of the multi-atom cavity QED system. Fig. 3(a) plots the transmitted intensity of the probe field $\frac{I_t}{I_{in}}$ versus the probe frequency detuning

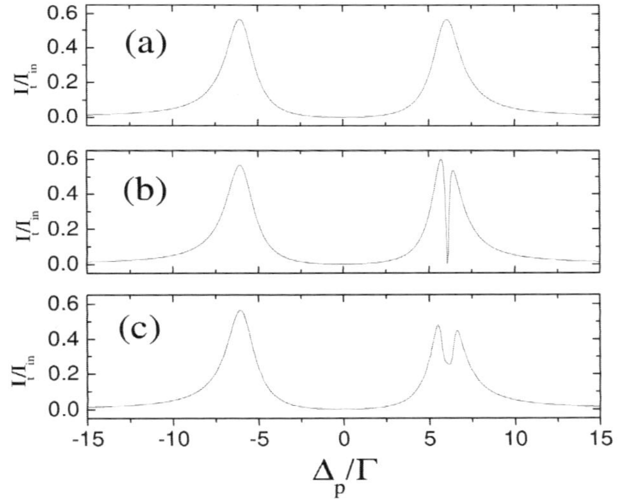

Fig. 3 The normalized transmission intensity I_t/I_{in} of the probe laser (I_{in} is the probe input intensity) through the cavity versus the probe detuning Δ_p/Γ. (a) The probe transmission spectrum without the control laser ($\Omega=0$) and the pump laser ($\Omega_s=0$). The two transmission peaks represents the two normal modes of the cavity QED system. (b) The probe transmission spectrum with the control laser ($\Omega=2\Gamma$), but without the pump laser ($\Omega_s=0$), (c) The probe transmission spectrum with both the control laser ($\Omega=0.5\Gamma$) and the pump laser ($\Omega_s=0.5\Gamma$). The relevant parameters are $\Gamma_3=\Gamma_4=\Gamma$, $g\sqrt{N}=6\Gamma$, $\kappa=1.5\Gamma$, $\gamma_{12}=0.001\Gamma$, and $\Delta_c=0$.

$\Delta p/\Gamma_3$ without the control laser and the pump laser. The probe spectrum exhibits two transmission peaks representing two normal modes separated by the vacuum Rabi frequency 2G. Fig. 3(b) plots the probe transmission spectrum with the control laser present (Ω=0.5Γ) and tuned to the normal mode resonance at $\Delta=\sqrt{N}g$, but without pump laser (Ω_s=0). Fig. 3(b) shows that the control laser induces the destructive interference and suppresses the normal mode excitation (the probe transmission at the normal mode resonance (Δ_p=G) is blocked) [23]. Fig. 3(c) plots the probe transmission when the pump laser is present (Ω_s=0.5Γ) and tuned to the atomic resonance (Δ_s=0). The spectrum shows that the pump laser reverses the destructive interference and enables the excitation of the normal mode and therefore the transmission of the probe light as seen by the increased transmission at the normal mode resonance Δ_p=G.

3. Experimental results

We carried out experimental studies of the coherently coupled cavity and four-level atoms system with cold ^{85}Rb atoms confined in a near confocal cavity consisting of two mirrors of 5 cm curvature with a mirror separation ~ 5 cm. The empty cavity finesse is measured to be ~ 150. A detailed description of our experimental set up can be found in elsewhere [13,21] and is only briefly outlined here. Three extended-cavity diode lasers were used as the control laser that drives the ^{85}Rb D_1 transition F=3 to F'=3, the pump laser that couples the ^{85}Rb D_2 transition F=3 to F'=4, and the probe lasers that couples the ^{85}Rb D_1 transition F=2 to F'=3. The circularly-polarized control laser and the linearly polarized pump laser were directed to overlap the cold atoms from the open side of the cavity and propagated in the directions nearly perpendicular to the cavity axis. The probe laser was linearly polarized parallel to the propagating direction of the control laser and then after sufficient attenuation, was coupled into the cavity. The transmitted probe light was collected by a photon counter (PerkinElmer SPCM-AQR-16-FC). The probe transmission spectrum was measured as the probe laser frequency was scanned across the ^{85}Rb D_1 transition F=2 to F'=3.

The experiment was run sequentially with a repetition rate of 10 Hz. All lasers were turned on or off by acousto-optic modulators (AOM) according to the time sequence described below. For each period of 100 ms, ~98 ms was used for cooling and trapping of the ^{85}Rb atoms, during which the trapping laser and the repump laser were turned on by two AOMs while the coupling laser, the pump laser, and the probe laser were off. The time for the data collection lasted ~ 2 ms, during which the repump laser was turned off first, and then after a delay of ~0.2 ms, the trapping laser was turned off (the current to the anti-Helmholtz coils of the MOT was always kept on), and the coupling laser, the pump laser, and the probe laser were turned on. After the coupling laser, the pump laser, and the probe laser were turned on by the AOMs for 0.2 ms, the probe laser frequency was scanned across the ^{85}Rb D_1 F=2→F=3 transitions and the probe light transmitted through the cavity was then recorded versus the probe frequency detuning.

Fig. 4 plots the measured cavity transmission intensity of the probe laser I_{out}/I_{in} (I_{in} is the resonant transmission of the probe light through an empty cavity) versus the probe frequency detuning Δ_p. The empty cavity is tuned to the atomic transition frequency $\Delta_c=\nu_c-\nu_{13}=0$ and both the control laser and the pump laser are on resonance ($\Delta\approx0$ and $\Delta_s\approx0$). The decay linewidth of the Rb transitions are Γ_3=5.7 MHz and Γ_4=5.9 MHz, respectively. Other parameters are $g\sqrt{N}=20$MHz, κ=14 MHz, Ω=12 MHz, γ_{12}=0.01Γ, and $\Delta_c=\Delta=\Delta_s=0$. The measured spectrum was the average of 50 scans. Fig. 4(a) plots the probe transmission spectrum without both the control laser and the pump laser. The two transmission peaks represent the two normal modes

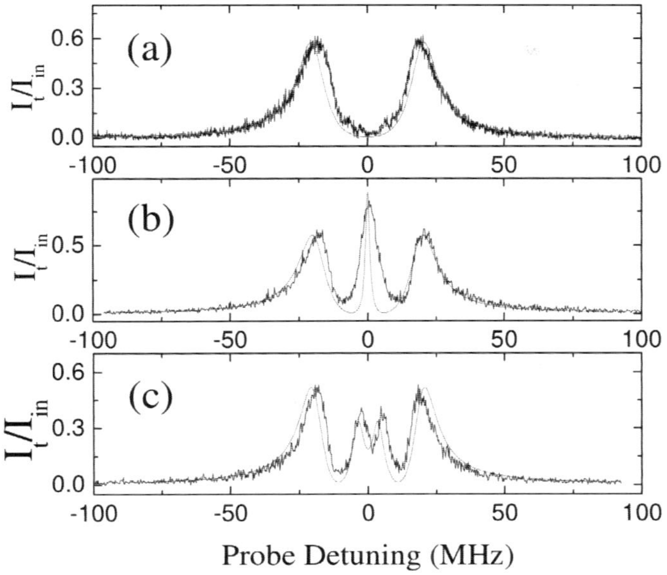

Fig. 4 The cavity transmission intensity I_t/I_{in} versus the probe detuning Δ_p. Black lines are experimental data and red lines are calculations. (a) Without both the control laser and the pump laser. (b) With the control laser ($\Omega \approx 12$ MHz) and without the pump laser. (c) With both the control laser ($\Omega \approx 10$ MHz) and the pump laser $\Omega_s \approx 8$ MHz.

separated in frequency by the vacuum Rabi frequency ($2g\sqrt{N} \approx 40MHz$). Fig. 4(b) plots the probe transmission spectrum with the control laser, but without the pump laser, and exhibits the three-peaked cavity EIT spectrum: two sideband peaks located at $\Delta_p = \pm\sqrt{\Omega^2 + g^2 N}$ represent the normal modes of the coupled cavity-atom system, and a central peak at Δ_p=0 is manifested by EIT (the intra-cavity dark state) [11,13]. Fig. 4(c) plots the probe transmission spectrum with both the control laser and the pump laser, in which the cavity EIT peak is split into two peaks. We observed that the splitting is nearly equal to $2\Omega_s$ and appear at sufficiently high pump intensities (Ω_3 >3 MHz). We are in the process of carrying out experimental studies of manipulating the excitation of the normal modes by the quantum interference in the cavity-coupled four-level atomic system. We observed destructive interference in the normal mode excitation induced by the control laser when it is tuned to the normal mode resonance (detuned from the atomic resonance by Δ=G). The experimental results are plotted in Fig. 5. Fig. 5(a) plots the probe transmission spectrum without the control laser. The measured vacuum Rabi splitting is $2G = 2g\sqrt{N} \approx 85$MHz (the experiment was done with the number of cold atoms about 4 times that of the experiment shown in Fig. 4). Fig. 5(b) plots the probe transmission spectrum when the control laser is present (Δ=42 MHz and Ω=12 MHz), but without the pump laser. It shows that the normal mode excitation at the resonance Δ_p=G is suppressed and a dip appears in the probe transmission. We plan to add the pump laser to form the coherently coupled four-level atom and cavity system and in subsequent experiments, expect to observe the phenomenon of reversed destructive interference and restoration of the normal mode excitation predicted in the theoretical calculation of Fig. 3(c).

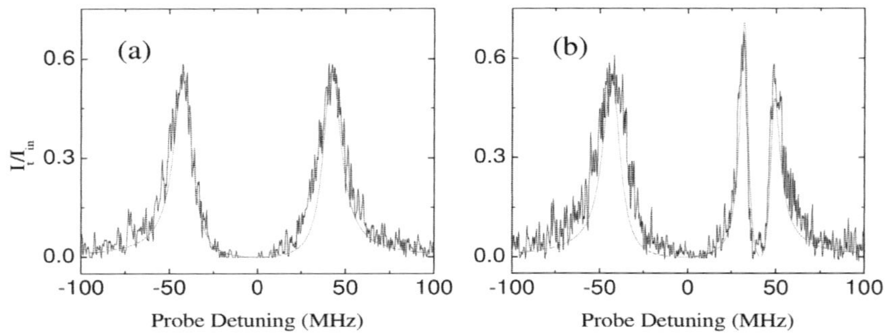

Fig. 5 The probe transmission intensity versus the probe detuning Δ_p. Black lines are experimental data and red lines are calculations. (a) Without both the control laser and the pump laser. (b) With the control laser ($\Omega \approx 12$ MHz) and without the pump laser. The parameters are $2G = 2g\sqrt{N} = 85$ $\kappa=14$ MHz, $\Omega=12$ MHz, $\gamma_{12}=0.01\Gamma$, $\Delta = g\sqrt{N}$, and $\Delta_c = \Delta_s = 0$

4. Conclusion

In conclusion, we have shown that the atomic coherence and interference induced by laser fields in four-level atoms can be used to manipulate and control quantum states of the coupled cavity and atom system. Specifically, the destructive interference can be induced by a free-space control laser in the excitation of the normal mode of the cavity-atom system and can be reversed by a free-space pump laser. Under appropriate conditions, the same control laser and the pump laser can be also used to manipulate the cavity EIT and create the dressed intra-cavity dark states. The coherent coupled cavity-atom system and the interference phenomena reported may be useful to a variety of applications in nonlinear optics and quantum physics. For example, the coherently coupled cavity-atom system can be used to explore the light-control-light phenomena such as all optical switching and cross-phase modulations at ultra-low light intensities [17,24], which may be useful for quantum gates applications. The cavity-atom system can be also used to realize the broadband cavity EIT and explore possibility of applications for multi-channel and multi-color light memories [10, 28]. Furthermore, our recent study shows that the transmitted photons and reflected photons from the coherently coupled cavity-atom system are highly correlated, which may render the system useful for studies of the photon correlation and quantum entanglement for the atoms and light fields.

Acknowledgement

This work is supported by the National Science Foundation under Grant No. 0757984.

References

1. *Cavity Quantum Electrodynamics*, edited by P. R. Berman (Academic, San Diego, 1994).
2. A. Boca, R. Miller, K. M. Birnbaum, A. D. Boozer, J. McKeever, and H. J. Kimble, Phys. Rev. Lett. **2004**, 93, 233603.
3. G. S. Agarwal, Phys. Rev. Lett. **1984**, 53, 1732-1735.
4. M. G. Raizen, R. J. Thompson, R. J. Brecha, H. J. Kimble, and H. J. Carmichael, Phys. Rev. Lett. **1989**, 63, 240 - 243.
5. Y. Zhu, D. J. Gauthier, S. E. Morin, Q. Wu, H. J. Carmichael, and T. W. Mossberg, Phys. Rev. Lett. 1990, **64**, 2499-2502.
6. S. E. Harris, Phys. Today **1997**, 50, 36-42.
7. E. Arimondo, in *Progress in Optics*, E. Wolf ed., Elsevier, Amsterdam, 1996; Vol. 31, pp.257-354.
8. M Fleischhauer, A. Imamoglu, & J. P. Marangos, Rev. Mod. Phys. **2005**, 77, 633-673.
9. M. D. Lukin, Rev. Mod. Phys. **2003**, 75, 457-472.
10. A. I. Lvovsky, B. C. Sanders, W. Tittel, Nature Photonics **2009**, 3, 706-714.
11. M. D. Lukin, M. Fleischhauer, M. O. Scully, and V. L. Velichansky, Opt. Lett. **1998**, 23, 295-297.
12. H. Wang, D. J. Goorskey, W. H. Burkett, and M. Xiao, Opt. Lett. 2000, 25, 1732-1734.
13. G. Hernandez, J. Zhang, and Y. Zhu, Phys. Rev. A **2007**, 76, 053814.
14. H. Wu, J. Gea-Banacloche, and M. Xiao, Phys. Rev. Lett. **2008**, 100, 173602.
15. M. Mücke, E. Figueroa, J. Bochmann, C. Hahn, K. Murr, S. Ritter, C. J. Villas-Boas, and G. Rempe, Nature **2010**, 465, 755-758.
16. L. Slodicka, G. Hétet, S. Gerber, M. Hennrich, and R. Blatt, Phys. Rev. Lett. **2010**, 105,153604.
17. Y. Zhu, Opt. Lett. **2010**, 35, 303-305.
18. J. Zhang, G. Hernandez, and Y. Zhu, Opt. Lett. 2008, **33**, 200846-200848.
19. J. Sheng, H. Wu, M. Mumba, J. Gea-Banacloche, and M. Xiao, Phys. Rev. A **2011**, 83, 023829.
20. M. Albert, A Dantan, and M. Drewsen, Nature Photonics **2011**, 5, 633-635.
21. G. Hernandez, J. Zhang, and Y. Zhu, Optics Express **2009**, 17, 4798-4805.
22. Y. Wang, J. Zhang, and Y. Zhu, "Observation of dressed intracavity darks states" Phys. Rev. A **2012**, 85, 013814.
23. J. Zhang, G. Hernandez, and Y. Zhu, Optics Express **2008**, 16, 7860-7868.
24. X. Wei, J. Zhang, and Y. Zhu, Phys. Rev. A, **2010**, 82, 033808.
25. "*Quantum Noise*", C. W. Gardiner (Springer, Berlin, Heidelberg, 1991).
26. "*Quantum Optics*", D. F. Walls and G. J. Milburn (Springer-Verlag, Berlin, Heidelberg, 1994).
27. C. Cohen-Tannoudji and S. Reynaud, J. Phys. B **1977**, **10**, 345-356.
28. X. Wei, Y. Wang, J. Zhang, and Y. Zhu, Phys. Rev. A **2011**, 84, 045806.

Journal of Physics: Conference Series **414** (2013) 012002 doi:10.1088/1742-6596/414/1/012002

Self-amplified spontaneous gamma-ray emission

A.A. Zadernovsky

Moscow State Technical University MIREA, Department of Physics,
78 Vernadsky Ave., 119454 Moscow, RUSSIA

e-mail: zadernovsky@mirea.ru

Abstract. We discuss a possibility for obtaining the self-amplified spontaneous emission of gamma radiation in an active medium with hidden population inversion of nuclear states. Hidden inversion appears in cooled nuclear ensembles due to assistance of the nuclear recoil in emission and absorption of gamma quanta. Establishing a hidden inversion allows an amplification of gamma radiation to occur in a particular frequency range without excess of the number of excited nuclei over the number of unexcited nuclei. We formulate a set of requirements to the pumping radiation and to the amplifying nuclear medium crucial to the development of gamma-ray lasing.

1. Introduction

In optical lasers, photon flux is amplified due to stimulated emission of radiation (this notion was first introduced by A. Einstein in 1916). Amplification is realized in an active medium with population inversion of quantum levels, if an increase in the number of photons due to stimulated emission exceeds the loss of the photons. The question arises as to whether this approach can be applied to create a gamma-ray laser on radiative nuclear transitions owing to the universal character of the Einstein emission laws.

For the photon flux propagating in an active medium, the gain coefficient is represented as

$$g = \sigma_{st}(n_2 - n_1 S) - \chi n, \tag{1}$$

where σ_{st} and χ are the cross sections for the stimulated emission of photons and for the nonresonant photon losses, n_2 and n_1 are the concentrations of emitters in the excited state (at the upper quantum level) and in the unexcited state (at the lower quantum level), n is the total concentration of emitters, $S=(2J_2+1)/(2J_1+1)$ is the degeneracy factor of quantum transition (J_2 and J_1 are the angular momenta of the upper and lower quantum levels, respectively).

At the center of emission line with wavelength λ_0 , the cross section for stimulated gamma radiation is

$$\sigma_{st} = \frac{\lambda_0^2}{2\pi}\frac{\Delta\omega_{rad}}{\Delta\omega_{tot}} = \frac{\lambda_0^2}{2\pi}\frac{1}{\Delta\omega_{tot}\tau(1+\alpha)}, \tag{2}$$

where $\Delta\omega_{tot}$ is the total width of the emission line, which takes into account all types of excessive homogeneous and inhomogeneous broadening, $\Delta\omega_{rad}$ - radiative width of the quantum transition (which is reciprocal value of the radiative lifetime τ_{sp} in excited state), τ is the lifetime in excited state, α is the transition quantum yield (for example, the internal electron conversion coefficient of the radiative nuclear transition).

If we pass from optics to gamma-rays, the wavelength λ_0 decreases by 4÷5 orders of magnitudes, the ratio of the radiative width to the total width of emission line, $\beta = \Delta\omega_{rad}/\Delta\omega_{tot}$, also decreases by 4÷5 orders of magnitudes (since the total width of emission line $\Delta\omega_{tot}$, being proportional to the nuclear transition energy, increases by 4÷5 orders of magnitudes), so that the cross section for stimulated emission σ_{st} decreases at least by 12÷15 orders of magnitudes. However, an amplification of stimulated gamma radiation is still possible. One can see that the amplification takes place if $g > 0$ and, in the case of the strong population inversion $n_2 >> n_1$ (we also assume $n=n_1+n_2$), this condition is reduced to the requirement that the cross section for stimulated emission σ_{st} should exceeds the cross-section for non-resonant photon losses χ.

Assuming the internal electron conversion coefficient α=10, the wavelength λ_0=1 Å (photon energy of about 10 keV) and taking into account that the nonresonant losses of the emitted photons are predominantly related to the photoeffect on the K–shell of atomic electrons with the corresponding cross section $\chi=10^{-21}$ cm^2 (for nuclei with intermediate atomic numbers Z) [1], we can transform the above amplification condition $\sigma_{st} > \chi$ to the following form $\Delta\omega_{tot}\tau < 10^3$.

The main contribution to the total width of emission line is provided by Doppler broadening. In the gamma range this broadening is relatively large, since the Doppler width $\Delta\omega_D$ is proportional to the energy of the nuclear transition. Using the estimate $\Delta\omega_D = 1.08\times10^{12} E_0\sqrt{T/A}$ (where the Doppler width $\Delta\omega_D$, energy E_0 of the nuclear transition, and temperature T are measured in inverse seconds, kiloelectronvolts, and kelvins, respectively), we obtain $\Delta\omega_{tot} = \Delta\omega_D \approx 10^{13}$ s^{-1} for the nuclei with mass number A=100 at T=300 K. From the amplification condition $\Delta\omega_{tot}\tau < 10^3$ we conclude that the nuclear lifetime in the excited state must be less than 0.1 ns.

A possibility for the application of such short-lived nuclear states is restricted by the power of pumping that induces population inversion of the nuclear levels. For example, if the upper nuclear level is populated due to neutron capture, the estimates of the pump neutron flux density needed for creation of population inversion in an ensemble of nuclei with the excited state lifetime of less than 0.1 ns yield extremely high values that can be only realized in nuclear explosions [2, 3]. Experimental results demonstrating the general process of excitation of nuclei by plasma radiation [4-6] and the first experimental evidence for x-ray driven nuclear population inversion [7] reinforce the harsh demands on pumping to reach the threshold for amplification of gamma-rays.

At first glance, a way to overcome this problem is the use of Mössbauer effect on active nuclei in crystal lattice. This approach was proposed at the very beginning of the search for a gamma-ray laser [8-16]. For Mössbauer phononless nuclear transitions, the radiation line width is close to the natural width and, hence, the product $\Delta\omega_{tot}\tau$ is close to its minimum value equal to unity. Thus, the amplification condition $\Delta\omega_{tot}\tau < 10^3$ is obviously satisfied and the pump power could be decreased by using the Mössbauer nuclei with relatively long lifetimes in the excited states.

For this purpose, L. Rivlin [8, 9] first proposed to separate (employing, for example, radiochemical methods) and collect long-lived nuclear isomers (the lifetimes of isomeric excited states range for various isomers from several fractions of a second to hundreds of years) and then introduce these isomers into a crystal lattice for the creation of an active medium with population inversion of nuclear states or to grow active crystals containing these excited isomeric nuclei.

However, Il'inskii and Khokhlov [17] have found that the Mössbauer condition $\Delta\omega_{tot}\tau \approx 1$ can be realized only for the nuclear states with the lifetime τ less than 10 μs and, thus, the total width of the Mössbauer line of the gamma transition can approach the natural width only for such relatively short-lived nuclei. This is due to the fact that, at longer lifetimes the natural emission line becomes so narrow that even insignificant distortions of the crystal lattice and/or the chemical and quadrupole shifts of the nuclear levels lead to an inevitable inhomogeneous broadening of the gamma emission line. In addition, the reason for the surplus inhomogeneous broadening that is inherent to even an ideal crystal is the magnetic dipole interaction with neighboring nuclei, since spins of the excited and unexcited nuclei are different and coordinates of the emitting nuclei are random. Owing to all these effects, the width of gamma emission line $\Delta\omega_{tot}$ at $\tau > 10$ μs becomes independent of the lifetime τ and can be estimated as 10^5 s^{-1}. The amplification condition $\Delta\omega_{tot}\tau < 10^3$ now leads to the upper-limit estimate $\tau < 10$ ms for the lifetime of excited nuclei.

Unfortunately, the above estimate does not allow rely neither on the growth of an active crystal consisting of such short-lived excited nuclei nor on direct pumping of nuclei in a crystal matrix for the realization of population inversion, since the flux densities needed for the pumping of such short-lived nuclear states are still too high and lead to the violation of the conditions at which the Mössbauer effect is observed or even to the destruction of the crystal. In spite of numerous and extremely refined proposals to improve the pumping process (see, for example, [18, 19]) as well as to narrow the emission line of long-lived isomers down to the natural value (see, for example, [17, 20]), the development of the standard Mössbauer scheme is almost stopped at present because of its impasse character. All these circumstances were discussed in details in the comprehensive reviews [21, 22] and they are often referred to as an unsolvable "gamma-ray laser dilemma".

The primary reason for the harsh demand on pumping in Mössbauer scheme is the resonant nuclear absorption, which leads to the population inversion requirement that at least a half of total number of nuclei of an active medium should be excited. Another approach that seems very promising as an alternative to the Mössbauer scheme does not require the real population inversion but only a *hidden population inversion* of nuclear states [23]. This approach utilizes entirely new possibilities revealing themselves only in radiative processes with hard photons in cooled ensembles of free nuclei. One of them is the possibility to use the nuclear recoil in the emission or absorption of gamma-photons to establish a hidden inversion of state populations. The hidden inversion may ensure a gain of stimulated gamma radiation in a particular frequency range without an excess of the number of excited nuclei over the number of unexcited ones. Since a total population inversion is not required for the gamma-ray lasing, this may result in a considerable reduction in the threshold flux of x-ray pumping radiation. Simultaneously, a sufficiently high the linewidth ratio β and the gain coefficient for amplified gamma-photon flux may be attained through deep cooling of atomic ensembles. The cooling may be accomplished by present-day or upcoming techniques [24] based on the light pressure of optical lasers. Together with some other attractive features [25] (the opportunity to use the so-called 'two-level' scheme for x-ray pumping of a cooled nuclear ensemble, the possibility of anisotropic amplification of a unidirectional gamma-photon flux without any reflecting mirrors, etc.) the concept of recoil-assisted gamma-ray amplification in deeply-cooled ensembles of free nuclei appears to be rather promising.

The question remains, however, whether destructive impact of x-ray pumping radiation on the active medium of cooled free nuclei brings to violation the conditions for hidden population inversion and breakdown the development of gamma-ray lasing process. We discuss in this paper some destructive features of x-ray pumping and formulate a set of requirements to the pumping radiation and to the amplifying nuclear medium crucial to the development of recoil- assisted gamma-ray lasing. We discuss, as well, the efficiency of pumping with modern sources of x-ray radiation.

2. Self-amplified spontaneous gamma-ray emission

Since in gamma region of wavelengths no mirrors exist to reflect the radiation effectively, the development of a gamma-ray laser source is likely to be based on self-amplified spontaneous emission principle. Spontaneous gamma radiation, propagating along the axis of cylindrically shaped nuclear active medium with hidden population inversion of nuclear states (which are assumed below to be the first excited and the stable ground ones), undergoes an amplification with the gain coefficient (1) with n_1=0.

In fact, it is well known that the centers of emission and absorption lines in ensemble of free nuclei are shifted relative to each other due to recoil momentum acquired by the nuclei in the radiative processes. At low temperatures, $k_B T < E_{rec}$ (where $E_{rec} = (E_0)^2/2Mc^2$ gives the recoil energy of a nucleus with mass M and k_B is the Boltzmann constant), sufficient for a radical suppression in an overlap between the emission and absorption lines, a hidden population inversion is established for a gamma-ray propagating oppositely to the direction of pumping x-ray [23, 25]. Since in this case the emitted photons cannot be resonantly absorbed by unexcited nuclei (but they are still capable of stimulating emission of the excited nuclei), the population difference merely equal to the concentration of excited nuclei and, accordingly, the gain coefficient (1) does not involve the concentration n_1 of unexcited nuclei.

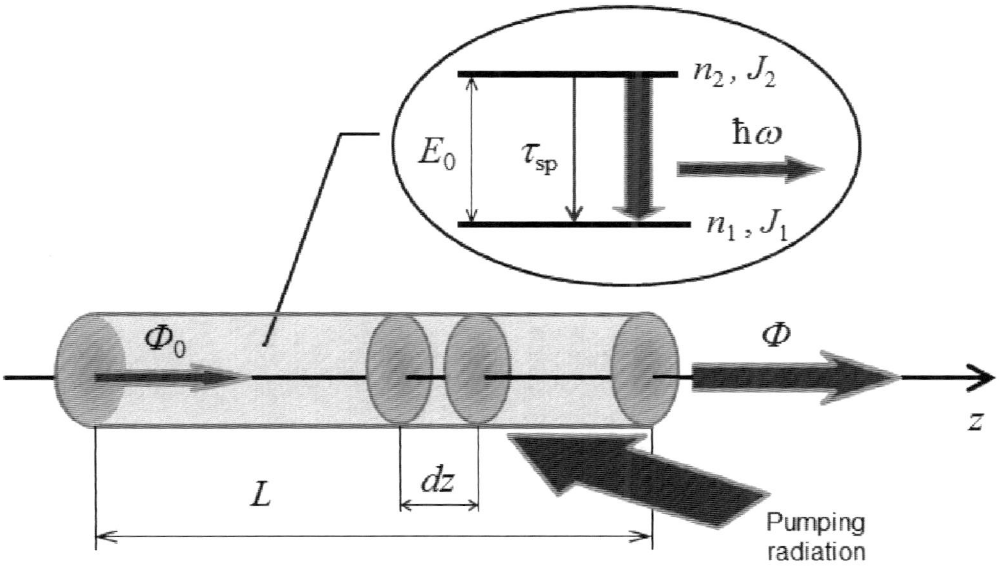

Figure 1. Amplification of spontaneous gamma radiation, propagating along the axis of cylindrically shaped nuclear ensemble with hidden population inversion

If we assume that the pump pulse duration exceeds the time for single pass of gamma-radiation, L/c, along the axis of a cylindrical nuclear medium with length L (see figure 1) or, alternatively, the pump is arranged in such a way that the traveling wave of nuclear excitation occurs in the direction of expected gamma-ray, then the output flux of self-amplified spontaneous emission into a solid angle $\Delta\Omega = \pi(R/L)^2$ (where R is the radius of the cylinder) could be collected from all over the length L. The output density of amplified photon flux integrated over all frequencies of emitted gamma quanta is

$$\Phi = \tfrac{\Delta\Omega}{4\pi}\int_0^\infty S_{sp}(\omega)\left[\int_0^L e^{g(L-z)}dz\right]d\omega, \tag{3}$$

where $S_{sp}(\omega) = (n_2/\tau_{sp})f(\omega-\omega_0)$ is the spectral density of the spontaneous gamma emission rate into unit volume. With approximation for the shape-factor of emission line $f(\omega-\omega_0)$ by a constant value of $(\Delta\omega_{tot})^{-1}$ within the spectral bandwidth $\Delta\omega_{tot}$ around the central frequency ω_0, the amplified photon flux density Φ (3) can be estimated, as compared to out-of-axis flux density of spontaneous gamma emission, Φ_0, by a factor of

$$\frac{\Phi}{\Phi_0} = \frac{G-1}{\ln G}, \tag{4}$$

where $G = \exp(gL)$ is the single-pass gain of stimulated gamma photons over the length L.

Truly exponential amplification of spontaneous gamma radiation occurs if $gL \geq 1$, the condition which is assumed to be a starting point to determine in the next sections a set of requirements crucial to the recoil assisted gamma-ray lasing process.

3. Active nuclear medium

The request for $gL \geq 1$ brings to the requirement to the relative concentration of excited nuclei

$$\frac{n_2}{n} \geq \frac{2\pi\left(\chi/\lambda_0^2 + 1/\lambda_0^2 nL\right)}{\Delta\omega_{\rm rad}/\Delta\omega_{\rm tot}}. \tag{5}$$

Taking into account that the concentration of excited nuclei n_2 is obviously less than the total concentration of nuclei n, we obtain from (5) the following requirement to the linewidth ratio

$$\frac{\Delta\omega_{\rm rad}}{\Delta\omega_{\rm tot}} \geq 2\pi\left(\frac{\chi}{\lambda_0^2} + \frac{1}{\lambda_0^2 nL}\right), \tag{6}$$

the condition which is, actually, to be considered as a demand on cooling degree of nuclear ensemble, since

$$\frac{\Delta\omega_{\rm rad}}{\Delta\omega_{\rm tot}} = \frac{1}{1+\alpha}\sqrt{\frac{T_{\rm cr}}{T}} \quad \text{at } T \geq T_{\rm cr}, \tag{7}$$

where $T_{\rm cr}$ is the critical temperature of nuclear ensemble at which the total width of emission line (which is basically dominated by the Doppler width) is reduced to the decay width of an upper nuclear level $\Delta\omega_2 = \Delta\omega_{\rm rad}(1+\alpha)$.

Even in deeply cooled nuclear ensembles the emission linewidth $\Delta\omega_{\rm tot}$ exceeds its minimum value equal to the decay width of the upper nuclear state $\Delta\omega_2$, which generally contains the contributions both from the radiative nuclear decay and the decay due to internal electron conversion. From the requirement (6) the relation $\Delta\omega_{\rm tot} > \Delta\omega_2$ gives for the dimensionless parameter of active medium, $\varepsilon = \lambda_0^2 nL$, involved into relations (5) and (6), the following restriction

$$\varepsilon = \lambda_0^2 nL \geq \frac{2\pi(1+\alpha)}{1 - 2\pi(1+\alpha)\,\chi/\lambda_0^2}. \tag{8}$$

For a numerical example we assume that the nuclear transition has an energy of E_0=10 keV. The total losses of the gamma radiation with the corresponding wavelength of $\lambda_0 = 1.24$ Å, are mainly due to the photoelectric effect with atomic cross section within the range of $\chi = (10^{-21} \div 10^{-20})$ cm^2 (for instance, $\chi = 1.14\times10^{-21}$ cm^2 for Al, 1.57×10^{-20} cm^2 for Fe, 4.32×10^{-20} cm^2 for Pb, 7.90×10^{-21} cm^2 for Rb, 5.55×10^{-21} cm^2 for Se, 2.82×10^{-20} cm^2 for W and *etc.* [1]). For nuclear transitions with internal electron conversion coefficient α =10 we obtain the estimate $2\pi(1+\alpha)\chi/\lambda_0^2 = 4.5\times(10^{-4} \div 10^{-3}) \ll 1$, which leads from (8) to the requirement $\varepsilon \geq \varepsilon_{\min} = 2\pi(1+\alpha) \approx 70$. Accordingly, if the dimensionless parameter ε =200, then gamma-ray lasing medium might have the total concentrations of nuclei of n=1.3×10^{16} cm^{-3} for the lengths of amplification L=1 m or n=2.6×10^{15} cm^{-3} for L=5 m.

Figure 2 shows the dependence of the relative concentration of excited nuclei n_2/n against the relative reciprocal temperature T_{cr}/T of an active medium with taken $\varepsilon = \lambda_0^2 nL$ as a dimensionless parameter of a family of curves. Gamma-ray lasing occurs for a set of nuclear medium parameters falling into a shaded domain in figure 2 with the boundaries putted by the restrictions (5), (6) and (8). In the case of $\varepsilon = \varepsilon_{min} = 2\pi(1+\alpha)$ this domain consists of only a single point, then spreading gradually together with growth of ε above the minimum value.

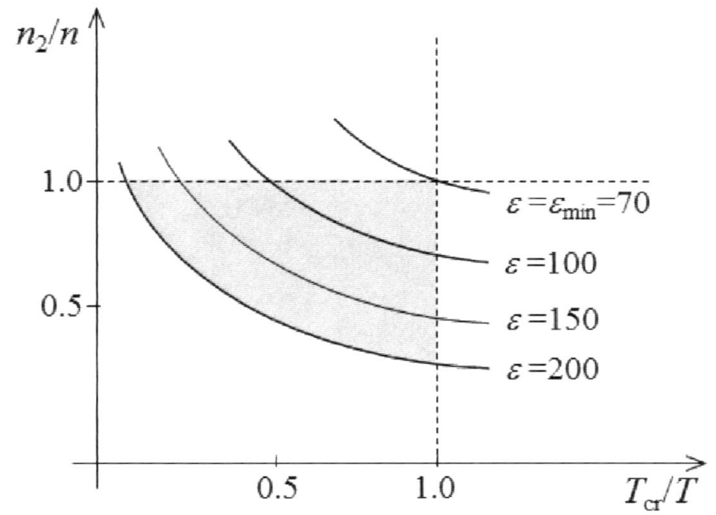

Figure 2. Dependence of the relative concentration of excited nuclei n_2/n against the relative reciprocal temperature T_{cr}/T of active medium with $\varepsilon = \lambda_0^2 nL$ taken as a dimensionless parameter of a family of curves

One can see from the figure 2 that the opportunity to use for gamma-ray lasing the lowest (for a given ε) concentration of excited nuclei,

$$\frac{n_2}{n} = \frac{\varepsilon_{min}}{\varepsilon} = \frac{2\pi(1+\alpha)}{\varepsilon}, \tag{9}$$

is conditioned by the need of deep cooling of the atomic ensemble down to the critical temperature T_{cr}.

The higher the temperature of an atomic ensemble, the greater concentrations of excited nuclei are required and at the highest (for a given ε) temperature, $T_{max}=T_{cr}(\varepsilon/\varepsilon_{min})^2$, still suitable for development of the recoil assisted gamma-ray lasing, all the available nuclei of the atomic ensemble must be pumped into the excited state.

The critical temperature, being measured in μK, can be estimated by $T_{cr} = 0.41A(E_0 T_{1/2})^{-2}$, where A is the nuclear mass number, the transition energy E_0 is measured in keV and the half-life in the nuclear excited state, $T_{1/2} = \tau \ln 2$, is in ns. For an ensemble of nuclei with A=100 and E_0=10 keV the estimate gives T_{cr}=0.41 μK for $T_{1/2}$ = 1 ns and T_{cr} =41 μK for $T_{1/2}$ = 0.1 ns.

4. X-ray pumping radiation

In attempting to propose any pump scheme, it should be noted that the pumping must interact with a previously-prepared cooled atomic ensemble. This is necessitated by the lifetime of the nuclear excited state, which lies in the nanosecond and even sub-nanosecond range, thereby precluding the reverse sequence of pumping and cooling operations. Therefore, the pumping radiation must not damage the initial monokinetic quality of the cooled atomic ensemble. This imposes some restrictions on the non-destructive pumping x-ray radiation.

There are a number of perturbations introducing by the radiative pumping into an active medium. Among them are: (1) an additional broadening of the initial velocity distribution in the group of excited nuclei caused by the broadband radiative pumping; (2) an additional broadening of the initial velocity distribution in the group of excited nuclei caused by an angular divergence of the pumping radiation; (3) the inter-atomic collisions destroying the monokinetic quality of the group of excited nuclei; and (4) the processes that are not directly relative to pumping of nuclei: coherent and incoherent x-ray scattering, absorption of x-rays by atomic electron shell, photoelectric x-ray absorption.

First of all, it is important to note that the pumping with a beam of resonant monochromatic and highly collimated x-ray photons preserves initial monokinetic quality of atomic medium, but through the recoil effect only adds a new narrow line into the velocity distribution. This line corresponds to a group of excited nuclei moving with the nuclear recoil velocity $v_{\rm rec} = E_0/Mc$. Frequency distribution of the pumping radiation within the bandwidth $\Delta\omega_{\rm pump}$ exceeding the Doppler width $\Delta\omega_{\rm D}$ leads to some dispersion of the nuclear recoil velocity

$$\Delta v_{\rm rec}^{(1)} = \hbar\Delta\omega_{\rm D}/Mc\,, \tag{10}$$

and, accordingly, to an additional broadening which can be neglected, however, since the corresponding linewidth

$$\Delta\omega_{\rm pump}^{(1)} = \frac{\omega_0}{c}\Delta v_{\rm rec}^{(1)} = \frac{E_0}{Mc^2}\Delta\omega_{\rm D} \tag{11}$$

is apparently far less than the initial Doppler linewidth $\Delta\omega_{\rm D}$.

Angular divergence of the pumping x-ray radiation also introduces an additional perturbation of the monokineticity into the atomic medium. Providing that the x-ray is confined to a narrow cone with the plane angle $\Delta\theta$ centered close to the axis of cylindrically shaped atomic medium (see figure 1), we obtain the following estimates for an additional dispersion of the longitudinal velocities of pumped nuclei

$$\Delta v_{\rm rec}^{(2)} = \frac{E_0}{2Mc}\left(\frac{\Delta\theta}{2}\right)^2 \tag{12}$$

and for the corresponding additional width

$$\Delta\omega_{\rm pump}^{(2)} = \frac{\omega_0}{c}\Delta v_{\rm rec}^{(2)} = \frac{E_{\rm rec}}{\hbar}\left(\frac{\Delta\theta}{2}\right)^2. \tag{13}$$

The pumping is non-destructive if the additional width $\Delta\omega_{\rm pump}^{(2)}$ is much less than the initial Doppler linewidth $\Delta\omega_{\rm D}$. This leads to the following requirement to the angular divergence of pumping radiation

$$\Delta\theta << 4\left(\frac{k_{\rm B}T\ln 2}{E_{\rm rec}}\right)^{1/4}. \tag{14}$$

Numerical estimate gives $\Delta\theta$ <<0.1 rad for the active medium of nuclei with mass number A=100, the nuclear transition energy $E_0 = 10$ keV and the temperature T=10 μK.

As concern the inter-atomic collisions, in spite of an appreciable value for the nuclear recoil velocity, which is estimated to be 32 m/s for a nucleus with A=100 and $E_0 = 10$ keV, the path covered by a pumped nucleus during its lifetime in excited state, τ=1 ns, is less than the mean free path of atoms for the concentrations up to 10^{16} cm^{-3}. Taking into account that the velocity of thermal atomic motion at T=10 μK is only 5 cm/s to be by three orders of magnitude less than the nuclear recoil velocity, one can conclude that the inter- atomic collisions cannot destroy the monokinetic quality of the considered group of excited nuclei. This group creates a hidden population inversion of nuclear states for the gamma-ray propagating opposite to the pumping x-ray.

Among other processes introducing perturbations into the nuclear active medium, the photoelectric absorption of pumping radiation is likely to be the most destructive one. The recoil momentum acquired by a free excited nucleus in the process of atomic ionization knocks it out of the participants of the gamma-ray lasing process. Therefore, the pumping radiation can be considered as non-destructive if the rate of creation of excited nuclei exceeds the rate of devastation of their concentration due to the photoelectric effect, that is

$$(S\lambda_0^2/4)(n/\tau_{sp})P(\omega_0) > n_2\chi\Delta\omega_{pump}P(\omega_0)\,, \tag{15}$$

where $P(\omega_0)$ is the spectral photon flux density of pumping radiation on the nuclear transition frequency, $\Delta\omega_{pump}$ is the frequency bandwidth of pumping radiation (it is assumed $\Delta\omega_{pump} >> \Delta\omega_D$), $S = (2J_2+1)/(2J_1+1)$ is the spin factor of the nuclear transition, τ_{sp} is the radiative lifetime in excited state. The condition (15) is consistent with the need to have concentrations of excited nuclei above the critical value (9) only for the x-ray pumping radiation confined to the following relative spectral bandwidth

$$\frac{\Delta\omega_{pump}}{\omega_0} < \frac{S\lambda_0^3\varepsilon}{[4\pi(1+\alpha)]^2\chi c\tau}\,. \tag{16}$$

In the case ε=200, $\lambda_0 = 1.24$ Å (E_0=10 keV), α=10, S=1, τ=0.1 ns and $\chi = 10^{-21}$ cm^2 we obtain for the upper-limit estimate of the degree of monochromaticity of pumping radiation the value of 10^{-5}.

At last we estimate intensity of pumping radiation. The growth of the concentration of excited nuclei can be describe as

$$n_2 \approx nS(P/P_{sat})(1-e^{-t/\tau}) \tag{17}$$

where $P_{sat} = 4(1+\alpha)/\lambda_0^2$ is the saturation spectral photon flux density.

The critical concentration (9) can be exceeded only with the help of intense pumping radiation of a high enough spectral photon flux density

$$P(\omega_0) \geq P_{cr} = S(\varepsilon_{min}/\varepsilon)P_{sat} \tag{18}$$

The spectral photon flux density of pumping radiation $P(\omega_0)$ in (18) is expressed in cm^{-2}, whereas the same value expressed in more relevant units of photons/(cm^2 s keV) is given by $P(\omega_0)/\hbar$ with Planck constant $\hbar$=6.58×10^{-19} keV s. Numerical estimate gives $P_{cr}/\hbar$=10^{35} photons/(cm^2 s keV). The critical spectral photon flux density of pumping radiation approaches the peak fluxes that could be supplied with the upcoming x-ray free electron lasers (HASYLAB/DASY, Hamburg, Germany).

5. Conclusions

We have formulated a set of requirements to the pumping radiation and to the amplifying nuclear medium crucial to the development of recoil assisted gamma-ray lasing. Performed analysis has revealed several most significant parameters to be fixed for the active nuclear medium and for the x-ray pumping radiation in order to obtain a self-amplified spontaneous gamma-ray emission. Unfortunately, the required values are far beyond that of available with current technology. First of all it concerns the pumping radiation. We have found that in order to preserve the initial monokinetic quality of the cooled nuclear ensemble, a highly collimated x-ray photon beam with $\Delta\theta \ll 0.1$ rad should be employed for pumping. Additionally, in order to diminish the destructive impact of pumping radiation through the atomic photoelectric effect, the x-ray photon beam should be of a high degree of monochromaticity with $\Delta\omega_{\text{pump}}/\omega_0 \ll 10^{-5}$. We have estimated the critical spectral photon flux density of pumping radiation. It falls into a hard-to-reach region of photon fluxes that could be supplied only with the upcoming x-ray free electron lasers. Summarizing, we may conclude that the lack of sources producing x-ray radiation with the required parameters is the primary impediment to realizing the gamma-ray lasing process in a cooled nuclear ensemble.

Acknowledgements
Author is grateful to L.A. Rivlin for fruitful discussions of the problem considered in this paper. This research was supported by the Ministry of Education and Science of the Russian Federation, the contract number 14.B37.21.0756.

References
[1] NIST Photon Cross Sections Database: http://www.nist.gov/pml/data/xcom/index.cfm
[2] Khokhlov R V 1972 *JETP Lett.* **15** 414
[3] Il'inskii Yu A and Khokhlov R V 1974 *Phys. Usp.* **16** 565
[4] Andreev A V, Volkov R V, Gordienko V M, Mikheev P M, Savel'ev A B, Tkalya E V, Chutko O V, Shashkov A A and Dykhne A M 1999 *JETP Lett.* **69** 371
[5] Ledingham K W D *et al* 2000 *Phys. Rev. Lett.* **84** 899
[6] Cowan T E *et al* 2000 *Phys. Rev. Lett.* **84** 903
[7] Stedile F *et al* 2001 *Phys. Rev.* C **63** 024320
[8] Rivlin L A 1961 *USSR Inventor's Certificate* No. 621256 on Jan. 10
[9] Rivlin L A 1979 *USSR Bull. Inventions* **23** 220
[10] Rivlin L A 1962 *Vopr. Radioelektron. ser. 1 Elektron.* **6** 60
[11] Rivlin L A 1963 *Vopr. Radioelektron., ser. 1 Elektron.* **6** 42
[12] Chirikov B V 1963 *Sov. Phys. JETP* **17** 1355
[13] Vali V and Vali W 1963 *Proc. IEEE* **51** 182
[14] Vali V and Vali W 1963 *Proc. IEEE* **51** 1248
[15] Baldwin G C, Neissel J P, Terhune J, and Tonks L 1963 *Trans. Am. Nucl. Soc.* **6** 178
[16] Baldwin G C, Neissel J P, Terhune J, and Tonks L 1963 *Proc. IEEE* **51** 1247
[17] Il'inskii Yu A and Khokhlov R V 1974 *Sov. Phys. JETP* **38** 809
[18] Goldansky V I and Kagan Yu M 1973 *Sov. Phys. JETP* **64** 90
[19] Letokhov V S 1973 *Sov. Phys. JETP* **64** 1555
[20] Andreev A V, Il'insky Yu A and Khokhlov R V 1975 *Sov. Phys. JETP* **40** 819
[21] Baldwin G C and Solem J C 1997 *Rev. Mod. Phys.* **69** 1085
[22] Rivlin L A 2007 *Quantum Electron.* **37** 723
[23] Rivlin L A 1999 *Quantum Electron.* **6** 467.
[24] Metcalf H J and van der Straten P 1999 *Laser Cooling and Trapping* (New York: Springer)
[25] Carroll J J, Karamian S A, Rivlin L A and Zadernovsky A A 2001 *Hyperfine Interact.* **135** 3

Recent Progress on Compact and Portable Terahertz Source Based on Parametric Conversion from Dual-Frequency Solid-State Laser Pulses

Yujie J. Ding*, Pu Zhao, and Da Li
Department of Electrical and Computer Engineering
Lehigh University, Bethlehem, PA 18015, USA
**Corresponding author:* yud2@lehigh.edu

Abstract

We review the progress made on power scaling and compact and portable THz sources. By reversely stacking GaP plates, we improved photon conversion efficiency from 25% to 40%, which is the maximum value. As the number of the plates was increased from 4 to 5, the output power was decreased due to back conversion. We also investigated THz generation by mixing two frequencies generated by a single Nd:YLF solid-state laser. The average output power reached 1 μW. By introducing two Nd:YLF crystals, we significantly improved the output power to 4.5 μW. Such a configuration allowed us to generate different output frequencies. We have also reviewed our effort of making the THz source further compact by exploiting passively Q-switched laser pulses.

Number of our report according to workshop program: 1.9.4

Topic: Interaction of Laser Radiation with Matter

PACS: 42.65.Ky, 42.65.-k, 42.65.Re

Keywords: Monochromatic THz pulses, power scaling, difference-frequency generation, nonlinear-optical materials, solid-state lasers, saturation of conversion efficiency, compact and portable terahertz sources.

I. Introduction

The terahertz (THz) region of the electromagnetic spectrum, having its frequencies being defined between 0.1 THz and 10 THz, many several potential applications. First, the transition frequencies between rotational states in gases are typically in this region [1]. Therefore, one can develop a rather compact spectrometer to fingerprint molecules. Since the linewidth for the rotational transitions can be 50 MHz (0.0017 cm^{-1}) or narrower for low-pressure gases, such a spectrometer must have a sufficiently high resolution to resolve each rotational transition. In comparison, in the domains from visible to mid-infrared, most molecules exhibit the congested and unresolved ro-vibrational spectra [2]. Therefore, it may not be feasible for anyone to truly fingerprint molecules in these domains. Second, identification and detection of biological agents based on THz spectroscopy have become increasingly important [3]. Third, due to the presence of the atmospheric windows, THz waves could be used for probing as well as imaging [4]. Based on our discussion made above, it is important for us to develop an active THz system that is capable of taking the multi-spectral images of targets and simultaneously analyzing the targets based on the THz spectra. One of the major components of the system is a THz source. A tunable and monochromatic THz source has its advantages over a broadband THz source, especially for revolving weak and congested absorption features.

Broadband THz pulses were generated based on optical rectifications [5], photoconduction [6], and Cherenkov radiation [7]. These schemes all require ultrafast laser pulses. In order to sense chemicals, time-domain spectroscopy was developed [8]. These broadband pulses were used to measure rotational spectra of gases [9], to identify explosives [10], and to take THz images [11].

Besides optical rectifications, other parametric processes in nonlinear crystals can be quite efficient for converting optical pulses to the THz waves. For example, based on THz parametric oscillation in a $LiNbO_3$ crystal [12], the output power on the level of ~ mW was generated in the range of 140-290 μm (1.03-2.14 THz). By mixing two coherent infrared laser beams in nonlinear-optical crystals such as GaSe [13-15], $ZnGeP_2$ [16-18], and GaP [19,20], we efficiently generated the THz waves with extremely-wide tuning ranges and very high peak powers as well as high conversion efficiencies. The widest tuning range achieved by us so far covers the wavelengths anywhere from 66.5 μm to 5.66 mm (from 4.51 THz down to 53 GHz) based on the type-*oe-e* phase-matched difference frequency generation (DFG). On the high-frequency side of the reststrahlen band, we were able to generate the radiation tunable in the range of 2.7–28.7 μm using a different phase-matching configuration [21]. One can see that the tuning range achieved by us almost covers the entire THz region (0.1-10 THz) and part of the millimeter-wave or microwave domain, i.e. 1 mm to 5.66 mm (53-300 GHz). The highest output peak power measured by us is 389 W at 203 μm (1.48 THz) and 4.7 W at the wavelength of 1 mm (300 GHz), corresponding to the conversion efficiencies of 0.098% and 0.0012% in terms of the pulse energies, respectively. The highest photon conversion efficiency was about 19% [15].

We have explored some important applications by using our widely-tunable THz sources for. Indeed, by measuring the absorption spectra of the commonly-used chemicals in the vapor phase, we demonstrated that THz absorption spectrometer developed by us could be used to fingerprint a class of molecular species [22,23]. We also measured the transmission spectra of the isotopic variants [23,24], biological species [14,25], and VX nerve agent [26].

Recently, by reversely stacking GaP plates, we were able to improve the photon conversion efficiency from 25% [27] to 40% [28]. We demonstrated that 40% corresponds to the maximum conversion efficiency [28]. As the number of the stacked GaP plates was increased from 4 to 5, the output power was decreased due to back conversion. In order to make our THz source truly compact and portable, we investigated THz generation by mixing two frequencies generated by a single Nd:YLF solid-state laser [29]. The average output power reached 1 μW at 1.64 THz [29]. Such a THz source can have a dimension of 12″×12″×6″. By introducing two Nd:YLF crystals in the laser cavity, we significantly improved the output power to 4.5 μW [30]. Such a configuration allowed us to generate different THz frequencies by combining two different laser crystals [30]. Through our further optimizations and utilizing novel configurations, we will reduce the dimension of the THz source down to first 12″×6″×4″ and eventually to the size of a laser pointer by exploiting passively Q-switched laser pulses [31].

In this article, we give a review over our effort on the power scaling of the THz waves generated by mixing two near-infrared coherent radiation beams in stacked GaP plates. We also summarize our recent results on the development of compact and portable THz sources by mixing a pair of frequencies generated by dual-frequency actively and passively Q-switched solid-state lasers.

II. Power scaling by stacking GaP plates

In this section, we summarize our results on power scaling of THz radiation based on DFG in high-resistivity and low-dislocation-density GaP plates [27]. We stacked GaP plates such that three nonzero elements of second-order nonlinear susceptibility have the same sign or opposite signs from one plate to the next. These two stacking configurations yielded completely different results in terms of the highest output peaks and corresponding wavelengths. The highest output peak power measured by us is about 2.36 kW, which is two orders of magnitude higher than our previous result [19,20].

Each of the semi-insulating undoped (110) GaP plate used in our experiment has a resistivity of > 1.0 MΩ-cm and a dislocation density of 2.25×10^5 cm^{-2}. It has a diameter of 48.5 mm and a thickness of 663 μm. Both of the (110) facets are polished. In order to generate monochromatic THz pulses, two coherent radiation beams were a Nd:YAG laser at 1.064 μm with pulse duration of 10 ns and pulse energy of 20.7 mJ (the highest used in our experiment) and the idler output from a β-BaB_2O_4-based optical parametric oscillator (OPO) pumped by the third-harmonic output of the Nd:YAG laser. The highest pulse energy for the idler beam is 18.7 mJ whereas its pulse duration is 5 ns. These two beams were collimated, combined, and then illuminated the GaP plates. The beam radius for both of the mixing beams at the GaP plates was measured to be 1.0 mm. Therefore, the highest peak intensity for the idler beam was 120 MW/cm^2, which is a factor of 5.4 below the highest intensity of 650 MW/cm^2 used without causing any damage [32]. They collinearly propagated in a direction perpendicular to the two large facets of the GaP plates with their polarizations parallel to the [1,-1,0] and [0,0,1] directions, respectively. Under such a configuration, the effective second-order nonlinear coefficient is much higher than that from a (100) GaP crystal [19,20]. Germanium and polyethylene windows were used to block the two mixing beams. The THz radiation generated by DFG was attenuated, collimated by a parabolic mirror, and then focused onto a power meter by using another parabolic mirror. The THz output wavelengths were calibrated by an etalon.

By measuring the transmission and reflection spectra of each GaP plate using Fourier transform infrared spectroscopy, we deduced the absorption spectrum. According to Fig. 1, the absorption coefficient is 2.2 cm^{-1} at 120 μm, which is 7 times lower than that from our previous sample. This implies that we could stack up to seven plates to effectively increase the output power. Based on the dispersion relation available [33], the coherence length for the THz generation is linearly increased with the wavelength from 581 μm at 76.5 μm. At 304 μm, e.g.

the coherence length becomes 7.37 mm. We measured the THz output powers at different output wavelengths. According to Fig. 2, the THz output radiation was tuned continuously from 3.91 THz (76.7 μm) to 454.9 GHz (659.5 μm). Such a tuning range was achieved by slightly adjusting the frequency (wavelength) of the idler beam generated by OPO. Within such a range, the plate thickness is either comparable to or much shorter than the coherence length calculated by us. Therefore, the THz generation is phase-matched. At 2.775 THz (108.1 μm), the output power reached the highest value, i.e. 722 W. Using the pulse width of ~5 ns, the highest average power of the generated THz radiation is 36.1 μW at the average input power of 207 mW at 1.0642 μm and 187 mW at 1.0748 μm, respectively. Such an output power is a factor of 46 larger than the highest peak power achieved on GaP so far [19,20].

By using a (110) GaP wafer, we can access a much higher value of the effective second-order nonlinear coefficient [20]. Second, the two parallel facets of the GaP crystal can induce Fabry-Perot effect, which can also contribute to the enhancement of the THz output power. In fact, from Fig. 2 one may notice that the THz output power changes periodically. On the low-frequency side, i.e. for the frequencies below 2.5 THz, the frequencies at most of the peaks and valleys can match those calculated based on Fabry-Perot effect. If we only assume such an effect for the THz wave, the ratio of a local maximum power and its adjacent minimum value should be about 3.3. For the frequencies higher than 2.5 THz, the frequencies at the peaks and valleys are no longer consistent with those predicted by Fabry-Perot effect. In addition, the modulation depths are much larger. We believe these modulations are primarily caused by the water-vapor absorption. Third, the absorption of the THz pulses generated by DFG is much less than that for the previous crystal [19,20]. As a result, the frequency at the highest output power from our crystal (2.775 THz) is much higher than that measured previously, i.e. 1.73 THz [19,20]. There was the subsequent absorption of the THz pulses by a rather thick crystal at relatively high frequencies [19,20]. Such an assumption is further supported by the fact that the high-frequency limit measured on our crystal is significantly higher than the previous value. Fourth, since the dislocation density of our GaP crystal is remarkably low, the damage threshold is significantly increased. As a result, we were able to use much higher peak powers for the two mixing beams.

We measured the dependence of the THz output power on the input power for one of the mixing beams. Based on the linear-least-square fit to the power dependences, we have determined the internal conversion efficiency to be 0.067% (the photon conversion efficiency of

6.8%). Without taking into consideration absorption losses, considering the Fabry-Perot effect for the THz wave, and assuming that the three nonzero elements of the second-order nonlinear optical susceptibility for GaP are identical, the maximum conversion efficiency for the THz generation in terms of peak powers is given by [34]:

$$\eta_{\max} = \frac{P_{THz}T_{THz}^{-1}}{P_1T_1} = \frac{16d_{eff}^2 L^2}{\varepsilon_0 c n_1 n_2 n_{THz} \lambda_{THz}^2} \frac{P_2}{\pi w_0^2} \frac{T_2}{T_{THz}^2} \tag{1}$$

where max is used to designate the local maximum values due to Fabry-Perot effect, d_{eff} is the effective nonlinear coefficient, P_{THz} is the THz output peak power, P_1 and P_2 are the input peak powers at 1.064 μm and at the idler wavelength, respectively, λ_{THz} is the output wavelength, L is the total thickness of the GaP wafer, w_0 is the beam radius for the pump beams, n_i are the refractive indices of the GaP crystal at the respective wavelengths of the three interacting waves, and T_i is the Fresnel transmission coefficients at single surface. Assuming $d_{\text{eff}} \approx 47$ pm/V, we have obtained $\eta_{\max} \approx 0.015\%$ from Eq. (1). This value is about a factor of 4.5 lower than our experimental value. We believe the difference between two may be caused by the uncertainty for the value of d_{eff} used in Eq. (1).

In addition to one GaP plate, we have attempted to stack two and three GaP plates. We have stacked them according to two different configurations, i.e. three elements of second-order nonlinear susceptibility for GaP have (A) same sign and (B) opposite signs between the adjacent GaP plates. Fig. 3 illustrates our result obtained for the stacking configuration A. One can see that the wavelength corresponding to the highest output power has been significantly red-shifted from 108.1 μm to 204.8 μm to 303.9 μm. The corresponding power is reduced from 722 W to 698 W to 611 W. 611 W corresponds to a photon conversion efficiency of 16% inside the stacked plates. These results indicate that there is an optimal interaction length for each specific output wavelength. We believe that the red-shift of the optimal output wavelength is primarily caused by the combination of the linear wavelength dependence of the coherence length and wavelength-dependent output power reflected by Eq. (1). Therefore, at the wavelengths of 108.1 μm, 204.8 μm, and 303.9 μm, the coherence lengths should be roughly close to 663 μm, 1.326 mm, and 1.989 mm, respectively.

Under the stacking configuration B our result is completely different. One can see from Fig. 4 that the wavelength corresponding to the highest output power is kept at 120.3 μm for the two and three plates. Therefore, quasi-phase matching is achieved at this wavelength. This implies that the coherence length at 120.3 μm is about 663 μm. On the other hand, using the dispersion relation [33], we have calculated coherence length to be 1.65 mm. We believe that the factor of 2.5 between the two is probably caused by the inaccuracy in the THz index of refraction calculated by using the dispersion relation [33]. Using 663 μm as the coherence length, we have determined the index of refraction at 120.3 μm to be 3.217. This value is just 3.8% lower than that calculated by using Ref. [33]. Such an estimate illustrates the fact that the coherence length is quite sensitive to the index of refraction at the output wavelength. At such a wavelength, the output peak power was significantly increased from 433 W to 1.36 kW and 2.36 kW, corresponding to two and three plates, respectively. The peak power of 2.36 kW corresponds to an average power of 118 μW. Compared with our previous result [19,20], we increased the peak power by a factor of 151, i.e. two orders of magnitude. The internal conversion efficiency deduced from our experiment is 0.22%, corresponding to the photon conversion efficiency of 25%.

III. Saturation of conversion efficiency and back conversion

In this section, we summarize our results following our previous investigation of power scaling of the THz generation based on DFG by stacking GaP plates [28]. In particular, we achieved the maximum photon conversion efficiency of 40%. The THz peak power generated by us approaches 4 kW. As the number of the GaP plates is increased from four to five, we observed the evidence on back conversion.

Similar to Section II above, we mixed two coherent input beams in stacked GaP plates. One beam has the highest input peak power of 1.5 MW at 1.064 μm (10 ns; 10 Hz). The second one generates an output wavelength tunable in 1.079 - 1.066 μm (5 ns; 10 Hz) with the highest peak power of 4.1 MW. The two collimated beams were directed to the GaP plates without focusing (the beam radii of 1 mm). The total intensity of the two beams at the GaP plates was 178 MW/cm^2. Each GaP plate had a diameter of 48.5 mm and thickness of 663 μm. They were cut and polished along the (110) plane and stacked between a pair of aluminum plates. The two input beams propagated along the [110] direction with their polarizations being parallel to [1-10] and [001], respectively. The output THz radiation was attenuated by filters and was then measured using a power meter.

In order to achieve quasi-phase-matching (QPM) for the THz generation, we rotated every the other plate by 180°. Such a configuration is named as alternatively-rotated stacked GaP plates (i.e. reversed plates). In such a case, the second-order nonlinear coefficient changes its sign between any adjacent ones. In order to find out the enhancement factor caused by QPM, we have compared with the second stacking configuration in which all the GaP plates are perfectly aligned such that their respective crystalline axes are along in the same directions (i.e. aligned plates). Therefore, in such a case the second-order nonlinear coefficients of all the plates have the same sign.

We measured the spectrum of the THz output generated by four reversed plates (the total thickness of 2.65 mm). As illustrated by Fig. 5, the generated THz radiation is continuously tuned from 78 μm to 300 μm. The highest output peak power was measured to be 2.73 kW (the corresponding power of 3.77 kW generated inside the plates) under the input powers of 1.5 MW and 4.1 MW at 1.064 μm and 1.074 μm, respectively. At such powers, the output wavelength was 120.3 μm. It was chosen as a compromise between the THz absorption dictating the effective interaction length and the dependence of the optimal power on the THz wavelength, see Eq. (3) below. The corresponding power conversion efficiency, defined as P_T/P_1 where P_T and P_1 are the THz peak power and input peak power at 1.064 μm, is determined to be 0.182%. Taking into the consideration the Fresnel reflections for the input beam at the entrance facet and THz output radiation at the exit facet, we have obtained the internal photon conversion efficiency of 39.6%. The optimal THz output wavelength of 120.3 μm corresponds to the 1st order QPM condition in the GaP plates having a period of 1.326 mm for the spatially modulated second-order nonlinear coefficients. The modulations appearing in Fig. 5 including those in 120-160 μm are caused by the water-vapor absorption and Fabry-Pèrot effect in the GaP parallel plate.

We measured the spectrum of the THz output radiation generated by four aligned plates. As one can see from Fig. 6 the generated THz radiation covers the range of 135-500 μm. Below 135 μm, the output powers were quite low, due to nearly zero values for the sine function, see Eq. (2) below. The highest output power reaches 510 W at the wavelength of 236 μm under the input power of 4.1 MW at 1.069 μm. The power conversion efficiency is determined to be 0.034%. The corresponding internal photon conversion efficiency is 14.6%, which is significantly lower than that for the reversed GaP plates. Around such a value, the optimal THz wavelength is primarily determined by the combination of the absorption and inverse-wavelength dependence of the THz power, see Eq. (1) below. Even though the aligned plates can be used to achieve phase-matching at sufficiently long THz wavelengths, the inverse dependence and increased optimal power, see Eq. (3) below, make the reversed plates significantly outperform the aligned plates.

Fig. 7 shows the dependence of the internal photon conversion efficiency on the average input power. For the four reversed plates, the conversion efficiency increases linearly below the average power of 50 mW for the input beam at 1.074 μm. However, as the average power is increased further, the saturation of the conversion efficiency becomes obvious. When the input power approaches 200 mW, the conversion efficiency reaches a maximum value. The highest conversion efficiency achieved in our experiment is 39.6% at the input power of 204 mW. Considering the repetition rate of 10 Hz, pulse width 5 ns and beam size of 1 mm, this high conversion efficiency was achieved at pump intensity of 130 MW/cm^2 at 1.074 μm. Assuming $P_2 >> P_1$, where P_2 is the peak power at 1.074 μm, the photon conversion efficiency for the THz generation inside the GaP plates can be obtained as follows [35]:

$$\eta_p = \eta_{\max} \sin^2[(\pi/2)\sqrt{P_{ave}/P_{opt}}] \tag{2}$$

where $\eta_{\max} = a_T/a_1$ is the maximum conversion efficiency with a_T and a_1 being the areas of the THz radiation and input beam at 1.064 μm, P_{ave} is the average input power at 1.074 μm. In Eq. (1), P_{opt} is an optimal power:

$$P_{opt} = \frac{c\varepsilon_0 n_1 n_2 n_T \lambda_1 \lambda_T a_2 \tau_2 R}{16 d_{eff}^2 L^2 T_2} \tag{3}$$

where d_{eff} is the effective second-order nonlinear coefficient, a_2, τ_2, and T_2 are the beam area, pulse duration, and transmittance at 1.074 μm, respectively, and R is the repetition rate. Following the nonlinear least-square-fit to the data points in Fig. 7 by using Eq. (2) above, we have obtained $\eta_{\max} \approx 40.0\%$ and $P_{opt} \approx 209$ mW. Therefore, we have indeed achieved the maximum photon conversion efficiency in our experiment. To the best of our knowledge, this is the first report on the complete saturation of the conversion efficiency for the THz pulses generated by DFG. Based on $\eta_{\max} \approx 40.0\%$, we have obtained $a_T \approx 0.400\ a_1$. Let us assume that the two input beams have Gaussian spatial distributions having exactly the same beam sizes, i.e. $a_1 = a_2$. After neglecting the effect caused by the saturation of the conversion efficiency, we have obtained $a_T \approx 0.5\ a_1$. The deviation of our result from the theory is attributed to the fact that the smaller effective area for the input beam at 1.074 μm. In addition, the weak reflections between the adjacent GaP plates could contribute to the deviation. In comparison, for the aligned plates no obvious saturation was observed, see Fig. 7. The highest photon conversion efficiency is 14.6%. If this value is significantly increased, we will need to introduce a more complicated expression by taking into the consideration the phase-mismatch factor [35] to fit the saturation behavior.

With increasing the number of the GaP plates from four to five, the THz output peak power is significantly decreased from 2.73 kW for the four plates to 2 kW for the five plates, see Fig. 8. This is due to back parametric conversion. Since the optimal power is inversely proportional to the square of the length, i.e. L^{-2}, see Eq. (3), we have used Eq. (2) to achieve a good fit of the data points in Fig. 8. Using thicker GaP plates we could increase the QPM wavelength such that a larger number of the plates may be stacked to reach the zero output power beyond the optimal point shown in Fig. 8 and predicted by Eq. (2).

IV. Compact and portable THz source

In this section, we summary our previous result following our demonstration of a compact and portable THz source which has been implemented by us based on DFG in a GaSe crystal [29]. An average THz output power can be as high as 1 μW at 1.64 THz (182 μm) in a linewidth of 65 GHz, corresponding to the peak power of 20 mW.

In our experiment, we have chosen a Nd:YLF crystal as the lasing medium since two different frequencies of 286.5 THz and 284.9 THz (wavelengths of 1047 nm and 1053 nm) can be generated from a single Nd:YLF laser. Indeed, since a Nd:YLF crystal is birefringent, it emits two transitions at 1047 nm and 1053 nm with the polarizations being orthogonal to each other, designated by π and σ, respectively. Since the two transitions have the stimulated emission cross sections of 1.8×10^{-19} cm^{-2} and 1.2×10^{-19} cm^{-2} at 1047 nm and 1053 nm, respectively, it is feasible for us to tailor the output powers at these two wavelengths from a single laser crystal to be close to each other. In such a case, the corresponding THz power reaches an optimal value under the same sum of the two input powers. Moreover, the THz output frequency generated by mixing these two laser beams is 1.643 THz (i.e. the wavelength of 182.42 μm), which is sufficiently close to 200 μm, and therefore, the THz output power is close to the maximum value from a GaSe crystal [13-15]. Furthermore, a Nd:YLF crystal is known for its low thermal effect and long upper-state lifetime. In the past, a single Nd:YLF crystal was used to simultaneously generate the output beams at 1047 nm and 1053 nm without introducing any optical element inside a single laser resonator [36]. However, the ratio of the powers generated at these two wavelengths could not be controlled at all. In addition, the laser output powers are also too low for the efficient generation of a THz output based on DFG. Although a Q-switched Nd:YLF laser was implemented at 1053 nm [37], the simultaneous lasing of the two wavelengths has not been achieved. Besides the Nd:YLF crystal, other combinations of the active ions and host materials can emit two different frequencies, see e.g. Ref. [38]. As demonstrated by us below, Q-switched solid-state lasers may be one of the optimal choices for implementing a compact THz source based on DFG.

Our experimental setup is shown in Fig. 9. The section bordered by a close-loop dashed line is a schematic for a dual-frequency Nd:YLF laser. In principle, one can simultaneously generate the coherent beams at 1047 nm and 1053 nm from a single Nd:YLF laser. However, under such a case balancing the output powers at the two wavelengths is out of question due to the strong competition between the gains for the two transitions within the same laser medium. In our experiment, we introduced an intracavity polarizer to separate the two perpendicularly-polarized beams into two laser cavities both of which shared the same lasing medium for alleviating the gain competition. In our design, input mirror (IM), output couplers (OC1), and polarizer (P) make up the cavity for lasing at 1047 nm (π, *o*-wave), whereas IM, P, and couplers OC2 make up the cavity for lasing at 1053 nm (σ, *e*-wave). An *a*-cut 1.0%-Nd-doped YLF ($4\times4\times10$ mm^3) crystal was shared arm by the two laser cavities. A plano input mirror (IM) has a coating for producing the high transmittance of T > 95% at 808 nm and simultaneously a high reflectivities of R > 99.5% at 1047 nm and 1053 nm. The polarizer (P) has a coating in such a way that when being placed at a Brewster's angle relative to the input beam, the π-polarized beam has a high transmittance of T $\approx$ 98% whereas the σ-polarized beam has the high reflectivity of R > 99.9. The Nd:YLF crystal was pumped by a diode laser via a fiber pigtail after the laser beam was collimated by a lens assembly. Such a diode laser has the dimension of 1.75″×1.25″, which is portable. Considering the fact that the stimulated emission cross section for the 1047-nm transition is 50% larger, one of the most effective approaches for balancing the output powers of the two transitions is to introduce an additional amount of the loss in the cavity for the 1047-nm beam. First, we have specifically chosen the different reflectivities for the two output coupler to be 75% and 80% at 1047 nm (OC1) and 1053 nm, respectively. As a result, the lasing thresholds for the two transitions become closer to each other. Second, the output coupler (OC1) was slightly tilted to introduce an additional amount of the loss at 1047 nm. In order to generate sufficiently high peak powers, an acoustic-optic Q-switch was placed next to the Nd:YLF crystal in such a way that it was shared by the two beams oscillating in the two cavities. This represents an important method for us to synchronize the output pulses generated from the two laser cavities. The typical dimension of the entire THz source illustrated by Fig. 9 is about 12″×12″×6″. Through further optimization, it can be reduced to 12″×6″×4″. In addition, such a compact THz source currently consumes the electrical power of $\leq$ 20 W. Therefore, it can be readily packaged into a portable system for field applications.

Fig. 10 shows the dependence of Q-switched output powers at 1047 nm and 1053 nm and the sum of the output powers on the output powers from the laser diode. The lasing thresholds were measured to be 1.17 W and 1.64 W at 1047 nm and 1053 nm, respectively. From Fig. 10, we can see that below the pump power of 5 W, both of the output powers increased linearly with the pump power. In addition, the powers at 1047 nm are significantly higher. Above 5 W, however, the output power at 1047 nm became slightly saturated whereas the output power at 1047 nm was increased at a higher rate. This is due to the fact that two transitions accessed the same population of the electrons in the upper lasing level. When the gain for one transition was saturated, the gain for the second one was increased. Therefore, the dependence for the sum of the output powers is still linear, which agrees well with the theory for the solid-state laser [39]. By measuring the output powers as a function the repetition rate, we found an optimal repetition rate of 3.9 kHz for the Q-switch in terms of the THz output powers. At the pump power of 9.69 W, we generated the output powers of 1.196 W and 0.608 W at 1047 nm and 1053 nm, respectively, corresponding to the net conversion efficiency of 18.6%. The ratio of the output powers at the two wavelengths is measured to be around 2:1. By further optimizing the design of the optical cavities, we can reduce the ratio to nearly 1:1. M^2 factors were measured to be 5.18 and 3.07 for 1047 nm and 1053 nm, respectively, at the pump power of 5.9 W. The poorer beam quality at 1047 nm is attributed to the tilting of its cavity mirror, which was used by us to reduce the ratio of the output powers for the two transitions. The pulse shapes of the two output beams are shown in Fig. 11. The two pulse trains were measured simultaneously by using two photodiodes. According to Fig. 11, the two pulses were synchronized. At the pump power of

9.69 W, the pulse widths were measured to be 15.47 ns and 18.59 ns at 1047 nm and 1053 nm, respectively.

Using two convex lenses with the same focal length, the two beams were collimated and then combined by a polarization cube, as shown in Fig. 9. It was adjusted in such a way that the optical paths at the two wavelengths were exactly the same. As a result, the two input beams were overlapped in both time and space and focused on the 15-cm-long GaSe crystal by using a convex lens (L3). After DFG, the residual pump beams were blocked by using a white polyethylene filter and the reflected pump beams were collected by a beam dump. The generated THz beam was collected by two off-axis parabolic mirrors and then focused onto a power meter. The input beams at 1047 nm and 1053 nm were ordinary and extraordinary waves inside the crystal whereas the THz wave was an extraordinary wave. The external phase matching angle was measured to be 11.5°, which is close to the theoretical value of 11° [34]. The azimuthal angle was optimized in our experiment in order to reach the optimal value of the effective nonlinear coefficient [34]. The output wavelength was measured by scanning a Si-based etalon, shown in inset to Fig. 12. The output wavelength was measured to be 183.3 μm, which agrees well with 182.4 μm, calculated from the two input wavelengths of 1.04666 μm and 1.0527 μm [39]. We plotted the dependence of the average and peak output powers on the sum of the output powers generated by the Nd:YLF laser as the input powers for DFG, see Fig. 12. The solid curve corresponds to the quadratic fit to the data points, which is consistent with the characteristics of DFG. Our experimental result indicated that the average output reached optimal values at the repetition rate of 3.9 kHz. When the sum of the input powers was 1.8 W, the average output power from the GaSe crystal was measured to be 0.948 μW. The pulse width for the THz beam was deduced to be 12.36 ns by measuring the temporal profile of the sum-frequency signal in a KTP crystal. Based on such a pulse width, the highest THz peak power was determined to be 19.7 mW. The linewidth of the THz wave was estimated to be 65 GHz by measuring the linewidth of the sum-frequency signal (i.e. 0.06 nm). A DLATGS pyroelectric detector operating at room temperature can have a noise equivalent power of as low as 230 pW. Therefore, the output power of 0.948 μW corresponds to a dynamic range of 4120, which is sufficiently high for realizing the key applications mentioned above. We can further improve the output powers to at least 100 μW by placing a nonlinear crystal inside the optimized Nd:YLF laser cavities.

Journal of Physics: Conference Series **414** (2013) 012003 doi:10.1088/1742-6596/414/1/012003

V. Power scalability and frequency agility

In this section, we summary our results of introducing two laser crystals to generate two different output frequencies [30]. Our experimental result illustrates that the output powers of the solid-state lasers based on two crystals are signficantly improved. Consequently, we have improved the THz output power by mixing the two laser frequencies in a nonlinear crystal. After the second laser crystal is introduced, the THz source still maintains its compactness. Moreover, such a configuration can be used to extend to the combinations of different laser crystals for generating different output frequencies.

The experimental setup for the new configuration of laser cavities is shown in Fig.13. The input mirror 1, polarizer, output coupler make up the first cavity for lasing at 1053 nm while the input mirror 2, polarizer, output coupler make up the second cavity for lasing at 1047 nm. This cavity configuration is completely different that in Section IV above [29]. Indeed, in the previous section [29], the Nd:YLF crystal was placed at the arm sharing the 1047 nm and 1053 nm cavities. As a result, the two radiation beams at 1047 nm and 1053 nm transition accessed the same population inversion within a single Nd:YLF crystal pumped by the same diode laser. Since the 1047 nm and 1053 nm beams have perpendicular polarization directions, we introduced an intracavity polarizer to separate the two beams into the two cavities. This approach alleviated the gain competition between the 1047 nm and 1053 nm radiation beams inside the Nd:YLF crystal. However, due to the competition, we observed the unbalanced powers between the two frequencies, see Section IV above [29]. In our current configuration, we introduced two Nd:YLF crystals (*a*-cut, Nd doped at 1.0%, and $4\times4\times10$ mm^3), labeled as laser crystals 1 and 2 in Fig. 13. The two laser crystals were placed at the two divided arms decoupled by the polarizer. As a result, the 1047 nm and 1053 nm transition beams now access the population inversions from two separate gain media, representing an ultimate solution to the gain competition. An acoustic-optic Q-switch was placed at the shared arm of the two laser cavities. Thus, the dual-frequency pulses are synchronized by simultaneously modulating the losses of the two cavities. The other improvement of the cavities is an output coupler shared by the two cavities, i.e. a concave mirror (curvature = 15 cm) having a reflectivity of R=75% at both 1047 nm and 1053 nm. Since the dual-frequency beams are emitted from the same output coupler, see Fig. 13, the two beams are collinearly propagating. In order to generate a THz output based on frequency-mixing, we can simply place a nonlinear crystal right after the output coupler. Therefore, the THz source is truly compact.

The two diode pump beams at 808 nm are collimated and focused onto the Nd:YLF crystals through a couple of convex lenses, respectively. The diode pump power can be tuned separately. This feature is quite critical for synchronizing the dual-frequency pulses when the lasers are operated at the pulsed mode. In such a case, the pulse build-up times for the two lasers must be exactly the same. After the acoustic-optic become highly transparent, the laser pulse is not generated instantly. The pulse build-up time is the time it takes to generate the laser pulse when the Q-switch is instantly open. This value is determined by the stimulated emission cross section of the laser gain medium, loss of the cavity, output coupling coefficient, and pumping level above threshold. In fact, once the laser crystals are chosen and the cavity is well aligned, the former three factors cannot be changed. As a result, the only way for changing the pulse build-up time is the pump power level. In our configuration, we can separately change the pump power levels of the two lasers simply by varying the driving currents of the lasers.

When the repetition rate of the Q-switch is set at 5 kHz, we have measured the dependence of the Q-switched output powers on the pump powers at both 1047 nm and 1053 nm, see Fig. 14. Both of the output powers at 1047 nm and 1053 nm have increased linearly. As a result, we have completely solved the issue of the gain competition in our previous section [29]. At the pump power of 10.55 W and 9.55 W, we have obtained the output powers of 2.8 W and 1.918 W at 1047 nm and 1053 nm, corresponding to net conversion efficiencies of 26.5% and 20.1%, respectively. The corresponding slope efficiencies are 29.8% and 22.5%, respectively. Compared with the results in the previous section [29], the total output power has been improved by more than twice. In addition, the ratio of slope efficiencies between 1047 nm and 1053 nm is 1.32. This value is close to the ratio of the corresponding stimulated cross sections (i.e. 1.5) [29].

We have used a 15-mm-long GaSe crystal to measure the THz output by mixing the two laser beams, see Fig. 13. The GaSe crystal is placed directly after the output coupler. This setup is different from our previous one where a beam splitter was used to combine the two dual-frequency beams onto a nonlinear crystal. Obviously, our configuration shown by Fig. 13 is much more compact. We have measured the power dependence, see Fig. 15. At the highest incident power of 4.24 W, we have achieved an output power of 4.464 μW at 1.643 THz (182.4 μm). According to Fig. 15, the dependence is nearly quadratic, which is characteristic for DFG.

Compared with the previous result, we have increased the output power by 4.7. Such an enhancement is attributed to the improvement of the laser powers at 1047 nm and 1053 nm. The pulse widths at 1047 nm and 1053 nm are measured to be 17.7 ns and 11.71 ns, respectively. The laser linewidths are measured to be 77.5 GHz and 76.5 GHz, respectively. Assuming the pulse shape is Gaussian, the pulse width of THz radiation is estimated to be 9.766 ns. We have also measured the polarization of the THz radiation as a function of the azimuthal angle of a THz polarizer, shown by inset to Fig. 15. Based on the sinusoidal oscillation, the THz polarization can be determined.

One of the advantages for our configuration lies in the fact that we can generate the THz output frequency simply by choosing two different laser crystals. Since more than 100 Nd-doped laser materials or ceramics have been studied [40] with their wavelengths covering from 1.03 μm to 1.1 μm, it is conceivable for us to generate any THz output wavelength. To demonstrate the frequency agility, we have replaced the Nd:YLF crystal with a 10-mm-long and 1.0%-Nd-doped YAG crystal. All the rest of the optical components remain the same. At the diode pump powers of 9.55 W and 8.653 W, we have generated the output powers of 1.981 W and 1.706 W at 1053 nm and 1064 nm, respectively. At the repetition rate of 5 kHz, the pulse widths are measured to be 9.92 ns and 12.48 ns, respectively. The laser linewidths are measured to be 76.5 GHz and 75.0 GHz, respectively. By mixing the two laser beams on the GaSe crystal, we have generated the THz radiation at 2.983 THz (100.5 μm). The external phase-matching angle for the type II DFG is measured to be 18.3°, which is consistent to the theoretical value [34]. The highest output power is 2.09 μW. The lower output power at such a frequency, compared with that at 1.64 THz, is due to the increased absorption of the THz radiation by the GaSe crystal. The estimated THz pulse width is 7.766 ns. We have measured the THz output wavelength by scanning a Si-based etalon, see Fig. 16. According to Fig. 16, the output wavelength is to be 98 μm, which is close to 100.5 μm, calculated from the two pump wavelengths. The THz linewidth is deduced to be 42 GHz from Fig. 16. This value is close to 53.5 GHz, estimated from the laser linewidths. Using surface-emitting geometry [41,42], it is feasible for us to extend the output wavelengths to the far-infrared region.

Journal of Physics: Conference Series **414** (2013) 012003 doi:10.1088/1742-6596/414/1/012003

VI. THz generation by passively Q-switched laser pulses

In this Section, we summarize our results on THz generation by mixing the passively Q-switched laser pulses in a nonlinear crystal. When two Nd:YLF crystals share a Cr:YAG crystal functioning as a single passive Q-switch, the timing jitter between each pair of dual-frequency pulses generated by the two crystals has been reduced by a factor of 20. Such a reduction in the timing jitter allows us to generate THz pulses by focusing such a passively Q-switched laser beam onto a nonlinear crystal. Such a result represents the first step for us to eventually implement a compact THz source based on ultra-compact microchip lasers. Our success in the THz generation is primarily attributed to our novel design of the cavity for the passively Q-switched dual-frequency Nd:YLF laser, see Fig. 17.

Passively Q-switched lasers [43] have completely revolutionized solid-state laser industry. However, the pulses generated by these lasers suffer from large timing jitters. Such a disadvantage severely limits their applications in differential optical absorption spectroscopy [44] and THz generation [29]. In the past, several approaches for reducing the timing jitter were investigated [45-49]. In order to synchronize each pair of the pulses generated by two passively Q-switched lasers, specific cavities were exploited [49-51].

The laser consists of two cavities being jointed together by a polarizer (P). The two cavities are used to generate two beams at 1047 nm and 1053 nm with the orthogonal polarizations (Π and σ). The polarizer being placed at the Brewster's angle has a high transmittance of $T \approx 98\%$ for the Π-polarized beam and a high reflectivity of $R > 99.9\%$ for the σ-polarized beam. This configuration is different from our previous work [30]. In this cavity, a passive Q-switch made of a 1-mm-thick Cr4+:YAG crystal with an initial transmittance of 90% (i.e. under a sufficiently low input intensity) is shared by the two cavities. Without introducing any complicated circuit and electronic driver, such a passive Q-switch can be fully integrated with the laser crystal [43]. Thus, the dimension of the entire laser can be reduced to 1-2".

Before the dual cavities simultaneously reached the lasing conditions, we tested the timing jitter of the laser pulses at 1047 nm. While the pulse width of the laser was measured to be about 80 ns, the timing jitter was higher than 400 ns. Since the corresponding pulses at the two wavelengths are no longer synchronized, such a dual-frequency laser system is not suitable to THz generation based on difference-frequency generation (DFG) or sum frequency generation (SFG). Due to the unique design of the cavities, we effectively reduced the timing jitter between the laser pulses at the two frequencies. First, the dual-frequency laser pulses are self-triggered

Journal of Physics: Conference Series **414** (2013) 012003 doi:10.1088/1742-6596/414/1/012003

through the occurrence of multiple pulsing [39]. The opening time of the Q-switch is shorter than the pulse build-up time. When the first pulse is generated, the loss of the Q-switch has not reached the minimum. However, the population inversion has not been depleted completely. As the loss the Q-switch is reduced further, an additional pulse is generated shortly after the first one. Via our unique design, multiple pulsing is used to our advantage. As the first pulse is generated in the 1047-nm laser, the passive Q-switch becomes saturated and changes the cavity loss for both of the 1047-nm and 1053-nm laser oscillators. The pulse at 1053 nm is generated shortly after the first pulse at 1047 nm. In the meantime, due to the multiple-pulsing effect, the second pulse at 1047 nm is also generated. As a result, the second pulse at 1047 nm and pulse at 1053 nm can be synchronized, and therefore, they can be used as the input mixing signals for DFG and SFG.

Besides, we were able to manipulate the instance when each pulse is generated. This was accomplished by controlling the pumping current for each output frequency, and therefore, we synchronized the pulses at the two different frequencies. In our experiment, we varied the pump current for the laser at 1053 nm while keeping the pump current for the laser at 1047 nm at a fixed value. The temporal profiles are shown in Fig. 18; all of the measurements were triggered by the laser pulses at 1047 nm.

When the current was low, the laser pulse at 1053 nm exhibited a random temporal profile, and therefore, the timing jitter between the dual-frequency pulses is expected to be large. As the pump current for the laser at 1053 nm was increased, the laser pulse at 1053 nm moved slowly toward the laser pulse at 1047 nm. At some current, the laser pulse at 1053 nm became stable, and therefore, the timing jitter was reduced to about 40 ns, see inset (a) (zone II). This represents the reduction of the timing jitter by a factor of 20. It is attributed to the presence of the passive Q-switch. Indeed, the laser pulse at 1047 nm was generated ahead of the laser pulse at 1053 nm due to a higher pumping level for the laser cavity at 1047 nm. The laser pulse at 1047 nm saturated the passive Q-switch, and therefore, reduced the loss for the Q-switch from 10% to a negligible amount. Since the laser cavity at 1053 nm shared the same Q-switch, the laser at 1053 nm started to emit pulses shortly after the laser pulses at 1047 nm.

Based on our experiment, such a mechanism can be used to significantly reduce the timing jitter for the laser pulses at 1053 nm relative to the laser pulses at 1047 nm. However, as the pump current was further increased, the laser cavity at 1053 nm can reach the lasing threshold without the loss modulation by the 1047 nm pulse. As a result, it was not possible for us to precisely control the instance when the laser pulses were generated at 1053 nm, see inset (b) (zone III). Besides, it was hard for us to determine the wavelength at which the first pulse was generated. But once the first pulse was generated, the loss in the Q-switch was saturated and the pulse at the other wavelength was correspondingly triggered. When the pump current reached a certain value, the laser pulses at 1053 nm were generated first. Since the loss in the Q-switch was reduced, the laser pulses at 1047 nm were subsequently generated. As a result, the temporal profile became relatively stable again, see inset (c) (zone IV). When the pump current became even higher, the temporal profile became relatively unstable, see inset (d) (zone V), similar to the profiles in zone I.

Due to the significant reduction of the timing jitter between each pair of the dual-frequency pulses, these pulses can be used for achieving THz generation based on frequency mixing. Indeed, we have successfully generated THz pulses from a 15-mm GaSe crystal. We believe that this is the first demonstration of THz generation from passively Q-switched dual-frequency laser pulses. While keeping the pump current for the laser cavity at 1047 nm as a constant, we have varied the pump current for the laser cavity at 1053 nm. According to Fig. 19, the behavior of the THz output power vs. the pump current is peculiar. Through our analysis, we have concluded that such a behavior is strongly correlated with the timing jitter between each pair of the dual-frequency laser pulses. Indeed, the first THz peak from the low-current side in Fig. 19 corresponds to the region between zones I and II in Fig. 18 where the second pulse at 1047 nm synchronizes well with the pulse at 1053 nm. Then, as the pump current increases, the pulse at 1053 nm moves to the center part between the two pulses at 1047 nm whereas none of the pulses at these two wavelength could be synchronized. This corresponds to the first dip in the THz power around the pump current 1.2 A. As the pump current is further increased, either 1053 nm or 1047 nm pulse may be generated first, shown in Fig. 18(b) (zone III). However, once being triggered by the first pulse at one wavelength, the second pulse can still be synchronized with the pulse at the other wavelength. As a result, the THz radiation can still be efficiently generated. This region corresponds to the second peak in the THz power in Fig. 19. Then, after the current reaches around 1.4 A, the pulse at 1053 nm is generated first and becomes dominant as the triggering pulse. Similarly, as the pulse at 1047 nm moves around the center point between the two 1053 nm pulses, few pulses at the two wavelength will be synchronized. Accordingly, the

Journal of Physics: Conference Series **414** (2013) 012003 doi:10.1088/1742-6596/414/1/012003

THz power has the second dip. The last peak in terahertz power is attributed to the mixing between the 1047-nm pulse and second pulse at 1053 nm.

Besides the THz generation, we have also demonstrated SFG by using this dual-frequency passively Q-switched laser as the two mixing beams. In order to understand the synchronization between the SF pulses and second-harmonic pulses, a dual-frequency laser beam was divided into two beams using a beamsplitter. Each SF pulse was generated by a KTP crystal by one of the split beams. The other split beam, used for SHG, was sent through a half-wave plate and a cubic polarizer. Due to the perpendicular polarizations of the beams at 1047 nm and 1053 nm, either the 1047-nm beam or 1053-nm beam can be blocked completely by rotating the half-wave plate. Another KTP crystal was used for achieving second-harmonic generation (SHG). In Fig. 20(a), the first pulse of the SH beam at 1047 nm was generated by the first pulse at 1047 nm that triggered the following pulses at 1047 nm and 1053 nm. Fig. 20(b) shows the pulse profiles of the SF and SH beams. Our measurement yielded a timing jitter of 5 ns, which is much smaller than the value for the timing jitter between each pair of the pulses at 1047 nm and 1053 nm. In this approach, we synchronized the pulses at two different wavelengths based on a passively Q-switched laser.

VI. Concluding remark

We have reviewed our recent results on THz generation based on difference frequency generation. In particular, we demonstrated that just using four reversely stacked GaP plates, we achieved the maximum conversion efficiency of 40% in terms of photon numbers. By adding the fifth GaP plate to the stack, the THz output power was significantly reduced from that for the four stacked plates. This is the evidence on back conversion. On the other hand, to make our THz source truly compact and portable, we took a completely different approach to the THz source. Indeed, by mixing two frequencies generated by a dual-frequency Nd:YLF laser we implemented a compact and portable THz source with an average output power reaching 1 μW. By introducing two Nd:YLF crystals instead of one inside the laser cavity, we significantly scaled up the output power to 4.5 μW. Through further optimizations and utilizing novel configurations, we will be able to improve the THz output power to the mW level. The current dimension of our THz source is about 12"×12"×6", which can be readily reduced to 12"×6"×4" through the further optimizations. Our future goal is to develop a THz source as compact as a laser pointer, operating at room temperature. Such a dimension can be achieved by exploiting the passively-Q-switched laser pulses.

Acknowledgements

We are indebted to Y. Jiang, S. Ragam, and I. B. Zotova for making valuable contributions to some of the projects reviewed in this article. This work has been supported by U.S. National Science Foundation.

References

[1]. W. Gordy and R. L. Cook, *Microwave molecular spectra,* 3rd Ed., Wiley, NY (1984).

[2]. R. A. McClatchey, R. W. Fenn, J. E. A. Selby, F. E. Volz, and J. S. Garing, "Optical properties of the atmosphere," in Handbook of optics, Ed. W. G. Driscoll, McGraw Hill, NY (1978).

[3]. A. Bykhovski, T. Globus, T. Khromova, B. Gelmont, and D. Woolard, "An analysis of the THz frequency signatures in the cellular components of biological agents," Intern. J. High Speed Electron. Sys. **17**, 225 (2007).

[4]. A. Lee, Q. Qin, S. Kumar, B. S. Williams, Q. Hu, and J. L. Reno, "Real-time terahertz imaging over a standoff distance (> 25 m)," Appl. Phys. Lett. **89**, 141125 (2006).

[5]. L. Xu, X.-C. Zhang, and D. H. Auston, "Terahertz beam generation by femtosecond optical pulses in electro-optic materials," Appl. Phys. Lett. **61**, 1784 (1992).

[6]. D. H. Auston, K. P. Cheung, and P. R. Smith, "Picosecond photoconducting Hertzian dipoles," Appl. Phys. Lett. **45**, 284 (1984).

[7]. D. H. Auston, K. P. Cheung, J. A. Valdmanis, and D. A. Kleinman, "Cherenkov radiation from femtosecond optical pulses in electro-optic media," Phys. Rev. Lett. **53**, 1555 (1984).

[8]. K. P. Cheung and D. H. Auston, "A novel technique for measuring far-infrared absorption and dispersion," Infrared Phys. **26**, 23 (1986).

[9]. M. van Exter, C. Fattinger, and D. Grischkowsky, "Terahertz time-domain spectroscopy of water vapor," Opt. Lett. **14**, 1128 (1989).

[10]. H.-B. Liu, Y. Chen, G. J. Bastiaans, and X.-C. Zhang, "Detection and identification of explosive RDX by THz diffuse reflection spectroscopy," Opt. Exp. **14**, 415 (2006).

[11]. D. M. Mittleman, S. Hunnsche, L. Boivin, and M. C. Nuss, "T-ray tomography," Opt. Lett. **22**, 904 (1997).

[12]. K. Kawase, M. Sato, T. Taniuchi, and H. Ito, "Coherent tunable THz-wave generation from $LiNbO_3$ with monolithic grating coupler," Appl. Phys. Lett. **68**, 2483 (1996).

[13]. W. Shi, Y. J. Ding, N. Fernelius, and K. L. Vodopyanov, "An efficient, tunable, and coherent 0.18–5.27 THz source based on GaSe crystal," Opt. Lett. **27**, 1454 (2002).

[14]. W. Shi and Y. J. Ding, A monochromatic and high-power THz source tunable in the ranges of 2.7-38.4 μm and 58.2-3540 μm for variety of potential applications," Appl. Phys. Lett. **84**, 1635 (2004).

[15]. W. Shi and Y. J. Ding, "Tunable coherent radiation from terahertz to microwave by mixing two infrared frequencies in a 47-mm-long GaSe crystal," Intern. J. High Speed Electron. Sys. **16**, 589 (2006).

[16]. W. Shi and Y. J. Ding, "Coherent and widely-tunable THz and millimeter waves based on difference-frequency generation in GaSe and $ZnGeP_2$," Optics in 2002, Optics and Photonics News, Dec. 2002, p. 57.

[17]. W. Shi and Y. J. Ding, "Continuously-tunable and coherent terahertz radiation by means of phase-matched difference-frequency generation in zinc germanium phosphide," Appl. Phys. Lett. **83**, 848 (2003).

[18]. W. Shi, Y. J. Ding, and P. G. Schunemann, "Coherent terahertz waves based on difference-frequency generation in an annealed zinc-germanium phosphide crystal: Improvements on tuning ranges and peak powers," Opt. Commun. **233**, 183 (2004).

[19]. W. Shi and Y. J. Ding, "Tunable terahertz waves generated by mixing two copropagating infrared beams in GaP," Opt. Lett. **30**, 1030 (2005).

[20]. Y. J. Ding and W. Shi, "Efficient THz generation and frequency upconversion in GaP crystals," Sol. State Electron. **50**, 1128 (2006).

[21]. W. Shi, X. Mu, Y. J. Ding, and N. Fernelius, "Tunable and coherent nanosecond radiation in the range of 2.7–28.7 μm based on difference-frequency generation in gallium selenide," Appl. Phys. Lett. **80**, 3889 (2002).

[22]. W. Shi and Y. J. Ding, "Chemical identification based on direct measurement of absorption spectrum by frequency-tuning monochromatic THz source," Laser Phys. Lett. **1**, 560 (2004).

[23]. H. Sun, Y. J. Ding, and I. B. Zotova, "THz spectroscopy by frequency-tuning monochromatic THz source: from single species to gas mixtures," IEEE Sensors J. **10**, 621 (2010).

[24]. H. Sun, Y. J. Ding, and Y. B. Zotova, "Differentiation of CO isotopic variants by frequency-tuning a terahertz source," Appl. Opt. **46**, 3976 (2007).

[25]. R. Song, Y. J. Ding, and Y. B. Zotova, "Fingerprinting insulins in the spectral region from mid-IR to THz," Intern. J. High Speed Electron. Sys. **17**, 251 (2007).

[26]. R. Song, Y. J. Ding, and I. B. Zotova, "Fingerprinting malathion vapor: a simulant for VX nerve agent," SPIE Proc. **6949**, 694903 (2008).

[27]. Y. Jiang, Y. J. Ding, and I. B. Zotova, "Power scaling of widely-tunable monochromatic THz radiation by stacking high-resistivity GaP plates," Appl. Phys. Lett. **96**, 031101 (2010).

[28]. Y. Jiang, D. Li, Y. J. Ding, and I. B. Zotova, "Terahertz generation based on parametric conversion: from saturation of conversion efficiency to back conversion," Opt. Lett. **36**, 1608 (2011).

[29]. P. Zhao, S. R. Ragam, Y. J. Ding, and I. B. Zotova, "Compact and portable terahertz source by mixing two frequencies generated simultaneously by single solid-state laser," Opt. Lett. **35**, 3979 (2010).

[30]. P. Zhao, S. Ragam, Y. J. Ding, and I. B. Zotova, "Power scalability and frequency agility of compact terahertz source based on frequency mixing from solid-state lasers," Appl. Phys. Lett. **98**, 131106 (2011).

[31]. P. Zhao, S. Ragam, Y. J. Ding, and I. B. Zotova, "Investigation of terahertz generation from passively Q-switched dual-frequency laser pulses," Opt. Lett. **36**, 4818-4820 (2011).

[32]. L. P. Gonzalez, S. Guha, and S. Trivedi, *CLEO Technical Digest* on CD-ROM (OSA, Washington, DC, 2004), CWA47.

[33]. F. L. Madarasz, J. O. Dimmock, N. Dietz, and. J. Bachmann, "Sellmeier parameters for ZnGeP2 and GaP," J. Appl. Phys. **87**, 1564 (2000).

[34]. D. N. Nikogosyan, *Nonlinear optical crystals*, Springer, NY, 2005.

[35]. A. Yariv, Quantum Electronics, Wiley, NY, 1989, p. 399-425.

[36]. B. Frei and J. E. Balmer, "1053-nm-wavelength selection in a diode-laser-pumped Nd:YLF laser," Appl. Opt. **33**, 6942-6946 (1994).

[37]. H. Zbinden and J. E. Balmer, "Q-switched Nd:YLF laser end pumped by a diode-laser bar," Opt. Lett. **15**, 1014-1016 (1990).

[38]. B. Wu, P. P. Jiang, D. Z. Yang, T. Chen, J. Kong, and Y. H. Shen, "Compact dual-wavelength Nd:$GdVO_4$ laser working at 1063 and 1065 nm," Opt. Express **17**, 6004-6009 (2009).

[39]. W. Koechner and M. Bass, *Solid-State Lasers*, Springer, New York, 2003, pp. 63-65.

[40]. F. Trager, *Handbook of Lasers and Optics*, Springer, New York, 2007.

[41]. Y. Avetisyan, Y. Sasaki, and H. Ito, "Analysis of THz-wave surface-emitting difference-frequency generation in periodically-poled lithium niobate waveguide," Appl. Phys. B **73**, 511 (2001).

[42]. Y. J. Ding, "Efficient generation of high-frequency THz waves from highly-lossy second-order nonlinear medium at polariton resonance under transverse-pumping geometry," Opt. Lett. **35**, 262 (2010).

[43]. J. J. Zayhowski, and C. Dill, "Diode-pumped passively Q-switched picosecond microchip lasers," Opt. Lett. **19**, 1427-1429 (1994).

[44]. J. F. Noxon, "Nitrogen dioxide in the stratosphere and troposphere measured by ground-based absorption spectroscopy," Science **189**, 547–549 (1975).

[45]. J. B. Khurgin, F. Jin, G. Solyar, C. C. Wang, and S. Trivedi, "Cost-effective low timing jitter passively Q-switched diode-pumped solid-state laser with composite pumping pulses," Appl. Opt. **41**, 1095-1097 (2002).

[46]. B. Cole, L. Goldberg, C. W. Trussell, A. Hays, B. W. Schilling, and C. McIntosh, "Reduction of timing jitter in a Q-Switched Nd:YAG laser by direct bleaching of a Cr^{4+}:YAG saturable absorber," Opt. Express **17**, 1766-1771 (2009).

[47]. T. Hakulinen, R. Koskinen, and O. G. Okhotnikov, "Low jitter Q-switched fiber laser using optically driven surface-normal saturable absorber modulator," Opt. Express **16**, 8720-8726 (2008).

[48]. Steinmetz, D. Nodop, A. Martin, J. Limpert, and A. Tunnermann, "Reduction of timing jitter in passively Q-switched microchip lasers using self-injection seeding," Opt. Lett. **35**, 2885-2887 (2010).

[49]. X. J. Wang, and Z. Y. Xu, "Timing jitter reduction and single-frequency operation in an acousto-optic Q-switched Cr,Nd : YAG laser," Appl. Opt. **45**, 8477-8483 (2006).

[50]. V. K. Bagdasarov, N. N. Denisov, A. A. Malyutin, and I. A. Chigaev, "Pulse synchronization in passively Q-switched lasers emitting at 1.053 and 1.064 μm," Quantum Electron. **39**, 887-890 (2009).

[51]. H. P. H. Cheng, P. Tidemand-Lichtenberg, O. B. Jensen, P. E. Andersen, P. M. Petersen, and C. Pedersen, "Experimental investigation of relative timing jitter in passively synchronized Q-switched lasers," Opt. Lett. **36**, 415-417 (2011).

[52]. P. Tidemand-Lichtenberg, J. Janousek, R. Melich, J. L. Mortensen, and P. Buchhave, "Synchronization of 1064 and 1342 nm pulses using passive saturable absorbers," Opt. Commun. **241**, 487-492 (2004).

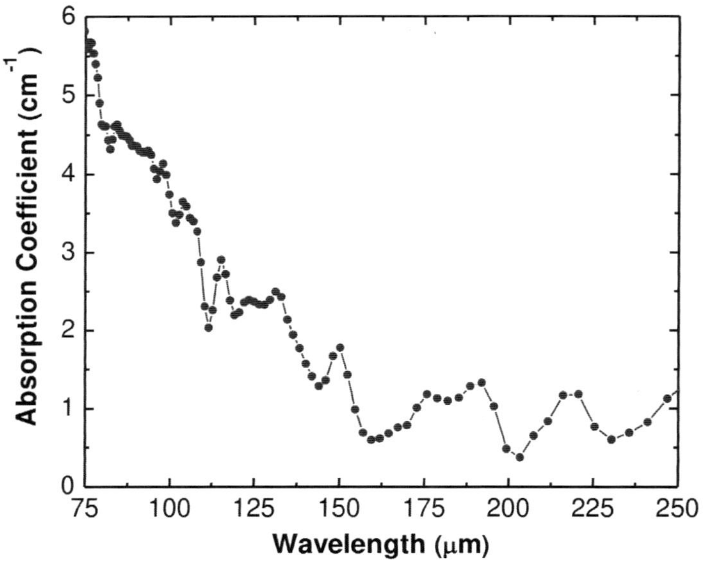

Fig. 1. Absorption coefficient is plotted vs. wavelength following the measurement of transmission and reflection spectra on a single GaP plate.

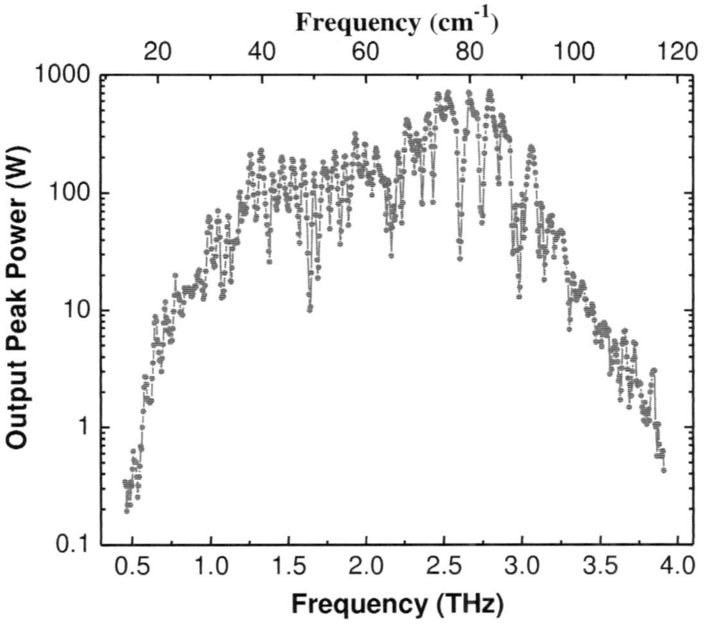

Fig. 2. THz output power was measured vs. output wavelength for a single GaP plate. As discussed in the text, the modulations were caused by convolution between Fabry-Perot effect for THz wave and water-vapor absorption.

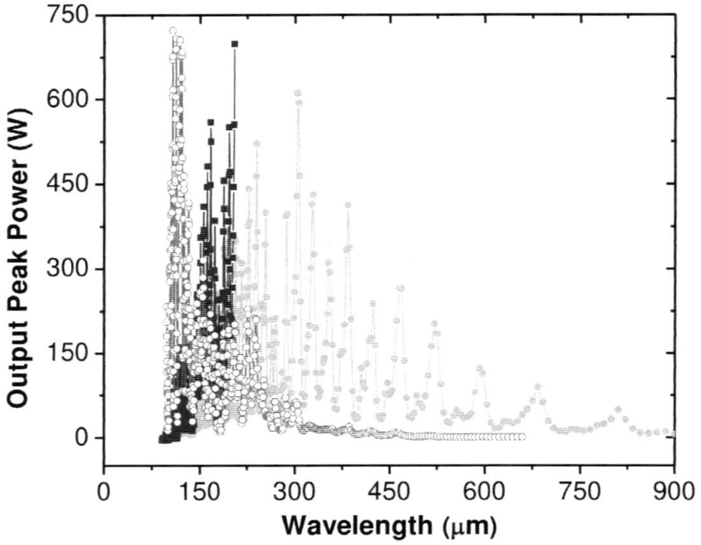

Fig. 3. Output peak power vs. wavelength: open circles – single GaP plate; squares – stacked two GaP plates; dots – stacked three GaP plates. For both of the stacked two and three plates, second-order nonlinear coefficients have same sign between adjacent plates.

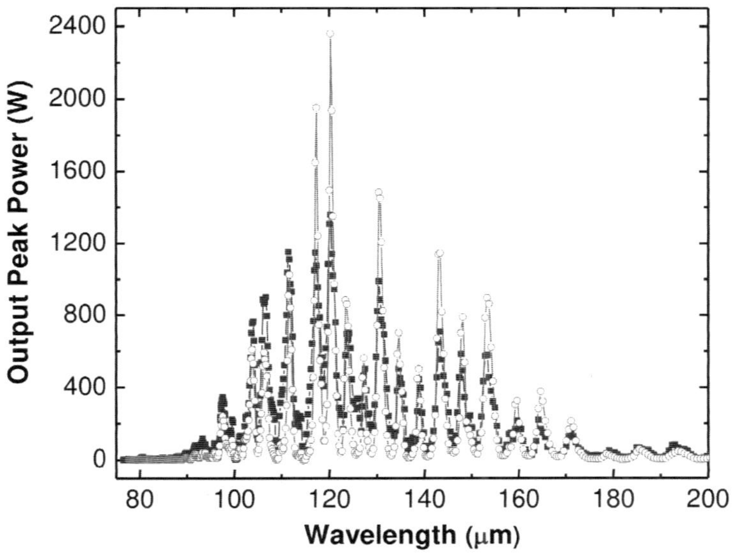

Fig. 4. Output peak power vs. wavelength: squares – stacked two GaP plates; open circles – stacked three GaP plates. For both of the stacked two and three plates, second-order nonlinear coefficients have the opposite signs between adjacent plates.

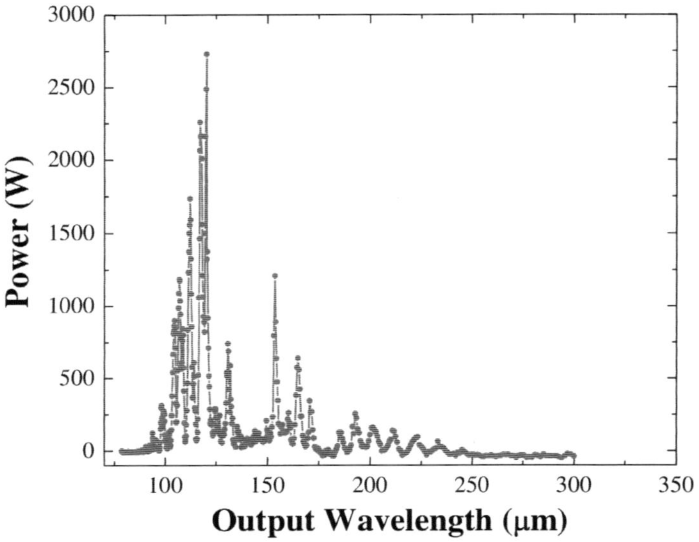

Fig. 5. Spectrum of THz output generated by four reversed GaP plates, covering range of 78-300 μm.

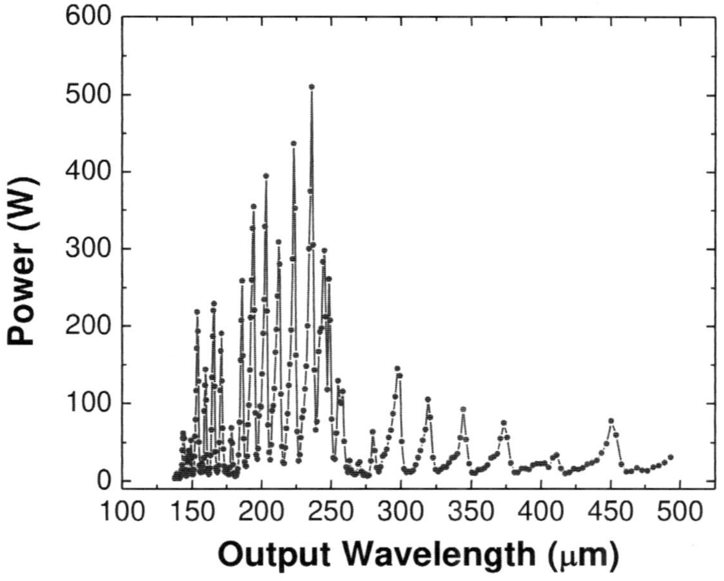

Fig. 6. Spectrum of THz output generated by aligned four GaP plates, covering tuning range of 135-500 μm.

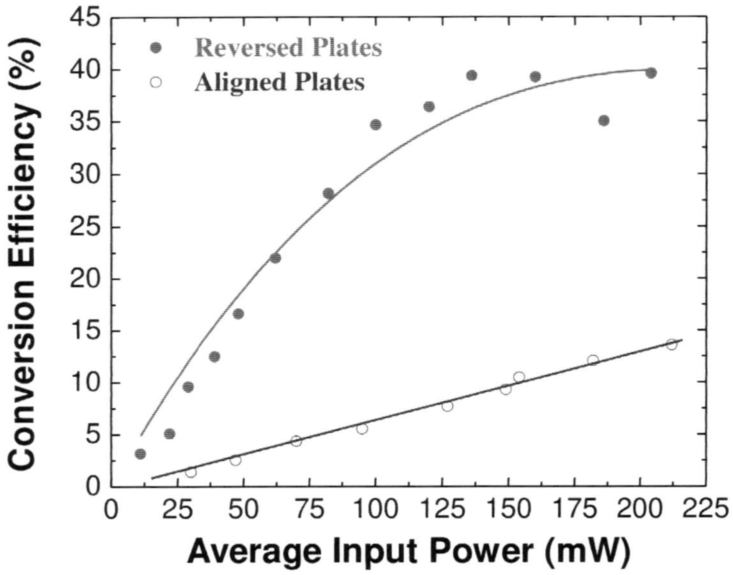

Fig. 7. Conversion efficiency for THz generation versus input power at 1.074 μm for four reversed GaP plates (red dots) and 1.069 μm for four aligned GaP plates (blue circles). Blue dashed curve (bottom) represents linear fit, whereas the red curve (top) is fitted using Eq. (2).

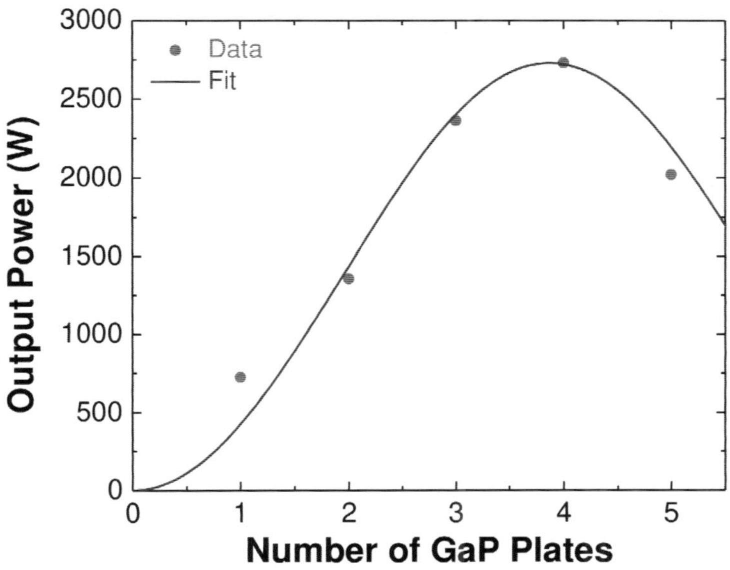

Fig. 8. Highest THz output power generated by reversed GaP plates versus the number of plates. Blue curve represents nonlinear least squares fit to data points using Eq. (2).

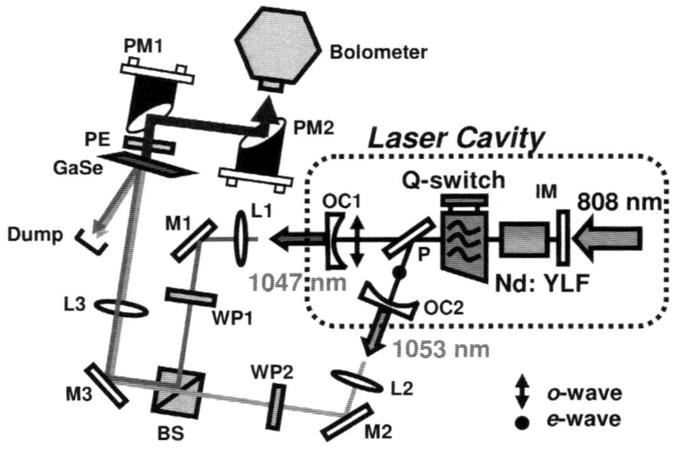

Fig. 9. Experimental setup for THz DFG based on compact and portable dual-frequency Nd:YLF laser having dual cavities: IM, laser input mirror; P, polarizer; OC1-OC2, laser output couplers; L1-L3, convex lenses; M1-M3, high reflection mirrors; WP1-WP2, half-wave plates; PE, high density white polyethene filter; and PM1-PM2, parabolic mirrors.

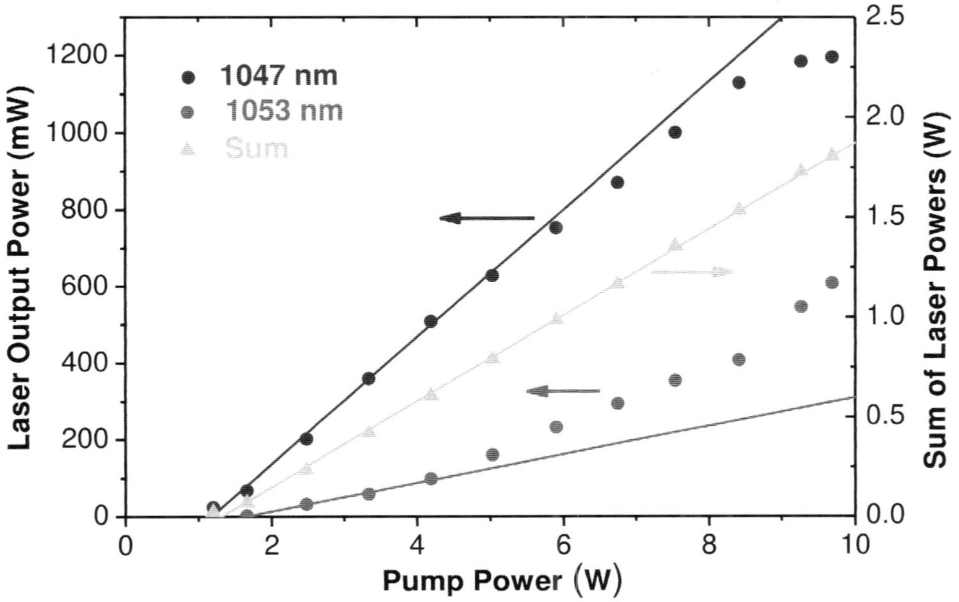

Fig. 10. Dependence of powers from dual-frequency Q-switched Nd:YLF laser on pump power from laser diode.

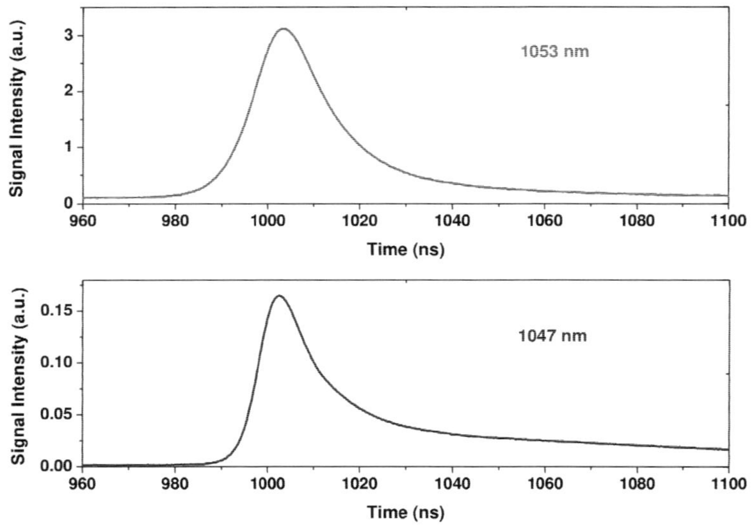

Fig. 11. Temporal profiles of dual-frequency output pulses generated by from Q-switched Nd:YLF laser. Ch1: 1053 nm and Ch2, 1047 nm.

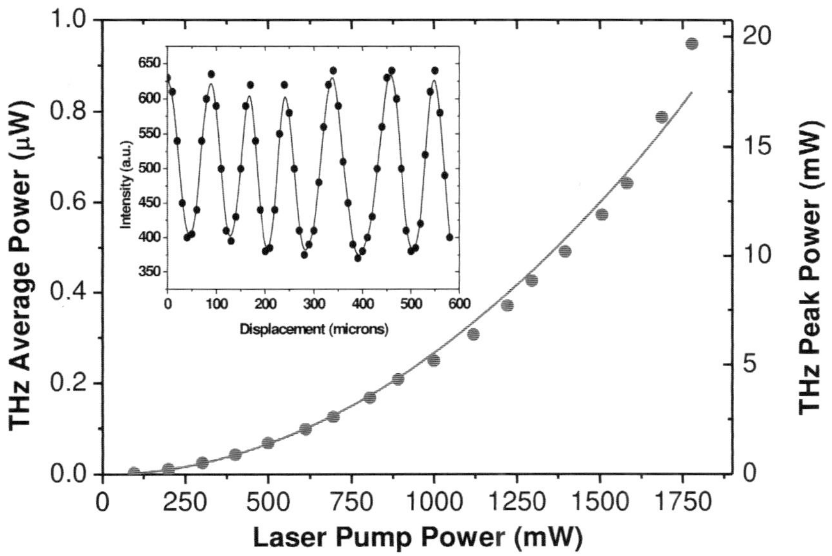

Fig. 12. Dependences of average and peak output powers on sum of the output powers generated by the single Nd:YLF laser. Solid curve corresponds to quadratic fit to data points. Inset: plot of power being transmitted through a Si etalon vs. distance between two Si plates in the etalon.

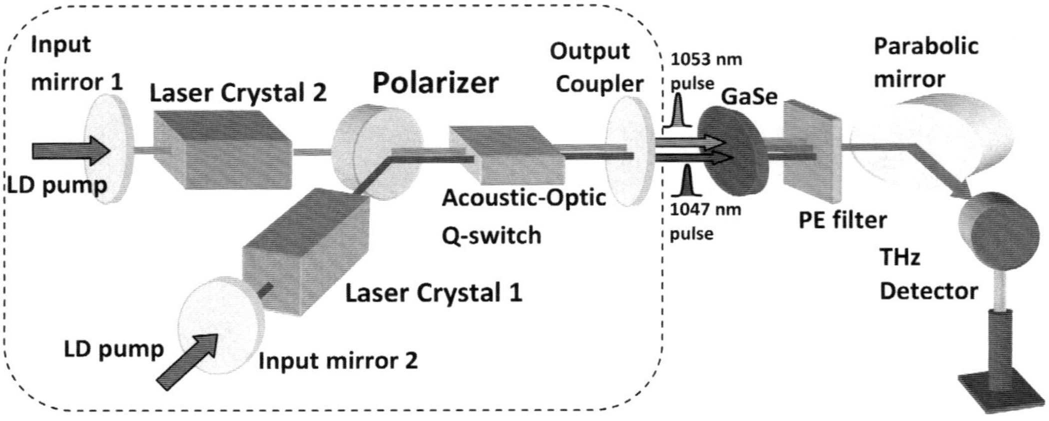

Fig. 13. Experimental layout for a configuration of the dual-frequency solid-state laser and compact THz source. The dual-frequency laser cavities are marked by the dashed border line.

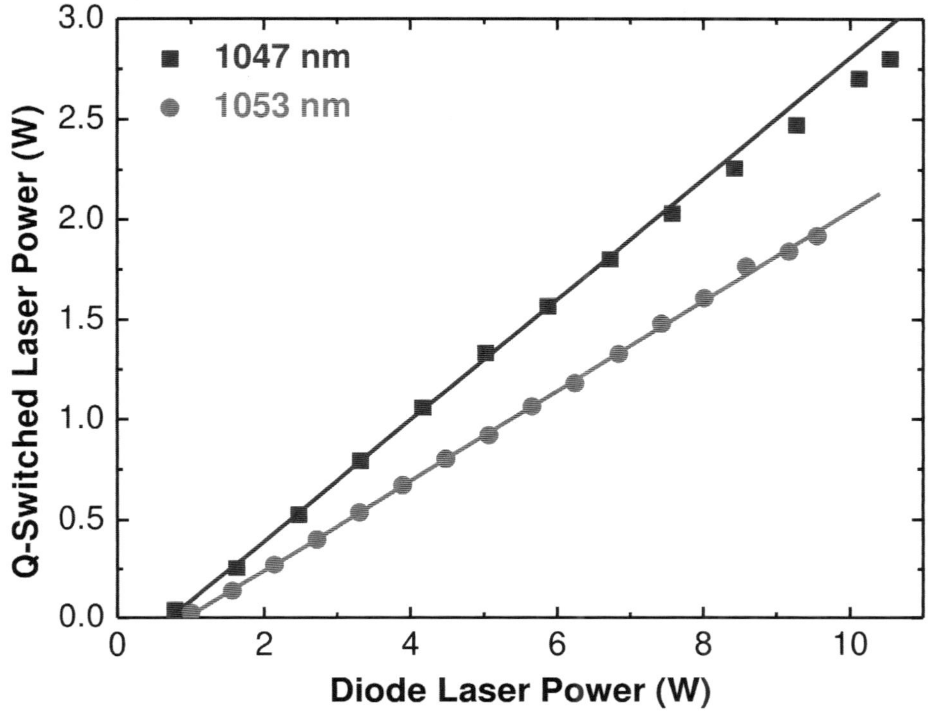

Fig. 14. Q-switched output power vs. pump power at 1047 nm and 1053 nm. Dots and squares correspond to data points; straight lines correspond to linear fits to data points.

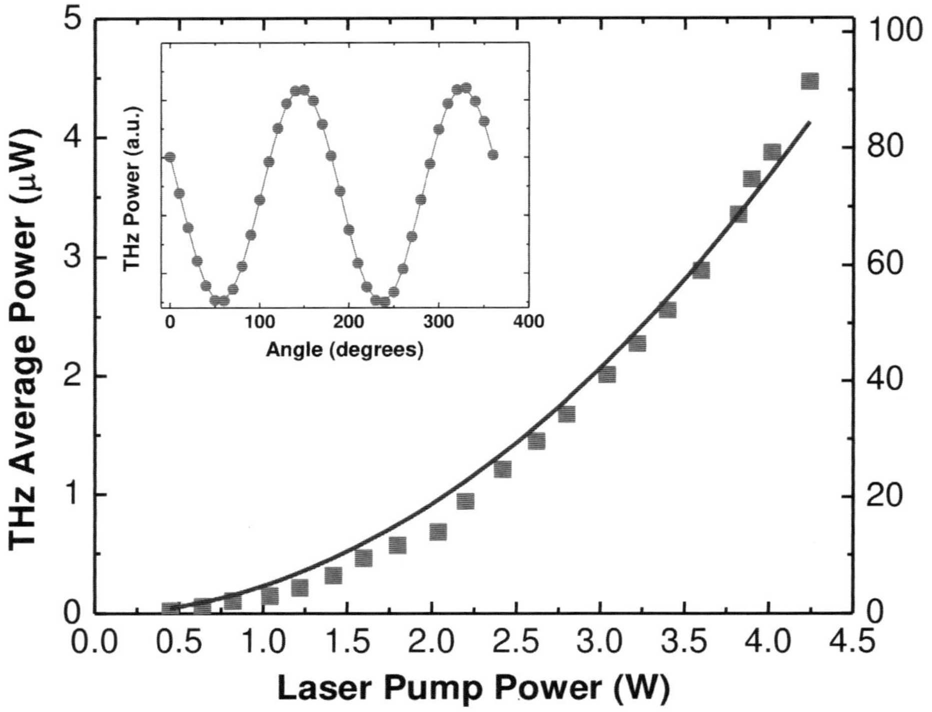

Fig. 15. THz average and peak power dependence on the combined power at 1047 nm and 1053 nm; straight lines correspond to quadratic fits to data points. Inset is the intensity of the THz radiation as a function of the azimuthal angle of a THz polarizer.

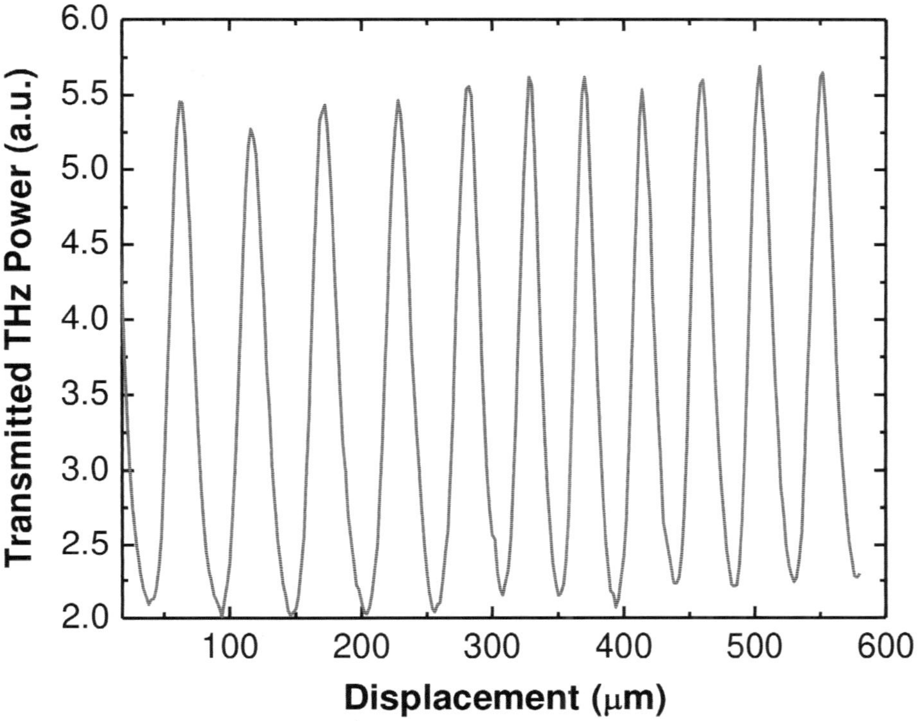

Fig. 16. THz power at 100.5 μm, generated by mixing dual frequencies from Nd:YLF and Nd:YAG lasers, being transmitted through a Si etalon as a function of the displacement between two Si wafers.

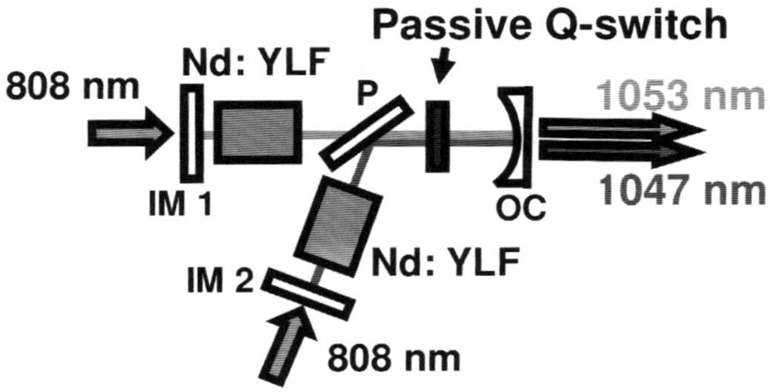

Fig. 17. Experimental setup of the dual-frequency passively Q-switched Nd: YLF solid-state laser. IM 1 and 2, input mirror; P, polarizer; OC, output coupler.

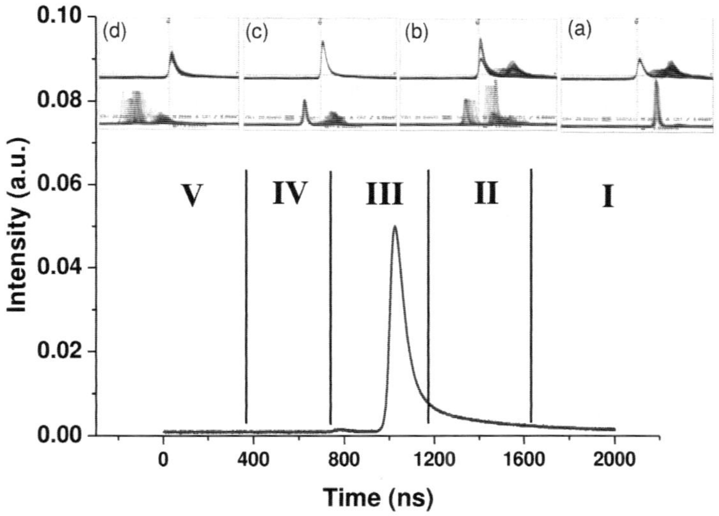

Fig. 18. Temporal profiles of dual-frequency laser pulses.

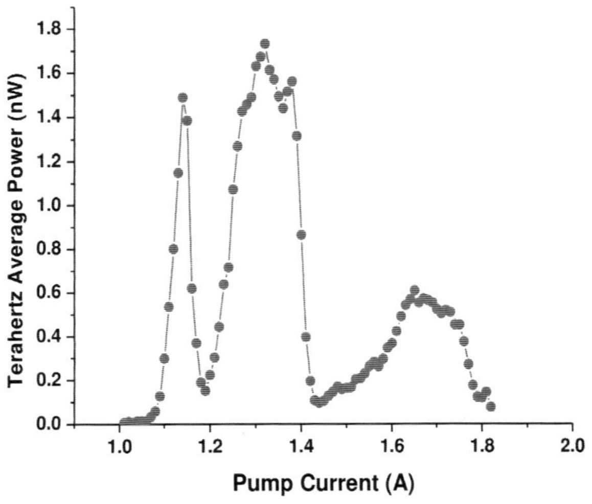

Fig. 19. THz intensity vs. pump current of 1053 nm laser.

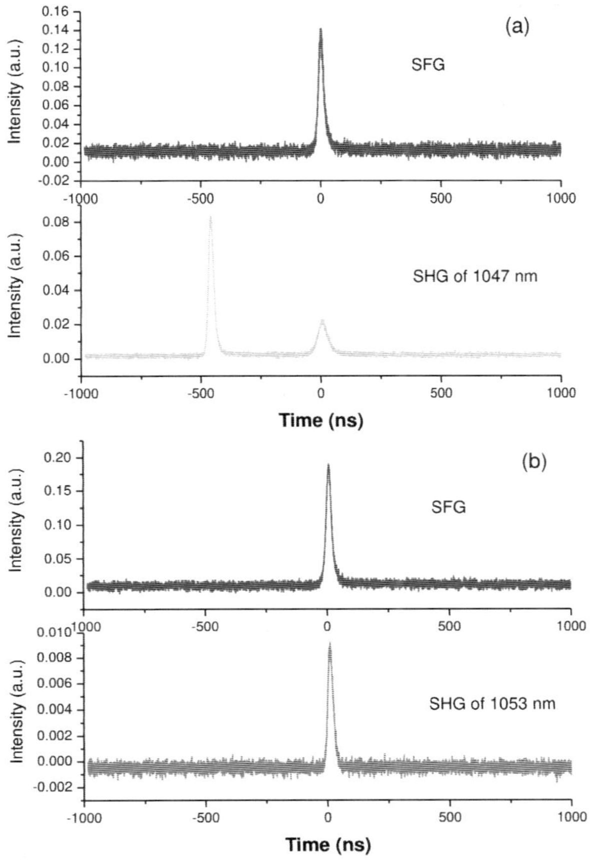

Fig. 20. a) Synchronized pulses of SFG and SHG of 1047 nm; (b) Synchronized pulses of SFG and SHG of 1053 nm.

Interplay of light and collisions in photon echo formation

N N Rubtsova[1], V G Gol'dort[1], V N Ishchenko[1], E B Khvorostov[1], S A Kochubei[1], V A Reshetov[2], and I V Yevseyev[3]

[1]A.V. Rzhanov Institute of Semiconductor Physics of Siberian Branch of Russian Academy of Sciences, 13, Academician Lavrentyev prospect, Novosibirsk, 630090, Russia
[2]Togliatti State University, 14, Belorusskaya str., Togliatti, 445667, Russia
[3]Moscow Physical Engineering State Institute, 31,Kashirskoye shosse, Moscow, 115409, Russia

E-mail: rubtsova@isp.nsc.ru

Abstract. Situations when a set of resonant radiation pulses can not create photon echo (PE) or stimulated photon echo (SPE) at relatively simple optically allowed transition in a gas are investigated. Some cases when dilution by atomic buffer can imply PE or SPE generation are analyzed. Discovered experimentally mechanisms, amplitudes, and polarization properties of these collision induced PE and collision induced SPE are in agreement with theoretical predictions. Dynamic Stark suppression of photon echo is also studied.

1. Introduction

The most common way to generate photon echo is to apply a sequence of at least two light pulses to resonant non-homogeneously broadened (in the case of gaseous media — due to Doppler effect) quantum transition. In the frames of the model which represents a matter as a gas of two-level atoms, a definite set of optimal conditions for an echo generation appears [1]. The first condition comes from evident requirement of low gas pressure: collision relaxation time should be considerably less than time delay between exciting light pulses. Besides, exciting pulses area should be optimal; this requirement is connected with the nature of interaction between quantum transition and resonant light, i.e. with Rabi oscillations.

In realistic atomic or molecular transitions, the levels degeneracy implies new phenomena. In particular, polarizations of exciting pulses acquire great importance [2]. Depending on the angular momenta of the resonant working levels, photon echo polarization and even its power become strongly dependent on the polarizations of exciting pulses.

The goal of this article is to highlight the conditions which imply apparently strange (from the viewpoint of photon echo formation) results which origin is related to specific features of light and collisions.

2. Experimental technique

The experimental scheme in Fig. 1 shows the excimer XeCl laser (wavelength is 308 nm, pulse duration is 5–6 ns, average pulse energy is of about 20 mJ) used as a pump source for the dye laser operating on a solution of Rhodamine 110 and the external optical scheme.

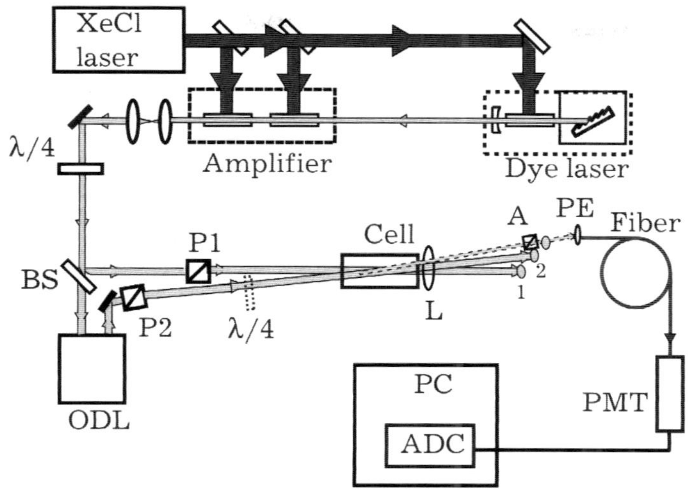

Figure 1. Experimental scheme for investigation of photon echo polarization.

The dye laser consisted of master oscillator (10 cm cavity shown in Fig. 1 by broken line box) and two sequential amplification stages. A slow circulation of alcohol dye solution was ensured for master oscillator and both amplifiers. The master oscillator cavity was formed by 1800 grooves/mm grating, operating in the second diffraction order in the auto-collimation regime and an output dielectric mirror. The grating placed inside a chamber with a controllable dry nitrogen pressure allowed us to make a fine tuning of the master oscillator frequency. An additional 1200 grooves/mm diffraction grating placed at the output of the master oscillator served as spectral selector for the rejection of spontaneous emission of the dye cell. Spatial-angular selectors consisted of two positive lenses with a diaphragm between them were placed between the master oscillator and the first amplification stage as well as between the amplification stages. The central part of the beam emerging from the final amplification stage was extracted through the diaphragm of 1.5 mm diameter and expanded by a Kepler telescope up to 10 mm diameter to be delivered to the external optical scheme. The application of spatial-angular selectors and a telescopic beam expander allowed us to reduce the background related to the super-luminescence of dye and to prepare a radiation beam with a sufficiently uniform transverse intensity distribution and with a divergence close to the diffraction limit. The resulting power of dye laser pulse after passing through two amplifiers was 250 μJ, the pulse duration was 2.5–5 ns. A neutral optical filter was placed at dye laser output to control the areas of exciting pulses. The filter transmission of 10 percent was applied for echo formation in our experimental condition — in such a case the pulses areas were close to the optimal values of $\pi/2$ and π for the first and second pulses correspondingly. Spectral properties of our dye laser were improved as compared to the previous work [3] by Fabry-Pérot interferometer inserted into the master oscillator between the grating and the dye cell. Now dye laser pulse spectrum contains only one strong mode with the spectral width less than 100 MHz in 90 cases out of 100. Other 10 percent of radiation pulses contain two modes with the spectral separation of 1.5 GHz. Optical responses from medium for these 10 percent of pulses were rejected at a stage of computer processing of registered data.

External optical scheme (see Fig. 1) included beam splitter BS and optical delay line ODL used to create the pair of coherent exciting pulses and two Glan-Taylor prisms P1 and P2 as independent polarizing elements for each of two exciting pulses. Quarter-wavelength plate (denoted as "$\lambda/4$" in Fig. 1) inserted between the output beam and the beam splitter BS gave the possibility of choosing the necessary orientation for the polarization plane of each beam by the corresponding orientation of Glan-Taylor prisms. If necessary, linear polarization of any exciting beam (or even both) could be converted into circular polarization using another quarter-wavelength plate (marked by symbol "$\lambda/4$" and drawn as broken line plate inserted into the second exciting beam in Fig. 1). The film polarizer A placed after the working cell (Cell in Fig. 1) provided the polarization analysis of photon echo detected by high sensitive short time response photomultiplier PMT. An additional photomultiplier (not shown in Fig. 1) was used to control the power of both exciting pulses. All the signals were converted into digital form by the analogue-to-digital converter (ADC in Fig. 1) triggered by the first exciting pulse. Data acquisition, averaging and treatment were provided by personal computer PC.

The scheme of angled photon echo was used for reliable photon echo detection. Part of the beam delayed by an optical delay line ODL (second pulse of exciting radiation) was directed to the working cell under the angle of about 2×10^{-3} rad to the first pulse. Such a scheme allowed us to detect a signal of photon echo without considerable losses of its amplitude and, simultaneously, without photomultiplier bleaching by the radiation of exciting pulses — due to spatial separation between echo and exciting pulses. The photon echo signal was focused onto the fiber end (Fiber in Fig. 1) and delivered to photomultiplier PMT for detection.

The state of polarization for exciting pulses and for the photon echo were analyzed by the detection of polarization diagrams obtained using the analyzer A (see Fig. 1) in front of photomultiplier PMT.

To suppress the influence of power and spatial distribution instabilities of excimer laser on the data, several technical solutions were undertaken, including the above-mentioned design of dye laser and external optical scheme, investigation of dye laser radiation properties and special technique of data acquisition and treatment (see for details [3, 4]).

In this experiment, a special attention was paid to the absence of any external stray fields, especially the magnetic ones, in the area of the working cell filled with ytterbium vapours. The magnetic field was measured inside the cold cell without vacuum and its value was not higher than 20 mG. The evaporation of ytterbium was induced by double heating wires passing along the cylindrical cell. This construction allowed us to minimize the stray longitudinal magnetic field from the heater. The absence of any stray magnetic field was finally controlled via the absence of non-Faraday rotation for the photon echo polarization plane. This could be done with accuracy of polarizer angle measurement of about one degree; above-mentioned upper limit for stray magnetic field (if this field was directed strongly along the cell axis) could imply the non-Faraday rotation by an angle of about 1.4 degree according to our measurements in [3]. Hence, both methods to estimate the stray magnetic field agree.

The temperature of the working cell with ytterbium vapour was kept between 800 and 840 K, which corresponds to Yb pressure 3.9–12.9 mTorr. In order to avoid the undesirable collision effects on the polarization properties of photon echo, the temperature in a working cell was maintained as low as possible; so most of experiments were made at 800 K. The working cell heater was driven by a special autonomic electronic scheme which warranted the temperature stability not worse than ± 2 K.

The sample with a natural content of ytterbium isotopes was used. The exciting radiation wavelength was thoroughly controlled by an electronic wavelength meter in order to get the operation spectral region coinciding with the Doppler center of ^{174}Yb isotope. The precision of wavelength meter was 10^{-4} nm which (for the wavelength of ^{174}Yb inter-combination transition under investigation) corresponds to 100 MHz in the frequency scale. Isotopic shifts for natural ytterbium were measured in [5]. The neighbouring isotopic lines for ^{174}Yb are ^{176}Yb and ^{172}Yb, which are shifted by -954 MHz and $+1$ GHz, correspondingly. All these even isotopes do not have a hyperfine structure, too. Ytterbium odd isotopes have higher isotopic shifts and do not overlap with the Doppler width of ^{174}Yb absorption line in our experimental conditions.

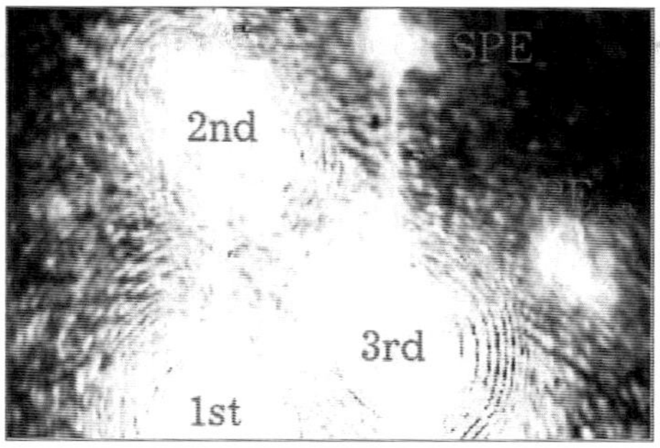

Figure 2. Exciting optical beams, SPE beam spot and PE generated by the first and the third exciting pulses PE_{1-3}.

The deviation of exciting pulses polarization from the ideal linear one was not greater than 10^{-3} in power. The circular polarization of exciting pulses deviated from the ideal circularity at about three percent. The origin of this inaccuracy of measurements is related to some amplitude instability of dye laser (Rhodamine 110) pumped by the excimer laser.

For SPE investigations, corresponding modification of the set up was made. External optical scheme used in our early works on SPE at the transition $(6s^2)\ ^1S_0 \leftrightarrow (6s6p)\ ^3P_1$ in ^{174}Yb vapour was described in [6, 7]. The pulsed Rhodamine 110 dye laser was a source of three resonant radiation exciting pulses formed by two optical delay lines. Each of the beams passed through the Glan-Thompson polarizers; the beams were directed into the working cell with ytterbium vapour in a non-coplanar geometry making small ($\approx 2 \cdot 10^{-3}$ rad) angle between any pair of exciting pulses propagation vectors. As a result, the angled SPE could be spatially selected from the exciting pulses.

Fig. 2 shows picture detected by web-camera in the focal plane of lens L (Fig. 1 for the case of SPE generation. Three exciting beams, SPE beam spot, and PE generated by the first and by the third exciting pulses can be seen here.

In this work, unlike [6, 7], the dye laser was pumped by the second harmonic of Nd^{3+}:YAG laser, which provided much better amplitude stability for exciting pulses.

The parasite longitudinal magnetic field was minimized with residual value of 20 mG to ensure correct results on SPE polarization. Linear polarization of each exciting pulse was checked through polarization diagram, and the deviation of polarization from the "ideal" linear one was of about 10^{-3} in power.

The temperature of the working cell was kept at 820 K with accuracy of 2 K. The buffer gas pressure was measured with accuracy of $5 \cdot 10^{-3}$ Torr in the range of 0–3 Torr.

3. When resonant exciting radiation does not form coherent response in pure gas

3.1. No PE at $0 \leftrightarrow 1$ transition

Polarization properties of photon echo (PE), stimulated photon echo (SPE) and other coherent transients is the first source of seemly paradoxical situation: we have exciting pulses of radiation which are resonant to a chosen quantum transition, they have optimal pulse areas, but they can not create photon echo.

Really, let us consider relatively simple transition of type $0 \leftrightarrow 1$. Experimental realization of such a transition can be done at the inter-combination transition $(6s^2)\ ^1S_0 \leftrightarrow (6s6p)\ ^3P_1$ of ^{174}Yb vapour. It is worth to note that polarization properties of photon echo generated in a gas at the transitions of type $0 \leftrightarrow 1$ or $1 \leftrightarrow 1$ do not depend on the area of exciting radiation pulses. This property predicted in [8] make two above mentioned transitions very attractive for experimental research. Photon echo generated in a gas by two pulses of resonant radiation with linear polarizations at such transitions should have linear polarization, and its polarization vector should be located along the polarization vector of the second exciting pulse. So, photon echo generated at these two types of transitions by linearly polarized exciting pulses "follows" the polarization of the second pulse. If polarization vectors of exciting pulse make an angle ψ, PE amplitude is proportional to $\cos(\psi)$. For the angle $\psi = 90°$ photon echo is absent [8]. Obvious explanation is that polarization created at the transition $0 \leftrightarrow 1$ by the first exciting pulse with linear polarization oriented, say, vertically can not be re-phased by the second exciting pulse with linear polarization oriented horizontally.

The combinations of circular and linear polarizations of exciting pulses were investigated for the transition $0 \leftrightarrow 1$ experimentally in ^{174}Yb vapour [9]. For all combinations of exciting pulses polarizations, for pure ytterbium vapour, the photon echo polarization "follows" the polarization of the second exciting pulse — in agreement with [8].

For exciting pulses with circular polarizations we meet another situation when PE can not be generated: that is for two exciting pulses with opposite circular polarizations. In this case, the first exciting pulse creates optical coherence between one definite pair of magnetic sub-levels (say, $m = 0$ of the lower and $m = +1$ of the upper levels). This optical coherence, in other words, circular polarization wave, can not by re-phased by the second exciting pulse which interacts with another pair of magnetic sub-levels (in this example, $m = 0$ of the lower and $m = -1$ of the upper levels).

3.2. No SPE at $0 \leftrightarrow 1$ *transition*

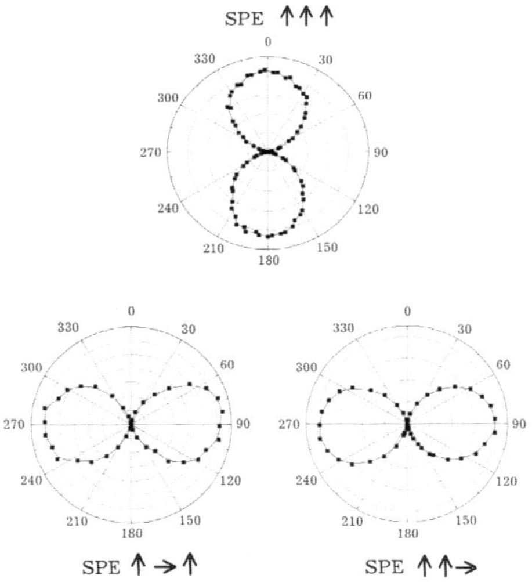

Figure 3. SPE polarization in a pure gas at $0 \leftrightarrow 1$ transition has linear polarization coinciding with exciting pulses polarization (upper graph, SPE↑↑↑) or its linear polarization "follows" polarization of crossed exciting pulse(lower row SPE↑→↑ and SPE↑↑→).

Now, let us consider stimulated photon echo (SPE) at the transition $0 \leftrightarrow 1$ in a pure gas.

If all three pulses of exciting radiation have identical linear polarization, the SPE at the transition $0 \leftrightarrow 1$ is polarized in the same manner. This polarization diagram denoted as SPE↑↑↑ is shown in the upper part of the Fig.3.

In the case when one of the exciting pulses was polarized linearly and orthogonally to another two exciting pulses, SPE polarization at the transition $0 \leftrightarrow 1$ in the pure ytterbium vapour showed linear polarization oriented in parallel to this exciting pulse of "crossed" polarization.

Journal of Physics: Conference Series **414** (2013) 012004 doi:10.1088/1742-6596/414/1/012004

This result is illustrated by two polarization diagrams in the lower row of the Fig.3. Left part of this row corresponds to case of the second pulse with crossed linear polarization denoted as SPE↑→↑. Lower right part of the Fig.3 denoted as SPE↑↑→ corresponds to the case when the third exciting pulse has linear polarization crossed to another two pulses.

It was surprising to find that the situation when the second and the third pulses have "crossed" polarization relative to the first pulse is not equivalent to the case when the third pulse has the "crossed" linear polarization relative the first two exciting pulses. Evidently, the first exciting pulse plays a special role in the SPE formation. In a pure ytterbium vapour at the transition $0 \leftrightarrow 1$ stimulated echo of type SPE→↑↑ does not appear.

4. Collision induced PE

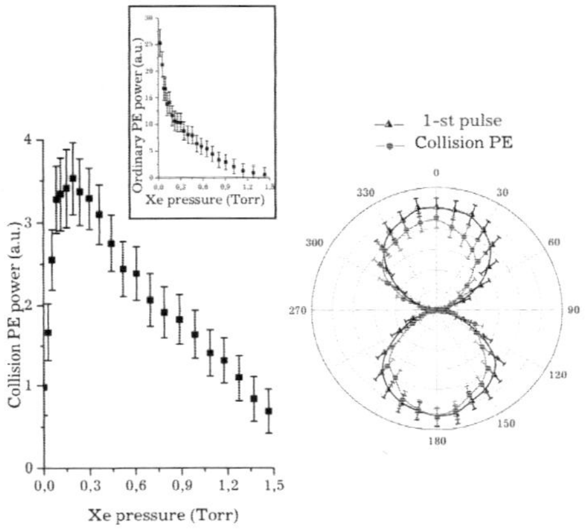

Figure 4. Ordinary (upper insert) and collision induced PE dependence on buffer gas pressure (left bottom graph) and polarization diagram together with 1-st exciting pulse (right graph).

The left bottom part of the Fig. 4 shows the power (in arbitrary units) of collision-induced photon echo generated by two mutually orthogonal linearly polarized exciting pulses of laser radiation resonant to the transition $(6s^2)\ ^1S_0 \leftrightarrow (6s6p)\ ^3P_1$ of ^{174}Yb in a mixture of Yb+Xe as a function of Xe pressure. Collision-induced echo is absent at zero buffer gas pressure. As buffer gas pressure grows, collision-induced photon echo power increases, then reaches its maximum value and starts to decay exponentially in a high pressure limit of buffer gas.

This non-monotonic behaviour of collision-induced echo power as a function of buffer gas pressure is in contrast to ordinary (generated in the same gas mixture by pulses of resonant radiation with linear parallel polarizations) echo behaviour. The error bars for echoes power shown in both parts of Fig. 4 arose from the statistics of power storage and averaging; the origin of these errors lies in fluctuations of pulse laser power. The buffer pressure was measured with high accuracy; these errors are less than experimental points size; that is why the error bars for pressure are not indicated in Fig. 4.

The maximum value of collision-induced echo is reached at Xe pressure of about 200 mTorr, and its value is several orders of magnitude lower than ordinary echo power at the same value of buffer gas pressure.

For comparison, the upper insert of Fig. 4 shows buffer gas behaviour for the "ordinary" photon echo generated at the same conditions but by application of exciting pulses with parallel linear polarizations. Here, the monotonous decay of PE power can be seen.

Polarization of collision-induced photon echo generated by two mutually orthogonal linearly polarized exciting pulses of laser radiation resonant to transition $(6s^2)\ ^1S_0 \leftrightarrow (6s6p)\ ^3P_1$ of ^{174}Yb in a mixture of Yb+Xe was investigated using analyzer A (see Fig.1). In the right part of Fig. 4, the power of collision induced echo (gray circles) is plotted as a function of rotation angle of analyzer A. The distances from the center of the picture are proportional to the radiation power transmitted by analyzer A at a given rotation angle. Both these polarization diagrams — for collision induced photon echo and for the first exciting pulse (black squares) — are represented in Fig. 4 for comparison. The error bars for echo and exciting pulse power are indicated in both curves. As it was already discussed, the polarizer angle measurement accuracy was about one degree; this value is less than experimental points size; so, the angle error bars are not shown in Fig. 4.

Unlike ordinary photon echo, which "follows" the polarization of the second exciting pulse, the collision-induced echo polarization coincides with the polarization of the first exciting pulse. This result well agrees both with the earlier prediction [10], and with the analytical calculation in the previous section.

As it was predicted in [8] and experimentally verified in [9], there is another combination of exciting pulses polarizations which gives zero power of ordinary photon echo, that is opposite circular polarizations. The calculations [11] predict that the collision-induced echo can not be generated at transition $0 \leftrightarrow 1$ even if we take into account depolarizing collisions anisotropy.

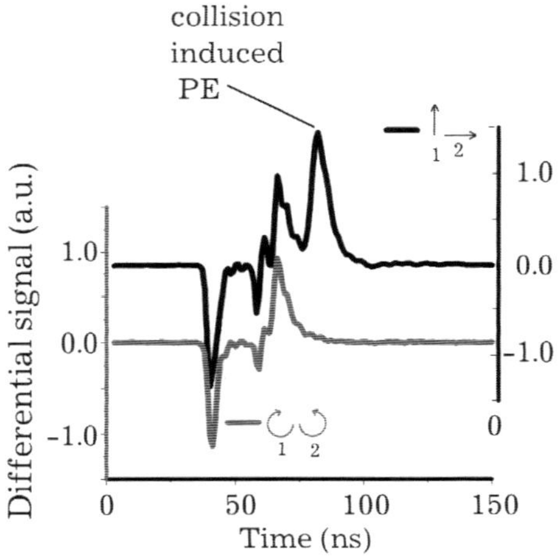

Figure 5. Collision induced PE appears for linear crossed polarization but does not exist for opposite circular polarizations.

The sensitivity of collision-induced echo detection was improved the following way. From the signal detected by PMT at the conditions of resonant excitation at wavelength $\lambda = 555.802$ nm (as measured by λ-meter at the atmospheric pressure), which corresponded to the center of Doppler contour of ^{174}Yb, the signal at non-resonant wavelength of $\lambda = 555.815$ nm was subtracted. This technique permitted the suppression by at least one order of magnitude of a residual stray signal arising from the wings of spatial distribution of exciting pulses. This differential signal is shown in Fig. 5 for two experimental cases: one corresponds to mutually orthogonal linear polarizations of exciting pulses (black curve, right vertical axis), and another one (grey curve, left vertical axis) — to the opposite circular polarizations. For the sake of clarity, the black curve and its axis are shifted up.

In Fig. 5, the following signals appear from the left to the right. Two peaks directed down were interpreted as absorption of exciting pulses in a mixture of Yb+Xe; the time delay between them corresponds to 42 ns, which equals to the time delay between exciting pulses provided by the optical delay line. At the tail of the second peak the signal changes its sign to the opposite one. This can be interpreted as the appearance of a free polarization decay signal. All these signals appeared both for linear orthogonal and opposite circular polarizations of exciting pulses. As for the photon echo signal, it appeared only for linear orthogonal polarizations of exciting pulses. It is worth emphasizing that other experimental conditions (pressure of Yb and Xe in a mixture, powers and radial distributions of exciting pulses, time delay between them, a small angle between wave vectors) were identical. The only one difference consisted in the quarter wave plate inserted in both exciting beams in order to convert linear polarizations into circular ones. As it is clear from Fig. 5, the collision echo is detected only for linear orthogonal polarizations of exciting pulses.

So, in the same Yb+Xe mixture at identical intensities of exciting pulses of resonant radiation, only pulses with linear perpendicular polarizations can generate collision induced photon echo. Pulses with opposite circular polarizations can not do that. In our experiments [11] this was shown with accuracy of about 10^{-2} of collision induced PE power for crossed linear polarizations.

5. Collision induced SPE

All above described cases of SPE generated in a pure ytterbium vapour at the transition $0 \leftrightarrow 1$ are in complete agreement with the theory [12] considered with an account of elastic depolarizing collisions. It was shown in [12] that with the polarization of the first pulse orthogonal to that of the second and third pulses, the appearance of the echo signal will be entirely determined by the action of collisions. The existence of this echo is determined by the difference of collision induced decay rates of two polarization moments of the upper working level — the alignment and the orientation.

As it was found experimentally, collision induced SPE can really appear for the case SPE$\rightarrow\uparrow\uparrow$ at the huge dilution of active atoms by buffer atoms Xe. The right graph of Fig.6 shows collision induced SPE power versus buffer gas pressure. SPE$\rightarrow\uparrow\uparrow$ is practically zero in the absence of buffer, its amplitude increases sharply as buffer gas pressure raises from zero. Then collision induced SPE$\rightarrow\uparrow\uparrow$ reaches its maximum value approximately at xenon pressure of about $200 \cdot 10^{-3}$ Torr. Let's remind that active atoms pressure was kept at $\approx 7.5 \cdot 10^{-3}$ Torr, so, Xe pressure was more than an order of magnitude higher. As Xe pressure overcomes value of $200 \cdot 10^{-3}$ Torr, collision SPE power starts decreasing, as is clear from the right graph of Fig.6.

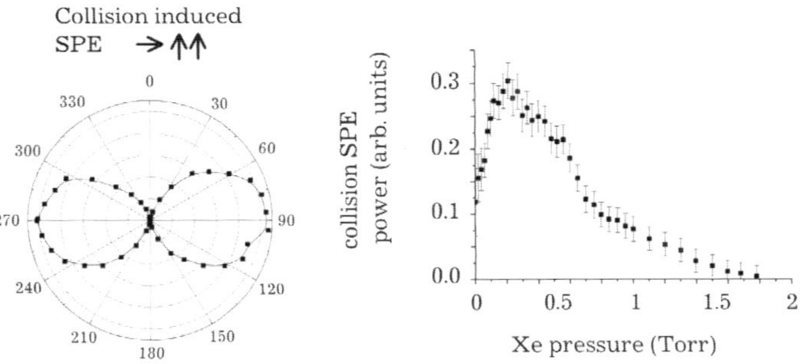

Figure 6. Polarization diagram (left graph) for collision induced SPE of type SPE$\rightarrow\uparrow\uparrow$ and collision induced SPE amplitude dependence on the buffer gas pressure (right graph).

The left graph of Fig.6 shows polarization diagram of SPE induced by collisions with buffer gas Xe at the transition $0 \leftrightarrow 1$ in ytterbium vapour. The diagram was registered at buffer gas Xe pressure $200 \cdot 10^{-3}$ Torr which is optimal for collision induced SPE formation. The collision induced SPE polarization is linear and it coincides with the polarization of the first exciting pulse "crossed" with another two exciting pulses.

Journal of Physics: Conference Series **414** (2013) 012004 doi:10.1088/1742-6596/414/1/012004

6. When resonant radiation blocks PE generation

Intense resonant CW electromagnetic radiation in the presence of weak irreversible relaxation changes the behaviour of a quantum system due to its dynamic interaction with the field. In the stationary case, quasi-energy levels appear which correspond to the so-called dressed states of an atom or a molecule: the initial levels are split into two sub-levels spaced by a distance determined by the Rabi frequency. The absorption spectrum usually detected by the probe field [13, 14] is significantly changed. The dynamics of nonstationary optical phenomena including coherent transient processes also changes significantly in the presence of a strong field. The known experimental works in this area for purely rotational microwave transitions in OCS molecules [15], in a Yb atomic beam [16], for Cs atoms trapped in an optical lattice [17], and in a dense Sr atomic gas [18] were carried out in the presence of the homogeneous broadening of a spectral line.

The inhomogeneous broadening of the line in the rarefied gas, which is caused by the Doppler effect, complicates the effect of the intense radiation on the kinetics of the coherent transient processes as demonstrated theoretically [19] for the photon echo in a three-level system in the presence of a dressing-field pulse.

The possibility to control coherent transient processes by intense resonant CW radiation was demonstrated in our work [20] in the rarefied gas, i.e., when the Doppler effect is significant.

In [20] coherent transient processes were generated through the Stark switching of levels in the presence of CW laser radiation. Experiments were carried out at the vibrational rotational transition R(4,3) $0 \leftrightarrow 1\ \nu_3$ $^{13}CH_3F$ in the radiation field of a CO_2 laser on the 9 P(32) generation line. The laser radiation was single mode with a line width of no more than 100 kHz and its frequency was tuned to the center of the absorption line of $^{13}CH_3F$.

Intense-field radiation was linearly polarized with polarization vector parallel to the Stark field strength, and its intensity was varied from 0.6 to 56 W/cm^2. The intensity of the weak field linearly polarized orthogonally to the strong field was equal to 0.2 W/cm^2. After the passage through a cell with the low-pressure $^{13}CH_3F$ gas, the weak field was separated by means of the reflection from a plate placed at the Brewster angle for the strong field, and then it was detected. The Stark pulses were applied on the internal electrodes of the working cell. The transverse distribution of the beams of high- and low-intensity radiation was close to a Gaussian distribution whose radius was about 2 mm for high-intensity radiation and was somewhat larger for low-intensity radiation. The intensity of the weak field was sufficient for the formation of coherent transient processes; i.e., this radiation gave rise to the saturation of the working transition; the term weak is used here only as compared to more intense radiation.

Focusing of radiation beam made it possible to reach a higher intensity; from other side, owing to this narrow beam, the relaxation due to the transit effect prevails over collision relaxation rate, and gave relaxation rate of about $\Gamma = 1$ MHz. This value is much lower than the Doppler width of ≈ 80 MHz. Thus, all of our experiments were performed in the presence of a significantly inhomogeneous broadening of the spectral transition.

The photon echo in the $^{13}CH_3F$ gas was formed and detected in low intensity, linearly polarized radiation. The molecular levels undergo a fast switching by two Stark voltage pulses with durations $T_1 = 0.1\mu s$ and $T_2 = 0.2\mu s$ spaced by the delay time $\tau = 1\mu s$. The gas was simultaneously irradiated by intense resonant CW radiation linearly polarized along the Stark field and orthogonally to the weak field forming the photon echo. An echo signal appears in the form of a weak heterodyne beating signal at the time approximately equal to the double delay time between the Stark pulses.

Each photograph in Fig. 7 exhibits from the left to the right two intense signals corresponding to the gas response to the action of two forming Stark pulses (a sort of optical nutation signals), and weaker response appearing through the time interval approximately equal to the delay time between the Stark pulses after the second Stark pulse, which is just the photon echo signals.

Journal of Physics: Conference Series **414** (2013) 012004 doi:10.1088/1742-6596/414/1/012004

The strong (control) field radiation power is marked by P_z and is indicated for each photograph in Fig. 7. As seen in the images Fig. 7, the behaviour of the photon echo amplitude is a non-monotonic function of the strong field. At low control field an increase of PE amplitude was observed; at higher control field, PE signal decrease and complete suppression of the photon echo was detected. In addition, a change in the shape of the photon echo signal is seen in Fig. 7 (we recall that the heterodyne receipt occurs). The maximum of the photon echo signal or, for a more complex shape, the center of gravity of the photon echo signal also varies non-monotonically with driving field power. These changes occur in the limits that do not exceed the duration of the forming pulses.

Observed photon echo variations as a function of dressing field (amplitude, PE-signal temporal shape and t_{min}) are non-monotonous functions of dressing field, in agreement with calculations.

As it can be seen in Fig. 7, photon echo can by suppressed by the intense radiation due to dynamic Stark effect.

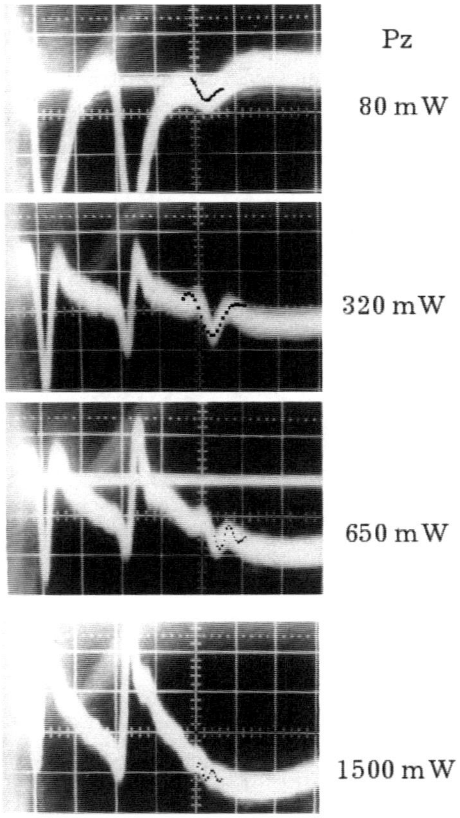

Figure 7. Collision induced SPE.

7. Discussion

Because of the levels degeneracy, we meet situations when exciting pulses can not create PE or SPE at the relatively simple transition $0 \leftrightarrow 1$.

In some cases, collision relaxation can change this situation. In particular, collision induced PE at the transition $0 \leftrightarrow 1$ appears due to anisotropy of collisions (i.e. due to different relaxation rates along and across velocity of active atoms) for the pair of exciting radiation pulses with crossed linear polarizations. Exciting pulses with opposite circular polarization can not create collision induced PE.

Explanation "at the finger tips" for the collision induced PE generated at the transition $0 \leftrightarrow 1$ looks as following. The first exciting pulse creates polarization for each active atom, and polarization vector for each atom is parallel to polarization vector of the first exciting pulse. The anisotropic collision relaxation "gives a turn" to dipole moments of individual atoms, depending on their velocity. As a result, medium polarization created by the first exciting pulse acquires orthogonal component; as a result, it can interact with the second pulse of resonant exciting radiation (which has crossed linear polarization) and obtains possibility for Doppler re-phasing. During the time interval after the second pulse and collision induced echo appearance, along with the Doppler rephrasing, the same kind of reverse atomic dipole "turns" due to collisions takes place, so that at the instance of echo appearance the atomic dipoles are aligned along the polarization vector of the first radiation pulse, and that is why collision induced photon echo has linear polarization along the first exciting pulse.

Collision induced SPE at the transition $0 \leftrightarrow 1$ appears under action of three linearly polarized exciting pulses with mutual orientation of type $\rightarrow\uparrow\uparrow$. Such a combination of polarizations can not create SPE in a pure gas. At the dilution of active atoms by great amount of heavy buffer atoms, SPE appears due to the difference in relaxation rates of polarization moments (orientation and alignment).

Both collision induced PE and collision induced SPE at the transition $0 \leftrightarrow 1$ have some similarity. Both have linear polarization coinciding with polarization of the first exciting pulse. Both reveal non-monotonous buffer gas pressure dependence. Both have small power, several orders of magnitude less that corresponding "ordinary" signal (created by two or three pulses of identical linear polarizations).

However, they have quite different nature. Mechanism of induced PE generation at the transition $0 \leftrightarrow 1$ is related to collision relaxation dependence on the *direction* of active atoms translational velocity. Collision induced SPE appears due to different collision relaxation rates for orientation and alignment of the upper working level.

There is possibility to suppress PE or other coherent response due to dynamic Stark effect produced by additional (driving) radiation.

8. Acknowledgments

This work was supported by the Russian Foundation for Basic Research (grants 10-02-00071 and 11-02-00141), and by Physical Sciences Department Programme "Fundamental optical spectroscopy and its applications" (grant III.9.8).

References

[1] Allen L, Eberly J H 1975 *Optical Resonance and Two-Level Atoms* New York: John Wiley & Sons
[2] Yevseyev I V, Yermachenko V M, and Samartsev V V 2004 *Depolarizing collisions in nonlinear electrodynamics*, CRC Press, Boca Raton, London, New York, Washington, D.C., 318 pp., 2004.
[3] Rubtsova N N, Ishchenko V N, Khvorostov E B, Kochubei S A, Reshetov V A, and Yevseyev I V 2004 *Physical Review* **A 70** 023403
[4] Rubtsova N N, Gol'dort V G, Yevseyev I V, Ishchenko V N, Kochubei S A, and Khvorostov E B 2008 *JETP Lett.* **87** 103
[5] Wijkngaarden van W A, and Li J 1994 *J. Opt. Soc. Am.* **B 11** 2163
[6] Yevseyev I V, Ishchenko V N, Kochubei S A, Reshetov V A, Rubtsova N N, Khvorostov E B 2006 *Laser Physics* **16** 535
[7] Yevseyev I V, Ishchenko V N, Kochubei S A, Reshetov V A, Rubtsova N N, Khvorostov E B 2007 *Laser Physics* **17** 324
[8] Alekseyev A I, Yevseyev I V 1969 *JETP* **56** 2118

[9] Khvorostov E B, Gol'dort V G, Ishchenko V N, Kochubei S A, Rubtsova N N, Salakhutdinov V D, Yevseyev I V 2011 *Laser Phys. Lett.* **8** 290
[10] Matskevich V K, Yevseyev I V, Yermachenko V M 1978 *Optika i Spektroskopiya* **45** 17
[11] Rubtsova N N, Gol'dort V G, Ishchenko V N, Khvorostov E B, Kochubei S A, Reshetov V A, Yevseyev I V 2011 *Phys. Rev.* **A 84** 033413
[12] Reshetov V A and Popov E N 2012 *J. Phys. B: At. Mol. Opt. Phys.* **45** 225502
[13] Rubtsova N N 2001 *Opt. Spectrosc.* **91** 53
[14] Rubtsova N N 2004 *Optical Spectroscopy and Frequency Standards. Molecular Spectroscopy* 44 Tomsk: IOA
[15] Glorieux P, Legrand J, Macke B, and Segard B 1973 *Mol. Phys.* **25** 739
[16] Bai Y S, Mossberg T W, Lu N, and P. R. Berman P R 1986 *Phys. Rev. Lett.* **57** 1692
[17] Triche C, Verkerk P, and Grynberg 1996 *OSA Topics on Ultracold Atoms and BEC* **7** 82
[18] Mazzoni 1999 *Opt. Commun.* **171** 99
[19] Berman P R 1984 *Opt. Commun.* **52** 225
[20] Rubtsova N N a and Reshetov V A 2008 *JETP Letters* **87** 276

Switching between superluminal to subluminal velocities and Tunable slow light in a four-level atomic system

Sankar Davuluri and Yuri V. Rostovtsev[1]

[1] *Department of Physics, University of North Texas, Denton, TX 76203*

A four-level atomic system interacting with two strong drive fields and a weak probe field has been studied. We demonstrate that the group velocity of probe field can be controlled to switch from subluminal to superluminal or even negative values around the probe resonance without any significant absorption. The basic idea is to create a gain with slow light on the probe beam resonance with one drive field while the other drive creates a dip in the gain profile of probe leading to superluminal and negative group velocities. A simple switching from slow light to negative group velocity light can be obtained by changing the relative intensities of drive fields. We also describe a control to tune the position of slow light frequency continuously simply by adjusting the relative intensities of both the driving fields.

I. INTRODUCTION

The velocity at which a peak of optical pulse travels is known as the group velocity. It is well known that the group velocity depends on susceptibility and its rate change. When spatial dispersion is ignored[10], eq(1) shows the relation between v_g and $\chi(\omega)$ for $\chi(\omega) << 1$

$$v_g = \frac{c}{n_g} = Re\frac{c}{1 + \frac{\chi(\omega)}{2} + \omega\frac{\partial\chi(\omega)}{2\partial\omega}} \tag{1}$$

where c is the velocity of light in vacuum, n_g is the group refractive index, ω is the angular frequency and $\chi(\omega)$ is the susceptibility of the medium. By modified the group refractive index we can make $v_g << c$ which is slow light or negative i.e., $v_g < 0$ for $n_g < 0$ or even faster than light, which is superluminal velocity, for $|n_g| < 1$.

All the above mentioned cases can be realized by changing the dispersion of pulse ($\frac{\partial\chi(\omega)}{\partial\omega}$) to positive or negative values. Consider an electromagnetic radiation interacting with a two level atom non-resonantly. Then we can never have fast light because denominator is always positive and greater than or equal to one. Now consider a resonant interaction case, at the resonance electromagnetic fields are either absorbed or gained depending on the initial state of the atom. This absorption or gain of the field will be accompanied by a non zero negative or positive dispersion terms. As a result the third term in the denominator of eq(1) makes a significant contribution to group velocity leading to either slow or super luminal or negative group velocities.

In a two-level atom fast light is accompanied by strong absorption making it impossible to study. Meanwhile, we take a multi level atom which is coupled to several laser fields and making use of quantum interference we discuss the generation and control for switching from slow to negative group velocities without any absorption at resonance. One of the most famous and simplest examples for modifying group velocity by using quantum interference is Electromagnetically induced transparency(EIT)[14] where probe has very high dispersion leading to slow light.In the last decade, making use of quantum interference in a multilevel atomic system coupled with several lasers, a considerable work is done to control fast and slow light as both of them has many potential applications. for example slow light can be used in applications like radars[4][18], spectrometers[19], magnetometers[10] etc while fast light can be used in precision measurements in interferometer. Slow light is also known to enhance both rotatory[11] and longitudinal[16] photon dragging effects. However, in most of that work either fast or slow light is accompanied by absorption. There are very few papers which discuss the possibility of generation of both slow and fast light without significant absorption and being able to switch from subluminal to superluminal group velocities.

An experimental demonstration of fast light with gain is presented in [7,8]. A nice description of fast light in terms of classical interference between several different wave is given in [1]. Non-violation of causality with fast light is explained in [2, 15]. Generation of light with different group velocities is discussed in[9]. In[9] control over sub luminal and superluminal group velocities with decreased absorption is achieved by controlling phase betwen the driving fields. Switching between subluminal and superuminal group velocities is discussed [17] in a three level lambda system by using an incoherent pump on the probe field. Decay induced interference properties are used in[5] to generate subluminal and superluminal group velocities. Simultanous generation of slow and fast light at different frequencies is demonstrated in N-type atomic system in(Dingan han). Until now, most of the work done in controlling fast and slow light has significant absorption in different conditions while very few papers,using incoherent pump or phase differences, demonstrated both sub luminal and superluminal group velocities without any absorption.

In this paper we study a four-level system for controlling group velocity of a probe pulse from subluminal to superluminal velocities. While having the ability to tune slow light continuously and switching between slow to fast light without any absorption are the major results of this paper. It is noteworthy to mention that we can do all the controlling simply by adjusting the relative intensities of the drive fields.

II. MODEL

Figure-1 shows the schematic diagram of our system. The levels a-b have the lasing with out inversion. Levels b-c and a-d are resonantly coupled by two strong laser fields of Rabi frequencies ω_q and ω_d respectively. A weak probe field of Rabi frequency Ω_{ab} couples levels a and b with detuning given by δ. Population decay channels between different levels is shown in the diagrams by γ_{ij}, where i and j represent the atomic levels(a,b,c,d). The levels a-b are coupled by a weak probe field of frequency ω_p. we assume that all the initial population is in level c i.e., $\rho^o_{cc} = 1$.

Hamiltonian of the system can be written as linear combination of unperturbed (H_o) and perturbed Hamiltonians (H_i). Hamiltonian $H = H_o + H_i$ where

$$H_o = \hbar\omega_a|a\rangle\langle a| + \hbar\omega_b|b\rangle\langle b| + \hbar\omega_c|c\rangle\langle c| + \hbar\omega_d|d\rangle\langle d| \tag{2}$$

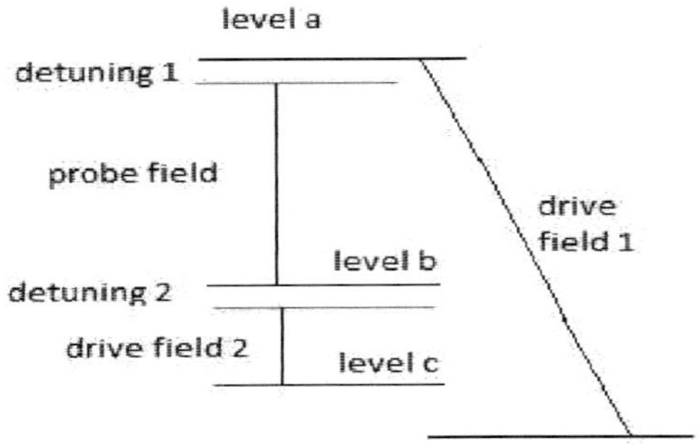

FIG. 1. detuning 1 represents the detuning of probe field $(\omega_{ab} - \omega_p)$ and detuning 2 is given by $(\omega_{bc} - \omega_q)$

$$H_i = \hbar|a\rangle\langle b|(\Omega_{ab}e^{-i\omega_p t} + \tilde{\Omega}_{ab}e^{i\omega_p t}) + |a\rangle\langle d|(\Omega_{ad}e^{-i\omega_d t} + \tilde{\Omega}_{ad}e^{i\omega_d t}) + |b\rangle\langle c|(\Omega_{bc}e^{-i\omega_q t} + \tilde{\Omega}_{bc}e^{i\omega_q t}) + c.c \tag{3}$$

Where $\hbar\omega_a, \hbar\omega_b, \hbar\omega_c, \hbar\omega_d$ represent energies of bare atomic states a,b,c,d respectively. $\Omega_{ij} = \langle i| - er_{ij}e^{ikz}|j\rangle$ and $\tilde{\Omega}_{ij} = \langle i| - er_{ij}e^{-ikz}|j\rangle$ are the Rabi frequencies between the corresponding levels i and j. ω_p,ω_d,w_q represent the angular frequencies of probe field , drive field1 and drive field2 respectively.

Time evolution of the coherence between different levels is given by the density matrix equation

$$i\hbar\dot{\rho} = [H, \rho] - i\hbar\gamma\rho \tag{4}$$

$$i\hbar\dot{\rho}_{ab} = \hbar\omega_{ab}\rho_{ab} + (\rho_{bb} - \rho_{aa})(\Omega_{ab}e^{-i\omega_p t} + \tilde{\Omega}_{ab}e^{i\omega_p t}) + \rho_{db}(\Omega_{ad}e^{-i\omega_d t} + \tilde{\Omega}_{ad}e_{i\omega_d t}) - \rho_{ac}(\tilde{\Omega}_{cb}e^{i\omega_q t} + \Omega_{cb}e^{-i\omega_q t}) - i\hbar\eta_{ab}\rho_{ab} \tag{5}$$

$$i\hbar\dot{\rho}_{db} = \hbar\omega_{db}\rho_{db} + \rho_{ab}(\tilde{\Omega}_{da}e^{i\omega_d t} + \Omega_{da}e^{-i\omega_d t}) - \rho_{da}(\Omega_{ab}e^{-i\omega_p t} + \tilde{\Omega}_{ab}e^{i\omega_p t}) - \rho_{dc}(\tilde{\Omega}_{cb}e^{i\omega_q t} + \Omega_{cb}e^{-i\omega_q t}) - i\hbar\eta_{db}\rho_{db} \tag{6}$$

$$i\hbar\rho_{da} = \hbar\omega_{da}\rho_{da} + (\rho_{aa} - \rho_{dd})(\Omega_{da}e^{-i\omega_d t} + \tilde{\Omega}_{da}e^{i\omega_d t}) - \rho_{db}(\Omega_{ba}e^{-i\omega_p t} + \tilde{\Omega}_{ba}e^{i\omega_p t}) - i\hbar\eta_{da}\rho_{da} \tag{7}$$

$$i\hbar\dot{\rho}_{dc} = \hbar\omega_{dc}\rho_{dc} + \rho_{ac}(\tilde{\Omega}_{da}e^{i\omega_d t} + \Omega_{da}e^{-i\omega_d t}) - \rho_{db}(\Omega_{bc}e^{-i\omega_q t} + \tilde{\Omega}_{bc}e^{i\omega_q t}) - i\hbar\eta_{dc}\rho_{dc} \tag{8}$$

$$i\hbar\dot{\rho}_{ac} = \hbar\omega_{ac}\rho_{ac} + \rho_{bc}(\Omega_{ab}e^{-i\omega_p t} + \tilde{\Omega}_{ab}e^{i\omega_p t}) + \rho_{dc}(\Omega_{ad}e^{-i\omega_d t} + \tilde{\Omega}_{ad}e^{i\omega_d t}) - \rho_{ab}(\Omega_{bc}e^{-i\omega_q t} + \tilde{\Omega}_{bc}e^{i\omega_q t}) - i\hbar\eta_{ac}\rho_{ac} \tag{9}$$

$$i\hbar\dot{\rho}_{bc} = \hbar\omega_{bc}\rho_{bc} + (\rho_{cc} - \rho_{bb})(\Omega_{bc}e^{-i\omega_q t} + \tilde{\omega}_{bc}e^{i\omega_q t}) + \rho_{ac}(\Omega_{ba}e^{-i\omega_p t} + \tilde{\Omega}_{ba}e^{i\omega_p t}) - i\hbar\eta_{bc}\rho_{bc} \tag{10}$$

$$i\hbar\dot{\rho}_{aa} = \hbar\rho_{ba}(\Omega_{ab}e^{-i\omega_p t} + \tilde{\Omega}_{ab}e^{i\omega_p t}) + \hbar\rho_{da}(\Omega_{ad}e^{-i\omega_d t} + \tilde{\Omega}_{ad}e^{i\omega_d t}) - \hbar\rho_{ab}(\Omega_{ba}e^{-i\omega_d t} + \tilde{\Omega}_{ba}e^{i\omega_d t}) - \hbar\rho_{ad}(\Omega_{da}e^{-i\omega_d t} + \tilde{\Omega}_{da}e^{-i\omega_d t}) - i\hbar(\gamma_{ab} + \gamma_{ac} + \gamma_{ad})\rho_{aa} \tag{11}$$

$$i\hbar\dot{\rho}_{bb} = \hbar\rho_{cb}(\Omega_{bc}e^{-i\omega_q t} + \tilde{\Omega}_{bc}e^{i\omega_q t}) + \hbar\rho_{ab}(\Omega_{ab}e^{-i\omega_p t} + \tilde{\Omega}_{ab}e^{i\omega_p t}) - \hbar\rho_{ba}(\Omega_{ab}e^{-iw_q t} + \tilde{\Omega}_{ab}e^{iw_q t}) - \hbar\rho_{bc}(\Omega_{cb}e^{-i\omega_q t} + \tilde{\Omega}_{cb}e^{i\omega_q t}) + i\hbar\gamma_{ab}\rho_{aa} - i\hbar\gamma_{bd}\rho_{bb} - i\hbar\gamma_{bc}\rho_{bb} \tag{12}$$

$$i\hbar\dot{\rho}_{cc} = \hbar\rho_{bc}(\Omega_{cb}e^{-i\omega_q t} + \tilde{\Omega}_{cb}e^{i\omega_q t}) - \hbar\rho_{cb}(\Omega_{bc}e^{-i\omega_q t} + \tilde{\Omega}_{bc}e^{i\omega_q t}) + i\hbar\gamma_{ac}\rho_{aa} - i\hbar\gamma_{cd}\rho_{cc} + i\hbar\gamma_{bc}\rho_{bb} \tag{13}$$

$$i\hbar\dot{\rho}_{dd} = \hbar\rho_{ad}(\Omega_{da}e^{-i\omega_d t} + \tilde{\Omega}_{da}e^{i\omega_d t}) - \hbar\rho_{da}(\Omega_{ad}e^{-i\omega_d t} + \tilde{\Omega}_{ad}e^{i\omega_d t}) + i\hbar\gamma_{ad}\rho_{aa} + i\hbar\gamma_{cd}\rho_{cc} + i\hbar\gamma_{bd}\rho_{bb} \tag{14}$$

The dipole de-phasing rates are given by η_{ij}. Neglecting the collisional de-phasing rates, we can relate dipole de-phasing rates and population decay rates as follows

$$\eta_{ab} = \frac{\gamma_{ab} + \gamma_{ac} + \gamma_{ad} + \gamma_{bd} + \gamma_{bc}}{2} \tag{15}$$

$$\eta_{da} = \frac{\gamma_{ab} + \gamma_{ac} + \gamma_{ad}}{2} \tag{16}$$

$$\eta_{ac} = \frac{\gamma_{ab} + \gamma_{ac} + \gamma_{ad} + \gamma_{cd}}{2} \tag{17}$$

$$\eta_{db} = \frac{\gamma_{bc} + \gamma_{bd}}{2} \tag{18}$$

$$\eta_{bc} = \frac{\gamma_{bc} + \gamma_{bd} + \gamma_{cd}}{2} \tag{19}$$

$$\eta_{dc} = \frac{\gamma_{cd}}{2} \tag{20}$$

applying rotating wave approximation by taking $\rho_{ab} = \sigma_{ab}e^{-i\omega_p t}$; $\rho_{db} = \sigma_{db}e^{-i(\omega_p - \omega_d)t}$; $\rho_{ac} = \sigma_{ac}e^{-i(\omega_p + \omega_q)t}$; $\rho_{dc} = \sigma_{dc}e^{-i(\omega_p + \omega_q - \omega_d)t}$; $\rho_{da} = \sigma_{da}e^{i\omega_d t}$; $\rho_{bc} = \sigma_{bc}e^{-i\omega_q t}$to density matrix equations and neglecting the terms oscillating at very high frequencies as they average to zero, the density matrix equations for coherences would transform into following equations.

$$i\dot{\sigma}_{ba} = \Gamma_{ab}\sigma_{ab} + (\rho_{bb} - \rho_{aa})\Omega_{ab} + \sigma_{db}\Omega_{ad} - \sigma_{ac}\tilde{\Omega}_{cb} \tag{21}$$

$$i\dot{\sigma}_{db} = \Gamma_{db}\sigma_{db} + \tilde{\Omega}_{da}\sigma_{ab} - \Omega_{ab}\sigma_{da} - \tilde{\Omega}_{cb}\sigma_{dc} \tag{22}$$

$$i\dot{\sigma}_{da} = \Gamma_{da}\sigma_{da} + \tilde{\Omega}_{da}(\rho_{aa} - \rho_{dd}) - \tilde{\Omega}_{ba}\sigma_{db} \tag{23}$$

$$i\dot{\sigma}_{ac} = \Gamma_{ac}\sigma_{ac} + \Omega_{ab}\sigma_{bc} + \Omega_{ad}\sigma_{dc} - \Omega_{bc}\sigma_{ab} \tag{24}$$

$$i\dot{\sigma}_{bc} = \Gamma_{bc}\sigma_{bc} + \Omega_{bc}(\rho_{cc} - \rho_{bb}) + \tilde{\Omega}_{ba}\sigma_{ac} \tag{25}$$

$$i\dot{\sigma}_{dc} = \Gamma_{dc}\sigma_{dc} + \tilde{\Omega}_{da}\sigma_{ac} - \Omega_{bc}\sigma_{db} \tag{26}$$

where $\Gamma_{bc} = \delta_{bc} - i\eta_{bc}$,$\Gamma_{ab} = \delta_{ab} - i\eta_{ab}$,$\Gamma_{db} = \delta_{db} - i\eta_{db}$,$\Gamma_{da} =$,$\Gamma_{dc} = \delta_{dc} - i\eta_{dc}$,$\Gamma_{ac} = \delta_{ac} - i\eta_{ac}$ As the time scale of evolution of electromagnetic field is much longer than time scale in which coherences evolve, we can solve the above equations in steady state to get the coherence between levels a and b. As the probe field is weak, we can use perturbation theory to solve only up to first order of probe field coherence while keeping the strong driving fields to all orders.

$$\sigma^1_{dc} = \frac{\tilde{\Omega}_{da}\sigma^1_{ac} - \Omega_{bc}\sigma^1_{db}}{-\Gamma_{dc}} \tag{27}$$

$$\sigma^1_{ac} = \frac{\Omega_{ab}\sigma^o_{bc} + \Omega_{ad}\sigma^1_{dc} - \Omega_{bc}\sigma^1_{ab}}{-\Gamma_{ac}} \tag{28}$$

$$\sigma^1_{db} = \frac{\tilde{\Omega}_{da}\sigma^1_{ab} - \Omega_{ab}\sigma^o_{da} - \tilde{\Omega}_{cb}\sigma^1_{dc}}{-\Gamma_{db}} \tag{29}$$

$$\sigma^o_{da} = \frac{\tilde{\Omega}_{da}(\rho^o_{aa} - \rho^o_{dd})}{-\Gamma_{da}} \tag{30}$$

$$\sigma^o_{bc} = \frac{\Omega_{bc}(\rho^o_{cc} - \rho^o_{bb})}{-\Gamma_{bc}} \tag{31}$$

solving the above equations give us the coherence between levels a and b. which is related to polarization and susceptibility in the following way

$$P_{ij} = \varpi_{ij}\rho_{ab} = \epsilon_o \chi E e^{i(kz-\omega t)} \Rightarrow \chi_{ab} = \frac{N\varpi^2\sigma^1_{ab}}{\hbar\epsilon_o\Omega_{ab}} \tag{32}$$

where N is the density of atoms, ϵ_o is the electric permeability of the medium,ϖ is the dipole moment. Susceptibility of this system is given the equation below. Before we observing how the absorption and gain profiles look we derive the sufficient condition for getting population non-inversion and then a necessary condition for observing gain. Combining both those conditions we get the LWI which is shown at the end of the paper.

$$\chi_{ab} = -\frac{N\varpi^2}{\hbar\eta_o}\frac{[n_{ba} - (n_{cb}|\Omega_{bc}|^2/(\Gamma_{ac}\Gamma_{bc}) + n_{ad}|\Omega_{ad}|^2/(\Gamma_{db}\Gamma_{da})) + \frac{|\Omega_{bc}|^2|\Omega_{ad}|^2(n_{cb}/(\Gamma_{ac}\Gamma_{bc})+n_{ad}/(\Gamma_{db}\Gamma_{da}))}{-\Gamma_{dc}+|\Omega_{bc}|^2/\Gamma_{db}+|\Omega_{ad}|^2/\Gamma_{ac}}(1/\Gamma_{db}+1/\Gamma_{ac})]}{[-\Gamma_{ab} - \frac{|\Omega_{bc}|^2|\Omega_{ad}|^2(1/\Gamma_{db}+1/\Gamma_{ac})^2}{(-\Gamma_{dc}+|\Omega_{bc}|^2/\Gamma_{db}+|\Omega_{ad}|^2/\Gamma_{ac})} + (|\Omega_{bc}|^2/\Gamma_{ac} + |\Omega_{ad}|^2/\Gamma_{db})]} \tag{33}$$

$$\chi_{ab} = -\frac{N\varpi^2}{\hbar\eta_o}\frac{f}{g} \tag{34}$$

$$f = (-\Gamma_{dc}\Gamma_{db}\Gamma_{ac} + \Gamma_{ac}|\Omega_{bc}|^2 + \Gamma_{db}|\Omega_{ad}|^2)(n_{ba}\Gamma_{ac}\Gamma_{bc}\Gamma_{db}\Gamma_{da}) + n_{cb}(\Gamma_{dc}\Gamma_{db} - |\Omega_{bc}|^2 + |\Omega_{ad}|^2)|\Omega_{bc}|^2\Gamma_{db}\Gamma_{da}\Gamma_{ac} + n_{ad}(\Gamma_{dc}\Gamma_{ac} - |\Omega_{ad}|^2 + |\Omega_{bc}|^2)|\Omega_{ad}|^2\Gamma_{ac}\Gamma_{bc}\Gamma_{db} \tag{35}$$

$$g = \{(-\Gamma_{dc}\Gamma_{db}\Gamma_{ac}+\Gamma_{ac}|\Omega_{bc}|^2+\Gamma_{db}|\Omega_{ad}|^2)(-\Gamma_{ab}\Gamma_{db}\Gamma_{ac}+\Gamma_{db}|\Omega_{bc}|^2+\Gamma_{ac}|\Omega_{ad}|^2) - |\Omega_{bc}|^2|\Omega_{ad}|^2(\Gamma_{ac}+\Gamma_{db})^2\}\Gamma_{bc}\Gamma_{da} \tag{36}$$

III. CONTROLLING AND TUNING GROUP VELOCITY OF PROBE

As pointed in [4], We can get slow or fast light by creating a dip in absorption and gain profiles respectively. This is idea which is being in this work for creation of super-luminal light and slow without any absorption. The drive1 helps in creating a gain on the probe resulting in slow light, at the same time the drive 2 can be used to induce absorption on the probe beam by optically pumping atoms from level c to level b. This absorption of probe photon induced through drive2 creates a dip in probes gain profile which would lead to superluminal or negative group velocity light. When the drive2 is switched off then a normal gain accompanied with slow light is observed on the probe beam.As the absorption of probe photons can be controlled by the strength of drive2 we can control the dip formation in probes gain profile by controlling the intensities of driving fields and hence we can control the switching between subliminal group velocity to super luminal group velocities. Note that we have fast light with absorption but that is not useful as the field will be simply absorbed by the medium. Hence we restrict our analysis and results only to the behavior of real part of susceptibility when there is no significant absorption.

We start with analysis of a simple condition where drive2 is zero, only drive1 and probe fields are present. In the steady state, without drive2, most of the population remains in levels a and d. Level's b and c gets populated according to the decay rate from level a. When the probe photon starts interacting with the medium in steady state there are two things that can happen one being the stimulated emission of probe photons by knocking out electrons from level a and the other one being absorption of probe photons from level b. As the population in level b tend to decay either to level d or c, the chances of a probe photon getting absorbed from level b is far less when compared with the chances of stimulated emission of probe from level a to b. As result we get gain on the probe beam which is accompanied by steep dispersion(or slow light) profile of probe. The fig below shows the absorption and dispersion profiles of probe beam when drive2 is zero. Now we focus on the case where we can create a dip in the gain profile of

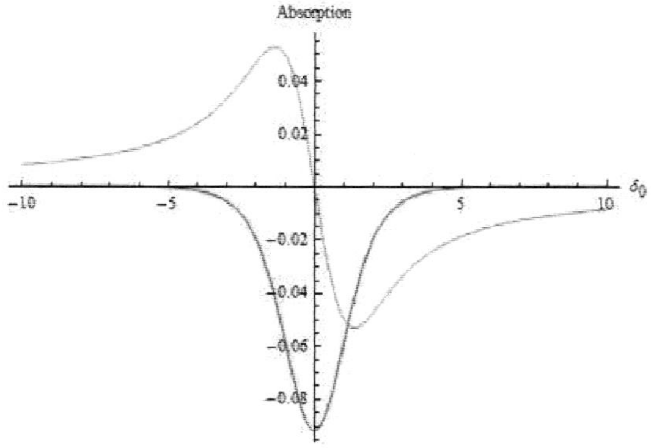

FIG. 2. $\gamma_{ab} = \gamma_{ac} = \gamma_{ad} = \gamma_{bd} = \gamma_{bc} = \gamma_{dc} = 1\gamma, \Omega_{ad} = 1\gamma$ and $\Omega_{bc} = 0$. Slow light with gain on probe profile

probe which would result in negative group velocity light. One simple way to is to increase the strength of drive1 to very high value so we can create Autler-Townes splitting in level a. As a result we get to gain profile from the two split levels of level a. However this requires a very strong drive1. Fig(3) shows this effect when drive2 is zero. Real part of susceptibility shows positive slope or negative group velocity. Instead of using a very strong drive1 to create a

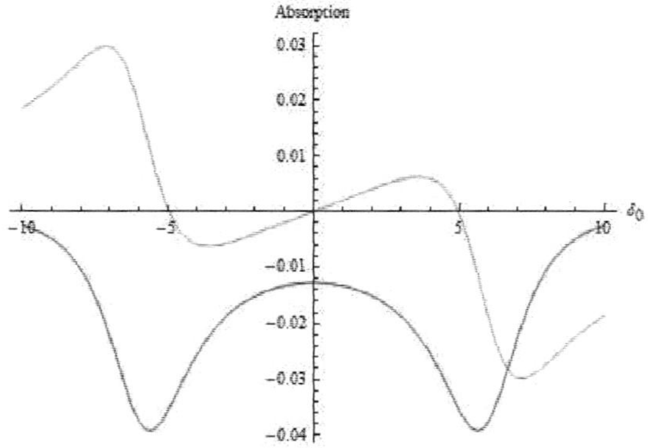

FIG. 3. $\gamma_{ab} = \gamma_{ac} = \gamma_{ad} = \gamma_{bd} = \gamma_{bc} = \gamma_{dc} = 1\gamma, \Omega_{ad} = 6\gamma$ and $\Omega_{bc} = 0$. Negative group velocity light with reduced gain

dip in the gain profile, we can use drive2 to do the same work but with less intensity requirement. As the intensity of the drive2 can be easily controlled, we have the ability control dispersion of probe light as well as the amount of gain or absorption the probe field goes through. Fig(4) shows the scenario when both the drive fields are used. It is clear that absorption at the probe resonance is completely nullified and the probe dispersion has a positive slope(resulting in negative group velocity). As the dip formed in the gain profile depends on the strength of drive2, It is possible to obtain negative group velocity with gain. Before proceeding further it is necessary to understand the role drive2 in influencing the gain profile of probe field. Without drive(2),as shown in FIG(2), we simply have gain on probe field and once we start increasing the strength of drive2 intensity more and more atoms that are de-excited from level a

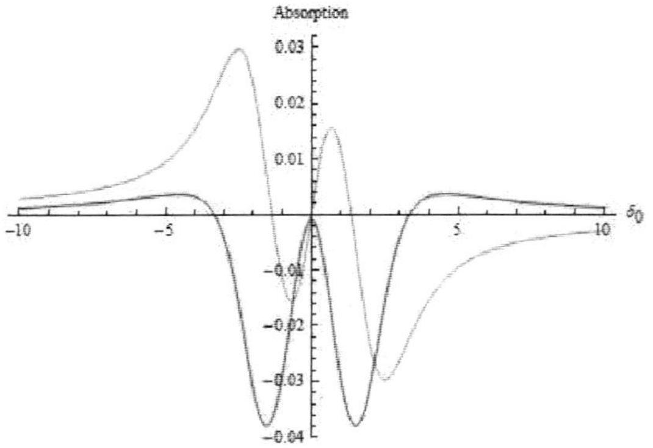

FIG. 4. $\gamma_{ab} = \gamma_{ac} = \gamma_{ad} = \gamma_{bd} = \gamma_{bc} = \gamma_{dc} = 1\gamma, \Omega_{ad} = \gamma$ and $\Omega_{bc} = 1.35\gamma$. Negative group velocity light with zero absorption

to c will be pumped back to level b and hence the probability of absorption of probe around the resonance increases. This increased chance of absorption of probe field will result in decrease of gain(creating a dip in absorption profile and negative light) at the resonance and further increasing drive2 would completely cancel the gain resulting in Fig(4) and eventually leading to absorption instead of gain on probe resonance.

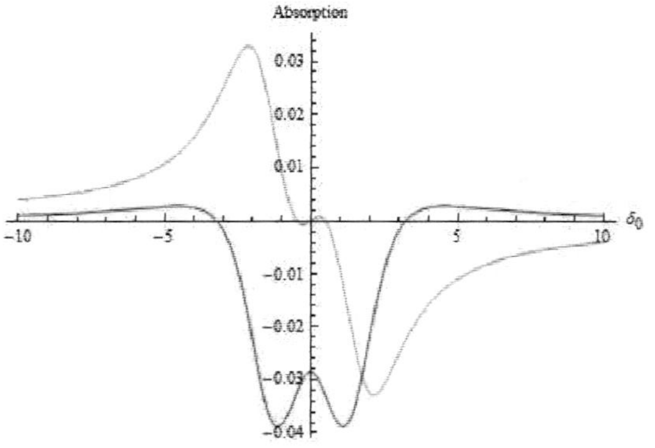

FIG. 5. Gradual change of probes group velocity from fast to slow with disappearance of dip in probes gain profile. $\gamma_{ab} = \gamma_{ac} = \gamma_{ad} = \gamma_{bd} = \gamma_{bc} = \gamma_{dc} = \gamma, \Omega_{ad} = \gamma$ and $\Omega_{bc} = 1\gamma$

Figure 6 shows the gradual disappearance of dip in probe fields gain profile when Ω_{bc} is relatively getting weaker and a gradual change of group velocity of light from negative to positive. It should be noted that probes absorption if zero or negative all the time. Another interesting possibility of this system is having ability ot tune the generation of slow light continuously over a broad range frequencies. Such a possibility is demonstrated in solids by(Yoshitomo) using Brillouin scattering. In this case tuning of slow light is provided by changing wavelength one of the lasers involved while in this work slow light tuning can be done changing strength of intensities of drive fields. This can be done by increasing the strengths of both the drive fields. By increasing the strength of driving fields we can continuously move the position of Autler-Townes components and as a result we can the gain, which is accompanied by slow light, at the wings of probe field. The fig illustrates such possibility

The Fig below shows the change of absorption as a function of drive2 strength It is clear that having a weak drive2 would result is slow light and a relatively stronger light would result in fast light. Now we go on to analyzing the requirement condition for changing from fast to slow light. The complete condition for getting fast light with zero absorption if given in the appendix. Here we will analyze a special case in where $\Omega_{ad} = \Omega_{ab} >> \gamma$. As we are interested in dispersion at the probe resonance only, we can set the probe detuning to zero in eq(33). Under this condition denominator of the susceptibility is always positive and non-zero(details are shown in the appendix). We will get zero absorption when numerator is zero and we get gain when numerator is less than zero. We can write

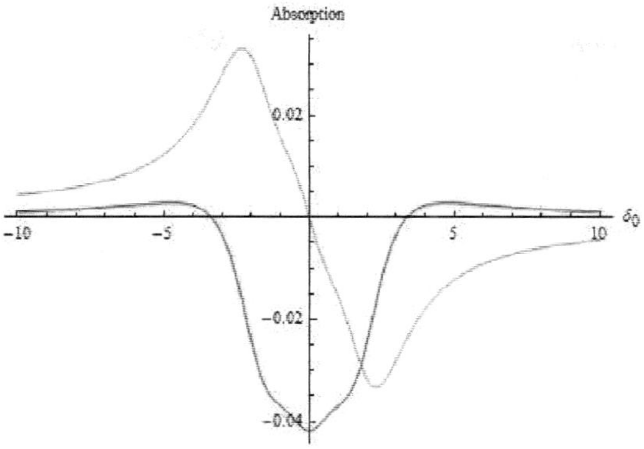

FIG. 6. Slow light with gain with drive1's intensity more than drive2. $\gamma_{ab} = \gamma_{ac} = \gamma_{ad} = \gamma_{bd} = \gamma_{bc} = \gamma_{dc} = 1\gamma, \Omega_{ad} = 1.1\gamma$ and $\Omega_{bc} = \gamma$

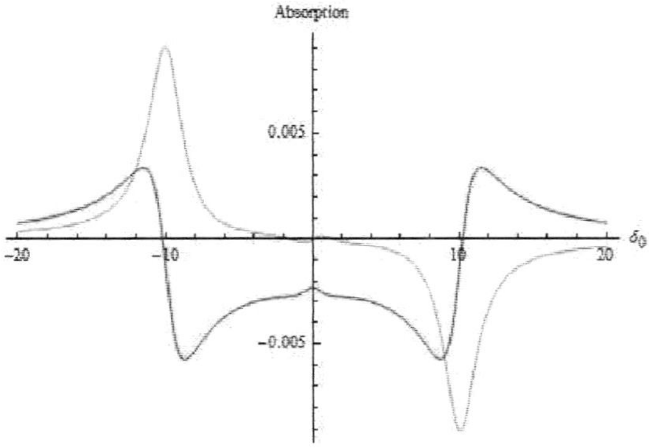

FIG. 7. tunable slow light. $\gamma_{ab} = \gamma_{ac} = \gamma_{ad} = \gamma_{bd} = \gamma_{bc} = \gamma_{dc} = 1\gamma, \Omega_{ad} = 5\gamma$ and $\Omega_{bc} = 5\gamma$

numerator as

$$i(|\Omega_{ad}|^2 - |\Omega_{bc}|^2)((\rho^o_{cc} - \rho^o_{bb})|\Omega_{bc}|^2\eta_{da} + (\rho^o_{dd} - \rho^o_{aa})|\Omega_{ad}|^2\eta_{bc}) \tag{37}$$

so when $\Omega_{ad} = \Omega_{ab} >> \eta_{ij}$ we always have a zero or very small absorption on the probe resonance,which means we have fast light, when both the drive fields are equal. As we have already discussed changing the intensity of drive2 with respect to the drive1 would give us gain or absorption so, having a low drive2 intensity would guarantee gain on probe beam along with slow light.

IV. CONCLUSION

We discussed a four-level atomic system, which is a combination of ladder and lambda system, coupled with two driving fields and a weak probe fields for varying the group velocity of the probe pulse. It is shown that probe pulses group velocity can be changed from subluminal to super luminal and negative values continuously by changing the intensities of driving fields. We also described the role of driving fields in controlling absorption of probe pulse. Another important possibility is to able to tune the frequency of slow light continuously. This can be done simply increasing the strength of relative intensities of both the driving fields.

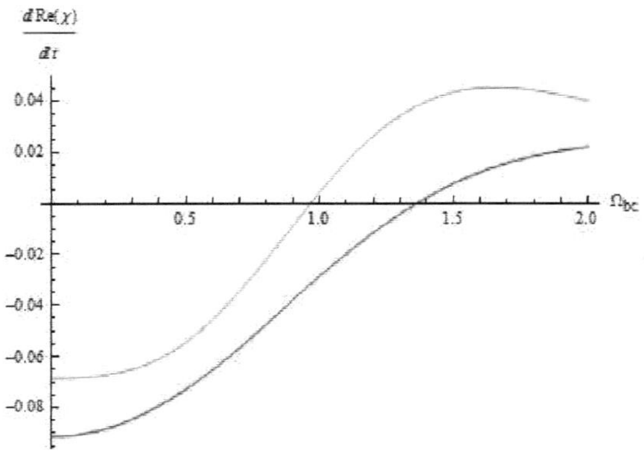

FIG. 8. Continuous change of probes group velocity when there is no absorption. $\gamma_{ab} = \gamma_{ac} = \gamma_{ad} = \gamma_{bd} = \gamma_{bc} = \gamma, \gamma_{cd} = 1\gamma, \Omega_{ad} = \gamma$ and $\Omega_{bc} = \gamma$

Appendix A: zero order solutions

The zero order steady state population equations are given by

$$\sigma_{da}\Omega_{ad} - \sigma_{ad}\tilde{\Omega}_{da} - i(\gamma_{ab} + \gamma_{ac} + \gamma_{ad})\sigma_{aa} = 0 \tag{A1}$$

$$\sigma_{cb}\Omega_{bc} - \sigma_{bc}\tilde{\Omega}_{cb} + i\gamma_{ab}\sigma_{aa} - i\gamma_{bd}\sigma_{bb} - i\gamma_{bc}\sigma_{bb} = 0 \tag{A2}$$

$$\sigma_{bc}\tilde{\Omega}_{cb} - \sigma_{cb}\Omega_{bc} + i\gamma_{ac}\sigma_{aa} - i\gamma_{cd}\sigma_{cc} + i\gamma_{bc}\sigma_{bb} = 0 \tag{A3}$$

$$\sigma_{ad}\tilde{\Omega}_{da} - \sigma_{da}\Omega_{ad} + i\gamma_{ad}\sigma_{aa} + i\gamma_{cd}\sigma_{cc} + i\gamma_{bd}\sigma_{bb} = 0 \tag{A4}$$

using eq(29) and eq(30) we can write the following relations

$$\sigma_{da}\Omega_{ad} - \sigma_{ad}\tilde{\Omega}_{da} = -|\Omega_{ad}|^2(\sigma_{aa} - \sigma_{dd})(1/\Gamma_{da} - 1/\Gamma^*_{da}) \tag{A5}$$

$$\sigma_{cb}\Omega_{bc} - \sigma_{bc}\tilde{\Omega}_{cb} = -|\Omega_{bc}|^2(\sigma_{cc} - \sigma_{bb})(1/\Gamma_{bc} - 1/\Gamma^*_{bc}) \tag{A6}$$

Substituting (A5) in (A1) directly gives the relation between level a and level d zero order populations as follows

$$\sigma_{aa} = \frac{\sigma_{dd}|\Omega_{ad}|^2(\Gamma^*_{da} - \Gamma_{da})}{|\Omega_{ad}|^2(\Gamma^*_{da} - \Gamma_{da}) + |\Gamma_{da}|^2(i\gamma_{ab} + i\gamma_{ac} + i\gamma_{ad})} \tag{A7}$$

From (A7) we can write $\sigma_{dd} = \Lambda\sigma_{aa}$ where Λ is the inverse of coefficient of σ_{dd}. Using this and eq(A6) we can write (A3) and (A4) as follows

$$\sigma_{cc}\{|\Omega_{bc}|^2(1/\Gamma^*_{bc} - 1/\Gamma_{bc}) - i\gamma_{cd}\} - \sigma_{bb}\{|\Omega_{bc}|^2(1/\Gamma^*_{bc} - 1/\Gamma_{bc}) - i\gamma_{bc}\} = -i\gamma_{ac}\sigma_{aa} \tag{A8}$$

$$\sigma_{cc}\{|\Omega_{bc}|^2(1/\Gamma^*_{bc} - 1/\Gamma_{bc})\} - \sigma_{bb}\{-|\Omega_{bc}|^2(1/\Gamma^*_{bc} - 1/\Gamma_{bc}) + i\gamma_{bc} + i\gamma_{bd}\} = -i\gamma_{ab}\sigma_{aa} \tag{A9}$$

$$\sigma_{cc} = i\frac{-\gamma_{ac}k_4 + \gamma_{ab}k_2}{k_1k_4 - k_3k_2}\sigma_{aa} \quad ; \quad \sigma_{bb} = i\frac{-\gamma_{ac}k_3 + \gamma_{ab}k_1}{k_1k_4 - k_3k_2}\sigma_{aa} \tag{A10}$$

we also know that probability of finding an atom in one of the four levels a,b,c,d is one. so using (A7) and (A10) with $\sigma_{aa} + \sigma_{bb} + \sigma_{cc} + \sigma_{dd} = 1$ gives the population evolution in steady state. The figure below shows the evolution of population in different levels as a function of drive 2

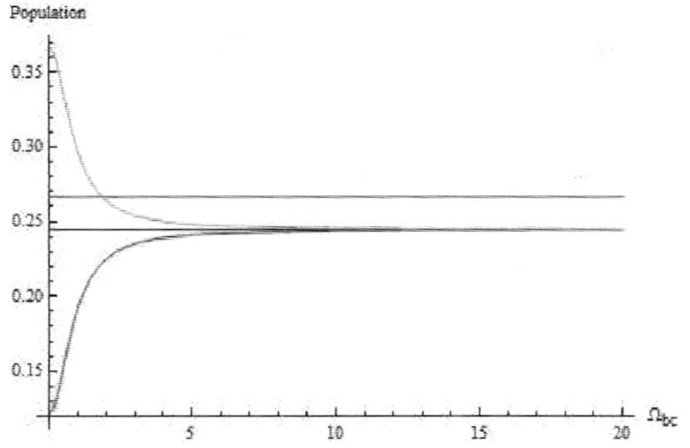

FIG. 9. tunable slow light. $\gamma_{ab} = \gamma_{ac} = \gamma_{ad} = \gamma_{bd} = \gamma_{bc} = \gamma_{cd} = \gamma, \Omega_{ad} = 5\gamma$ and $\delta = 0$

[1] A. Dogariu, A. Kuzmich, and L. J. Wang, Phys. Rev. A 63, 053806 (2001).
[2] A. Kuzmich, A. Dogariu, and L. J. Wang, Phys. Rev. Lett. 86,3925 (2001).
[3] G. S. Agarwal, T. N. Dey, S. Menon, Phys. Rev. A 64, 053809 (2001).
[4] R. W. Boyd, Journal of Mod. Opt. 56, Nos. 1819, 19081915 (2009).
[5] D. Bortman-Arbiv, A. D. Wilson-Gordon, and H. Friedmann, Phys.Rev. A 63, 043818 (2001).
[6] I. Gonzalo, M.A.Anton, F. Carreno, O. G. Calderon, Phys. Lett. A 372,6334-6339 (2008).
[7] K. Kim, H. S. Moon, C. Lee, S. K. Kim, and J. B. Kim, Phys. Rev. A 68, 013810 (2003).
[8] L. J. Wang, A. Kuzmich , A. Dogariu, Nature 406 (2000).
[9] M. Sahrai, H. Tajalli, K. T. Kapale, and M. S. Zubairy, Phys. Rev. A 70, 023813 (2004).
[10] A. B. Matsko, 0. Kocharovskaya, Y. Rostovtsev, G. R. Welch, A. S. Zibrov, M. O. Scully,Advances in Atomic, Molecular and Optical Physics 46 (2001).
[11] S. Franke-Arnold, G. Gibson, R. W. Boyd, M. J. Padgett, Science 333 (2011).
[12] M. O. Scully and M. S. Zubairy, Science 301 (2003).
[13] Y. Okawachi, M. S. Bigelow, J. E. Sharping, Z. Zhu, A. Schweinsberg,D. J. Gauthier, R. W. Boyd, A. L. Gaeta Phys. Rev. Lett, 94, 153902 (2005).
[14] L. V. Hau, S. E. Harris, Z. Dutton , C. H. Behroozi, Nature 397, 594-598 (1999).
[15] N. Brunner, V. Scarani, M. Wegmller, M. Legre, and N. Gisin, Phys. Rev. Lett 93,203902 (2004)
[16] S. Davuluri, Y. Rostovtsev, Phys. Rev. A 86, 013806 (2012).
[17] M. Mahmoudi, S. W. Rabiei, L. E. Zohravi, M. Sahrai , Opt. Comm. 281,4681-4686 (2008).
[18] A. Schweinsberg, Z. Shi, J. E. Vornehm and R. W. Boyd, Opt. Exp. 19, 17, 15760 (2011).
[19] Z. Shi, R. W. Boyd, Proc. of SPIE Vol. 8007 (2011).

Journal of Physics: Conference Series **414** (2013) 012006 doi:10.1088/1742-6596/414/1/012006

Room-temperature single-photon sources with definite circular and linear polarizations based on single-emitter fluorescence in liquid crystal hosts

Justin M. Winkler[1], Svetlana G. Lukishova[2], Luke J. Bissell[3]

[1] Department of Physics and Astronomy, University of Rochester, Rochester, New York 14627, USA

[2] The Institute of Optics, University of Rochester, Rochester, New York 14627, USA

[3] U.S. Air Force Research Laboratory, Wright-Patterson Air Force Base, Ohio 45433, USA

E-mail: sluk@lle.rochester.edu

Abstract. Definite circular and linear polarizations of room-temperature single-photon sources, which can serve as polarization bases for quantum key distribution, are produced by doping planar-aligned liquid crystal hosts with single fluorescence emitters. Chiral 1-D photonic bandgap microcavities for a single handedness of circularly polarized light were prepared from both monomeric and oligomeric cholesteric liquid crystals. Fluorescent emitters, such as nanocrystal quantum dots, nitrogen vacancy color centers in nanodiamonds, and rare-earth ions in nanocrystals, were doped into these microcavity structures and used to produce circularly polarized fluorescence of definite handedness. Additionally, we observed circularly polarized resonances in the spectrum of nanocrystal quantum dot fluorescence at the edge of the cholesteric microcavity's photonic stopband. For this polarization we obtained a ~4.9 enhancement of intensity compared to the polarization of the opposite handedness that propagates without photonic bandgap microcavity effects. Such a resonance is indicative of coupling of quantum dot fluorescence to the cholesteric microcavity mode. We have also used planar-aligned nematic liquid crystal hosts to align DiI dye molecules doped into the host, thereby providing a single-photon source of linear polarization of definite direction. Antibunching is demonstrated for fluorescence of nanocrystal quantum dots, nitrogen vacancy color centers, and dye molecules in these liquid crystal structures.

1. Introduction

Single-photon sources are distinct from faint light sources in that *all* emitted photons are separated in time (photon antibunching). In the case of photon antibunching, the second-order correlation function $g^{(2)}(0) < 1$, where $g^{(2)}(\tau)$ is defined at time τ as [1]

$$g^{(2)}(\tau) = \frac{\langle I(t)I(t+\tau)\rangle}{\langle I(t)\rangle^2}. \quad (1)$$

In equation (1), $I(t)$ is the light intensity and, for the practical case of stationary fields, averaging is performed over time. In the ideal case for antibunching, $g^{(2)}(0) = 0$. For laser light, $g^{(2)}(0) = 1$, and for thermal light, $g^{(2)}(0) = 2$.

Journal of Physics: Conference Series **414** (2013) 012006 doi:10.1088/1742-6596/414/1/012006

Single-photon sources are of interest for applications in the field of quantum communication and quantum metrology. A particularly intriguing usage for single photons is as vectors of quantum information in quantum key distribution (QKD) schemes [2-6]. QKD allows for the secure distribution of randomly generated cryptographic keys between two parties, whereby any passive eavesdropper performing measurements on quantum states transmitted via a QKD scheme will alter the state in a manner detectable by legitimate parties [7-10]. Many QKD schemes specifically require usage of single photons to transmit information, as multi-photon packets can be split, potentially allowing eavesdroppers to siphon off information about the key without risking detection [11,12].

Approaches to generating on-demand single photons usually rely on individual fluorescent emitters, which will emit a single photon at a time when excited. Major challenges to this approach include practicality of implementation, as many single photon emitters require cryogenic temperatures. Another issue of note is that many QKD approaches encode information via photon polarization to generate a key. A quantum cryptography system based on an unpolarized single-photon source, however, uses polarizers to encode information and therefore has a maximum efficiency of 50%. Thus, an ideal single-photon source would emit photons on-demand, with high efficiency and definite polarization, and would operate at room temperature.

This paper concerns our efforts in producing room-temperature single-photon sources that emit photons of definite polarization [13-20]. We have achieved this by fabricating liquid-crystal planar-aligned structures doped with single emitters, including nanocrystal quantum dots, fluorescent molecules, and color centers in nanodiamonds. Cholesteric liquid crystal chiral 1-D photonic bandgap structures were used to create circularly polarized single photons. Nematic liquid crystals were used to align the dipole moment of emitters, allowing for definite linear polarization of emitted photons. It should be noted that single-photon sources with definite linear polarization have also been obtained using heterostructured quantum dots at cryogenic temperatures in elliptical-shape microcavities by the authors of Refs [21-25].

We have also observed coupling of nanocrystal quantum dot fluorescence to the modes of a chiral photonic bandgap microcavity made from an aligned cholesteric liquid crystal structure. Due to the Purcell effect [26], coupling of emitted light to the microcavity causes enhanced rates of spontaneous emission at the edge of the photonic stopband [18]. Planar-aligned cholesteric liquid crystal microcavities have previously been doped with dye molecules (1-3% concentration by weight) for obtaining lasing in cholesterics at the bandedge of a photonic stopband [27-30]. Indeed, a great deal of research has been devoted to the control of spontaneous emission using a variety of photonic crystal architechtures, see, e.g., Refs [24,31-35]. Spontaneous emission enhancement is relevant to QKD as enhancing the rate of single photon emission might potentially be used to increase the bit rate of a practical QKD implementation.

2. Experimental Setup

Fluorescence antibunching proves the single-photon nature of the source. Measuring the time interval between two consecutive photons and building the histogram of coincidence counts $c(\tau)$ allows us to measure $g^{(2)}(\tau)$, which is proportional to $c(\tau)$ for sufficiently small count rates and short interphoton times [36,37]. Such measurements are carried out using a Hanbury Brown and Twiss (HBT) intensity interferometer, which consists of a beamsplitter and two single-photon detectors.

For examining our samples, we used two confocal fluorescence microscopes (see Figure 1) equipped with HBT interferometers: a homemade microscope and a Witec alpha-SNOM in a confocal mode. The homemade confocal microscope had an oil-immersion, 1.35 NA objective to focus laser light into the sample, with more detailed schematics given in Figure 2. Information on the Witec alpha-SNOM can be found at Witec's website (see Ref [38]).

Figure 1. Experimental setups. Left – homemade confocal fluorescence microscope. Right – Witec alpha-SNOM microscope.

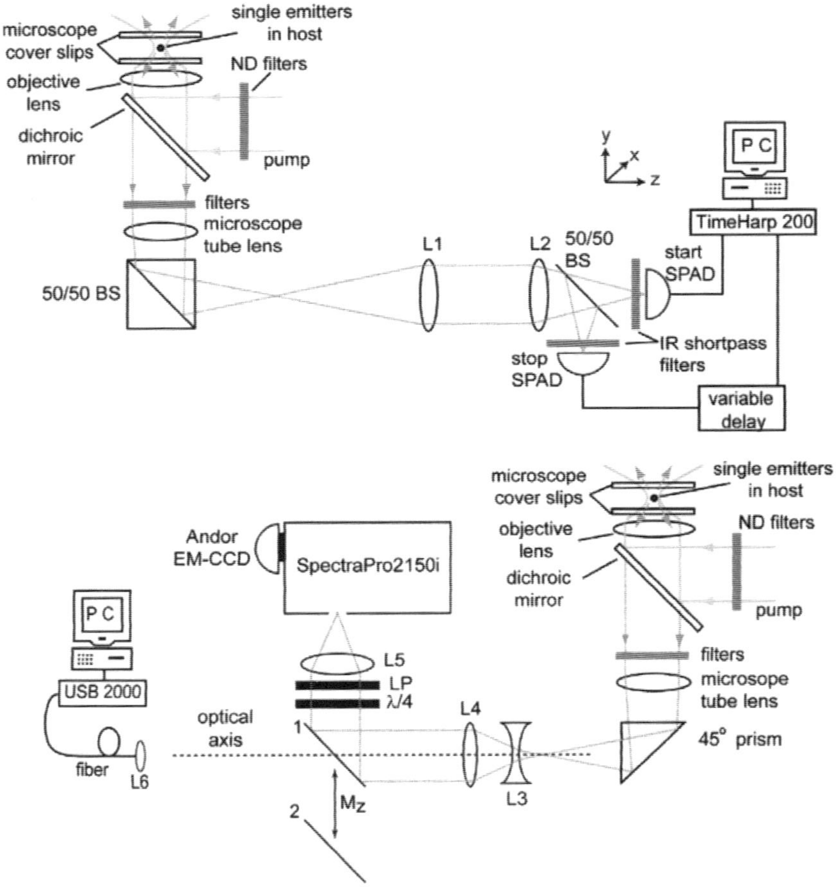

Figure 2. Schematics of experimental setup with a homemade confocal microscope. Top – single-emitter confocal fluorescence imaging and photon antibunching measurements. Bottom – spectral and polarization measurements. L1-L6, lenses; BS, beamsplitter; SPAD, single-photon counting avalanche photodiode modules; λ/4, achromatic quarter wave plate; LP, linear polarizer, M_z, mirror on a translation stage; ND filters, neutral density filters.

Single emitters were excited using either a cw, 633 nm HeNe laser, a pulsed, 532 nm Nd:Vanadate laser (76 MHz repetition rate, 6 ps pulse duration), or a cw, 976 nm diode laser depending on the emitter being examined. Sample fluorescence was collected by the same microscope objective, with any remaining excitation light filtered out using a dichroic mirror and interference filters. The fluorescence light could then be directed to one of two microscope ports, which were connected to diagnostics such as:

- An HBT interferometer, utilizing two SPCM-AQR-14 single-photon counting avalanche photodiode modules (SPADs), from Perkin Elmer. SPADs were connected to a TimeHarp 200 time-correlated single-photon counting PCI board (PicoQuant). We used this setup to image the fluorescence of several single emitters, then focused the laser light on one of these single emitters and characterized the fluorescence (photon antibunching or fluorescence lifetime measurements). See Figure 2, top.
- A Princeton Instruments Acton SP2150i spectrometer, used in conjunction with an iXon DV 887 electron multiplying (EM) CCD camera from Andor Technology. Collected fluorescence could be sent through an achromatic quarter waveplate and a linear polarizer, allowing for distinction between right-handed circularly polarized (RHCP) and left-handed circularly polarized (LHCP) fluorescence. In some measurements, we used an Ocean Optics spectrometer with a long acquisition time. See Figure 2, bottom.

3. Single-Photon Sources with Circular Polarization

3.1. Monomeric Cholesteric Liquid Crystals

We used planar-aligned cholesteric liquid crystal (CLC) to fabricate 1-D chiral photonic bandgap microcavities. CLC consists of rod-shaped chiral anisotropic molecules that form periodic helical structures of pitch p when planar-aligned [39]. In CLC chiral photonic bandgap structures, circularly polarized light of handedness with electric-field vector following the rotation of the liquid crystal molecules is reflected within a spectral band centered at $\lambda_c = n_{av}p$, with bandwidth of roughly $\Delta\lambda = \lambda_c\Delta n/n_{av}$. The variables n_{av} and Δn are defined from the extraordinary index of refraction, n_e, and the ordinary index of refection, n_o, where $n_{av} = (n_o + n_e)/2$ and $\Delta n = n_e - n_o$. Light with polarization of opposite handedness is unaffected by the structure and propagates without any changes. When doped with single emitters, such chiral photonic bandgap structures allow for the creation of single-photon sources with a high purity of circular polarization, with the potential for resonance effects of the microcavity being used to enhance the rate of single photon emission.

To create CLC microcavities, we used a mixture of E7 nematic liquid crystal and chiral additive CB15 [16]. When using this liquid crystal mixture to produce a photonic bandgap structure, the stopband center wavelength was tuned by the ratio of CB15 to E7, with a CB15 concentration by weight c_m being related to the stopband position by $c_m = n_{av}/(\lambda_c \times \text{HTP})$, where the chiral twisting power of CB15 HTP is equal to ~ 7.3 μm^{-1}. Thus it was possible to tune the resulting bandgap based on the relative concentrations of these compounds, allowing us to create photonic bandgap structures with predetermined stopband features. Figure 3 provides examples of the selective transmission curves of prepared CLC hosts, with quantum dot fluorescence spectral curves superimposed, demonstrating the viability of this approach in creating chiral microcavity structures tuned to the fluorescence maxima of chosen emitters.

Once a CLC mixture was created, the liquid crystal was mixed with a solution containing ~nM concentrations of quantum dots, and the solvent was evaporated. The resulting monomeric CLC doped with quantum dots was placed between two glass coverslips and planar-aligned via a mechanical shearing motion of the coverslips (for further details see Ref [14]). This relatively simple process created a chiral microcavity doped with single emitters.

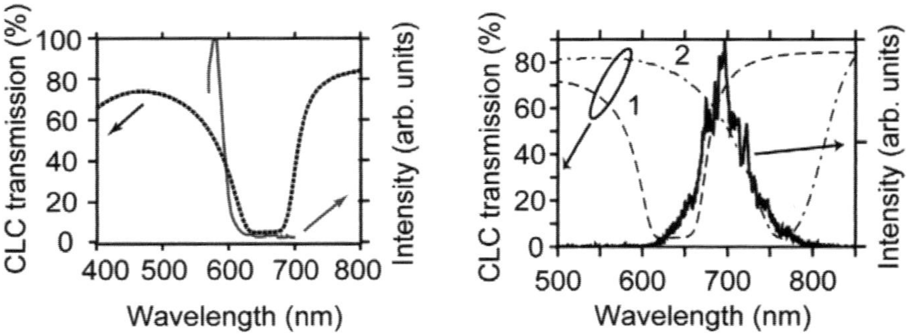

Figure 3. Selective transmission curves of right-handed circularly polarized light through chiral photonic bandgap structures formed from planar-aligned monomeric CLC hosts (dashed curves), with different concentrations c_m (by weight) of CB15 in E7 adjusted to give different stopband positions. Superimposed on these curves are the fluorescence spectra (solid lines) of the nanocrystal quantum dots that these transmission curves were tuned to. Left – c_m = 36.6%, CdSe quantum dots, maximum fluorescence wavelength λ_0 = 580 nm; Right – curve 1 corresponds to c_m = 36.0% and curve 2 corresponds to c_m = 29%, CdSeTe quantum dots, maximum fluorescence wavelength λ_0 = 700 nm.

Figure 4, left, shows the emission spectra of CdSe quantum dots in one of these chiral microcavities, with curve 1 (black) showing RHCP fluorescence and curve 2 (red) showing LHCP fluorescence. The degree of circular polarization is measured by the circular polarization dissymmetry factor [40]

$$g_e = 2(I_L - I_R) / (I_L + I_R), \tag{2}$$

where I_L and I_R are the intensities of left and right-handed circularly polarized light respectively. The observed dissymmetry factor was g_e = -1.6 at 580 nm. For unpolarized light, g_e = 0.

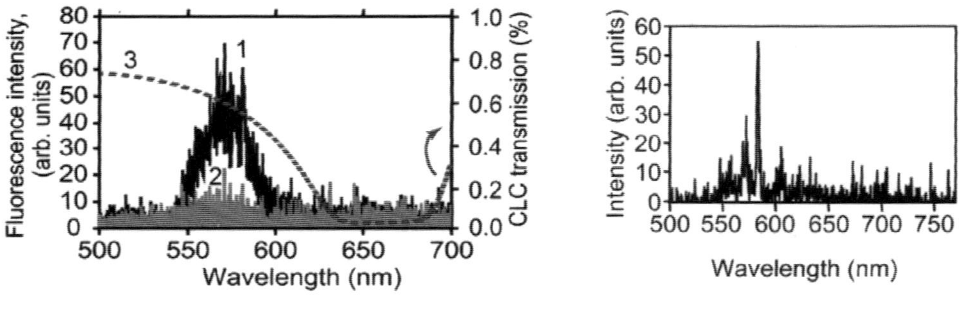

Figure 4. Left – Comparison of right-handed circularly polarized fluorescence (black, curve 1) and left-handed circularly polarized fluorescence from CdSe quantum dots (red, curve 2) in a planar-aligned monomeric CLC host (E7 with CB15). The blue dashed curve (curve 3) shows the selective transmission of the CLC host. Right – Fluorescence and Raman spectra of the same monomeric CLC without quantum dots.

Our HBT interferometer was used for examining the photon statistics of these samples. Detector coincidence counts were histogrammed based on the time delay between photon detection events (interphoton times), with $g^{(2)}(\tau)$ found by normalizing these coincidence histograms [36,37]. Figure 5 presents the $g^{(2)}(\tau)$ histograms we obtained from our microcavities under pulsed, 532 nm excitation.

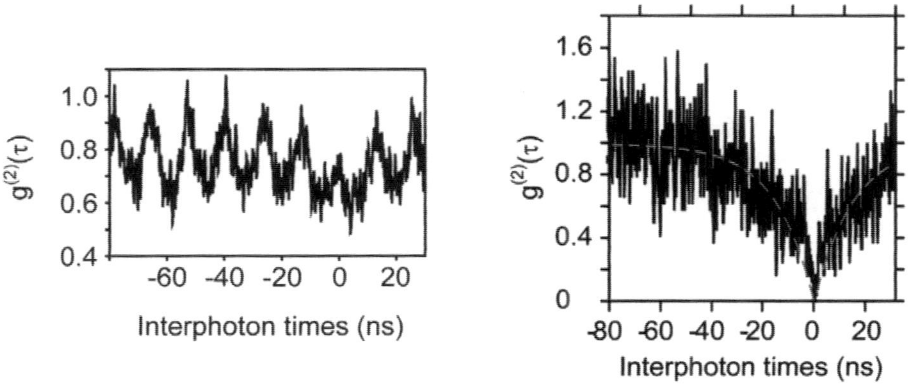

Figure 5. Histograms showing the second-order correlation function $g^{(2)}(\tau)$ (obtained by normalization of coincidence counts) versus interphoton times found from the fluorescence of quantum dots in a CLC host under pulsed, 532 nm excitation. The dip at the zero interphoton time indicates photon antibunching. Left – Antibunching from CdSe quantum dots from Figure 3, left. Right – Antibunching from CdSeTe quantum dots from Figure 3 (right, curve 1) with superimposed red dashed line showing fit of form $g^{(2)}(\tau) = 1-(1/N)\mathrm{e}^{(-\tau/\tau_0)}$ (see explanation in the text).

Figure 5, left, shows the $g^{(2)}(\tau)$ histogram from the fluorescence of the CdSe quantum dots from Figure 3, left. The peaked structure occurs due to the pulsed excitation. A distinctive dip in the height of the peak at the zero interphoton time is clearly present, indicating photon antibunching and corresponding to a value of $g^{(2)}(0) = 0.76 \pm 0.04$. Such a large value of $g^{(2)}(0)$ can be explained by the background from the liquid crystal host, located at the same spectral region as the quantum dot fluorescence, with the spectrum of the raw liquid crystal host shown in Figure 4, right.

Figure 5, right, in turn, presents the $g^{(2)}(\tau)$ histogram from the fluorescence of the CdSeTe quantum dots from Figure 3, right, with the sample's selective transmission presented by curve 1, where the quantum dot fluorescence spectrum is outside the range of liquid crystal background. While a pulsed, 532 nm laser is still used to excite this sample, the peaked structure seen in Figure 5, left, is not present in Figure 5, right, due to these CdSeTe quantum dots having fluorescence lifetimes larger than the time between two laser pulses (a 76 MHz pulse repetition rate corresponds to ~13 ns between pulses). A clear dip at the zero interphoton time is present, regardless, indicating antibunching. A fit of form [41] $g^{(2)}(\tau) = 1-(1/N)\mathrm{e}^{(-\tau/\tau_0)}$ was found for this histogram, where $\tau_0 = 1 / (k_{12} + k_{21})$, k_{12} is the excitation pump rate, k_{21} is the spontaneous decay rate, and N is the number of emitters producing the fluorescence spot under examination (at $N = 1$, $g^{(2)}(0) = 0$, at $N = 2$, $g^{(2)}(0) = 0.5$). This fit is shown by the red dashed curve in Figure 5, right, and yielded a $g^{(2)}(0)$ value of 0.001 ± 0.034 and a decay time constant of $\tau_0 \sim 15$ ns.

3.2. Glassy Cholesteric Liquid Crystal Oligomers: Microcavity Resonances

We used left-handed cyclosiloxane oligomeric CLC powder from Wacker Chemie [42,43] and produced a planar-aligned glassy CLC structure doped with CdSeTe quantum dots (Qdot 800 ITK organic, Invitrogen, fluorescence maximum at 790 nm) [18]. Doping was accomplished by heating the CLC to ~135° C (the oligomer's melting temperature) and then mixing the melted CLC with quantum dots dispersed in toluene at a concentration of ~1 μM, with heating allowed to continue until the toluene evaporated. Subsequently, cells were prepared using two polyimide buffed glass coverslips.

The CLC doped with quantum dots was placed on a buffed coverslip and heated beyond the oligomer clearing temperature of 180° C. After the sample was cooled to ~135° C, the second buffed coverslip was placed on the first and sheared along the direction of polyimide buffing. A slow cooling process back to a glassy (solid) state preserved the CLC order. This resulted in a photonic bandgap microcavity with a center wavelength of 910 nm.

Prepared samples were analyzed using our homemade confocal microscope setup. We excited the sample with cw, 633 nm laser light from a HeNe laser and observed the spectrum of the sample's fluorescence. By placing an achromatic quarter waveplate and linear polarizer in front of the spectrometer, we were able to filter for different handedness of circularly polarized fluorescence.

The resulting fluorescence spectra can be seen in Figure 6. LHCP light experienced the photonic bandgap and therefore the black LHCP curve in Figure 6 (curve 1) shows microcavity resonance, indicating that the LHCP light coupled to the cavity mode. The LHCP fluorescence had a center wavelength of 833 nm and a FWHM of 16 nm (Q ~ 50), as compared to a FWHM of 76 nm for RHCP fluorescence. The center wavelength of this resonance roughly matches the edge of the photonic stopband, centered at 910 nm, and shown in Figure 6 by the blue curve 3. The observed RHCP fluorescence is shown in red in Figure 6 (curve 2) and was less intense due to not experiencing the CLC microcavity, showing no sign of line narrowing. The maximum intensity ratio between LHCP and RHCP was a factor of 4.9, and the circular polarization dissymmetry factor, from equation (2), was $g_e = 1.3$.

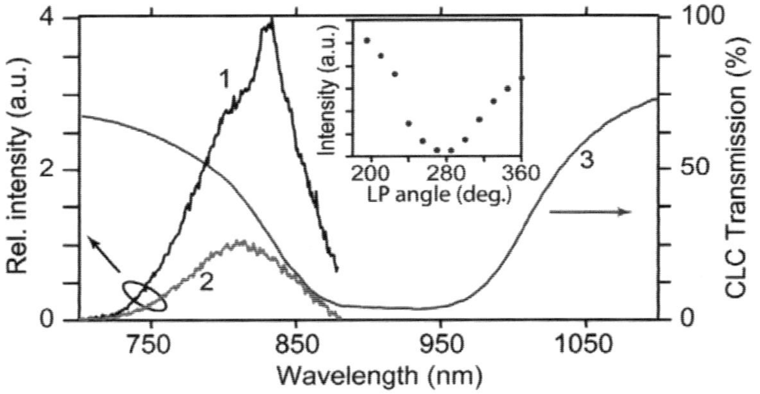

Figure 6. Circularly polarized fluorescence spectrum from CdSeTe quantum dots doped in a planar-aligned glassy cholesteric liquid crystal microcavity. Curve 1 shows LHCP fluorescence and has a resonance at 833 nm. Curve 2 shows RHCP fluorescence from the same source. Curve 3 shows the selective transmission of LHCP light through the aligned liquid crystal host, showing a stopband centered at 910 nm. Inset: Dependence of resonance peak intensity on the rotation of a linear polarizer (LP) after a fixed quarter-wave plate.

Figure 7, left, shows a confocal fluorescence raster scan taken of a sample prepared using a relatively low concentration of quantum dots dispersed in toluene (~10 nM), with the higher intensity spots indicating the location of fluorescing quantum dots in the glassy CLC microcavity. Focusing on a spot (circled in white on Figure 7, right), we checked for photon antibunching and obtained the coincidence histogram shown in Figure 7, right.

Figure 7, right, displays the histogram of coincidence counts $c(\tau)$ in blue, with $g^{(2)}(\tau)$ derived by normalizing $c(\tau)$. The measured $g^{(2)}(0)$ value from the fit shown in Figure 7, right (green, solid curve) is $g^{(2)}(0) = 0.382 \pm 0.037$. As $g^{(2)}(0) < 0.5$, this indicated that we have managed to isolate the fluorescence of a single quantum dot, serving as a source of antibunched light in a glassy CLC microcavity.

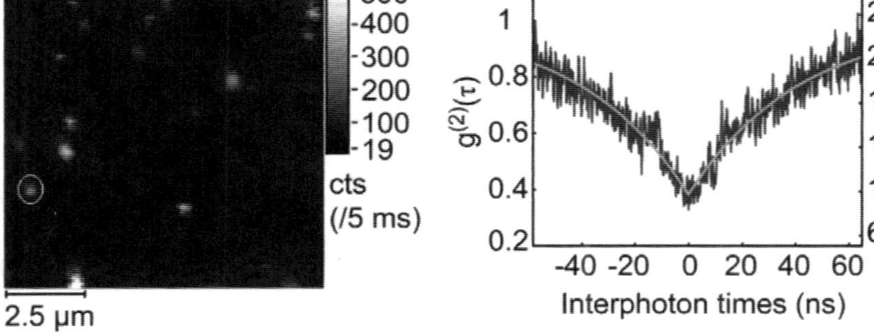

Figure 7. Left – Confocal fluorescence microscopy scan of CdSeTe quantum dots doped in planar-aligned glassy oligomeric cholesteric liquid crystal. Right – Raw coincidence counts $c(\tau)$ (right axis) of fluorescence from the circled quantum dot in the confocal scan on left, as measured by an HBT interferometer. The second-order correlation function, $g^{(2)}(\tau)$ (left axis) is found by normalizing $c(\tau)$ and indicates antibunched fluorescence.

3.3. Rare-Earth-Doped Nanocrystals and Color Centers in Nanodiamonds

Additional work with monomeric CLC based microcavities included doping with emitters such as nitrogen vacancy color centers in nanodiamonds and rare-earth ions in nanocrystals. These alternative emitters can provide a source of single photons that is less likely to bleach.

Figure 8, left, shows antibunching of nitrogen vacancy color centers in nanodiamonds doped in a CLC (E7 and CB15) microcavity with a stopband centered at 725 nm, under cw, 514 nm excitation. The $g^{(2)}(0)$ value determined via fit (shown in Figure 8, left, as the dashed red line) was $g^{(2)}(0) = 0.74 \pm 0.08$. The fluorescence spectrum of nitrogen vacancy centers in nanodiamonds doped in this monomeric CLC when excited using cw, 514 nm excitation from an argon ion laser is shown in Figure 8, right.

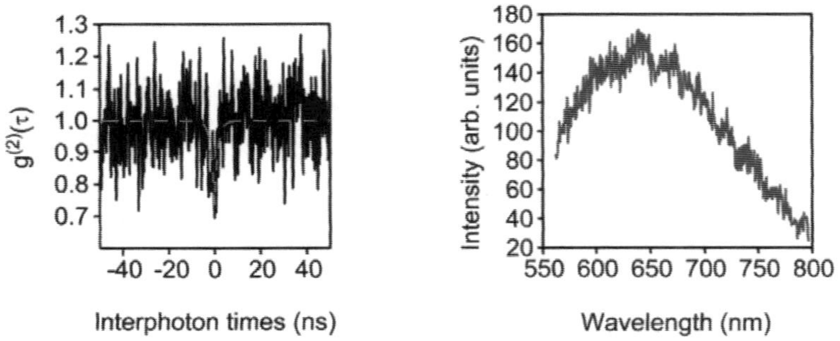

Figure 8. Left – Photon antibunching observed in fluorecence from nitrogen vacancy centers in a CLC host. Black curve shows the raw data. Red dashed curve shows the fit. Right – Fluorescence spectrum of nitrogen vacancy centers in nanodiamonds doped in CLC host when excited by cw, 514 nm laser light.

A monomeric CLC microcavity made from a mixture of E7 and CB 15 was also doped with rare-earth Er^{3+} and Yb^{3+} ions in 20 nm to 30 nm sized $NaYF_4$ nanocrystals with 20% Yb and 2% Er. When these ions were excited using a cw, 976 nm diode laser at incident powers of ~500 μW, we were able to observe upconverted fluorescence of Er^{3+}, as shown in Figure 9. The emission lines observed were attributed to the transitions $^2H_{11/2}$, $^4S_{3/2} \rightarrow {}^4I_{15/2}$ (green) and $^4F_{9/2} \rightarrow {}^4I_{15/2}$ (red) of the Er^{3+} ions [44]. The populations of upper levels in Er^{3+}occur due to an efficient energy transfer from the Yb^{3+} to the Er^{3+}. This fluorescence was measured to have a circular polarization dissymmetry factor of $g_e = -0.77$ at 680 nm. These nanocrystals doped with rare-earth ions were prepared at the University at Buffalo (The Institute for Lasers, Photonics and Biophotonics).

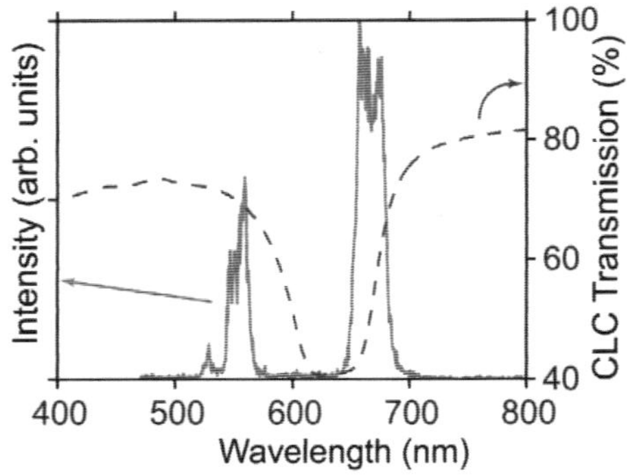

Figure 9. Red solid lines show the fluorescence spectrum of Er^{3+} ions doped in $NaYF_4$ nanocrystals dispersed in a chiral microcavity formed from planar-aligned monomeric CLC (E7 and CB15). The blue dashed curve shows the spectral transmission of the doped microcavity, measured using unpolarized light.

4. Single-Photon Sources with Linear Polarization

Another intriguing application of liquid crystal as a host for single emitters is using the planar-alignment of nematic liquid crystals to provide doped molecular dipoles definite alignment along a preferred direction for efficient excitation [15]. This allows for the creation of a single-photon source with fluorescence of definite linear polarization.

Samples were prepared using $DiIC_{18}(3)$ dye (DiI) molecules from Molecular Probes doped in a glassy nematic liquid crystal oligomer synthesized by S. H. Chen's group of University of Rochester [15]. Planar-aligned films of this doped liquid crystal host of ~100 nm thickness were prepared using photoalignment of the liquid crystal.

Photoalignment was performed by first spin-coating a Staralign-2100 linearly photopolymerizable polymer (Rolic Technologies Ltd.) to a cleaned cover glass slip, which was then cured at ~135° C. This film was then irradiated by a polarized UV light for 10 to 15 minutes, with further irradiation used to bleach the polymer impurity fluorescence.

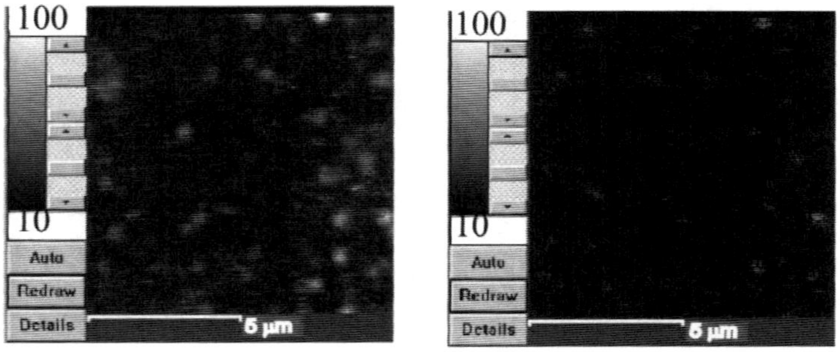

Figure 10. Confocal fluorescence microscopy scans of DiI dye molecules in a planar-aligned nematic liquid crystal host. Left – Fluorescence with polarization perpendicular to the alignment direction of the host. Right – Fluorescence with polarization parallel to the alignment direction of the host.

An oligomer solution doped with dye and diluted in chloroform was subsequently spin-coated onto these Staralign coated glass slips. After the chloroform evaporated, we heated the sample to ~80°C, slightly above where the oligomer transition to a nematic state occurs, after which the sample was slowly cooled to a glassy state, preserving the planar-aligned nematic order.

To characterize prepared samples, we used the Witec alpha-SNOM (see Figure 1, right) in a confocal mode, exciting the sample using a Nd:YAG laser for cw, 532 nm irradiation. Single-photon counting avalanche photodiodes were used as photodetectors for confocal fluorescence scans, with a polarizing beamsplitter used so that each photodetector collected light of orthogonal polarization. Figure 10 shows confocal fluorescence microscopy images taken of a planar-aligned sample of glassy nematic liquid crystal doped with DiI dye, with one photodetector used to collect fluorescence polarized parallel to the sample alignment direction, while the other photodetector collected fluorescence polarized perpendicular to the sample alignment direction.

Molecules were identified by the peak pixel intensity values from Figure 10 and a linear polarization measure ρ was determined for each molecule by comparison of the perpendicular and parallel polarization images. Our measure of linear polarization was

$$\rho = (I_{par} - I_{perp}) / (I_{par} + I_{perp}), \quad (3)$$

where I_{par} and I_{perp} are, respectively, the fluorescence intensities parallel and perpendicular to the direction of alignment. The only difference of ρ with the degree of polarization [45] is that it allows inclusion of the direction of polarization as part of the measure, depending on whether ρ is positive or negative.

The different ρ values that we found are histogrammed in Figure 11. A clear asymmetry in Figure 11 demonstrates a preference towards fluorescence polarized perpendicular to sample alignment. This is in contrast to the expected fluorescence from an unoriented sample, which would yield a symmetric ρ histogram [46].

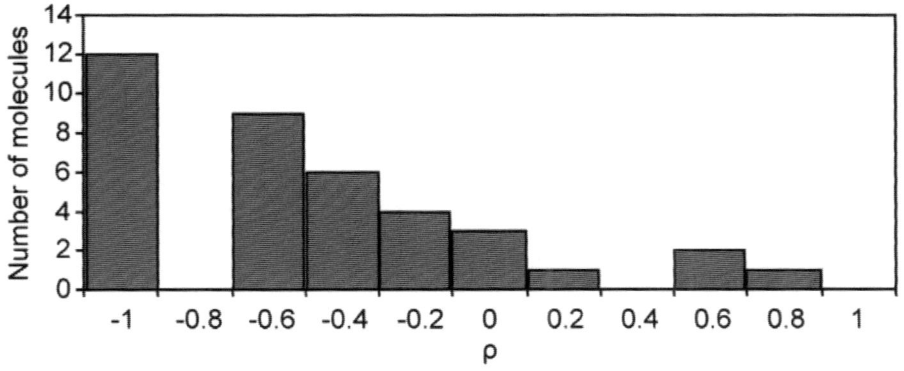

Figure 11. Histogram of polarization measure ρ from the fluorescence of 38 different DiI dye molecules in a planar-aligned nematic liquid crystal host.

That DiI molecules doped in this planar-aligned liquid crystal host tend to fluoresce with polarization perpendicular to the alignment of the liquid crystal can be explained by the molecular structure of these molecules. As illustrated in Figure 12, it is likely that two alkyl chains orient themselves parallel to the rod-like molecules of the nematic liquid crystal host. The absorbing and emitting dipoles, however, are parallel to the bridge between these alkyl chains. Therefore, these dipoles end up oriented perpendicular to the direction of liquid crystal alignment, hence the fluorescence having a polarization orthogonal to the alignment of the liquid crystal.

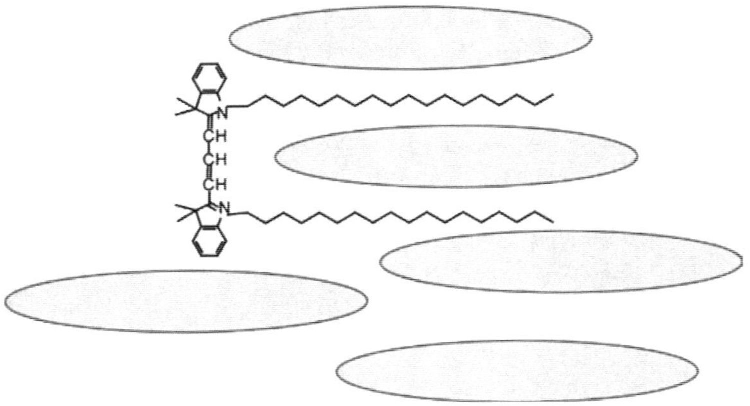

Figure 12. Schematic view of a DiI dye molecule in a nematic liquid crystal host. The long axes (alkyl chains) of the DiI molecules tend to orient themselves along the rod-like nematic liquid crystal molecules, while the dipole, which is parallel to the bridge between the alkyl chains, orients perpendicular to the direction of liquid crystal alignment.

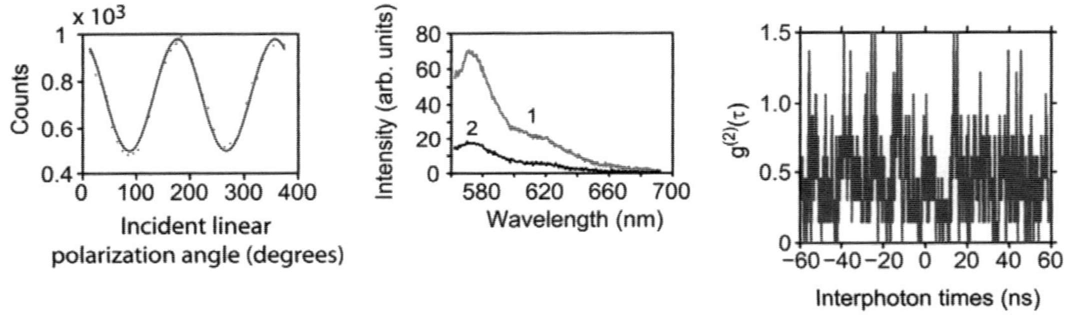

Figure 13. Polarized fluorescence of DiI dye in planar-aligned nematic liquid crystal hosts. Samples were excited using a 532 nm laser. Left – Intensity of DiI fluorescence in E7 as linear polarization of exciting light was rotated over 360°, with red dots showing experimental measurements and the blue solid curve providing a sinusoidal fit. The zero angle of polarization corresponds to the direction perpendicular to liquid crystal alignment. Center – Red (curve 1) and black (curve 2) show fluorescence spectrum with polarization perpendicular and parallel to the host alignment direction respectively. Right – Antibunching histogram taken from the fluorescence of a single molecule of DiI dye in an E7 nematic host under pulsed, 532 nm excitation.

To confirm these results, additional experiments were carried out both with monomeric nematic liquid crystal and nematic glassy liquid crystal oligomers. Figure 13, left, shows intensity changes of fluorescence of DiI molecules in a planar-aligned monomeric nematic liquid crystal (E7), as the linear polarization of the exciting 532 nm beam was rotated over 360° degrees. Figure 13, left, shows a clear dependence of the intensity on the exciting angle of polarization, where the maximum fluorescence occurred when the DiI molecules were excited by light with linear polarization perpendicular to the alignment of the liquid crystal.

Figure 13, center, shows spectrofluorimeter measurements made for polarization perpendicular and parallel to the sample alignment, this time having used a sample of planar-aligned glassy nematic liquid crystal oligomer doped with DiI molecules of more than 1% concentration by weight. The value of ρ was measured from Figure 13, center, by comparing the peak intensities of the curves, yielding ρ = -0.5. The results of Figure 13, left and center, indicated a clear preference for fluorescence with definite linear polarization perpendicular to the sample alignment.

Figure 13, right, shows an antibunching histogram for a single molecule of DiI dye doped in an E7 nematic host, taken using pulsed, 532 nm excitation [47]. The value for $g^{(2)}(0)$ is $g^{(2)}(0) = 0.77 \pm 0.10$. This relatively high value of $g^{(2)}(0)$ may be due to the fact that the Raman spectrum of the E7 overlaps with the fluorescence spectrum of DiI dye.

5. Conclusion

We have employed liquid crystals as the hosts for single emitters to create room-temperature single-photon sources that emit photons of definite circular and linear polarization. Circularly polarized light of definite handedness was observed from the fluorescence of nanocrystal quantum dots, nitrogen vacancy centers in nanodiamonds, and rare-earth ions in nanocrystals doped in a photonic bandgap cholesteric liquid crystal microcavity. We also observed circularly polarized resonance from quantum dot fluorescence in a cholesteric microcavity. The resonance was located at 833 nm, at the edge of a stopband centered at 910 nm, and had a FWHM of 16 nm as compared to a FWHM of 76 nm for fluorescence with polarization of opposite handedness, demonstrating coupling of left-handed circularly polarized light to a microcavity mode. Circularly polarized fluorescence from the microcavity had a circular polarization dissymmetry factor of 1.3 and an intensity ratio between left and right hand circularly polarized light of 4.9 at the resonant wavelength. Antibunching was demonstrated in this architecture, showing the feasibility of this approach for creating circularly polarized single-photon sources. Additional work was done on the usage of nematic liquid crystal for aligning the orientation of DiI dye molecules, providing a source of photons of definite linear polarization.

6. Acknowledgement

The authors acknowledge the support given by the US Army Research Office under Award no DAAD19-02-1-0285 and the US National Science Foundation (awards ECS-0420888, DUE-063362, and DUE-0920500). Justin Winkler's work was supported by a NASA Office of the Chief Technologist's Space Technology Research Fellowship. Luke Bissell was supported by an Air Force SMART Fellowship. We would also like to thank K. Marshall, A.W. Schmid, C.R. Stroud, R.W. Boyd, L. Novotny, A. Lieb, T. Krauss, C.H. Chen, P. Freivald and R. Knox for their advice and assistance and T. Ohulchanskyy for providing nanocrystals doped with rare-earths.

References

[1] Gerry C C and Knight P L 2005 *Introductory quantum optics* (Cambridge University Press) p 128

[2] Grangier P, Sanders B and Vuckovic J 2004 Focus on single photons on demand *New J. Phys.* **6**

[3] Kumar P, Kwiat P, Migdall A, Nam S W, Vuckovic J and Wong F N C 2004 Photonic technologies for quantum information processing *Quantum Inf. Process.* **3** 215–31

[4] Lounis B and Orrit M 2005 Single-photon sources *Rep. Prog. Phys* **68** 1129–79

[5] Yamamoto Y, Santori C, Solomon G, Vuckovic J, Fattal D, Waks E and Diamanti E 2005 Single photons for quantum information systems *Prog. Inform.* 5

[6] Eisaman M D, Fan J, Migdall A and Polyakov S V 2011 Invited Review Article: Single-photon sources and detectors *Rev. Sci. Instrum.* **82** 071101–071101–25

[7] Bennett C H and Brassard G 1984 Quantum cryptography: public key distribution and coin tossing *Proc. of IEEE Int. Conf. on Computers, Systems and Signal Processing* (Bangalore, India: IEEE Press) p 175

[8] Ekert A K 1991 Quantum cryptography based on Bell's theorem *Phys. Rev. Lett.* **67** 661–3

[9] Bennett C H 1992 Quantum cryptography using any two nonorthogonal states *Phys. Rev. Lett.* **68** 3121–4

[10] Gisin N, Ribordy G, Tittel W and Zbinden H 2002 Quantum cryptography *Rev. Mod. Phys.* **74** 145–95

[11] Brassard G, Lütkenhaus N, Mor T and Sanders B C 2000 Limitations on practical quantum cryptography *Phys. Rev. Lett.* **85** 1330–3

[12] Lütkenhaus N and Jahma M 2002 Quantum key distribution with realistic states: photon-number statistics in the photon-number splitting attack *New J. Phys.* **4** 44

[13] Lukishova S G, Schmid A W, McNamara A J, Boyd R W and Stroud C R 2003 Room temperature single-photon source: single-dye molecule fluorescence in liquid crystal host *IEEE J. Sel. Topics Quantum Electron.* **9** 1512–18

[14] Lukishova S G, Schmid A W, Supranowitz C M, Lippa N, McNamara A J, Boyd R W and Stroud C R 2004 Dye-doped cholesteric-liquid-crystal room-temperature single-photon source *J. Mod. Opt.* **51** 1535–47

[15] Lukishova S G, Schmid A W, Knox R, Freivald P, Bissell L J, Boyd R W, Stroud C R and Marshall K L 2007 Room temperature source of single photons of definite polarization *J. Mod. Opt.* **54** 417–29

[16] Lukishova S G, Bissell L J, Menon V M, Valappil N, Hahn M A, Evans C M, Zimmerman B, Krauss T D, Stroud C R and Boyd R W 2009 Organic photonic bandgap microcavities doped with semiconductor nanocrystals for room-temperature on-demand single-photon sources *J. Mod. Opt.* **56** 167–74

[17] Lukishova S G, Bissell, L J, Stroud C R and Boyd R W 2010 Room temperature single photon sources with definite circular and linear polarizations *Opt. Spectrosc.* **108** 417–24

[18] Lukishova S G, Bissell L J, Winkler J and Stroud C R 2012 Resonance in quantum dot fluorescence in a photonic bandgap liquid crystal host *Opt. Lett.* **37** 1259–61

[19] Lukishova S G 2012 Liquid crystals under two extremes: (1) high-power laser irradiation, and (2) single-photon level *Mol. Cryst. Liq. Cryst.* **559** 127–57

[20] Lukishova S, Knox R, Freivald P, McNamara A, Boyd R, Stroud C, Schmid A and Marshall K 2006 Single-photon source for quantum information based on single dye molecule fluorescence in liquid crystal host *Mol. Cryst. Liq. Cryst.* **454** 1–14

[21] Gayral B, Gérard J M, Legrand B, Costard E and Thierry-Mieg V 1998 Optical study of GaAs/AlAs pillar microcavities with elliptical cross section *Appl. Phys. Lett.* **72** 1421

[22] Daraei A, et al 2006 Control of polarized single quantum dot emission in high-quality-factor microcavity pillars *Appl. Phys. Lett.* **88** 051113–3

[23] Unitt D C, Bennett A J, Atkinson P, Ritchie D A and Shields A J 2005 Polarization control of quantum dot single-photon sources via a dipole-dependent Purcell effect *Phys. Rev.* B **72** 033318–4

[24] Englund D, Fattal D, Waks E, Solomon G, Zhang B, Nakaoka T, Arakawa Y, Yamamoto Y and Vuckovic J 2005 Controlling the spontaneous emission rate of single quantum dots in a two-dimensional photonic crystal *Phys. Rev. Lett.* **95** 013904

[25] Chang W-H, Chen W-Y, Chang H-S, Hsieh T-P, Chyi J-I and Hsu T-M 2006 Efficient single-photon sources based on low-density quantum dots in photonic-crystal nanocavities *Phys. Rev. Lett.* **96** 117401

[26] Purcell E M 1946 Spontaneous emission at radio frequencies *Phys. Rev.* **69** 681

[27] Kopp V I, Fan B, Vithana H K M and Genack A Z 1998 Low-threshold lasing at the edge of a photonic stop band in cholesteric liquid crystals *Opt. Lett.* **23** 1707–9

[28] Il'chishin I P, Tikhonov E A, Tishchenko V G and Shpak M T 1980 Generation of a tunable radiation by impurity cholesteric liquid crystals *JETP Lett.* **32** 24

[29] Wei S K H, Chen S H, Dolgaleva K, Lukishova S G and Boyd R W 2009 Robust organic lasers comprising glassy-cholesteric pentafluorene doped with a red-emitting oligofluorene *Appl.*

Phys. Lett. **94** 041111–3

[30] Dolgaleva K, Wei S K H, Lukishova S G, Chen S H, Schwertz K and Boyd R W 2008 Enhanced laser performance of cholesteric liquid crystals doped with oligofluorene dye *J. Opt. Soc. Am.* B **25** 1496–504

[31] Lodahl P, Floris van Driel A, Nikolaev I S, Irman A, Overgaag K, Vanmaekelbergh D and Vos W L 2004 Controlling the dynamics of spontaneous emission from quantum dots by photonic crystals *Nature* **430** 654–7

[32] Fushman I, Englund D and Vuckovic J 2005 Coupling of PbS quantum dots to photonic crystal cavities at room temperature *Appl. Phys. Lett.* **87** 241102–3

[33] Wu Z, Mi Z, Bhattacharya P, Zhu T and Xu J 2007 Enhanced spontaneous emission at 1.55 μm from colloidal PbSe quantum dots in a Si photonic crystal microcavity *Appl. Phys. Lett.* **90** 171105–3

[34] Bose R, Yang X, Chatterjee R, Gao J and Wong C W 2007 Weak coupling interactions of colloidal lead sulphide nanocrystals with silicon photonic crystal nanocavities near 1.55 μm at room temperature *Appl. Phys. Lett.* **90** 111117–3

[35] Barrelet C J, Bao J, Lončar M, Park H-G, Capasso F and Lieber C M 2006 Hybrid single-nanowire photonic crystal and microresonator structures *Nano Lett.* **6** 11–5

[36] Messin G, Hermier J P, Giacobino E, Desbiolles P and Dahan M 2001 Bunching and antibunching in the fluorescence of semiconductor nanocrystals *Opt. Lett.* **26** 1891–3

[37] Treussart F, Clouqueur A, Grossman C and Roch J-F 2001 Photon antibunching in the fluorescence of a single dye molecule embedded in a thin polymer film *Opt. Lett.* **26** 1504–6

[38] http://www.witec.de/products/snom/alpha300-s/

[39] Chandrasekhar S 1992 *Liquid Crystals* (Cambridge University Press)

[40] Chen S H, Katsis D, Schmid A W, Mastrangelo J C, Tsutsui T and Blanton T N 1999 Circularly polarized light generated by photoexcitation of luminophores in glassy liquid-crystal films *Nature* **397** 506–8

[41] Hollars C W, Lane S M and Huser T 2003 Controlled non-classical photon emission from single conjugated polymer molecules *Chem. Phys. Lett.* **370** 393–8

[42] Bunning T J and Kreuzer F-H 1995 Cyclosiloxane-based liquid crystalline materials *Trends Polym. Sci.* **3** 318–23

[43] Lukishova S G and Schmid A W 2006 Near-field optical microscopy of defects in cholesteric oligomeric liquid crystal films *Mol. Cryst. Liq. Cryst.* **454** 15

[44] Wang Z, Tao F, Cai W, Yao L and Li X 2007 Controlled-synthesis and up-conversion luminescence of $NaYF_4$: Yb, Er phosphors *Solid State Commun.* **144** 255–8

[45] Born M and Wolf E 1999 *Principles of Optics: Electromagnetic Theory of Propagation, Interference and Diffraction of Light* (Cambridge University Press)

[46] Chung I, Shimizu K T and Bawendi M G 2003 Room temperature measurements of the 3D orientation of single CdSe quantum dots using polarization microscopy *Proc. of the National Academy of Sciences of the USA* **100** 405–408

[47] Bissell L J 2011 Experimental realization of efficient, room temperature single-photon sources with definite circular and linear polarizations *PhD Dissertation* (Rochester, NY: University of Rochester)

Journal of Physics: Conference Series **414** (2013) 012007 doi:10.1088/1742-6596/414/1/012007

Generation of Polarization-Entangled Photons in a Standard Polarization-Maintaining Fiber

V. O. Lorenz, Bin Fang, Offir Cohen*

Department of Physics and Astronomy, University of Delaware, Newark, DE 19716, USA
*Present address: Joint Quantum Institute, National Institute of Standards and Technology & University of Maryland, Gaithersburg, MD 20849

E-mail: vlorenz@udel.edu

Abstract. We demonstrate the capability of standard, commercially available polarization-maintaining fibers (PMFs) to generate polarization-entangled photon-pairs by inserting a PMF source into a Sagnac interferometer. We perform a quantum state tomography to reconstruct the density matrix, yielding, without background subtraction or spectral filtering, a polarization-entangled photon-pair state with $92.37 \pm 0.14\%$ fidelity to the maximally entangled Bell state. With its high coupling efficiency into SMFs and ability to produce spectrally uncorrelated photons, we expect this source to be useful for free-space and fiber-based quantum information protocols.

1. Introduction

Entangled photon pairs are an important resource for many applications such as quantum communication protocols [1], quantum computation schemes [2] and fundamental tests of quantum mechanics [3]. Many of these applications often require the distribution or processing of entangled photons through single-mode fiber (SMF) networks. However, coupling into SMFs has been a challenge due to the spatial mode mismatch between the created photons and the guided mode in the SMF. Recently, it was demonstrated that efficient generation of photon-pairs at visible wavelengths is possible using standard, commercially available polarization-maintaining fibers (PMFs) [4] by employing the so-called birefringent phase-matching, in which the photons are produced with polarization orthogonal to the pump beam. Also, it was demonstrated that such a source can generate the photons with no spectral correlations between them, and that the coupling efficiency of the generated photons into single-mode fiber networks was as high as 85% [5]. Here we demonstrate the capability of the source to generate polarization-entangled photon-pairs by inserting a PMF source into a Sagnac interferometer.

2. Experiment

The experimental setup is shown in Fig. 1(a). An 80 MHz Ti:sapphire femtosecond laser provides pump pulses at 708 nm central wavelength with 6 nm full-width at half-maximum (FWHM) bandwidth. A bandpass filter (BPF1) (Semrock FF01-692/40) cleans up the spectrum and a half-wave-plate (HWP1) sets the polarization of the pump at 45°. The horizontal and vertical polarization components of the pump are then coupled into opposite ends of a 20 cm-long PMF (Nufern PM630-HP), which is angle-cleaved to reduce back-reflections. Inside the Sagnac loop,

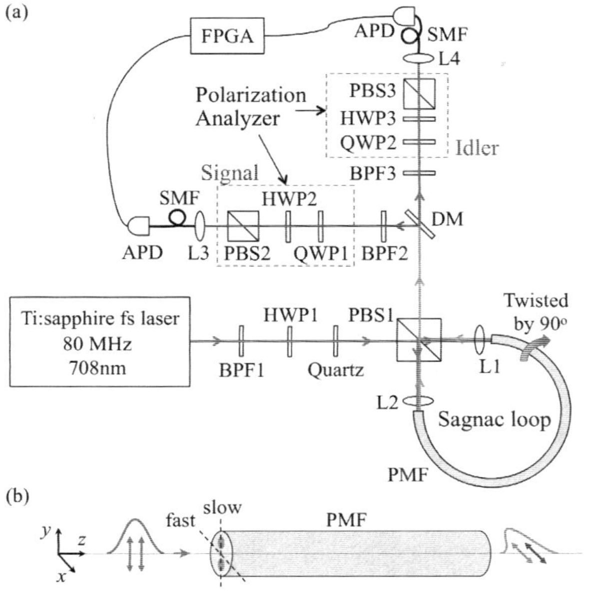

Figure 1. (a) Schematic of photon-pair generation in standard polarization-maintaining fiber. (b) Experimental setup. M, mirror; L1-4 lens; HWP1-3, half-wave-plate; QWP1-2, quarter-wave-plate; PBS1-3, polarizing beam splitter; PMF, polarization-maintaining fiber; DM, dichroic mirror; BPF1-3, band-pass filter; SMF, single-mode fiber; APD, avalanche photodiode; FPGA, field programmable gate array.

two sideband photons are created through spontaneous four-wave mixing. In the spontaneous four-wave mixing process, two pump photons at angular frequency ω_p are annihilated by a $\chi^{(3)}$ medium to generate two side-band photons at angular frequencies ω_s and ω_i, denoted as signal and idler, respectively ($\omega_s > \omega_i$). Energy conservation and phase-matching result in the conditions, assuming self- and cross-phase modulation are negligible, [4]

$$2\omega_p = \omega_s + \omega_i, \tag{1a}$$

$$\Delta k = 2k(\omega_p) - k(\omega_s) - k(\omega_i) + 2\Delta n\frac{\omega_p}{c} = 0, \tag{1b}$$

where $\omega_{p,s,i}$ are the angular frequencies of the pump, signal and idler, respectively, $k(\omega)$ is the wave vector of the angular frequency ω given by the dispersion of pure silica glass, Δk is the phase mismatch and Δn is the birefringence of the PMF. For our experiment these conditions result in the signal at 620 nm and the idler at 825 nm. As shown in Fig. 1(b), the pump propagates on the slow axis of the fiber ($\Delta n > 0$), and signal/idler photons are created with polarization orthogonal to the pump, traveling on the fast axis of the fiber. The fiber is twisted by 90° such that an entangled state is generated at the output port of PBS1. For an ideally symmetric Sagnac loop, the outputs of the two counter-propagating paths are spatially and temporally indistinguishable such that the polarization-entangled state $|\Psi\rangle = (|H_sH_i\rangle + e^{i\varphi}\,|V_sV_i\rangle)/\sqrt{2}$ is created. A dichroic mirror (DM) (Semrock FF685-Di02) then splits the signal and idler photons into two arms to perform joint polarization measurements. Bandpass filters (BPF2,3) (Semrock FF01-609/54, FF01-832/37) suppress the pump and other background. Each arm has a polarization analyzer composed of a quarter-wave-plate (QWP), a half-wave-plate (HWP) and a PBS. The transmitted photons are coupled into SMFs connected to silicon-based avalanche photodiodes (APDs) that detect the photons (Excelitas SPCM AQ4C). The electronic output of the APDs is analyzed by a field-programmable gate array (FPGA) [6] coincidence counter connected to a personal computer.

Any source of distinguishability of the two pathways through the source can result in degradation of the entanglement. For femtosecond sources, small temporal differences in the two paths is a particular issue [7] as the photons' coherence time is short. As will be explained in more detail in the next section, in this experiment we find that a temporal difference may arise due to the fact that one pathway is reflected twice from the PBS, while the other is transmitted twice, meaning that a small separation between the two reflective surfaces in the PBS may impose a temporal mismatch between the two pathways. One way to address temporal distinguishability is to filter the photons through a narrow bandpass filter, thus imposing temporal and spectral matching between the H and V photons that pass through. However, the filtering technique degrades both the efficiency and reliability of the source. Instead, we pre-compensate for the presence of temporal delay between the two paths by inserting a quartz plate to delay one of the pump's polarizations with respect to the other. This method is preferable to the use of spectral filtering as it does not result in loss of photons and allows for loop-hole free tests of Bell's inequality.

3. Results and Discussion

To characterize the photon-pair state we perform a quantum state tomography [8] through 36 different settings of HWP2, HWP3, QWP1 and QWP2 that project the photons onto different joint combinations of the $\{H, V\}$; $\{D = (H+V)/\sqrt{2}, A = (H-V)/\sqrt{2}\}$; and $\{L = (H - iV)/\sqrt{2}, R = (H + iV)/\sqrt{2}\}$ bases and obtain coincidences statistics. We count coincidences for 25 seconds for each setting of the wave-plates. We reconstruct the density matrix of the state using the maximum likelihood method and determine the tangle and linear entropy.

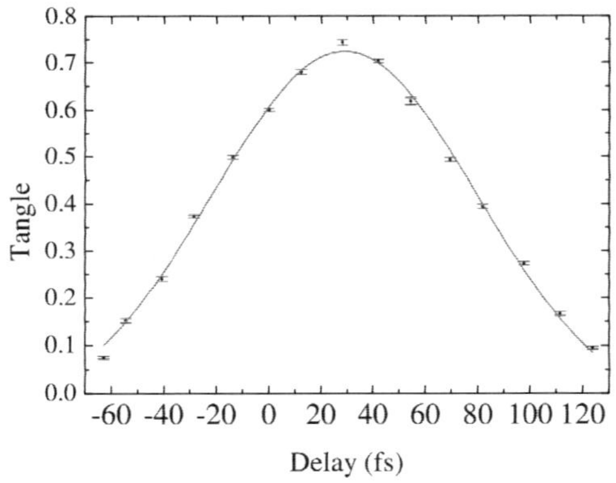

Figure 2. Tangle (black dots) versus delay imposed on the vertical polarization of the pump relative to the horizontal polarization. Error bars are estimated assuming Poissonian statistics of the counts. The FWHM of the Gaussian fit (solid red) is $\Delta\tau_{FWHM} \approx 120\,\text{nm}$.

To determine the degree to which the entanglement is degraded by the presence of temporal distinguishability, we measure the tangle for various combinations of three quartz crystals which produce 13, 41, and 68 fs delays between orthogonal polarizations of the pump. As shown in Fig. 2, we achieve the highest tangle $T_{\max} = 0.7362 \pm 0.0049$ when the vertically polarized pump is delayed by 28 fs with respect to the horizontally-polarized pump (all errors are estimated assuming Poissonian statistics of the counts). For this case we also obtain the lowest linear entropy $L_{\min} = 0.1791 \pm 0.0034$. For comparison, without any delay compensation we find tangle and linear entropy $T = 0.6010 \pm 0.0032$ and $L = 0.2710 \pm 0.0023$, respectively. These results suggest that, without quartz crystals, the optical path along which the vertically polarized photons are created is shorter than the path in which the horizontally polarized photons are created.

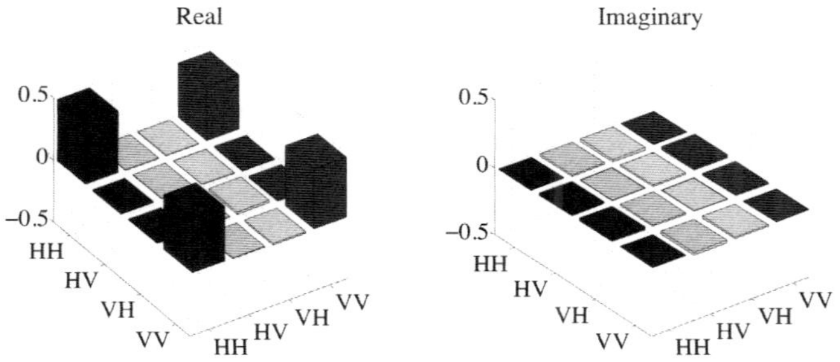

Figure 3. Real and imaginary parts of the reconstructed density matrix with 28 fs temporal compensation. Without any corrections or background subtractions the fidelity to the Bell state $|\Phi^+\rangle = (|H_sH_i\rangle + |V_sV_i\rangle)/\sqrt{2}$ is calculated as $92.37 \pm 0.14\%$.

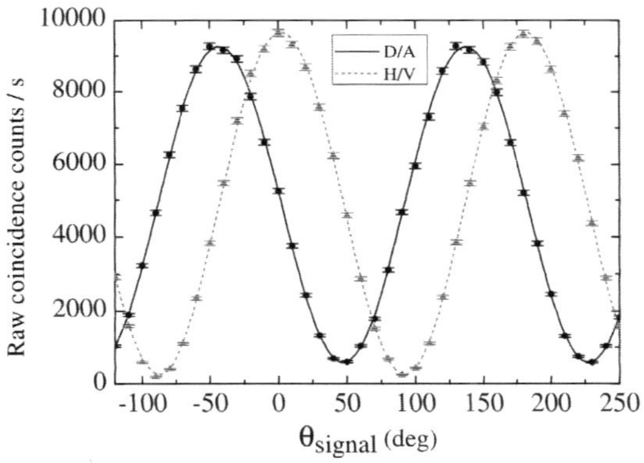

Figure 4. Quantum interference visibility in the H/V (red triangles) and D/A (black circles) bases. θ_{signal} is the angle by which the signal arm polarization is rotated.

For the setup with maximal tangle, we tilt one of the quartz crystals to control and set the phase to $\varphi = 0$. Using 30 mW total pump power, we obtain $\sim 9,000$ coincidences/s. The reconstructed density matrix , shown in Fig. 3, suggests a 92.37±0.14% fidelity with the Bell state $(|H_sH_i\rangle + |V_sV_i\rangle)/\sqrt{2}$. We validate the generated state by an independent quantum-interference visibility measurement, in which QWP1 and QWP2 are set to 0° so that they do not affect the photons, and HWP3 is fixed at either 0° for the H/V basis or 22.5° for the D/A basis interference measurement. Coincidences between signal and idler are recorded as a function of the rotation angle θ of the signal polarization, controlled using HWP2. The results, shown in Fig. 4, show a visibility of $95.00 \pm 1.16\%$ in the H/V basis and $88.21 \pm 0.80\%$ in the D/A basis (again, without any corrections or background subtractions). This result agrees with the expected visibilities given the density matrix in Fig. 3, which are 94.31% and 89.04%, respectively.

4. Conclusion

In summary, we have presented a polarization-entangled photon pair source based on standard polarization-maintaining fiber and have shown that entanglement was created with quality comparable to other demonstrated sources. Due to the large birefringence of the fiber, the photons are far-detuned from the pump, practically eliminating Raman contamination even at room temperature. We showed that temporal distinguishability can be compensated without the use of spectral filters, avoiding photon loss and allowing for loop-hole free tests of Bell's inequality. With its capabilities for spatial mode-matching, the creation of spectrally-uncorrelated photons [5], low cost, and high wavelength tunability [4] we expect such a source to be useful for free-space and fiber-network-based quantum information protocols.

Acknowledgements

This work was supported in part by the University of Delaware Research Foundation and the NSF Physics Division, Grant No. 1205812.

References

[1] Ekert A K 1991 *Phys. Rev. Lett.* **67**(6) 661–663 URL `http://link.aps.org/doi/10.1103/PhysRevLett.67.661`
[2] Pellizzari T, Gardiner S A, Cirac J I and Zoller P 1995 *Phys. Rev. Lett.* **75**(21) 3788–3791 URL `http://link.aps.org/doi/10.1103/PhysRevLett.75.3788`
[3] Tittel W and Weihs G 2001 *Quantum Information & Computation* **1** 3–56
[4] Smith B J, Mahou P, Cohen O, Lundeen J S and Walmsley I A 2009 *Opt. Express* **17** 23589–23602 URL `http://www.opticsexpress.org/abstract.cfm?URI=oe-17-26-23589`
[5] Söller C, Cohen O, Smith B J, Walmsley I A and Silberhorn C 2011 *Phys. Rev. A* **83**(3) 031806 URL `http://link.aps.org/doi/10.1103/PhysRevA.83.031806`
[6] *Fabricated based on a design by the group of A. Steinberg, University of Toronto* URL `http://www.physics.utoronto.ca/~astummer/pub/mirror/Projects/Archives/2008%20Coincidence%20Counter/Coincidence%20Counter.html`
[7] Kim Y H, Kulik S P and Shih Y 2000 *Phys. Rev. A* **62**(1) 011802 URL `http://link.aps.org/doi/10.1103/PhysRevA.62.011802`
[8] James D F V, Kwiat P G, Munro W J and White A G 2001 *Phys. Rev. A* **64**(5) 052312 URL `http://link.aps.org/doi/10.1103/PhysRevA.64.052312`

21st International Laser Physics Workshop IOP Publishing
Journal of Physics: Conference Series **414** (2013) 012008 doi:10.1088/1742-6596/414/1/012008

X-ray emission from relativistic electrons in a transverse high intensity optical lattice

I.A. Andriyash

E-mail: igor.andriyash@gmail.com

E. d'Humières

V.T. Tikhonchuk

Ph. Balcou

Univ. Bordeaux, CNRS, CEA, CELIA (Centre Lasers Intenses et Applications), UMR 5107, F-33400 Talence, France

Abstract. A new scheme of all-optical free electron laser (FEL), operating in the X-ray range and triggered by interaction of a moderately relativistic electron bunch with a transverse high intensity optical lattice is presented. The ponderomotive potential of the optical lattice forms guiding channels, where the electrons are trapped and oscillate transversely. The collective oscillations of the guided electron bunch scatter the laser beams and amplify the signal at the Stokes frequency, thereby yielding a coherent beam at the Doppler-shifted frequency in the XUV or X-ray domain. Analytic descriptions of a small signal amplification in such a system are presented, and the optimal resonance conditions are defined. The gain scaling laws are confirmed by numerical simulations using a simplified kinetic model in the co-propagating reference frame.

1. Introduction

We present here a new scheme of an all-optical free electron laser (FEL) operating in the X-ray range and triggered by interaction of a moderately relativistic electron bunch with a transverse high intensity optical lattice. The lattice is created by two co-polarized counter-propagating laser beams overlapping in the interaction region with the electrons. The ponderomotive potential of the optical lattice forms guiding channels where the electrons are trapped and oscillate transversely [1, 2]. The collective oscillations of the guided electron bunch [3] can amplify the scattered signal wave at the Stokes frequency, thereby yielding a coherent beam at the Doppler-shifted frequency in the XUV or X-ray domain. Such a scheme combines some properties of conventional XFELs [4] and the XFEL based on the stimulated Raman scattering [5] of electron plasma waves; however, the physics of amplification is significantly different due to the guiding effect of the optical lattice [6]. It is expected to be less demanding with respect to the electron beam quality.

In section 2 we describe the injection and the state of electron beam, trapped in the lattice potential. In the sections 3 and 4, we derive the gain scaling laws in linear fluid and kinetic descriptions, and analyze the typical amplification regimes. The theoretical estimates are confirmed by the comparison with numerical simulations in section 5, using a simplified kinetic model in the co-propagating reference frame. Numerical estimates in the conclusion show that this scheme holds the potential to achieve X-ray amplification over millimeter-scale lengths, while coupled to the Laser Wakefield Acceleration of electrons section 6.

2. Injection and the ground state of electron beam in the optical lattice

Let us consider a beam of relativistic electrons, which propagates along the z-axis into an optical lattice with a wave vector oriented along the x-axis (the direction of laser pulse propagation) and polarized along the y-axis. The lattice is formed by two counter-propagating plane waves:

$$\mathbf{a}_L = \mathbf{a}_1 + \mathbf{a}_2 = 2a\,\mathbf{e}_y\,\sin k_0 x\,\cos\omega_0 t\,,$$

where the normalized vector potential $a(z)$ charactrizes the field amplitude in the relativistic units $m_e\omega_0 c/e$. Within the interaction region the lattice amplitude is considered as uniform along x and y axes, and does not depend on time. The profile of the lattice along the z-axis has a ramp $a(0 < z < l_{ramp}) = a_0 z/l_{ramp}$ and is assumed to be constant a_0 for $z > l_{ramp}$.

Entering the ramp, electrons become trapped in the ponderomotive potential, $U_p \propto \sin^2 k_0 x$, of the lattice channels. Propagation along the ramp modifies the electron distribution in the (x, p_x) plane and deforms the beam spatial profile. The maximum density is localized on the channel axes, where $U_p = 0$, and the electrons are less affected by injection. Without entering in the details, it is reasonable to assume that the maximal density of the trapped beams is the same as the average beam density before injection. The maximal transverse kinetic energy of trapped electrons is limited as, $\epsilon_\perp < m_e c^2 (a_0/\gamma_b)^2$, where $\gamma_b = 1/\sqrt{1-\beta_b^2}$ is the beam Lorentz factor.

The study is performed in the reference system moving with the average velocity electron beam $\beta_b c$. In such a system the electron beam parameters and the laser field are modified due to the Lorentz transformation. The wave vectors of both laser pulses obtain a component along the electron beam, $k_{0\|} = \gamma_b\beta_b k_0^l$, while the transverse projections remain invariant $k_{0\perp} = k_0^l$. The superscript "l" hereafter indicates the parameters in the laboratory reference system. The electron velocity spread in the beam frame is assumed to be non-relativistic, it can be estimated as $\delta\beta_\perp = \epsilon_N^l/(\beta_b\sigma_\perp^l)$, $\delta\beta_\| = \delta\gamma_\|^l/(\gamma_b\beta_b)$, where ϵ_N^l is a normalized emittance of the beam, $\sigma_\perp^l$ its radius and $\delta\gamma_\|^l$ characterizes its energy spread. The laser wavelength λ_0^l and the electron density are decreased by the factor of γ_b, and assuming a time independent beam density one may calculate the ratio of the electron beam density to the critical density for the laser field as follows:

$$\frac{n_e}{n_c} = \frac{\lambda_0^{l\,2}}{\pi\sigma_\perp^{l\,2}\gamma_b^3}\,\frac{J^l}{J_A}\,,$$

where J is the electron current and $J_A = m_e c^3/e \simeq 17$ kA is the Alfven current.

Considering a perturbative approach to the lattice–beam interaction, we assume the electron distribution function (EDF) to be a sum of the EDFs of the ground state f_0 and a small perturbation f, while the vector potential includes the laser field a_L and a small signal wave a_s. Thus, the ponderomotive potential $U_p = (m_e c^2/4)\nabla|\bar{a}|^2$ includes the zeroth-order and first-order perturbation terms,

$$U_p^{(0)} = m_e (c/2)^2\,|\bar{a}_L|^2\,, \tag{1a}$$

$$U_p^{(1)} = m_e (c/2)^2\,(\bar{a}_s\bar{a}_L^* + c.c.)\,. \tag{1b}$$

We describe the interaction process using the Vlasov equation averaged over the laser period $\tau_0 = \lambda_0/c$, and the enveloped wave equation for the electromagnetic vector potential. In the zeroth order there is no signal wave, and the system reads:

$$\left(\partial_t + v_x\partial_x + v_z\partial_z - \partial_x U_p^{(0)}\partial_{v_x}\right) f_0 = 0\,, \tag{2a}$$

$$(\partial_t^2 - c^2\nabla^2)\bar{a}_L = -\bar{a}_L \frac{\omega_p^2}{n_0}\int d\mathbf{v}\, f_0\,, \tag{2b}$$

where $\omega_p = \sqrt{4\pi e^2 n_0/m_e}$ is the electron plasma frequency on the channel axis, where the electron density reaches its maximum value, $n_0 = n_e(x=0)$. The enveloped field of the optical lattice reads $\bar{a}_L = 2a_0 \sin(k_{0\perp}x)e^{ik_{0\parallel}z}$. For the practically interesting cases, the relative electron density in the proper frame is very low (below 10^{-5} for the available sources). Thus, we may neglect the dispersion term in the right hand side of eq. (2b) and consider $\omega_0 = k_0 c$.

Considering a single trapped beam, we may approximate the sinusoidal channel potential with a parabolic one, $U_p^{(0)} = \Omega^2 x^2/2$, where $\Omega = \sqrt{2}a_0 k_{0\perp}c$ is a frequency of small-amplitude oscillations. The electron trajectories in such a potential are the characteristics of eq. (2a), and they read:

$$\begin{aligned} &X(x_0, v_{x0}, t) = x_0 \cos\Omega t + (v_{x0}/\Omega)\sin\Omega t\,,\\ &V_x(x_0, v_{x0}, t) = v_{x0}\cos\Omega t - x_0\Omega\sin\Omega t\,,\\ &Z(z_0, v_{z0}, t) = z_0 + v_{z0}t\,, \quad V_z(z_0, v_{z0}, t) = v_{z0} \end{aligned} \tag{3}$$

where x_0, v_{x0}, z_0 and v_{z0} are coordinates of the particle at $t=0$. Its easy to see, that along these trajectories the equation, $df_0/dt = 0$, is satisfied, so the ground state EDF is constant with respect to the motion integrals $v_z, v_x^2 + \Omega^2 x^2$.

Let us consider a model EDF of non-relativistic particles homogeneously distributed along z axis with no z velocity component. In the plane (x, v_x) they occupy the Kapchinsky-Vladimirsky (K-V) ellipse $v_x^2 + \Omega^2 x^2 = \epsilon_\perp$, (for more on the K-V distribution see [7]):

$$f_0 = n_0/(2\sqrt{\epsilon_\perp})\eta(\epsilon_\perp - v_x^2 - \Omega^2 x^2)\,. \tag{4}$$

where the maximal kinetic energy of the electrons transverse motion is limited by the ponderomotive potential amplitude, $\epsilon_\perp = 2a_0^2$. Using the definition eq. (4), we may calculate the distributions of electron density and pressure:

$$n_e = n_0\sqrt{1-\xi^2}, \tag{5a}$$

$$P_{xx} = n_e^3 v_\perp^2/3n_0^2\,,\ P_{zz} = n_e v_\perp^2\,, \tag{5b}$$

where $\xi = x/L$ is a transverse coordinate normalized to the "effective" beam half-width $L = \sqrt{\epsilon_\perp}/\Omega$. The system eq. (5b) represents an equation of state, that may be used subsequently to describe the perturbation dynamics in a fluid approach.

3. Fluid description of the stimulated scattering of electrons in the optical lattice

We first study the collective electron dynamics in a fluid approach, i.e. electron motion follows the hydrodynamic equations, interacting with the electromagnetic field via the first order of ponderomotive potential eq. (1b). In this simplified model the electrostatic potential and the beam longitudinal temperature are neglected and the transverse electron temperature does not exceed the trapping limit. The interaction is described with the following set of equations:

$$\partial_t n_e + \nabla\cdot(n_e\,\mathbf{u}) = 0\,, \tag{6a}$$

$$\partial_t u_x + \nabla_x P_{xx}/m_e n_e + \nabla_x U_p = 0\,, \tag{6b}$$

$$\partial_t u_z + \nabla_z P_{zz}/m_e n_e + \nabla_z U_p = 0\,, \tag{6c}$$

$$\left((\partial_t - i\omega_0)^2 - c^2\nabla^2\right)\bar{a} = -(4\pi e^2/m_e)n_e\,\bar{a}\,, \tag{6d}$$

where $\mathbf{u}$ is an electron fluid velocity. For the signal wave propagating in the positive direction of z axis with frequency and wavenumber close to those of the pump $(\omega_s, k_s) \simeq (\omega_0, k_0)$, we may consider the electromagnetic field in the enveloped form $\bar{a}_s$, defined as $a_s = \mathrm{Re}[\bar{a}_s(t,z)\,\mathrm{e}^{-i(\omega_0 t - k_0 z)}]$. Equation (6d) may be further simplified by assuming the signal wave to be a plane wave, $\partial_\perp^2 a_s = 0$.

We describe the collective electron modes as a perturbation of the beam position Δ_x, defined as $\xi = (x + \Delta_x)/L$. Without a signal wave, $\partial_z U_p = 0$, and from equations (6) it follows that:

$$\partial_t^2 \Delta_x = -\Omega^2 \Delta_x \,.$$

This equation describes the collective beam mode with a frequency $\omega_b = \Omega$. For a small amplitude signal, considering eq. (6) in the Fourier domain, we obtain the dispersion equation:

$$\left(\omega^2 - \Omega^2\right)\left[(\omega + \omega_0)^2 - (k - k_{0\parallel})^2 c^2 - \tilde{\omega}_p^2\right] = 4\alpha\omega_0^4 \,, \tag{7}$$

where the coefficient

$$\alpha = K(a_0 \omega_p k c / 2\omega_0^2)^2 \,,$$

defines the coupling between the beam and signal modes, and the factor $K = (k_{0\perp}L)^2\,\xi^2\sqrt{1-\xi^2}$, accounts for the filling factor $k_{0\perp}L$ and the coupling localization within the beam, $|\xi| < 1$. For realistic beam parameters the filling factor is $k_{0\perp}L \simeq 1$, and we may estimate K by averaging the coupling localization over ξ. In the case of a plane signal wave, the localization shape is $\propto \xi^2\sqrt{1-\xi^2}$, and we evaluate the averaged factor as $K \simeq 0.2$. If the scattered wave is diffracted, its profile must be considered, and the value of K will be estimated numerically.

Amplification of the Stokes mode is described by the complex frequency in eq. (7) with a real part $\omega \simeq -\Omega$ and imaginary part Γ corresponding to an exponential growth rate. Considering the resonant wave vector, $k_s = k - k_{0\parallel} \simeq (\omega_0 - \Omega)/c$, and assuming $\Omega \ll \omega_0$, we estimate the imaginary part as:

$$\Gamma/\omega_0 = \omega_0\sqrt{\alpha/\Omega(\omega_0 - \Omega)} \,. \tag{8}$$

4. Kinetics of the beam and field perturbations

The amplification growth rate may be also estimated in a linear kinetic model of the perturbation dynamics. To first order, the enveloped electromagnetic equation reads :

$$(\partial_t + c\partial_z)\bar{a}_s - i\,\frac{c}{k_0}\,\partial_\perp^2 \bar{a}_s = -i\,a_0 \sin(k_{0\perp}x)\mathrm{e}^{-ikz}\,\frac{\omega_p^2}{n_0\omega_0}\int d\mathbf{v} f \,, \tag{9}$$

where $k = k_0 + k_{0\parallel}$. As in section 3, we assume the plane signal wave, so $\partial_\perp^2 a_s = 0$, and reduce the order of eq. (9). The electron dynamics is described by the Vlasov equation:

$$\left(\partial_t + v_x\partial_x + v_z\partial_z - \partial_x U_p^{(0)}\partial_{v_x}\right) f = \left(\partial_x U_p^{(1)}\,\partial_{v_x} + \partial_z U_p^{(1)}\,\partial_{v_z}\right) f_0 \,, \tag{10}$$

where two terms in the right hand side represent the sources of EDF perturbations along x and z axes respectively. These perturbations are related to the components of the ponderomotive force $\partial_z U_p^{(1)}$ to $\partial_x U_p^{(1)}$, and their ratio can be estimated as $k/k_{0\perp} \sim \gamma_b \gg 1$, so that the x component may neglected. Integrating eq. (10) over the time along the electron trajectories eq. (3) and over the electron velocity, we calculate the electron density perturbation:

$$\delta n/n_0 = \sum_{\pm}(N_\pm \mathrm{e}^{ikz} + c.c.) \,, \tag{11}$$

where,

$$N_\pm = ia_0 \frac{kc^2}{4} k_{0\perp} L\, \xi \sqrt{1-\xi^2} \int_{t_0}^{t} \mathrm{d}t' \bar{a}_s(t') \mathrm{e}^{\pm i\Omega(t-t')} .$$

Here the components $N_+ \mathrm{e}^{ikz}$, $N_- \mathrm{e}^{ikz}$, $N_+^* \mathrm{e}^{-ikz}$, $N_-^* \mathrm{e}^{-ikz}$ may be resonantly coupled to the up- or down-shifted waves, propagating either along the positive or negative direction of z axis. For the signal propagating along the z axis we may neglect the short wavelength terms $\propto \mathrm{e}^{-i2kz}$, and describe the coupled of electromagnetic (9) and beam (11) modes as follows:

$$(\partial_t + c\partial_z)\bar{a}_s = -i\, a_0 k_{0\perp} x \frac{\omega_p^2}{\omega_0} \sum_\pm N_\pm . \qquad (12)$$

We consider the harmonic solutions of eq. (12), $N_\pm\, , \bar{a}_s \propto \mathrm{e}^{-i\omega t + i\Delta k z}$, where $\omega = \omega_0 - \omega_s$ and $\Delta kc = \omega_0 - k_s c$. The values of ω and Δk define the frequency and the wave number detuning of the signal field with respect to the laser wave. They are related by the dispersion equation:

$$\omega - \Delta kc = -\alpha\, \omega_0^3 \sum_\pm (\omega \pm \Omega)^{-2} \, , \qquad (13)$$

Two terms in the right hand side of eq. (13) represent the Stokes (+) and anti-Stokes (-) modes. It can be shown, that there are no unstable solutions in the anti-Stokes domain. Considering the instability near the resonance $\omega \simeq -\Omega$, one may neglect the anti-Stokes term, and find:

$$(\omega - \Delta kc)\,(\omega + \Omega)^2 = \alpha\, \omega_0^3 . \qquad (14)$$

where factors in the left hand side characterize the electromagnetic and electron beam modes and the right hand side defines the coupling. Assuming only the temporal signal amplification (that is, a periodic system), the detuning term has to be real $\Delta kc \in \Re$.

The maximal growth rate corresponds to detuning $\Delta kc = -\Omega$, where the dispersion equation reads: $(\omega + \Omega)^3 = \alpha\omega_0^3$. The imaginary part of the frequency is evaluated as

$$\Gamma/\omega_0 = (\sqrt{3}/2)\alpha^{1/3} \, , \qquad (15)$$

which corresponds to the signal frequency, $\omega_s = \omega_0(1 - \alpha^{1/3}/2) - \Omega$. For the case, $\alpha^{1/3} \ll \Omega/\omega_0$, it is close to the Stokes frequency $\omega_s \simeq \omega_0 - \Omega$. Note, that a the scaling eq. (15) is similar to the SRS in a strong coupling regime, while eq. (8) corresponds to the weak coupling [5]. For the non-detuned wave, the instability develops in a weak coupling regime, and eq. (13) gives the following growth rate:

$$\Gamma/\omega_0 = \sqrt{\alpha\omega_0/\Omega} . \qquad (16)$$

Our linear models are based on the perturbation theory in the approximation of a small amplitude signal and electron density perturbations, and a constant electron ground state temperature. In numerical simulations, we observe that electrons are rapidly heated along the propagation direction and trapped in the longitudinal potential due to the interference of the signal and pump waves. Trapping continues till all resonant (interacting with the wave) electrons are trapped. The density perturbations reach the maximum value $|N_\pm| \simeq 1$, which results in the signal saturation at the level that follows from eq. (12):

$$|a_s/a_0| \simeq \omega_p^2/(\omega_0\Omega) . \qquad (17)$$

These theoretical models of the beam trapping and the signal amplification are compared with the numerical simulations presented in the next section. They account for the dynamics of electrons in the ground state and the signal wave saturation.

5. Numerical modeling of the signal amplification in the optical lattice

For the numerical simulations we use the two-dimensional code EWOK, where the electron motion is averaged over the laser period τ_0 and is described with help of the particle-in-cell (PIC) method. The electromagnetic field is calculated from the single-component, reduced-order Maxwell equation eq. (9), and its transverse distribution is presented by the Fourier harmonics.

Here we study a temporal amplification of the signal wave in a periodic system. The vertical size of the simulation domain presents the width of one potential channel $\gamma_b\lambda_0/2$. The length in the horizontal dimension is limited by the condition that the signal and the beam modes are periodic. It is chosen as an integer number of two wavelengths: the longitudinal potential $a_s a_L* \sim e^{i(k_{0\parallel}+k_s)z}$ and the enveloped signal wave $\bar{a}_s \sim e^{i(k_s-k_0)z}$. This choice of the length of the simulation domain may limit the detuning of the signal wave and result in its artificial diffraction. The electrostatic force violates the periodicity and is neglected in what follows.

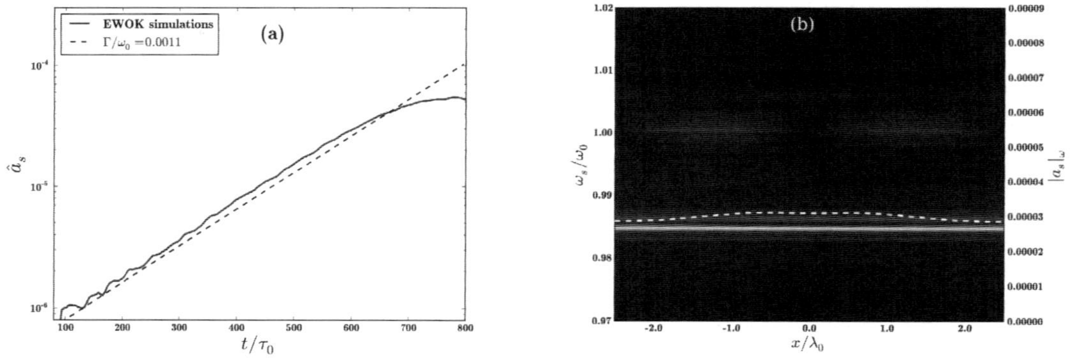

Figure 1. (a) Amplitude of the scattered field in the center of right boundary, $(x, z) = (0, 70.2)\lambda_0$. (b) Spatial-spectral distribution of the amplified wave along the right boundary.

Let us consider the interaction of an electron beam having the Lorentz factor $\gamma_b = 10$ with the optical lattice, created by two laser waves with the amplitudes $a_0 = 0.1$. The dimensions of the simulation domain are $5\lambda_0 \times 70.2\lambda_0$, and the injection duration is $t_{inj} = 20\lambda_0/c$. The transverse momentum spread is close to the trapping conditions $\beta_\perp = 0.1$, while the longitudinal momentum spread and the electron density vary in the calculations.

For the electron density $n_e/n_c = 0.8 \times 10^{-5}$, the signal amplitude at the center of the right boundary of simulation box, demonstrates the exponential growth with the rate $\Gamma/\omega_0 = 1.1 \times 10^{-3}$ as it is shown in fig. 1a. For these conditions $\alpha^{1/3} \ll \Omega/\omega_0$, the instability develops in the weak coupling regime, and the growth rate value is in good agreement with eqs. (8) and (16). The strong coupling scaling eq. (15) predicts a few times higher growth rate $\Gamma/\omega_0 = 3.7 \times 10^{-3}$.

The field spectrum at the right boundary $z/\lambda_0 = 70.2, x/\lambda_0 \in [-2.5, 2.5]$ presented in fig. 1b gives the frequency of the amplified wave and its transverse profile. The amplified wave has a narrow spectrum around the resonant frequency $\omega_{\text{res}} = 0.985\omega_0 \simeq \omega_0 - \Omega$, which corresponds to the Stokes mode. The cut of this picture at the resonant frequency $\omega_s = \omega_{\text{res}}$ along the x coordinate gives the transverse distribution of the amplitude of the amplified wave a_s. It is shown with the white dashed curve in fig. 1b. The frequency shift increases by 8% at the edges of the amplification zone.

The amplification of the signal wave implies the growth of the electron beam modes. The density of electrons that are trapped in this mode, $n_e \propto \xi\sqrt{1-\xi^2}\exp(i(k_{0\parallel}+k_s)z)$, is presented in fig. 2. Trapping of all electrons corresponds to the instability saturation at the level given by

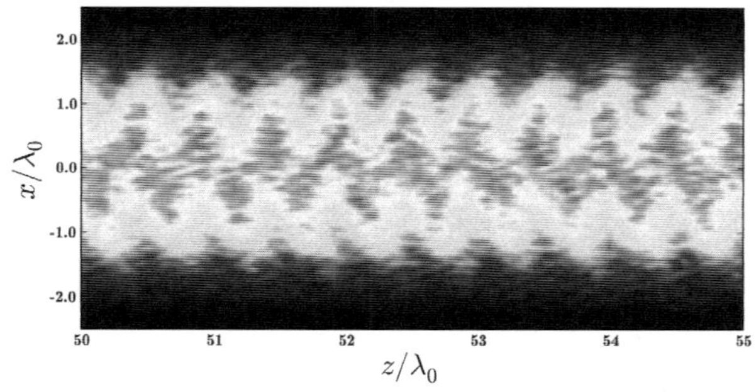

Figure 2. Fragment of the bunched electron density distribution near the saturation time.

eq. (17). For the parameters presented in fig. 1a, this field is estimated as $a_s^{\mathrm{sat}} = 5.3 \cdot 10^{-5}$. It is in a good agreement with the field measured in the simulations : $\max[a_s] = 5.8 \cdot 10^{-5}$.

6. Conclusions

We presented two successive models of the stimulated Raman scattering of a relativistic electron beam in a high intensity optical lattice. A fluid and a kinetic perturbation theory were developed that describe the interaction of electrons trapped in the ponderomotive potential of the lattice with the electromagnetic field. The collective oscillations of electrons in the lattice potential enable a resonant transfer of the laser beam energy to the signal wave. The longitudinal momentum of the signal wave is transferred to electrons. The rate of the signal growth is obtained analytically and compared to the results of PIC simulations. A qualitatively good level of agreement is obtained.

The proposed interaction scheme may be used for amplification of XUV or X-ray coherent radiation. Let us consider for example a 4 kA electron beam produced in a laser wake field accelerator with $\gamma = 100$ and interacting with an optical lattice with an amplitude $a_0 = 0.5$. The maximal gain length for that case is estimated to $300 - 400\,\mu$m. In the saturated regime, a microjoule pulse at a wavelength of 0.4 Å can be generated, of few femtosecond duration . Even though these theoretical estimates may be reduced in real experimental conditions, this XFEL-SRS scheme will enable a high gain of X-rays, along with an efficient stabilization of the electron beam. This all-optical Raman X-ray laser may thus result in ultra-compact, fully coherent source up to the hard X-ray range, with innumerable applications to science, technology and medicine.

References

[1] Fedorov M V, Oganesyan K B and Prokhorov A M 1988 *Appl. Phys. Lett.* **53** 353–354
[2] Balcou Ph 2010 *Eur. Phys. J.* D **59**(3) 525–537
[3] Andriyash I A, Ph Balcou and Tikhonchuk V T 2011 *Eur. Phys. J.* D **65**(3) 533–540
[4] Freund H P and Antonsen T M 1996 *Principles of Free-Electron Lasers* (Chapman & Hall)
[5] Sprangle P and Drobot A T 1979 *J. Appl. Phys.* **50** 2652–2661
[6] Andriyash I A, d'Humières E, Tikhonchuk V T and Balcou Ph 2012 *Phys. Rev. Lett.* **109**(24) 244802
[7] Reiser M 2008 *Theory and Design of Charged Particle Beams* 2nd ed (Wiley-VCH)

Journal of Physics: Conference Series **414** (2013) 012009 doi:10.1088/1742-6596/414/1/012009

The applicability range of different forms of the radiation friction force in the ultrarelativistic electron interaction with electromagnetic wave (exact solutions)

S. S. Bulanov[1], T. Zh. Esirkepov[2], M. Kando[2], J. K. Koga[2], and S. V. Bulanov[2]

[1]University of California, Berkeley, CA 94720, USA
[2]Kansai Photon Science Institute, JAEA, Kizugawa, Kyoto 619-0215, Japan

Abstract. When the effects of radiation reaction dominate the interaction of electrons with intense laser pulses, the electron dynamics changes qualitatively. The adequate theoretical description of this regime becomes crucially important with the use of the radiation friction force either in the Lorentz-Abraham-Dirac form, which possess unphysical runaway solutions, or in the Landau-Lifshitz form, which is a perturbation valid for relatively low electromagnetic wave amplitude. The goal of the present paper is to find the limits of the Landau-Lifshitz radiation force applicability in terms of the electromagnetic wave amplitude and frequency.

1. Introduction

The recent development of ultrahigh-intensity laser systems has generated a great amount of interest in a class of well-known theoretical problems involving the interaction of strong electromagnetic fields with with charged particles that have not been experimentally demonstrated. Among them the problem of the radiation friction effects on the charged particle dynamics, which has been attracting attention for more than a century [1, 2, 3, 4]. The radiation friction imposes constraints on the highest attainable energy of charged particles accelerated by standard accelerators [5] and in space [6], in particular, on the energy of the ultra high energy cosmic rays [7]. The effects of radiation reaction on electrons in a magnetically confined plasma lead to the phase space contraction [8]. The radiation generated by present day [9, 10] lasers approaches limits when the radiation friction force will change the scenario of the electromagnetic (EM) wave interaction with matter, i.e. at $I > I_{rad} = 10^{23}\mathrm{W/cm}^2$. The electron dynamics will become dissipative with fast conversion of the EM wave energy to hard EM radiation, which for typical lasers parameter is in the gamma-ray range [11, 12, 13, 14]. There are discussions of the modification of the electron acceleration in the laser wake field acceleration regime [15] and the ion acceleration in the radiation pressure dominated regime [16] due the radiation friction, which are mainly obtained with computer simulations [11, 17]. If the laser intensity substantially exceeds I_{rad} novel physics of abundant electron-positron pair creation will come into play [18] (see also [13] and [19]) when the electron (positron) interaction with the EM field is principally determined by the radiation friction effects. The persistent interest towards the radiation friction effects stems from all these reasons [20, 21].

In order to self-consistently find the trajectory of the emitting electron, the so called

Minkovsky equations [3] should be solved with the radiation friction force taken into account

$$m_e c \frac{du^\mu}{ds} = \frac{e}{c} F^{\mu\nu} u_\nu + g^\mu, \tag{1}$$

$$\frac{dx^\mu}{ds} = u^\mu. \tag{2}$$

Here $u_\mu = (\gamma, \mathbf{p}/m_e c)$ is the four-velocity, $F_{\mu\nu} = \partial_\mu A_\nu - \partial_\nu A_\mu$ is the EM field tensor with A_μ being the EM four-vector and $\mu = 0, 1, 2, 3$, and $s = c \int dt/\gamma$. The radiation friction force in the Lorentz-Abraham-Dirac (LAD) form [22, 23, 24] is given by

$$g^\mu = \frac{2e^2}{3c} \left[\frac{d^2 u^\mu}{ds^2} - u^\mu \left(\frac{du^\nu}{ds} \right) \left(\frac{du_\nu}{ds} \right) \right]. \tag{3}$$

As is well known, the equation (1) with the radiation friction force in the LAD form (3) possesses unphysical self-accelerating solutions (e.g. see Refs. [2, 3]). When the radiation friction force is taken to be in the Landau-Lifshitz (L-L) form,

$$g^\mu = \frac{2e^3}{3m_e c^3} \left\{ \frac{\partial F^{\mu\nu}}{\partial x^\lambda} u_\nu u_\lambda - \frac{e}{m_e c^2} \left[F^{\mu\lambda} F_{\nu\lambda} u^\nu - \left(F_{\nu\lambda} u^\lambda \right) \left(F^{\nu\kappa} u_\kappa \right) u^\mu \right] \right\}, \tag{4}$$

the electron motion equations does not have pathological solutions, although they are not always consistent with energy-momentum conservation for an abruptly changing electromagnetic field [25]. Since the L-L radiation friction force is derived as a perturbation to the equations of motion, this approximation is valid provided there exists a frame of reference, where the L-L radiation friction force is small compared to the Lorentz force, $eF^{\mu\nu}u_\nu$, as noticed in Ref. [2]. Proving this frame of reference existence and finding the range of validity of the friction force in the L-L form is far from trivial. Below, using several exact analytical solutions to the electron motion equations in the EM field for the radiation friction force in the LAD and L-L forms, we discuss the validity range of the later approximation.

The electron motion equations with the LAD friction force admit exact solution for the stationary problem describing the electron motion in the rotating electric field (see Refs. [26, 12, 13]). This problem can also be solved for the case of the L-L force. Generalizing the electromagnetic field configuration, we consider the electric and magnetic field to be a superposition of the rotating with the frequency ω homogeneous in space and time-independent components

$$\mathbf{E} = \mathbf{e}_1 E_1 + \mathbf{e}_2 [D x_2 + E \cos(\omega t)] + \mathbf{e}_3 [D x_3 + E \sin(\omega t)], \tag{5}$$

$$\mathbf{B} = \mathbf{e}_1 B_1 + \mathbf{e}_2 J x_3 - \mathbf{e}_3 J x_2, \tag{6}$$

where $\mathbf{e}_1$, $\mathbf{e}_2$ and $\mathbf{e}_3$ are the unit vectors along the 1,2,3 axis. The EM field tensor is equal to

$$F^{\mu\nu} = \begin{pmatrix} 0 & -E_1 & -Dx_2 - E\cos(\omega t) & -Dx_3 - E\sin(\omega t) \\ E_1 & 0 & -Jx_2 & Jx_3 \\ Dx_2 + E\cos(\omega t) & Jx_2 & 0 & -B_1 \\ Dx_3 + E\sin(\omega t) & -Jx_3 & B_1 & 0 \end{pmatrix}. \tag{7}$$

In the case when E_1, B_1, J and D vanish the electric field can be realized in the antinodes, where the magnetic field vanishes, of a standing EM wave formed by two counter-propagating circularly polarized EM waves. Such an EM field configuration plays an important role in theoretical considerations of various nonlinear effects in quantum electrodynamics, e.g. see Refs. [13, 18, 27, 28]. This EM configuration corresponds also to the circularly polarized EM wave propagating in the underdense plasma for the frame of reference moving with the wave group velocity [29, 30]. In this frame of reference the EM wave frequency is equal to the Langmuir frequency, $\omega_{pe} = (4\pi n_0 e^2/m_e)^{1/2}$, where n_0 is the plasma density and the wave has no magnetic field component. The static component of the magnetic field, B_1, can be generated in the laser plasmas due to the inverse Faraday effect. Its effect on the charged particle motion has been studied in Ref. [26]. The radial component of the electric field, $\mathbf{e}_2 D x_2 + \mathbf{e}_3 D x_3$, and azimuthal component of magnetic field, $\mathbf{e}_2 J x_3 - \mathbf{e}_3 J x_2$, correspond to the plasma wave in the boosted frame of reference with E_1 being the longitudinal component of the wake field.

It is convenient to write the electron momentum $\mathbf{p} = \mathbf{e}_1 p_1(t) + \mathbf{e}_2 p_2(t) + \mathbf{e}_3 p_3(t)$ and coordinates $\mathbf{x} = \mathbf{e}_1 x_1(t) + \mathbf{e}_2 x_2(t) + \mathbf{e}_3 x_3(t)$ as a combination of non-rotating and rotating with the angular frequency ω vectors,

$$\begin{pmatrix} \tilde{u}_1 \\ \tilde{u}_2 \\ \tilde{u}_3 \end{pmatrix} \equiv \frac{1}{m_e c} \begin{pmatrix} p_1 \\ p_{\|} \\ p_{\perp} \end{pmatrix} = \frac{1}{m_e c} \begin{pmatrix} 1 & 0 & 0 \\ 0 & \cos(\omega t) & \sin(\omega t) \\ 0 & -\sin(\omega t) & \cos(\omega t) \end{pmatrix} \begin{pmatrix} p_1 \\ p_2 \\ p_3 \end{pmatrix} \tag{8}$$

and

$$\begin{pmatrix} \tilde{x}_1 \\ \tilde{x}_2 \\ \tilde{x}_3 \end{pmatrix} = \begin{pmatrix} 1 & 0 & 0 \\ 0 & \cos(\omega t) & \sin(\omega t) \\ 0 & -\sin(\omega t) & \cos(\omega t) \end{pmatrix} \begin{pmatrix} x_1 \\ x_2 \\ x_3 \end{pmatrix}. \tag{9}$$

2. Stationary solution to the electron equations of motion with the radiation friction force in the Lorentz-Abraham-Dirac form

Stationary solution to equations (1) and (2), for which the vectors $\tilde{\mathbf{u}} = (\tilde{u}_1, \tilde{u}_2, \tilde{u}_3)$ and $\tilde{\mathbf{x}} = (\tilde{x}_1, \tilde{x}_2, \tilde{x}_3)$ do not depend on time, with the radiation friction force in the LAD form (3) can be cast as

$$0 = a_1 - \varepsilon_{rad} \tilde{u}_1 \gamma \left(\gamma^2 - 1 - \tilde{u}_1{}^2\right), \tag{10}$$

$$\tilde{u}_2 = \left(d - j\frac{\tilde{u}_1}{\gamma}\right)\frac{\tilde{u}_2}{\gamma} + b\frac{\tilde{u}_3}{\gamma} + \varepsilon_{rad} \tilde{u}_3 \gamma \left(\gamma^2 - \tilde{u}_1{}^2\right), \tag{11}$$

$$\tilde{u}_3 = \left(d - j\frac{\tilde{u}_1}{\gamma}\right)\frac{\tilde{u}_3}{\gamma} - b\frac{\tilde{u}_2}{\gamma} + a - \varepsilon_{rad} \tilde{u}_2 \gamma \left(\gamma^2 - \tilde{u}_1{}^2\right), \tag{12}$$

where we use the relationship between $\tilde{x}_i$ and $\tilde{u}_i$, with $i = 1, 2, 3$, given by Eq. (2), which is

$$\tilde{x}_1 = s\gamma\tilde{u}_1, \quad \tilde{x}_2 = \frac{\tilde{u}_3 c}{\gamma\omega_0}, \quad \text{and} \quad \tilde{x}_3 = -\frac{\tilde{u}_2 c}{\gamma\omega_0}. \tag{13}$$

Here the dimensionless parameter,

$$\varepsilon_{rad} = \frac{2e^2\omega_0}{3m_e c^3}, \tag{14}$$

characterizes the radiation damping effect, $a_1 = eE_1/m_e\omega_0 c$, $a = eE/m_e\omega_0 c$, $d = eD/m_e\omega_0 c$, $j = eJ/m_e\omega_0 c$, and $b = eB_1/m_e\omega_0 c$ are normalized longitudinal and transverse components of the electric and magnetic field, and γ is the electron relativistic Lorentz-factor equal to $(1 + \tilde{u}_1{}^2 + \tilde{u}_2{}^2 + \tilde{u}_3{}^2)^{1/2}$. The parameter ε_{rad} can also be written as $\varepsilon_{rad} = 4\pi r_e/3\lambda_0$ or $\varepsilon_{rad} = 2\omega_0 t_e/3$, where $r_e = e^2/m_e c^2$ is the classical electron radius, $t_e = r_e/c$, and $\lambda_0 = 2\pi c/\omega_0$.

At first we analyze the most simple case with $B_1 = J = D = 0$. The stationary solution to equations (1) and (2), for which the vectors $\tilde{\mathbf{u}} = (\tilde{u}_1, \tilde{u}_2, \tilde{u}_3)$ and $\tilde{\mathbf{x}} = (\tilde{x}_1, \tilde{x}_2, \tilde{x}_3)$ do not depend on time, with the radiation friction force in the LAD form (3) can be cast as

$$0 = a_1 - \varepsilon_{rad} \tilde{u}_1 \gamma \left(\gamma^2 - 1 - \tilde{u}_1{}^2\right), \tag{15}$$

$$\tilde{u}_2 = \varepsilon_{rad}\tilde{u}_3\gamma\left(\gamma^2 - \tilde{u}_1{}^2\right), \tag{16}$$

$$\tilde{u}_3 = a - \varepsilon_{rad}\tilde{u}_2\gamma\left(\gamma^2 - \tilde{u}_1{}^2\right). \tag{17}$$

Multiplying Eq. (15) by $\tilde{u}_1$, Eq. (16) by $\tilde{u}_2$, and Eq. (17) by $\tilde{u}_3$, and adding them, we obtain

$$a_1\tilde{u}_1 + a\tilde{u}_2 = \varepsilon_{rad}\gamma^3\left(\gamma^2 - 1 - \tilde{u}_1{}^2\right). \tag{18}$$

The left hand side of this equation is proportional to the work produced by the electric field in the units of time and the right hand side is proportional to the energy dissipation rate due to the radiation losses.

Multiplying Eq. (16) by $\tilde{u}_3$ and Eq. (17) by $\tilde{u}_2$, and adding them, we obtain

$$\tilde{u}_2{}^2 + \tilde{u}_3{}^2 = a\tilde{u}_3. \tag{19}$$

3. Electron in the rotating electric field

3.1. Lorentz-Abraham-Dirac form of the radiation friction force

If, in addition, the longitudinal component of electric field vanishes, $a_1 = 0$ with $\tilde{u}_1 = 0$, we obtain from Eqs. (10 - 12)

$$p_{||} = \varepsilon_{rad}p_\perp\gamma^3 \quad \text{and} \quad p_\perp = m_e c a - \varepsilon_{rad}p_{||}\gamma^3, \tag{20}$$

where the parallel and perpendicular to the electric field components of the electron momentum defined by Eq. (8) are equal to

$$p_{||} = \frac{(\mathbf{p}\cdot\mathbf{E})}{|\mathbf{E}|} = m_e c\tilde{u}_2 \quad \text{and} \quad p_\perp = (p^2 - p_{||}^2)^{1/2} = m_e c\tilde{u}_3, \tag{21}$$

respectively. In this case equation (19) yields a relationship between $p_{||}$ and $p_\perp$:

$$p_{||}^2 + p_\perp^2 = m_e c a p_\perp. \tag{22}$$

The electron gamma-factor γ is equal to $(1 + \tilde{u}_2{}^2 + \tilde{u}_3{}^2)^{1/2} \equiv (1 + p_{||}{}^2 + p_\perp{}^2)^{1/2}$.

As we see, from the relationship $p_{||} = [p_\perp(p_\perp - m_e ca)]^{1/2}$ it follows that the perpendicular to the electric field component of the electron momentum is always equal or less than a. Multiplying the first equation in (20) by $p_\perp$ and the second equation by $p_{||}$ and subtracting them, we find

$$p_{||}a = \varepsilon_{rad}p_\perp^2\gamma^3, \tag{23}$$

which corresponds to the energy balance equation (18) for $a_1 = \tilde{u}_1 = 0$.

If the EM field amplitude is relatively small, i.e. $a \ll \varepsilon_{rad}^{-1/3}$ Eqs. (22) and (23) yield for the components of the electron momentum perpendicular and parallel to the electric field

$$p_\perp \approx m_e c\left(a - \varepsilon_{rad}^2 a^7\right) \quad \text{and} \quad p_{||} \approx m_e c\varepsilon_{rad}a^4. \tag{24}$$

In the opposite limit, when $a \gg \varepsilon_{rad}^{-1/3}$, we obtain

$$p_\perp \approx \frac{m_e c}{(\varepsilon_{rad}a)^{1/2}} \quad \text{and} \quad p_{||} \approx m_e c\left(\frac{a}{\varepsilon_{rad}}\right)^{1/4}. \tag{25}$$

In Fig. 1a we show a dependence of $p_\perp$ and $p_{||}$ on the EM field amplitude, a, for the dimensionless parameter $\varepsilon_{rad} = 10^{-8}$, obtained by numerical solution of Eqs. (20). Here the horizontal axis is normalized by $\varepsilon_{rad}^{-1/3}$ and the vertical axis is normalized by $(a_m/\varepsilon_{rad})^{1/4}$.

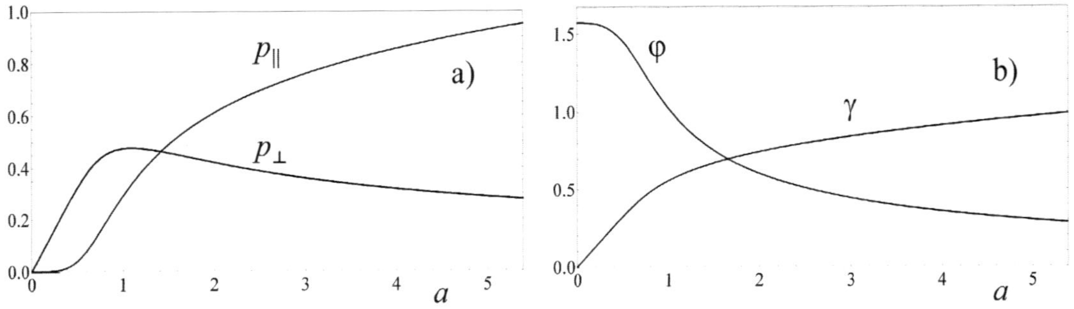

Figure 1. a) Dependence of perpendicular, $p_\perp$, and parallel, $p_{||}$, to the electric field components of the electron momentum (normalized on $m_e c(a_m/\varepsilon_{rad})^{1/4}$) on the normalized EM field amplitude, $a\varepsilon_{rad}^{1/3}$, and b) dependence of φ and normalized on $(a_m/\varepsilon_{rad})^{1/4}$ electron gamma-factor, γ, on $a\varepsilon_{rad}^{1/3}$ for $a_m = 2500$ and $\varepsilon_{rad} = 10^{-8}$.

As we see, the dependences of the components of the electron momentum perpendicular and parallel to the electric field correspond the asymptotics given by Eqs. (24) and (25). The perpendicular momentum reaches the maximum at $a \approx \varepsilon_{rad}^{-1/3}$ and then decreases. The parallel momentum component monotonuously increases with the EM amplitude growth.

It is also convenient to represent the momentum components in the complex form

$$p_{||} + ip_\perp = \mathfrak{p}\exp(-i\varphi) \qquad (26)$$

with $\mathfrak{p} = (p_\perp^2 + p_{||}^2)^{1/2}$ and φ being the momentum value and the phase between the rotating electric field and the momentum vector. Eqs. (20) can be rewritten as

$$m_e ca = \mathfrak{p}(1 + \varepsilon_{rad}^2\gamma^6)^{1/2} \quad \text{and} \quad \tan\varphi = -\frac{1}{\varepsilon_{rad}\gamma^3}, \qquad (27)$$

where the electron gamma-factor γ is equal to $(1 + \mathfrak{p}^2)^{1/2}$. These equations are the same as Eqs. (6) in Ref. [13].

In Fig. 1b we present the electron gamma factor γ normalized by $(a_m/\varepsilon_{rad})^{1/4}$ and the angle φ versus the EM field amplitude a for $\varepsilon_{rad} = 10^{-8}$. The angle φ changes from $\pi/2$ at $a = 0$, when the electron momentum is perpendicular to the electric field vector, to 0 at $a \to \infty$, when the electron momentum becomes antiparallel to the electric field. The horizontal axis is normalized in the same way as in Fig. 1a.

3.2. Landau-Lifshitz form of the radiation friction force

We look for the solutions describing a stationary electron orbit in a rotating homogeneous electric field, i.e. E_1, D, B_1, J vanish in Eqs. (5 - 7). From Eq. (4) we obtain for the $p_{||}$ and $p_\perp$ momentum components the algebraic equations

$$p_{||} = \varepsilon_{rad}\frac{p_\perp}{\gamma}a^2\left[1 + \left(\frac{p_\perp}{m_e c}\right)^2\right] \quad \text{and} \quad p_\perp = m_e ca - \varepsilon_{rad}\frac{p_{||}}{\gamma}a^2\left[1 + \left(\frac{p_\perp}{m_e c}\right)^2\right]. \qquad (28)$$

Using the variables $\mathfrak{p}$ and φ defined by Eq. (26) we can present these equations in the form

$$m_e ca = \varepsilon_{rad}\frac{\mathfrak{p}}{\gamma}\left[a^2 + \left(\frac{\mathfrak{p}}{m_e c}\right)^2\right] \quad \text{and} \quad \sin\varphi = -\frac{\mathfrak{p}}{m_e ca}. \qquad (29)$$

In the range of the EM field amplitude, $0 < a \ll \varepsilon_{rad}^{-1}$ solution to these equations has the same asymptotic dependences as given by Eqs. (20) and (22). However, when the EM field amplitude approaches the value of ε_{rad}^{-1}, the solution qualitatively changes. According to Eq. (29), the electron momentum decreases as also shown in Fig. 2. In Fig. 2a we present the components of the electron momentum parallel and perpendicular to the instantaneous electric field as functions of the electric field amplitude. Fig. 2b shows the dependences of the angle, φ, and the electron gamma-factor, γ, on the electric field. The momentum and gamma-factor are normalized by $(a_m/\varepsilon_{rad})^{1/4}$, and the dimensionless electric field amplitude by ε_{rad}.

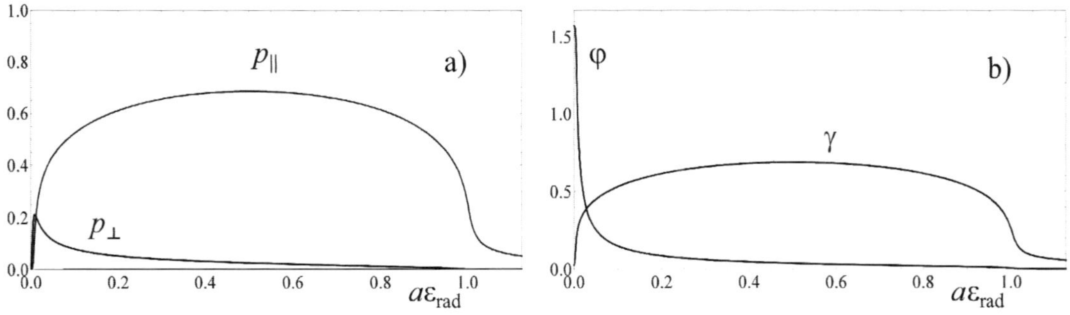

Figure 2. Solution of the electron motion equation with the radiation friction force in the LL form in the case of rotating homogeneous electric field: a) Dependence of the components of the electron momentum (normalized by $m_e c(a_m/\varepsilon_{rad})^{1/4}$) perpendicular, $p_\perp$, and parallel, $p_{||}$, to the electric field on the normalized EM field amplitude, $a\varepsilon_{rad}$, and b) dependence of φ normalized by $(a_m/\varepsilon_{rad})^{1/4}$ and the electron gamma-factor, γ, on $a\varepsilon_{rad}$ for $a_m = 1500$ and $\varepsilon_{rad} = 7.5 \times 10^{-4}$.

In Fig. 3 we present the results of the solution of the electron motion equation in a rotating homogeneous electric field. Here the dependences of the electron gamma-factors on the electric field, $\gamma_{\rm LAD}$ and $\gamma_{\rm LL}$, correspond to the radiation friction force taken in the LAD and L-L form, respectively. The normalization is the same as in Fig. 2.

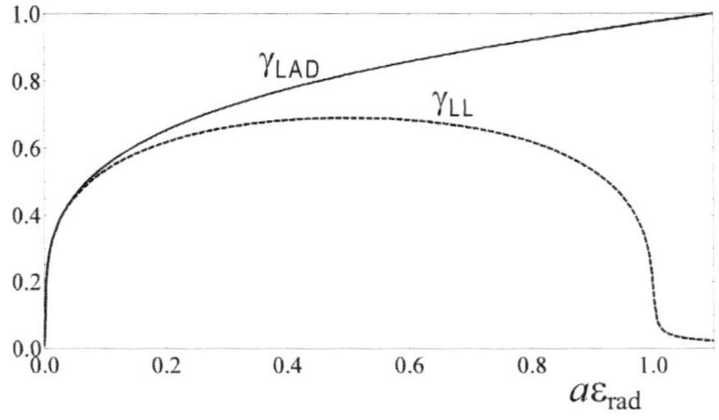

Figure 3. Solution of the electron motion equation in a rotating homogeneous electric field for $a_m = 1500$ and $\varepsilon_{rad} = 7.5 \times 10^{-4}$. Dependences of the electron gamma-factors on the electric field: $\gamma_{\rm LAD}$ and $\gamma_{\rm LL}$ correspond to the radiation friction force taken in the LAD and L-L form, respectively. The normalization is the same as in Fig. 2.

4. Electron in the superposition of rotating and radial electric fields

4.1. Lorentz-Abraham-Dirac form of the radiation friction force

Electron dynamics in the superposition of rotating and radial electric field corresponds to the case of the electron direct acceleration by the laser pulse propagating inside the self-focusing channel. In the frame of reference moving with the laser pulse group velocity equations of electron motion are Eqs. (10 - 12) with $b = j = a_1 = 0$:

$$\tilde{u}_2 = d\frac{\tilde{u}_2}{\gamma} + \varepsilon_{rad}\tilde{u}_3\gamma^3, \tag{30}$$

$$\tilde{u}_3 = d\frac{\tilde{u}_3}{\gamma} + a - \varepsilon_{rad}\tilde{u}_2\gamma^3. \tag{31}$$

For variables p and φ defined by Eq. (26) we can rewrite Eqs. (30 , 31) as

$$m_e c a = \mathsf{p}\left[\left(1 - \frac{d}{\gamma}\right)^2 + \varepsilon_{rad}^2\gamma^6\right]^{1/2} \quad \text{and} \quad \tan\varphi = -\frac{\gamma - d}{\varepsilon_{rad}\gamma^4} \tag{32}$$

with $\gamma = (1 + \mathsf{p}^2)^{1/2}$.

4.2. Landau-Lifshitz form of the radiation friction force

The equations of the electron motion in the superposition of rotating and radial electric fields ($b = j = a_1 = 0$) with the radiation friction force in the L-L form are

$$(\gamma - d)\tilde{u}_2 = \varepsilon_{rad})\tilde{u}_3\left[(d\gamma + a)\tilde{u}_3)^2 + a^2 - d^2\right], \tag{33}$$

$$(\gamma - d)\tilde{u}_3 = a\gamma - \varepsilon_{rad})\tilde{u}_2\left[(d\gamma + a)\tilde{u}_3)^2 + a^2 - d^2\right]. \tag{34}$$

We can rewrite Eqs.(33,34) as

$$a^2 = d^2 - \frac{(d - \gamma + \gamma^3)^2}{\gamma^2} + \frac{\gamma^2 - \sqrt{\gamma^4\left[1 - 4\varepsilon_{rad}^2\left(\gamma^2 - 1\right)^2\right]}}{2\varepsilon_{rad}^2\left(1 - \gamma^2\right)} \tag{35}$$

$$\tan\phi = \frac{2\varepsilon_{rad}(\gamma - d)(\gamma^2 - 1)}{\gamma^2 - \sqrt{\gamma^4\left[1 - 4\varepsilon_{rad}^2\left(\gamma^2 - 1\right)^2\right]}} \tag{36}$$

As in the above considered case of rotating electric field, the dependence of the electron momentum on the electric field amplitude shows similar behavior in the cases of LAD and L-L forms of the radiation friction force, provided $a \gg 1$, $\gamma \gg 1$, and $2\varepsilon_{rad}\gamma^2 \ll 1$.

5. Discussions and conclusions

As follows from consideration of the above presented exact solutions to the electron motion equations with the radiation friction force taken either in the LAD or in the L-L form, in the limit of relatively low electric field amplitude they show the same behavior, as seen in Fig. 3. When the electric field is strong, i.e. the normalized field amplitude a approaches the value of ε_{rad}^{-1}, the solutions are drastically different.

The condition $a = \varepsilon_{rad}^{-1}$ corresponds to the electric field equal to the critical electric field of classical electrodynamics, $E_{cr} = m_e^2 c^4/e^3$. This electric field can produce a work equal to $m_e c^2$ over the distance of the classical electron radius r_e.

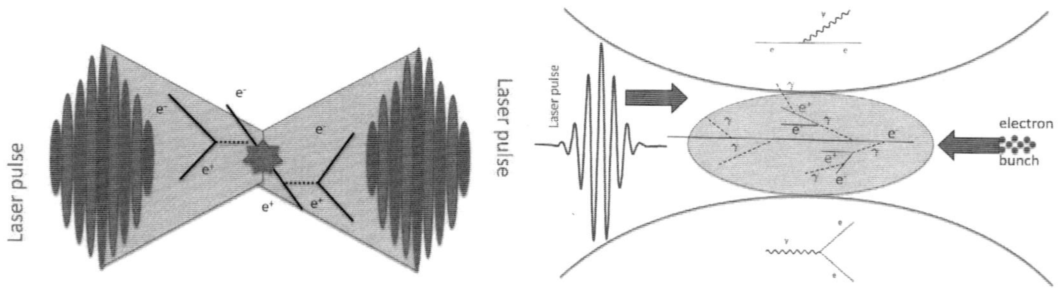

Figure 4. Principle experimental schemes aimed at the study of (i) colliding laser pulses (all optical setup); (ii) laser - e-beam interaction.

The radiation friction force in the Landau-Lifshitz form assumes the smallness of the EM field amplitude compared to the critical field of classical electrodynamics. Another parameter which should be small is the ratio of the EM field inhomogeity scale-length to the classical electron radius, r_e. The time dependent EM fields should be slowly evolved on a timescale compared to $t_e = r_e/c$, as discussed in Ref. [25] devoted to the problem of classical electrodynamics applicability.

Obviously, the limit of the EM field amplitude of the order of E_{cr} (and of the space- and time scales of the order of r_e and $t_e = r_e/c$) is of pure academic interest, because the quantum mechanical effects become important at electric field amplitudes substantially lower (and at the spatial scale of the order of the electron Compton wavelength, $\hbar/m_e c$). The critical electric field of quantum electrodynamics (QED), $E_S = m_e^2 c^3/e\hbar$, is a factor $\alpha = e^2/\hbar c \approx 1/137$ smaller. Here α is the fine structure constant. For the electron motion in colliding EM waves, as shown in Ref. [12] the QED effects, due to the recoil from the photon emission, should be incorporated into the description of the electron interaction with the EM field for an even smaller EM wave amplitude. This field is of the order of αE_S, which is equal to $\alpha^2 E_{cr}$, below which both the LAD and L-L forms for the radiation friction force give the same result.

Let us derive the extreme field limits, which mark the subsequent onset of the classical radiation reaction regime, the regime when quantum recoil becomes important, and the e^+e^- pair production from vacuum, for two principle experimental schemes aimed at the study of particle physics effects at high laser intensity: (i) colliding laser pulses (all optical setup) and (ii) laser - e-beam interaction (See Figure 4).

In what follows we consider the behavior of an electron in the focus of two colliding circularly polarized laser pulses, *i.e.*, in the antinode of a standing light wave with null magnetic field, and in the collision with the counterpropagating laser pulse and determine the thresholds for the classical radiation dominated regime of interaction and for the onset of the quantum regime. We also mention the threshold for the Schwinger process in the case of two colliding laser pulses.

The power emitted by the electron in the circularly polarized electric field in the ultra-relativistic limit is proportional to the forth power of its energy $P_{C,\gamma} = \varepsilon_{rad}\omega mc^2\gamma_e^2(\gamma_e^2-1) \sim \gamma_e^4$. In the non-radiative approximation, the electron can acquire the energy from the EM field with the rate $\approx \omega mc^2 a$. The condition of the balance between the acquired and emitted energy is $a^3 \approx \varepsilon_{rad}^{-1}$. Thus the radiation reaction (RR) effects become dominant at $a > a_{rad} = \varepsilon_{rad}^{-1/3}$. For a head-on collision of an energetic e-beam and an intense laser pulse with $\varepsilon_{rad} a\gamma_e^2 \gg 1$ the longitudinal momentum is given by $p_x = -p_0\left[1+\varepsilon_{rad}\omega(p_0/m)\int_0^t a^2(-2\eta)d\eta\right]^{-1}$. In this case the interaction is purely dissipative and all the EM forces can be neglected except the radiation reaction. This means that the classical radiation effects become dominant at

$a > a_{rad} = (\varepsilon_{rad}\omega\tau_{laser}\gamma_{e,0})^{-1/2}$, where τ_{laser} is the duration of the laser pulse and $\gamma_{e,0}$ is the initial energy of the e-beam.

The quantum effects become important when the energy of a photon emitted by an electron is of order of the electron energy, *i.e.*, $\hbar\omega_m = \gamma_e mc^2$. In the case of the electron circulating in the focus of two colliding laser pulses the electron emits photons with the energy $\hbar\omega_m = \hbar\omega\gamma_e^3$. Thus the quantum recoil comes into play when $a > a_Q = (2\alpha/3)^2\varepsilon_{rad}^{-1}$. Here we took into account the fact that in the limit $I > 10^{23}$ W/cm^2 due to strong radiation damping effects the electron energy scales as $mc^2(a/\varepsilon_{rad})^{1/4}$. For an electron colliding with a laser pulse a characteristic photon energy is $\hbar\omega_m \approx \hbar\omega a\gamma_e^2$ [19], which corresponds to the condition $\chi_e = e\hbar\sqrt{(F_{\mu\nu}p_\nu)^2/m^3c^4} \sim 1$. The parameter χ_e has the meaning of the EM field strength in the rest frame of the particle. It is responsible for the magnitude of the quantum nonlinear effects. Thus the quantum recoil comes into play when $a > a_Q = (2\alpha/3)\gamma_e^{-1}\varepsilon_{rad}^{-1}$. This condition is analogous to two conditions: $\chi_e > 1$ and $\alpha a > 1$ derived from the analysis of the $e \rightarrow e\gamma$ process probability [31]. The probability of pair creation acquires its optimum value over the characteristic scale of the process when the electric field strength is of the order of the "critical" for quantum electrodynamics (QED) value $a > a_S = (2\alpha/3)\varepsilon_{rad}^{-1}$.

Table 1. The extreme field limits for two colliding 0.8 μm laser pulses and a 10 GeV electron beam colliding with a 0.8 μm laser pulse.

	Classical RR	Quantum regime	Schwinger limit
two colliding pulses	$a_{rad} = \varepsilon_{rad}^{-1/3}$ $a_{rad} = 400$ $I_{rad} = 3.5 \times 10^{23}$ W/cm^2	$a_Q = \left(\frac{2}{3}\alpha\right)^2\varepsilon_{rad}^{-1}$ $a_Q = 1.6 \times 10^3$ $I_Q = 5.5 \times 10^{24}$ W/cm^2	$a_S = \left(\frac{2}{3}\alpha\right)\varepsilon_{rad}^{-1}$ $a_S = 3 \times 10^5$ $I_S = 2.3 \times 10^{29}$ W/cm^2
e-beam and laser pulse	$a_{rad} = (\varepsilon_{rad}\omega\tau_{laser}\gamma_{e,0})^{-1/2}$ $a_{rad} = 10$ $I_{rad} = 2.2 \times 10^{20}$ W/cm^2	$a_Q = \left(\frac{2}{3}\alpha\right)\gamma_e^{-1}\varepsilon_{rad}^{-1}$ $a_Q = 20$ $I_Q = 8.7 \times 10^{20}$ W/cm^2	

Although conclusions following from the above presented consideration do not have the character of a rigorously proved mathematical theorem, they give an indication that for

$$a < a_Q = \left(\frac{2}{3}\alpha\right)^2 \epsilon_{rad}^{-1} \tag{37}$$

the solutions of the electron equations of motion with the radiation friction force in the L-L form coincide with those obtained from solution of the equations of motion with the radiation friction force in the LAD form.

We thank G. Korn, N. B. Narozhny, A. S. Pirozhkov, H. Ruhl, and A. G. Zhidkov for discussions. We acknowledge support of this work from the NSF under Grant No. PHY-0935197, the Office of Science of the US DOE under Contract No. DE-AC02-05CH11231 and the Ministry of Education, Culture, Sports, Science and Technology (MEXT) of Japan, Grant-in-Aid for Scientific Research, No. 20244065.

[1] Jackson J D 1999 *Classical Electrodynamics* (New York: John Willey and Sons)
[2] Landau L D and Lifshitz E M 1975 *The Classical Theory of Fields* (Oxford: Pergamon Press)
[3] Barut A O 1980 *Electrodynamics and classical theory of fields and particles* (New York: Dover)
[4] Ginzburg V L 1989 *Applications of Electrodynamics in Theoretical Physics and Astrophysics* (New York: Gordon and Breach)
[5] Iwanenko D and Pomeranchuk I 1944 *Phys. Rev.* **65** 343; Bruck H 1974 *Circular particle accelerators* (Los Alamos Lab)

[6] Berezinskii V S, Bulanov S V, Dogiel V A, Ginzburg V L and Ptuskin V S 1990 *Astrophysics of Cosmic Rays* (Amsterdam: Elsevier); Bulanov S V, Esirkepov T Zh, Habs D, Pegoraro F and Tajima T 2009 *Eur. Phys. J.* D **55** 483

[7] Medvedev M V 2003 *Phys. Rev.* E **67** 045401; Neronov A Yu, Semikoz D V and Tkachev I I 2009 *New J. Phys.* **11** 065015; Ptitsyna K V and Troitsky S V 2010 *Physics-Uspekhi* **53** 691; Kotera K and Olinto A V 2010 *Annu. Rev. Astron. Astrophys* **49**

[8] Hazeltine R D and Mahajan S M 2004 *Phys. Rev.* E **70** 046407

[9] Mourou G, Tajima T and Bulanov S V 2006 *Rev. Mod. Phys.* **78** 309

[10] Yanovsky V, Chvykov V, Kalinchenko G, Rousseau P, Planchon T, Matsuoka T, Maksimchuk A, Nees J, Cheriaux G, Mourou G and Krushelnick K 2008 *Opt. Express* **16** 2109

[11] Zhidkov A, Koga J K, Sasaki A, Uesaka M 2002 *Phys. Rev. Lett.* **88** 185002

[12] Bulanov S V, Esirkepov T Zh, Koga J K and Tajima T 2004 *Plasma Phys. Rep.* **30** 196

[13] Bulanov S S, Esirkepov T Zh, Koga J K, Thomas A G R and Bulanov S V 2010 *Phys. Rev. Lett.* **105** 220407

[14] Thomas A G R, Ridgers C P, Bulanov S S, Griffin B J and Mangles S P D 2012 *Phys. Rev.* X **2** 041004

[15] Tajima T and Dawson J M 1979 *Phys. Rev. Lett.* **43** 267; Esarey E, Schroeder C B and Leemans W P 2009 *Rev. Mod. Phys.* **81** 1229

[16] Esirkepov T Zh, Borghesi M, Bulanov S V, Mourou G and Tajima T 2004 *Phys. Rev. Lett.* **92** 175003; Pegoraro F and Bulanov S V 2007 *Phys. Rev. Lett.* **99** 065002; Bulanov S V, Echkina E Yu, Esirkepov T Zh, Inovenkov I N, Kando M, Pegoraro F and Korn G 2010 *Phys. Rev. Lett.* **104** 135003; 2010 *Phys. Plasmas* **17** 063102

[17] Michel P, Schroeder C B, Shadwick B A, Esarey E and Leemans W P 2006 *Phys. Rev.* E**74** 026501; Naumova N, Schlegel T, Tikhonchuk V T, Labaune C, Sokolov I V and Mourou G 2009 *Phys. Rev. Lett.* **102** 025002; Schlegel T, Naumova N, Tikhonchuk V T, Labaune C, Sokolov I V and Mourou G 2009 *Phys. Plasmas* **16** 083103; Tamburini M, Pegoraro F, Di Piazza A, Keitel C H, Liseykina T V and Macchi A 2010 *New J. Phys.* **12** 123005; Schroeder C B, Esarey E, Geddes C G R, Benedetti C and Leemans W P 2010 *Phys. Rev.* ST - AB **13** 101301; Chen M, Pukhov A, Yu T and Sheng Z-M 2011 *Plasma Phys. Control. Fusion* **53** 014004

[18] Bell A R and Kirk J G 2008 *Phys. Rev. Lett.* **101** 200403; Fedotov A M, Narozhny N B, Mourou G and Korn G 2010 *Phys. Rev. Lett.* **105** 080402; Nerush E N, Kostyukov I Yu, Fedotov A M, Narozhny N B, Elkina N V and Ruhl H 2011 *Phys. Rev. Lett.* **106** 035001

[19] Bulanov S V, Esirkepov T Zh, Hayashi Y, *et. al.* 2011 *Nuclear Instruments and Methods in Physics Research* A **660** 31

[20] Ginzburg V L 1969 *Sov. Physics-Uspekhi* **12** 565; Steiger A D and Woods C H 1972 *Phys. Rev.* D **5** 2912; Zel'dovich Ya B 1975 *Sov. Phys. Usp.* **18** 97; Stewart P 1975 *Astron. Astrophys.* **41** 169; Aseo E, Kennel C F and Pellat R 1978 *Astron. Astrophys.* **65** 401; Kuz'menkov L S 1978 *Sov. Phys. Dokl.* **23** 469; Keitel C H, Szymanowski C, Knight P L and Maquet A 1998 *J. Phys.* B **31** L75; Hartemann F V, Troha A L, Baldis H A, Gupta A, Kerman A K, Landahl E C, Luhmann N C and Van Meter J R 2000 *Ap. J. Suppl.* **127** 347; Rohrlich F 1965 *Classical Charged Particles* (Reading: Perseus Books Publishing); Rohrlich F 2001 *Phys. Lett.* A **283** 276; Hartemann F V, Gibson D J and Kerman A K 2005 *Phys. Rev.* E **72** 026502; Koga J K, Esirkepov T Zh and Bulanov S V 2006 *J. Plasma Phys.* **72** 1315; Berezhiani V I, Mahajan S M and Yoshida Z 2008 *Phys. Rev.* E **78** 066403

[21] Spohn H 2000 *Europhys. Lett.* **50** 287; Rohrlich F 2002 *Phys. Lett.* A **303** 307; Koga J K, Esirkepov T Zh and Bulanov S V 2005 *Phys. Plasmas* **12** 093106; Di Piazza A 2008 *Lett. Math. Phys.* **83** 305; Di Piazza A, Hatsagortsyan K Z and Keitel C H 2009 *Phys. Rev. Lett.* **102** 254802; Sokolov I V, Naumova N M, Nees J A, Mourou G A and Yanovsky V P 2009 *Phys. Plasmas* **16** 093115; Griffiths D J, Proctor T C and Schroeter D F 2010 *Am. J. Phys.* **78** 391; Hadad Y, Labun L, Rafelski J, Elkina N, Klier C and Ruhl H 2010 *Phys. Rev.* D **82** 096012

[22] Abraham M 1905 *Theorie der Elektrizitat, Vol. II: Elektromagnetische Theorie der Strahlung* (Leipzig: Teubner)

[23] Lorentz A 1909 *The Theory of Electrons* (Leipzig: Teubner)

[24] Dirac P A M 1938 *Proc. R. Soc. London* Ser. A **167** 148

[25] Baylis W E and Huschilt J 2002 *Phys. Lett.* A **301** 7

[26] Zel'dovich Ya B and Illarionov A F 1972 *Sov. JETP* **34** 467

[27] Popov V S 2004 *Phys.-Usp.* **47** 855; Marklund M and Shukla P 2006 *Rev. Mod. Phys.* **78** 591; Salamin Y I, Hu S X, Hatsagortsyan K Z and Keitel C H 2006 *Phys. Rep.* **427** 41

[28] Bulanov S S, Mur V D, Narozhny N B, Nees J and Popov V S 2010 *Phys. Rev. Lett.* **104**, 220404

[29] Chian A C-L 1981 *Phys. Rev.* A **24** 2773

[30] Bulanov S S 2004 *Phys. Rev.* E **69** 036408

[31] Di Piazza A, Hatsagortsyan K Z and Keitel C H 2010 *Phys. Rev. Lett.* **105** 220403

Photo-proton emission from halo nuclei in the presence of a linearly polarized laser field

A Dadi[1] **and C Müller**[2]

[1] Max-Planck-Institut für Kernphysik, Saupfercheckweg 1, 69117 Heidelberg, Germany
[2] Institut für Theoretische Physik I, Heinrich-Heine-Universität, 40225 Düsseldorf, Germany

E-mail: c.mueller@tp1.uni-duesseldorf.de

Abstract. We investigate the emission of protons from halo nuclei which are subject to the combined electromagnetic fields of a γ-ray photon and an intense laser beam of linear polarization. Photo-proton energy spectra resulting from this laser-assisted nuclear photoeffect are calculated within an S-matrix approach based on the strong-field approximation. We analyze the influence of the applied γ-ray energy, laser field polarization and nuclear isotope on the proton emission spectra. The process represents an example from the field of laser-assisted photo-nuclear physics which may be explored at the upcoming Extreme Light Infrastructure.

1. Introduction

The nuclear photoeffect was discovered by Chadwick and Goldhaber in 1934. Using γ rays from a radioactive source, they observed the photodisintegration of the deuteron into a proton and a neutron [1]. In the years after, further pioneering contributions to this new research area were made by Bothe and Gentner in Heidelberg, whose systematic experimental investigations relied on high-energy γ rays stemming from an accelerator [2]. Since that time, studies of the nuclear photoeffect in the energy region up to a few tens of MeV have proven to be particularly important for our understanding of nuclear structure and nuclear reactions. While modern experiments on photonuclear reactions typically employ high-energy photons produced, for example, by synchrotron radiation sources, powerful lasers are nowadays emerging into an alternative tool for photonuclear studies [3]. A laboratory devoted particularly to the research area of laser-nuclear physics is currently being built as one of the pillars of the Extreme Light Infrastructure [4, 5].

The large mismatch between the typical energy scales of nuclei on the one side and laser fields on the other side generally renders the probability of inducing direct photonuclear reactions by laser radiation very small [6]. Therefore, indirect laser-nucleus interactions – mediated by secondary particles such as electrons in laser-produced plasmas – have mostly been investigated. In particular, photofission and photoneutron production arising from bremsstrahlung photons emitted by fast electrons have been observed experimentally [7, 8, 9].

Laser-*assisted* nuclear processes are generally less demanding than directly laser-induced nuclear processes in terms of the required laser field intensity and frequency. Such processes can already occur in the absence of any laser field but may be modified in a characteristic manner when a laser field is present. Corresponding studies have been conducted on laser-assisted internal conversion [10] and laser-assisted nuclear excitation by electron transition [6, 11]. Related studies have treated laser-triggered nuclear Coulomb excitation in atoms (see

Journal of Physics: Conference Series **414** (2013) 012010 doi:10.1088/1742-6596/414/1/012010

[12, 13, 14, 15] and references therein) and laser-driven proton streaking [16]. We point out that the ongoing increase of available laser intensities and frequencies has also inspired theoreticians to investigate directly laser-induced nuclear reactions during the last few years [17, 18, 19].

Very recently, we have considered the nuclear photoeffect assisted by a laser field of circular polarization [20]. In this process, a proton is emitted from a nucleus which is exposed to the combined fields of a γ photon and an intense laser beam (see also Fig. 1):

$$^{A}_{Z}[\mathrm{Xp}] + \omega_\gamma + n\omega_0 \quad \longrightarrow \quad ^{A-1}_{Z-1}[\mathrm{X}] + ^{1}_{1}\mathrm{p}\ . \tag{1}$$

The γ photon energy ω_γ is assumed to exceed the proton separation energy E_b, so that proton ejection can already occur without a laser field present. $^{A}_{Z}[\mathrm{Xp}]$ denotes a nucleus with A nucleons and Z protons. It is taken to be a halo nucleus, consisting of a nuclear core X surrounded by a rather loosely bound proton. This special structure, which some isotopes possess [21], reduces the γ photon energy required and also facilitates the theoretical treatment of the process. Simultaneous with the γ-absorption, also a certain number n of laser photons of energy ω_0 can be emitted or absorbed during the process, due to the laser assistance. This additional photon exchange with the laser field may lead to distinct modifications of the photo-proton energy spectra. We note that a pronounced impact of an assisting laser field has also been found in the photoelectric effect in atoms (see, e.g., [22, 23, 24, 25]).

In the current contribution we extend our results from [20] to the case of a linearly polarized laser field. Besides, while in [20] we examined the process dependencies on the applied laser intensity and frequency and the γ-ray polarization, here we study the influence of the γ-photon energy, the laser polarization and the nuclear isotope. An S-matrix description of the process based on the strong-field approximation [26] is given in Section 2. The cross section is expressed in analytical form as an infinite sum over the number n of participating laser photons. In Section 3 we discuss the various dependencies of the photo-proton spectra on the applied laser and nuclear parameters. A short summary and conclusion is given in Section 4.

Natural units with $\hbar = c = \varepsilon_0 = 1$ are used throughout unless explicitly stated otherwise.

2. Theoretical framework

The laser-assisted nuclear photoeffect of Eq. (1) can be treated theoretically as follows. The laser field is assumed to be a linearly polarized, monochromatic wave with vector potential

$$\vec{A}_L(t) = A_0 \cos(\omega_0 t)\, \vec{e}\ , \tag{2}$$

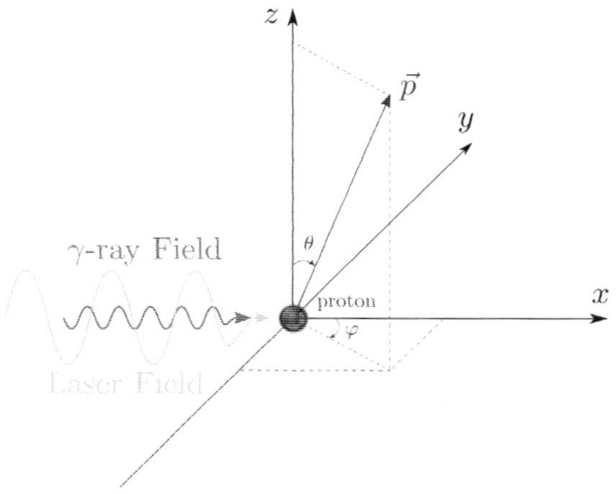

Figure 1. Scheme of the laser-assisted nuclear photoeffect. Due to the presence of the combined fields of a γ-photon and a laser beam, a proton is ejected from a halo nucleus.

where A_0 is the amplitude. Note that, for the laser parameters under consideration, the dipole approximation applies. The vector potential of the γ photon reads

$$\vec{A}_\gamma(\vec{r},t) = \sqrt{\frac{2\pi}{V_\gamma \omega_\gamma}}\, e^{i\left(\vec{k}_\gamma\cdot\vec{r}-\omega_\gamma t\right)}\,\vec{\varepsilon}, \tag{3}$$

with momentum $\vec{k}_\gamma$, polarization vector $\vec{\varepsilon}$ and normalization volume V_γ.

The nonrelativistic interaction Hamiltonian responsible for γ-photon absorption by the nucleus in the presence of the laser field is

$$\hat{H}_{\rm int} = -\frac{e}{m}\left(\hat{\vec{p}} - e\vec{A}_L(t)\right)\cdot\vec{A}_\gamma(\vec{r},t)\,, \tag{4}$$

with the proton charge e and mass m and the momentum operator $\hat{\vec{p}} = -i\vec{\nabla}$. The S-matrix of the process has the general form

$$S_{fi} = -i\int_{-\infty}^{\infty} \langle\Psi_f|H_{\rm int}|\Psi_i^{(+)}\rangle\, dt\,. \tag{5}$$

Here, $\Psi_i^{(+)}$ is a state of the full Hamiltonian $\hat{H}$, which includes the interactions of the proton with the photon fields and the nuclear potential. Ψ_f denotes an unperturbed state belonging to the Hamiltonian $\hat{H}-\hat{H}_{\rm int}$. Since neither of these states is known exactly, it is necessary to perform suitable approximations. Assuming that in the initial state the nuclear forces are dominant, we may write $\Psi_i^{(+)}$ approximately as a stationary state,

$$\Psi_i^{(+)}(\vec{r},t) \approx \phi_0(\vec{r})\, e^{iE_b t}, \tag{6}$$

which is subject solely to the nuclear potential. For the space-dependent part in Eq. (6), a Yukawa form may be used [27]

$$\phi_0(\vec{r}) = \frac{c_0}{\sqrt{4\pi}}\cdot\frac{e^{-\beta|\vec{r}|}}{\beta|\vec{r}|}\,, \tag{7}$$

with $\beta = 1/(\sqrt{2}R_{\rm rms})$ and $c_0 = \sqrt{2}\beta^{3/2}$, where $R_{\rm rms}$ denotes the rms-radius of the proton halo. With regard to the final proton state in the continuum, we note that it will be affected only rather moderately by the nuclear core potential, provided the energy of the ejected proton is relatively high. This state is mainly affected by the laser wave and may, thus, be approximated by a Volkov state. For the linearly polarized field of Eq. (2), the latter reads (in velocity gauge)

$$\psi_{\vec{p}}(\vec{r},t) = \frac{e^{i\vec{p}\cdot\vec{r}}}{\sqrt{V}}\exp\left[-i\left(\frac{p^2}{2m}+U_{\rm p}\right)t\right] f(t)\,, \tag{8}$$

with the proton momentum $\vec{p}$ and a normalization volume V. Besides, $U_{\rm p} = e^2A_0^2/(4m)$ denotes the ponderomotive energy and

$$f(t) = e^{i(\alpha\sin\omega_0 t - \beta\sin 2\omega_0 t)}\,, \quad \text{with} \quad \alpha = \frac{eA_0}{m\omega_0}\vec{p}\cdot\vec{e} \quad \text{and} \quad \beta = \frac{e^2A_0^2}{8m\omega_0}\,. \tag{9}$$

Within this strong-field approximation, the S-matrix of Eq. (5) becomes

$$S_{fi} \approx \frac{ie}{m}\sqrt{\frac{2\pi}{VV_\gamma\omega_\gamma}}\int dt\int d^3r\; e^{-i\vec{p}\cdot\vec{r}} e^{i\left(\frac{p^2}{2m}+U_{\rm p}\right)t} f(t)^*\left(\vec{p}-e\vec{A}_L\right)\cdot\vec{\varepsilon}\, e^{i\left(\vec{k}_\gamma\cdot\vec{r}-\omega_\gamma t\right)}\phi_0(\vec{r})\, e^{iE_b t}\,. \tag{10}$$

As in [20], it is also assumed here that $A_0 r_{\rm eff} \ll 1$ holds, where $r_{\rm eff}$ denotes the relevant radial extension of integration in Eq. (10). This condition will be fulfilled as we shall restrict our consideration to field amplitudes $A_0 \lesssim 1\,{\rm MeV}$. Since the relevant range of the space integral is limited to $r \lesssim 10$ fm due to the presence of the bound nuclear state, we have $A_0 r_{\rm eff} \lesssim 10^{-2}$.

In order to make progress, we expand the functions $f(t)^*$ and $\cos(\omega_0 t) f(t)^*$, which appear in the term $f(t)^*(\vec{p} - e\vec{A}_L)\cdot\vec{\varepsilon}$ in Eq. (10), into Fourier series according to

$$f(t)^* = \sum_{n=-\infty}^{+\infty} B_n\, e^{-in\omega_0 t}\ , \quad \cos(\omega_0 t) f(t)^* = \sum_{n=-\infty}^{+\infty} C_n\, e^{-in\omega_0 t}\ . \tag{11}$$

The Fourier coefficients $B_n = \widetilde{J}_n(\alpha,\beta)$ and $C_n = \frac{1}{2}\left[\widetilde{J}_{n+1}(\alpha,\beta) + \widetilde{J}_{n-1}(\alpha,\beta)\right]$ can be expressed in terms of generalized Bessel functions of integer order. The latter are given by infinite sums over ordinary Bessel functions, according to [28]

$$\widetilde{J}_n(\alpha,\beta) = \sum_{\ell=-\infty}^{+\infty} J_{n-2\ell}(\alpha) J_\ell(\beta)\ . \tag{12}$$

As a consequence, we obtain

$$S_{fi} \approx \frac{ie}{m}\sqrt{\frac{2\pi}{VV_\gamma\omega_\gamma}}\sum_{n=-\infty}^{+\infty}\mathcal{M}_n \int d^3r\, e^{-i(\vec{p}-\vec{k}_\gamma)\cdot\vec{r}}\,\phi_0(\vec{r})\int dt\, e^{i\left(\frac{p^2}{2m}+U_{\rm p}+E_b-\omega_\gamma-n\omega_0\right)t}\ , \tag{13}$$

with $\mathcal{M}_n = \vec{\varepsilon}\cdot\vec{p}\,B_n - eA_0\vec{\varepsilon}\cdot\vec{e}\,C_n$. The space and time integrations can be carried out analytically to yield

$$S_{fi} \approx \frac{i2\sqrt{2}\pi^2 ec_0}{m\sqrt{VV_\gamma\omega_\gamma}}\sum_{n=n_0}^{+\infty}\mathcal{M}_n\, G(\vec{p}-\vec{k}_\gamma)\,\delta\left(\frac{p^2}{2m} + U_{\rm p} + E_b - \omega_\gamma - n\omega_0\right). \tag{14}$$

Here, n_0 denotes the smallest integer which is in accordance with the energy conservation condition, as dictated by the δ-function. Besides,

$$G(\vec{p}-\vec{k}_\gamma) = \frac{2}{\beta[\beta^2 + (\vec{p}-\vec{k}_\gamma)^2]} \tag{15}$$

is the Fourier transform of the bound halo state.

From the absolute square of the S-matrix in Eq. (14) we obtain the total cross section for the laser-assisted nuclear photoeffect by integrating over the final proton momentum and by dividing out the γ-photon flux $j = 1/V_\gamma$ and a unit time T:

$$\sigma = \frac{1}{jT}\int\frac{V\,d^3p}{(2\pi)^3}\,|S_{fi}|^2 = \sum_{n=n_0}^{+\infty}\sigma_n\ . \tag{16}$$

The total cross section naturally decomposes into a sum over partial cross sections σ_n, in accordance with the structure of Eq. (14) which involves a summation over the number of exchanged laser photons. These partial cross sections read

$$\sigma_n = \frac{e^2c_0^2}{2m\omega_\gamma}\int_0^\pi \sin\theta\, d\theta\int_0^{2\pi} d\varphi\; p_n\,(\vec{\varepsilon}\cdot\vec{p}_n\,B_n - eA_0\vec{\varepsilon}\cdot\vec{e}\,C_n)^2\;G(\vec{p}_n - \vec{k}_\gamma)^2\ . \tag{17}$$

The absolute value of the momentum of the outgoing proton is fixed by the δ-function to be $p_n = [2m\,(\omega_\gamma + n\omega_0 - U_{\rm p} - E_b)]^{1/2}$. The remaining integrals over the polar and azimuthal emission angles of the proton in Eq. (17) can be performed by numerical means.

3. Numerical results and Discussion

We have applied our theory to laser-assisted photo-proton ejection from ^{8}B [21, 27, 29] and ^{26}P [30]. These nuclear isotopes possess low proton separation energies of $E_b \approx 137$ keV and $E_b \approx 860$ keV, respectively. The rms radius of the proton halo amounts to $R_{\rm rms} \approx 4.73$ fm in ^{8}B and $R_{\rm rms} \approx 3.11$ fm in ^{26}P. The γ photon is incident with a relatively high energy of a few MeV and is polarized along the z axis, $\vec{\varepsilon} = \vec{e}_z$. The high γ-photon energy guarantees that the nuclear potential has only a moderate influence on the proton after ejection.

In the following we demonstrate how the properties of the laser-assisted nuclear photoeffect depend on the incident γ-photon energy, the laser polarization and the nuclear isotope.

3.1. Dependence on γ-photon energy

Figure 2 displays the partial cross sections σ_n, as a function of the number n of emitted ($n < 0$) or absorbed ($n > 0$) laser photons. The laser field is assumed to have an x-ray frequency of $\omega_0 = 2$ keV at an intensity of $I = 1.0 \times 10^{23}$ W/cm^2 [31]. Like the γ-photon, it is polarized along the z axis, $\vec{e} = \vec{e}_z$. Note that the σ_n distribution corresponds to the photo-proton energy spectrum since, for a given value of n, the proton energy amounts to $E_n = \omega_\gamma + n\omega_0 - U_{\rm p} - E_b$. As the figure illustrates, the assistance of the laser field leads to sidebands in the photo-

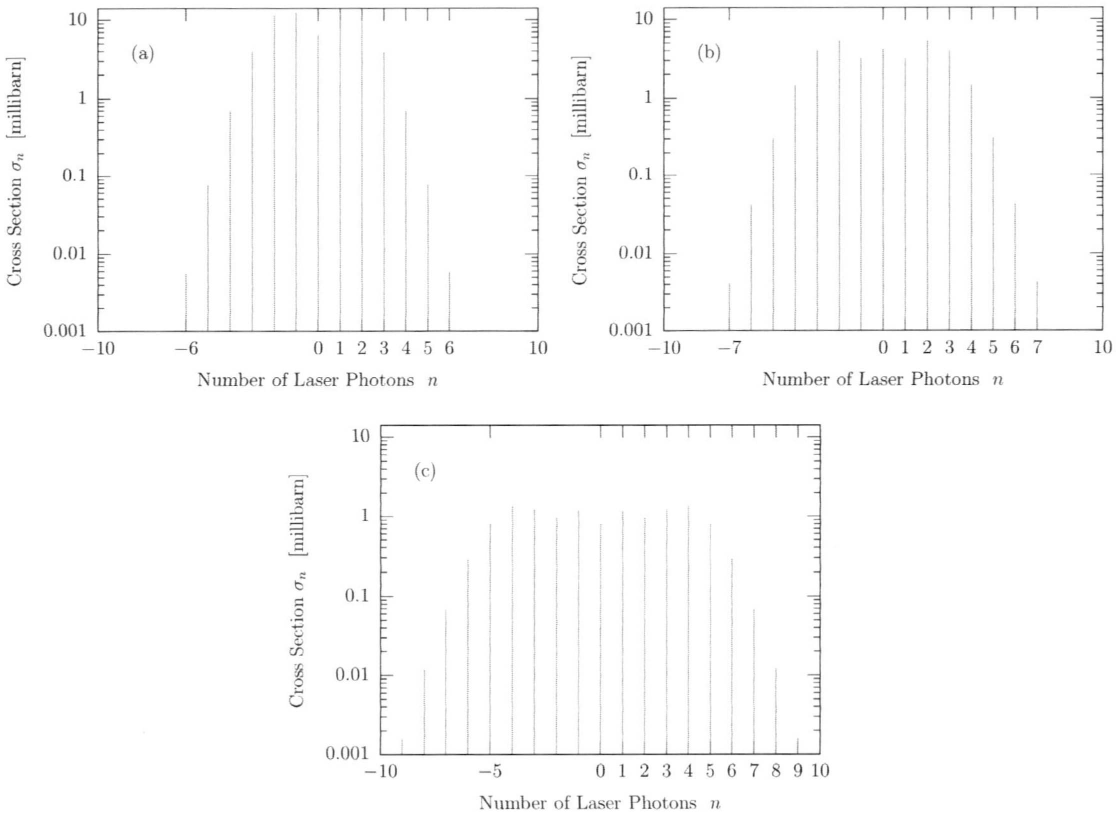

Figure 2. Distributions of the partial cross sections σ_n of laser-assisted photo-proton emission from ^{8}B for various energies of the incident γ-photon: (a) $\omega_\gamma = 3$ MeV, (b) $\omega_\gamma = 5$ MeV, (c) $\omega_\gamma = 10$ MeV. The parameters of the linearly polarized laser field are $\omega_0 = 2$ keV and $I = 1.0 \times 10^{23}$ W/cm^2. The polarization vectors of the γ-photon and laser field are parallel.

proton energy distribution. The sidebands are equally spaced by a laser photon energy and symmetrically surround the central line located at $n = 0$.

Moreover, as panels (a)-(c) of Fig. 2 show, the number of relevant sidebands grows when the γ-photon energy increases. The reason is that, due to the $\hat{\vec{p}} \cdot \vec{A}_L$ interaction between the proton and the laser field, larger proton momentum components along the laser polarization direction lead to enhanced proton-laser coupling. This fact is also reflected in the momentum dependence of the first argument [32] of the generalized Bessel functions in Eq. (9). When the γ-photon energy increases, also the emitted proton energy increases. As a consequence, also the relevant momentum component of the proton increases. Note in this regard that the proton is emitted preferentially along the polarization direction of the γ-photon.

The total cross sections $\sigma = \sum_n \sigma_n$ in Figs. 2(a)-(c) amount to about 63 mb, 33 mb and 13 mb. We point out that the same values would result from our model when the laser field is switched off. I.e., while the presence of the laser modifies the energy distributions of the ejected protons, it does not affect the total ejection probability. For comparison we note that an estimate of the total cross section via the well-known Bethe-Peierls formula gives values which are somewhat smaller (by a factor ≈ 3) than the cross sections mentioned above. Inclusion of the repulsive Coulomb interaction between the ejected proton and the core nucleus into our model would lead to a reduction of the total cross sections predicted [33].

3.2. Influence of laser polarization

In Fig. 3 we show the distributions of partial cross sections for laser-assisted photo-proton emission from ^{8}B by a 3-MeV γ-photon. Panels (a) and (b) refer to a laser field of linear polarization and circular polarization, respectively. The γ-photon as well as the linearly polarized laser field are polarized along the z axis, whereas the electric field vector of the circularly polarized laser wave rotates within the y-z plane. Both laser fields are assumed to have the same intensity. From the figure one can see that the linearly polarized laser field has a somewhat larger influence on the proton emission energies than the circularly polarized field. Indeed, the "plateau" in panel (a) comprising partial cross sections of similar magnitude is slightly more extended than in panel (b). This can be understood by noting that a linearly polarized field has an amplitude which is larger by a factor $\sqrt{2}$ than an equally intense field of circular polarization

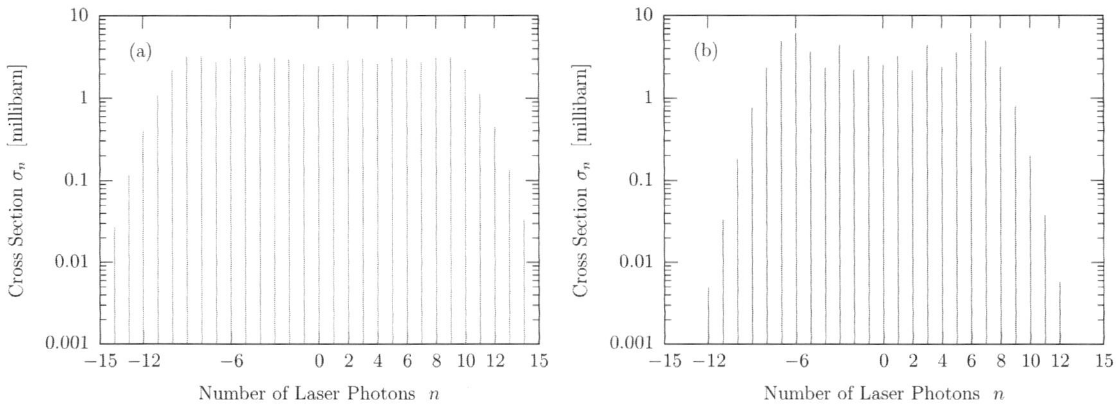

Figure 3. Comparison of the distributions of partial cross sections σ_n for photo-proton emission from ^{8}B assisted by a laser field of (a) linear polarization and (b) circular polarization. The laser frequency is $\omega_0 = 2\,\text{keV}$ and laser intensity $I = 1.25 \times 10^{24}\,\text{W/cm}^2$. The γ-photon has energy $\omega_\gamma = 3\,\text{MeV}$ and is polarized along the same axis as the linearly polarized laser field.

has. This larger amplitude leads to an enhanced proton-laser coupling, as may again be seen from the Bessel function argument α in Eq. (9). Note that for a circularly polarized laser wave, a very similar term enters as argument into an ordinary Bessel function J_n [20].

In terms of absolute numbers, the plateau in Fig. 3(a) extends over the range $-10 \lesssim n \lesssim 10$, corresponding to an energy width of $20\omega_0 = 40\,\text{keV}$. The proton energies thus lie approximately between 2.84 MeV and 2.88 MeV. The energy width arising for the assisting laser field of circular polarization in Fig. 3(b) is somewhat smaller and amounts to about 30 keV. Therefore, a linearly polarized laser field is advantageous for detecting signatures of the laser-assisted nuclear photoeffect. The total cross sections are the same for both polarizations. It is worth mentioning that similar energy widths of the proton spectra would arise if an optical laser with intensity $\sim 10^{18}\,\text{W/cm}^2$ was used instead of the x-ray laser beam [20].

3.3. Variation of nuclear isotope and field geometry

So far, we have considered photo-proton ejection from ^{8}B. Figure 4 shows distributions of partial cross sections for the laser-assisted nuclear photoeffect in ^{26}P. (Note that they are shown on a linear scale, in contrast to Figs. 2 and 3.) Since the proton separation energy in this isotope is much higher, a γ-photon energy of 10 MeV has been assumed here in order to lift the proton high up into the continuum.

Because of the larger value of E_b, a proton removed from ^{26}P has a smaller kinetic energy than a proton from ^{8}B, provided the γ-photon energy is the same. Therefore, at a laser intensity of $10^{23}\,\text{W/cm}^2$ the energy distribution of the photo-proton from ^{26}P would be slightly more narrow than the distribution in Fig. 2(c). However, when the laser intensity is enhanced, the number of sizeable sidebands in the case of ^{26}P grows accordingly. This is illustrated in Fig. 4(a). The total cross section amounts to 18 mb.

Besides, a comparison of panels (a) and (b) of Fig. 4 reveals that also the relative orientation of the γ-ray and laser fields has an impact on the photo-proton energy distribution. When both fields have parallel polarizations [see Fig. 4(a)], the proton can couple much more effectively to the laser field. The reason is that the proton is emitted preferably along the polarization vector of the γ-photon and, thus, has a large momentum component along this direction. When the laser vector potential points into the same direction, the $\vec{p} \cdot \vec{A}_L$ interaction is strong. This

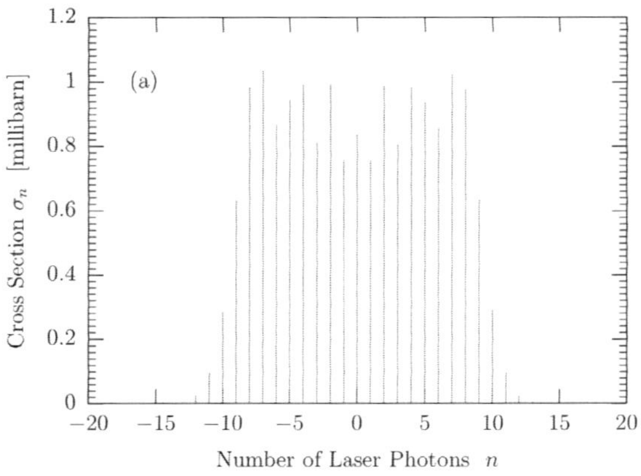

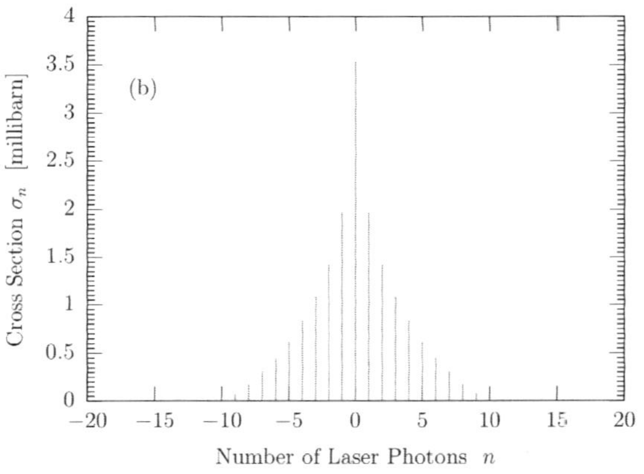

Figure 4. (a) Distributions of partial cross sections σ_n for laser-assisted photo-proton emission from ^{26}P. The field parameters are $\omega_\gamma = 10\,\text{MeV}$, $\omega_0 = 2\,\text{keV}$ and $I = 3.13 \times 10^{23}\,\text{W/cm}^2$ (i.e., $A_0 = 0.5\,\text{MeV}$). Both the γ-photon and the laser field are polarized along the z axis. (b) Same as (a) but the laser field is polarized perpendicularly to the polarization direction of the γ-photon.

leads to a relatively large number of emitted or absorbed laser photons and, accordingly, to a rather broad energy spectrum of the ejected proton. In contrast, when the polarization vectors of the γ-photon and the linearly polarized laser field are perpendicular [see Fig. 4(b)], then the proton's momentum component along the laser field is small, so that it cannot couple strongly to the laser field. The consequence is that the energy sidebands are largely suppressed.

4. Conclusion

We have examined proton emission from halo nuclei via the laser-assisted nuclear photoeffect. In contrast to an earlier study of the same subject, the laser field was assumed to be linearly polarized. An S-matrix theory based on the strong-field approximation was presented which resembles the theoretical treatment of laser-assisted photo-ionization of atoms.

The assistance of the laser field was shown to substantially modify the photo-proton energy spectra where emission and absorption of laser photons during the process lead to characteristic energy sidebands. The latter were found to be more pronounced for higher energies of the incident γ-photon and for laser fields which are linearly polarized along the same axis as the γ-photon. The total cross section of the process has not been modified for the laser parameters under consideration. While nuclear isotopes with a proton halo have the advantage of low proton separation energies, we point out that similar effects may also be expected for laser-assisted photo-proton emission from ordinary nuclei.

In view of a possible experimental investigation of the laser-assisted nuclear photoeffect in the future, we note that few-MeV γ-photons can nowadays be produced efficiently via bremsstrahlung of laser-accelerated electrons [34]. Thus, future experiments on laser-assisted photo-nuclear reactions may rely on setups involving two intense laser beams, with one of them being utilized to generate high-energy bremsstrahlung photons. This might evolve into an interesting research area for the nuclear pillar of the Extreme Light Infrastructure [4, 5].

Acknowledgments

A. D. gratefully acknowledges the scholarship support from the International Max Planck Research School for Quantum Dynamics in Physics, Chemistry and Biology (IMPRS-QD).

References

[1] Chadwick J and Goldhaber M 1934 *Nature* **134** 237
[2] Bothe W and Gentner W 1937 *Z. Phys.* **106** 236
[3] Ledingham K W D and Galster W 2010 New J. Phys. **12** 045005
[4] Zamfir N V, Habs D, Negoita F and Ursescu D 2011 *Extreme Light Infrastructure: Nuclear Physics*, *Proc. SPIE* 8080, 80800X (doi:10.1117/12.890139)
[5] For current information, see http://www.eli-laser.eu.
[6] Matinyan S 1998 *Phys. Rep.* **298** 199
[7] Ledingham K W D *et al.* 2000 *Phys. Rev. Lett.* **84** 899
[8] Cowan T E *et al.* 2000 *Phys. Rev. Lett.* **84** 903
[9] Schwoerer H *et al.* 2001 *Phys. Rev. Lett.* **86** 2317
[10] Kálmán P and Bergou J 1986 *Phys. Rev.* C **34** 1024
[11] Typel S and Leclercq-Willain S 1996 *Phys. Rev.* A **53** 2547
[12] Berger J F, Gogny D M and Weiss M S 1991 *Phys. Rev.* A **43** 455
[13] Mocken G R and Keitel C H 2004 *J. Phys.* B **37** L275
[14] Kornev A S and Zon B A 2007 *Laser Phys. Lett.* **4** 588
[15] Shahbaz A, Müller C, Bürvenich T J and Keitel C H 2009 *Nucl. Phys.* A **821** 106
[16] Habs D, Tajima T, Schreiber J, Barty C P J, Fujiwara M and Thirolf P G 2009 *Eur. Phys. J.* D **55** 279
[17] Bürvenich T J, Evers J and Keitel C H 2006 *Phys. Rev. Lett.* **96** 142501
[18] Bürvenich T J, Evers J and Keitel C H 2006 *Phys. Rev.* C **74** 044601
[19] Liao W-T, Pálffy A and Keitel C H 2011 *Phys. Lett.* B **705** 134
[20] Dadi A and Müller C 2012 *Phys. Rev.* C **85** 064604 (arXiv:1204.3736)
[21] Minamisono T *et al.* 1992 *Phys. Rev. Lett.* **69** 2058

[22] Véniard V, Taïeb R and Maquet A 1995 *Phys. Rev. Lett.* **74** 4161
[23] Glover T E, Schoenlein R W, Chin A H and Shank C V 1996 *Phys. Rev. Lett.* **76** 2468
[24] Meyer M *et al.* 2006 *Phys. Rev.* A **74** 011401(R)
[25] Madsen L B 2005 *Am. J. Phys.* **73** 57
[26] Reiss H R 1992 *Prog. Quantum Electron.* **16** 1
[27] Fukuda M *et al.* 1999 *Nucl. Phys.* A **656** 209
[28] Reiss H R 1980 *Phys. Rev.* A **22** 1786
[29] Smedberger M H *et al.* 1999 *Phys Lett.* B **452** 1
[30] Ren Z, Chen B, Ma Z and Xu G 1996 *Phys. Rev.* C **53** R572
[31] Brilliant x-ray beams with intensities up to 10^{18} W/cm^2 can presently be produced by free-electron lasers (see Young L *et al.* 2010 *Nature* **466** 56), and a further increase of the attainable x-ray intensity is envisaged. Another potential source of high-intensity x-rays are high-order harmonics emitted from plasma surfaces (see Dromey B *et al.* 2009 *Nature Phys.* **5** 146).
[32] For the present field parameters, the first argument is relevant since $\alpha \sim 1$–10 whereas $\beta \ll 1$.
[33] Berestetskii V B, Lifshitz E M and Pitaevskii L P 1971 *Relativistic Quantum Theory* (New York: Pergamon)
[34] Giulietti A *et al.* 2008 *Phys. Rev. Lett.* **101** 105002

Towards high intensity few-cycle pulses using plasma wakefield self-compression effect

A. Pipahl[1], E.A. Anashkina[2], M. Toncian[1], T. Toncian[1], S.A. Skobelev[2], A.V. Bashinov[2], A.A. Gonoskov[2], O. Willi[1], A.V. Kim[2]

[1] Institut für Laser- and Plasmaphysik, Heinrich-Heine-Universität Düsseldorf, Universitätsstraße 1, 40225 Düsseldorf, Germany
[2] Institute of Applied Physics, Russian Academy of Sciences, 603950 Nizhny Novgorod, Russia

E-mail: kim@ufp.appl.sci-nnov.ru

Abstract. We present experimental results for laser pulse compression due to wakefield excitation in the relativistic regime using the SPIDER technique for complete pulse characterization. The results indicate that the pulse was compressed to 10 fs duration which has been verified by 3D PIC simulations showing the possibility of getting laser pulses of few-cycle duration. These results were also verified by using a hydrodynamic code. The scalability of the latter allows to extend the wakefield compression scheme to the petawatt level, thus making it a route to PW-class few-cycle pulse generation.

1. Introduction

Tremendous progress has recently been made in laser technology primarily based on the chirped pulse amplification (CPA) [1] allowing ultrashort laser pulse generation with a duration as short as 30 fs and peak power up to one petawatt (PW) [2]. Now several projects worldwide are towards the 10 PW level in progress. However, the generation of high power few-cycle pulses, particularly below 10 fs for a Ti:sapphire laser at a wavelength of about 800 nm, is still a formidable task. The great demand of high intensity few-cycle pulses for various applications ranging from attosecond pulse production [3] to particle acceleration make [4] it challenging for present day laser physics. One of the ways relies on the optical-parametric chirped-pulse amplification (OPCPA) technology [5, 6, 7], however the progress of this technique has relative been slowly in obtaining high energy pulses. It possesses however a high potential, including ultrabroad amplification along with high gain. An alternative way is a recently proposed method for high energy few-cycle pulse generation using the ionization-induced self-compression effect [8] however this method has not been verified experimentally at high energy yet.

In this Letter, we propose to use the plasma wakefield self-compression effect for high power ultrashort laser pulses produced by conventional laser systems, e.g., such as based on Ti:sapphire laser technology, in order to obtain high intensity pulses with a duration of a few-cycles. We first carefully look at the temporal characteristics of the pulses compressed to 10 fs duration by using the SPIDER technique after passing a 100 TW, 30 fs pulse through a two millimeter gas jet. Then, these experimental data will be directly compared to 3D particle-in-cell (PIC) modeling showing that self-compression of the driver pulse takes place along with the self-focusing effect and is actually based on plasma wave excitation. To get a deeper insight into the underlying physics we developed a simplified 3D hydrodynamic model based on a quasi-one-dimensional description of the plasma waves and showed its qualitative agreement with PIC modeling and experiments as well. Finally, of particular interest is the extension of the wakefield compression scheme to the PW pulse level using the simplified model, since 3D PIC simulations for PW pulses cannot be done on presently available supercomputers.

Previous experiments in the relativistic regime [9, 10] have confirmed the general predictions [11, 12, 13] for pulse shortening in a plasma wave excited by the pulse. Going one step further towards few-cycle duration we employed a SPIDER (spectral phase interferometry for direct electric-field reconstruction) technique for complete pulse characterization, which is proven to be a perfect characterization method to the 10 fs temporal range [14]. However, it is worth noting that its application can become challenging when the spectrum is highly modulated or ultra-broadband [15].

2. Experiment

The experiment was conducted with the table-top Ti:Sa laser system (ARCTURUS facility; 100 TW, 30 fs, 800 nm, 10 Hz). The linearly polarized laser pulse with an energy of 1.8J was focused by an $f/7$ off-axis parabola onto a helium gas jet to a focal spot with a diameter of about 10 μm that leads to an intensity of 4×10^{19} Wcm^{-2} in vacuum, corresponding to a normalized vector amplitude of $a_0 = 4$. The gas jet length is about 2 mm and the electron densities vary in the range of $n_e = (3 \div 15) \times 10^{18}$ cm^{-3} obtained by changing the gas pressure.

The transmitted pulse was measured by collimating the central part of the laser beam. Figure 1 shows SPIDER interferograms, pulse spectra (right graphs), retrieved pulse intensity profiles and phases (central graphs) in the time domain for the electron densities of 6.6×10^{18} cm^{-3} (a) and 1.3×10^{19} cm^{-3} (b), respectively.

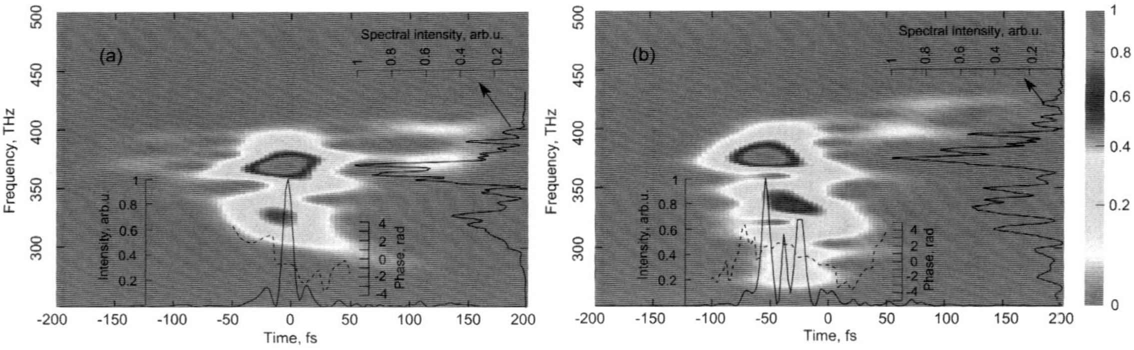

Figure 1. SPIDER interferograms, pulse spectra (right graphs), retrieved pulse intensity profiles and phases (central graphs) in the time domain for the electron densities of 6.6×10^{18} cm^{-3} (a) and 1.3×10^{19} cm^{-3} (b), respectively.

In the low density case of Fig. 1(a) the corresponding plasma wavelength $\lambda_p = 2\pi c/\omega_p = 13$ μm $[\omega_p = (4\pi e^2 n_e/m)^{1/2}$ is the plasma frequency] exceeds the pulse length $c\tau = 9$ μm. The retrieved output pulse duration is about 10 fs, i.e., the compression factor is ~ 3 after the pulse has propagated about 2 mm. Indeed, as shown below, this is also confirmed by simulations indicating that optimal compression takes place exactly at the distance of 2 mm. However, to increase the compression factor and get sub-10 fs pulses we can use higher plasma densities, since the interaction efficiency length is inversely proportional to the plasma density. Indeed, in the higher density case of Fig. 1(b), the pulse spectrum is essentially broader, which corresponds to a transform-limited pulse duration of 5 fs, but the retrieved experimental output pulse consists of three subpulses with pulse durations between 7 and 10 fs. But, it is important to note that the temporal resolution of the SPIDER technique we used was about 10 fs. In order to properly treat these results it is natural to assume that either an optimal compression length is less than 2 mm or the Raman instability destroys the pulse compression for the input pulse length comparable to the plasma wavelength which is 9.3 μm.

3. 3D PIC modeling

First, to understand the specific features of relativistic pulse compression we use 3D particle-in-cell (PIC) modeling for the parameters relevant to the experiment. The fully relativistic PIC code, developed earlier and applied to proton acceleration [16], accounts exactly for electromagnetic dispersion as it uses a fast Fourier transformation for field calculations. This allows in some cases even a fair quantitative comparison with experimental data. The simulation box co-moving with the speed of light is $125\lambda\times125\lambda\times100\lambda$ and contains $512\times512\times1024$ cells. Each cell is filled with 8 quasiparticles. A linearly polarized laser pulse with a Gaussian envelope $a = a_0 exp[-(x^2+y^2)/b^2]exp[-t^2/\tau^2]$ in both transverse (x, y) and longitudinal (z) directions is focused normally onto the front edge of the plasma slab from the left side ($z = 0$ μm). Since in the experiment the length of the gas jet was comparable with the beam Rayleigh length, we set the following parameters: $a_0 = 4$, $b = 20$ μm, $\tau = 30$ fs, and a uniform helium plasma slab of 2 mm in length, to mimic the experiment. The initial temperature of the electrons and protons is 20 keV.

In Fig. 2(a,b) we present the simulation results for a plasma density of 6.6×10^{18} cm^{-3} which agrees qualitatively and quantitatively with the experimental data shown in Fig. 1(a). Indeed, the laser pulse is continuously shortened during its propagation in the plasma, reaching 10 fs at a distance of 2 mm. Since the pulse compression is accompanied with relativistic self-focusing, the laser beam diameter shrinks from 20 μm to 10.6 μm and the output intensity increases to 5.7×10^{19} Wcm^{-2}. The most important feature of the compression is that it occurs so that the plasma wave is excited in a quasi-one-dimensional manner, i.e., unlike in the blowout or bubble regime [17]. This is very consistent with the weakly relativistic theory [18] and assures that the electron density immediately behind the laser pulse is controlled. Indeed, at the beginning of the pulse propagation the maximum excited plasma wave density is about $0.5n_c$, which is consistent with the 1D excitation in the relativistic regime. However, during propagation it decreases but not less than to $0.23n_c$, i.e., the electrons under the action of the ponderomotive force do not go around the laser pulse as in the bubble regime but mostly pass through it. At higher plasma densities we can expect more efficient pulse compression, if it is not destroyed by the Raman or self-focusing instabilities [13]. Although there are no systematic studies of the role of these instabilities at $a > 1$ for pulse compression, however from a general point of view it is obvious that in the strongly relativistic regime these obstacles maybe overcome by properly choosing the laser parameters, specifically its amplitude, beam diameter and pulse duration. In fact, for higher amplitudes, the effective plasma frequency becomes lower due to relativistic mass correction, which decreases the nonlinear interaction efficiency and may restrain the development of the instabilities. This is seen in Fig. 2(c) clearly demonstrating, on the one hand, pulse compression to 6.1 fs at a distance of 1.28 mm for a plasma density of 1.3×10^{19} cm^{-3} and, on the other hand, suppression of laser distortions due to relativistic self-focusing or Raman type instabilities. It is important to note that at a distance of 1.64 mm the pulse becomes broader (about 8.5 fs) and structured, both these factors increasing with further propagation. Thus, comparing this simulation with the experiment shown in Fig. 1(b) we can conclude that the pulse compression occurs at shorter distances of about 1 mm. The ultrabroad pulse spectrum actually reflects the possibility of such compression.

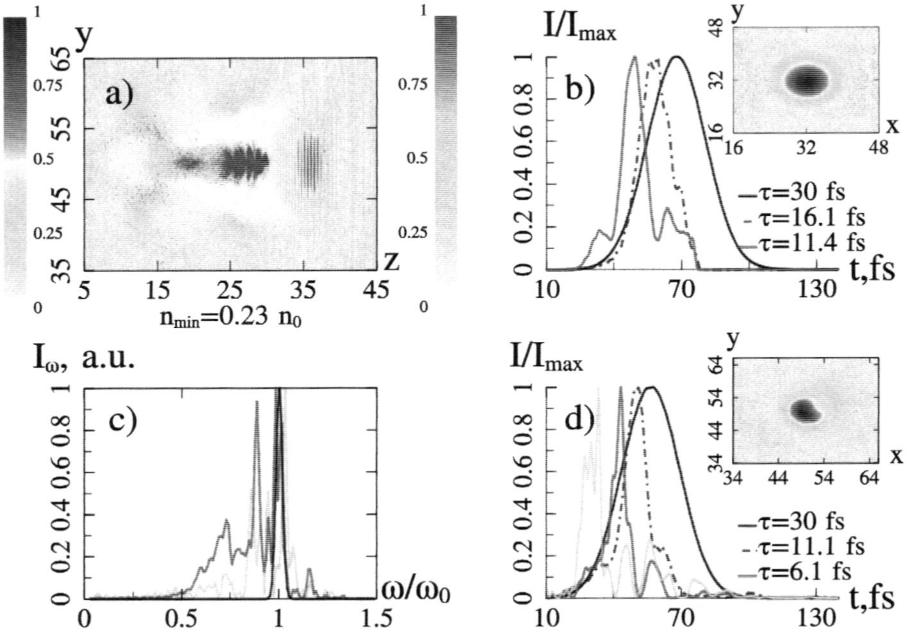

Figure 2. (a) Snapshot of the electron density distribution (green scale) and laser field (red scale) at the time of the pulse leaving the 2 mm long plasma with a density of 6.6×10^{18} cm^{-3} and (b) pulse profiles corresponding to this case: input (black line), in the middle $z = 1.2$ mm (blue-dotted line), and output (red line). (c,d) The result of modeling for a 2 mm long plasma with a density of 1.3×10^{19} cm^{-3}: input (black), in the middle $z = 1.28$ mm (red) and at $z = 1.642$ mm (light blue) pulse spectra, and the corresponding pulse profiles (additionally: green line at $z = 628$ μm), respectively. In the insets: the corresponding beam profiles of 10.6 μm (b) and 6.5 μm (d) in diameter (FWHM).

4. Hydrodynamic modeling

Despite the satisfactory agreement of the PIC simulations with the experiment it should be noted that 3D simulations, which are necessary for our case, cannot provide an exact description of the real laser-plasma interaction because of the high fluctuation level caused by the small number of quasiparticles in the cells, limited by present day supercomputers. Moreover, as we are interested in PW-class lasers it is impermissible to make any quantitative prediction from such comparison. To get one step further we have developed a simplified hydrodynamic code which allows, first, to assess the ultimate potential of the wakefield compression scheme and, second, to apply it to PW lasers. To do so we have employed the well-proven slowly evolving wave approach (SEWA) [19]. As the plasma density n_e is much less than the critical value $n_c = m\omega^2/(4\pi e^2)$ ($n_e \ll n_c$) [ω is the pulse carrier frequency], assuming that the field changes only slightly on the wavelength scale and the transverse size of the beam is large compared to the characteristic longitudinal scale of the field (quasi-optical beam), the Maxwell equations and plasma hydrodynamic equations can be reduced to the following set of self-consistent equations in the corresponding reference frame

$\tau = t - z/c$

$$\frac{2}{c}\frac{\partial^2 \mathbf{a}}{\partial z \partial \tau} - \Delta_\perp \mathbf{a} + \frac{\omega_p^2}{c^2}\frac{\mathbf{a}}{1+\Phi} = 0, \tag{1}$$

$$\frac{\partial^2 \Phi}{\partial \tau^2} + \omega_p^2 \frac{(1+\Phi)^2 - (1+ \mid \mathbf{a} \mid^2)}{2(1+\Phi)^2} = 0, \tag{2}$$

where $\omega_p = (4\pi e^2 n_e/m)^{1/2}$ is the plasma frequency, c is the speed of light in vacuum, $\Delta_\perp = \partial^2/\partial r^2 + (1/r)\partial/\partial r$. Here, we introduced the following dimensionless vector and scalar potentials: $\mathbf{a} = e\mathbf{A}/mc^2$, $\Phi = e\varphi/mc^2$. Equation (2) is the usual equation of state for a plasma wave [18, 20] when the electrons evolve mainly in the longitudinal direction (see also [21]). This occurs for laser pulses with a longitudinal dimension much smaller than the transverse one, which is especially true when the group velocity is close to the speed of light, i.e. for low density plasmas.

To begin with, we first compare the results of hydrodynamic modeling, as shown in Fig. 3, with the PIC simulations and experiment. We can definitely say that the results of the pulse compression dynamics for both plasma densities are in good qualitative agreement. For a lower plasma density of $n_e = 6 \times 10^{18}$ cm^{-3} and $a_0 = 4$ the pulse is compressed, being one-hump localized in all dimensions, to the duration comprising 1.5 optical oscillations, i.e. to 5 fs, at a distance of 2 mm. Whereas for the higher density of $n_e = 1.3 \times 10^{19}$ cm^{-3} and $a_0 = 3$ the minimum pulse duration is 1.2 optical periods, i.e. 4 fs, at a distance of 1.15 mm only. But in this case after passing the point of compression, at a distance of 2 mm the pulse splits into subpulses, as can clearly be seen in Fig. 3(b). For comparison we also present in Fig. 3(c,d) the temporal profiles of the input pulse and pulses with minimum durations along the axis, and their spectra. It is interesting that at the higher plasma density the quality of compression is better, i.e., the laser pulse (red line) is well localized and practically does not contain wings on the back side as compared to the blue line. It should also be mentioned that during propagation the pulse's peak shifts forward and its leading edge becomes steeper, thus making the longitudinal dimension smaller than the transverse one. From this modeling we can state with confidence that the wakefield compression is capable of producing laser pulses with few-cycle durations. In fact, in both cases of Fig. 3 we actually obtained single-cycle pulses with an energy efficiency of about 20%.

A very intriguing question is that of producing few-cycle pulses with petawatt powers. The remarkable feature of Eqs. (1), (2) is that they contain scalable solutions. In fact, with the following transformations for spatial and temporal variables: $\omega_p\tau \to \tau$, $\omega_p z/c \to z$, and $\omega_p r/c \to r$, Eqs. (1), (2) can be written in dimensionless form. This means that the simulations presented, for example, in Fig. 3 are suitable for other laser-plasma parameters with the corresponding scalability, i.e., laser pulse parameters such as laser frequency ω, pulse duration and beam diameter which can be scaled by a factor of ω_p/ω. But this scalability cannot help us if we have one type of laser, i.e., if ω is fixed. However, to get an insight into the physics of interest we can employ an approximate scalability valid for comparatively long pulses, when the envelope approximation holds. In fact, to obtain compression we usually start from long pulses containing many optical cycles. Thus, assuming $\mathbf{a} \sim e^{i\omega\tau}$ and reducing Eq. (1) to the form $(2i\omega/c)\partial\mathbf{a}/\partial z - \Delta_\perp\mathbf{a} + (\omega_p^2/c^2)\mathbf{a}/(1+\Phi) = 0$ [18] the following scalability

$$n_0^{1/2}\omega\tau \to \tau, n_0 kz \to z, n_0^{1/2} kr \to r \tag{3}$$

(where $k = \omega/c$ and $n_0 = \omega_p^2/\omega^2$) can now be applied with respect to the envelope of the laser field; the difference is that the length of the compression is proportional to n_0, i.e. to the plasma density. For compression of PW-class laser pulses for an input amplitude $a \sim 1$ we have to use larger laser beams and therefore, as it follows from Eq. (3), lower plasma densities.

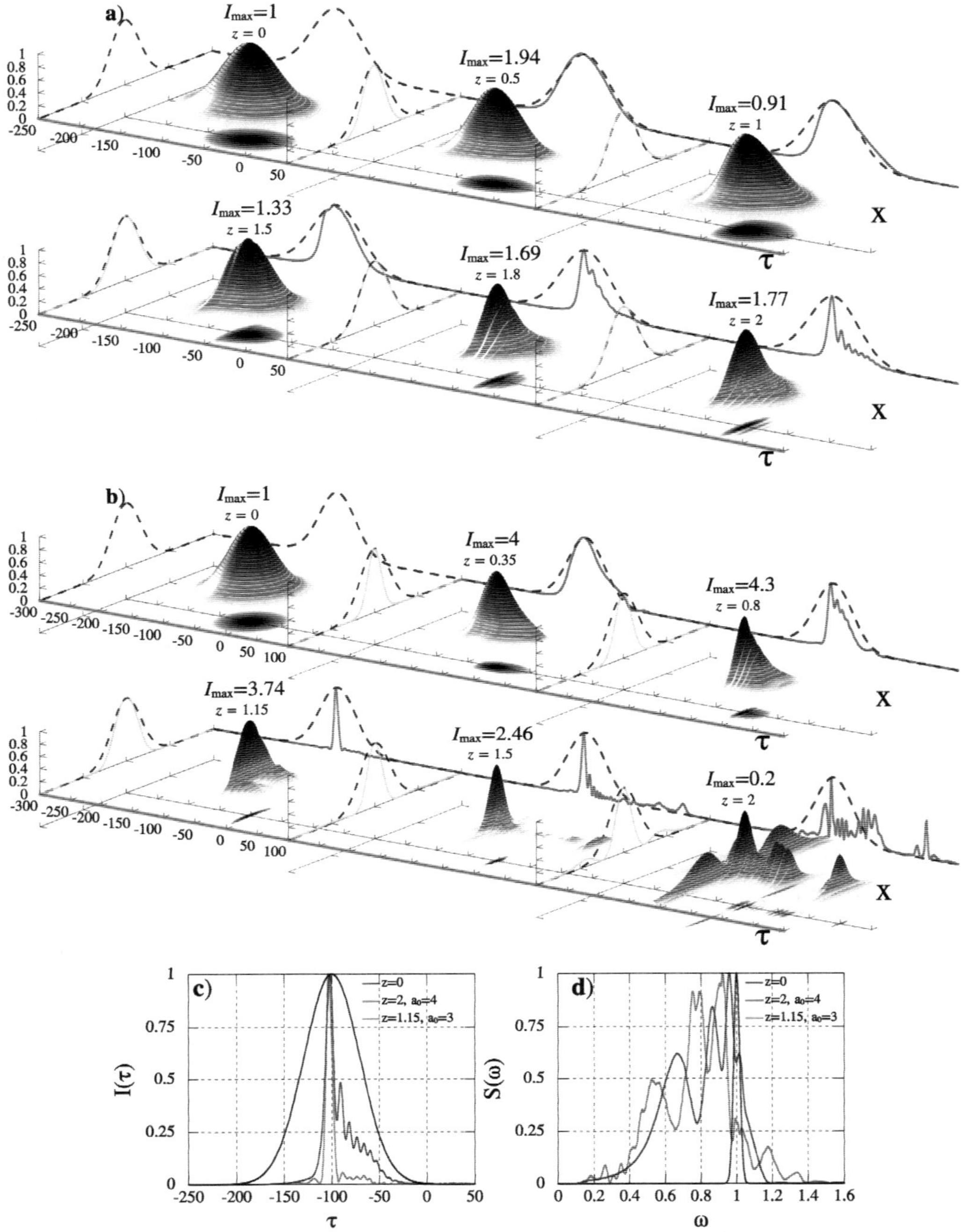

Figure 3. Spatial-temporal pulse dynamics along the propagation direction at plasma densities of 6.6×10^{18} cm^{-3} (a) and 1.5×10^{19} cm^{-3} (b) respectively; additionally input (dashed curve) and current (red) laser pulse profiles on the axis, and transverse (blue) beam profiles at the plane of maximum intensity are depicted. (c) Intensity profiles of the input pulse and compressed pulses with a minimum duration, and (d) corresponding spectra: blue line for $n_e = 6.6\times10^{18}$ cm^{-3} - 5 fs at FWHM and red line for $n_e = 1.3\times10^{19}$ cm^{-3} - 4 fs at FWHM.

As an example we make estimations for a 1 PW laser pulse with a duration of 500 fs (pulse energy of 500 J). Following the scalability given by Eq. (3) and the simulations presented in Fig. 3 we may suggest a two-stage wakefield compression scheme to get a few-cycle pulse. At the first stage, by using plasma densities of about 5×10^{17} cm^{-3} the pulse is compressed by a factor of 7, i.e., down to the duration of 70 fs at a plasma length of 4 cm with an energy efficiency of about 20%. At the second stage, this pulse should enter into the plasma jet with $(3 \div 4) \times 10^{18}$ cm^{-3} density. After passing a length of about 4 mm the pulse will be compressed down to 10 fs. Thus, with the same efficiency of 20% at this stage, we obtain an output pulse with the energy of 20 J that corresponds to the pulse power of two petawatts, i.e., even more than the input power but for the shorter pulse.

Another but more rigorous example we present by directly checking the possibility of compression of petawatt-class pulses to few-cycle durations. In Fig. 4 we present results for a plasma density of 10^{18} cm^{-3}, when the input pulse has the following parameters: pulse duration 70 fs, beam width 36 μm, and $a_0 = 4$, which correspond to 70 J pulse energy and 1 PW power, respectively. The pulse is smoothly compressed, reaching at a distance of 5 cm its minimal duration of 5 fs. As is clearly seen, the pulse temporal and beam profiles have good quality, the energy efficiency is about 18 %, so that at the output we obtain a 5 fs, 2 PW pulse.

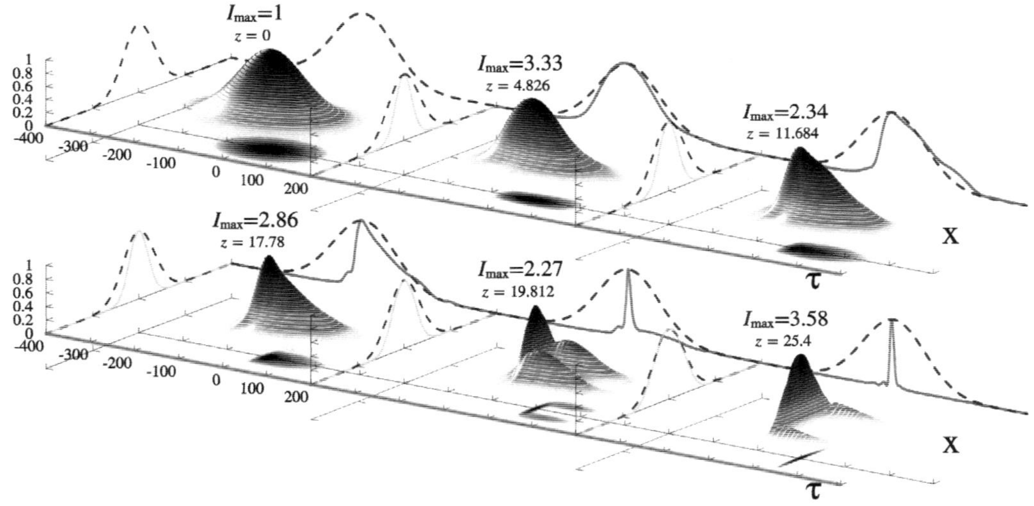

Figure 4. Spatial-temporal pulse dynamics along the propagation direction for the PW-class laser pulse (70 fs, 70 J, $a_0 = 4$) in a plasma with a density of $1{\times}10^{18}$ cm^{-3}. Notations as in Fig. 3.

5. Conclusion

We have examined wakefield pulse compression in a combined experimental and numerical study. It is shown that in the relativistic regime ($a_0 \sim 1 \div 5$) laser pulses can be effectively compressed to a duration of to few-cycles. Moreover, the scheme is scalable to the petawatt level, which means that for present day conventional laser systems delivering comparatively long pulses the wakefield compression opens up a new way of producing PW-class few-cycle pulses.

Acknowledgments

This research has been supported by the DFG SFB/TR 18 and the GRK 1203 programmes. Two of the authors (S.A.S. and E.A.A.) also acknowledges partially support from the RF President

Grant No. MK-4902.2011.2 and Dynasty Foundation.

References

[1] Strickland D and Mourou G 1985 *Opt. Commun.* **56** 219
[2] Sung J H, Lee S K, Yu T J, Jeong T M and Lee J 2010 *Opt. Lett.* **35** 3021
[3] Krausz F and Ivanov M 2009 *Rev. Mod. Phys.* **81** 163
[4] Schmid K et.al. 2009 *Phys. Rev. Lett.* **102** 124801
[5] Chekhlov O V et.al. 2006 *Opt. Lett.* **31** 3665
[6] Lozhkarev V V et.al. 2007 *Laser Phys. Lett.* **4** 421
[7] Herrmann D, Veisz L, Tautz R, Tavella F, Schmid K, PervakV and Krausz F 2009 *Opt. Lett.* **34** 2459
[8] Skobelev S A, Kim A V and Willi O 2012 *Phys. Rev. Lett.* **108** 123904
[9] Faure J, Glinec Y, Santos J J, Ewald F, Rousseau J-P, Kiselev S, Pukhov A, Hosokai T and Malka V 2005 *Phys. Rev. Lett.* **95** 205003
[10] Schreiber J, Bellei C, Mangles S P D , Kamperidis C, Kneip S, Nagel S R, Palmer C A J, Rajeev P P, Streeter M J V and Najmudin Z 2010 *Phys. Rev. Lett.* **105** 235003
[11] Ren C, Duda B J, Hemker R G, Mori W B, Katsouleas T, Antonsen T M, Jr. and Mora P 2001 *Phys. Rev.* E **63** 026411
[12] Lontano M and Murusidze I G 2003 *Opt. Express* **11** 248
[13] Gordon D F, Hafizi B, Hubbard R F, Penano J R, Sprangle P and Ting A 2003 *Phys. Rev. Lett.* **90** 215001
[14] Gallmann L, Sutter D H, Matuschek N, Steinmeyer G, Keller U, Iaconis C and Walmsley I A 1999 *Opt. Lett.* **24** 1314
[15] Wyatt A S and Walmsley I A 2006 *Opt. Lett.* **31** 1914
[16] Gonoskov A A, Korzhimanov A V, Eremin V I, Kim A V and Sergeev A M 2009 *Phys. Rev. Lett.* **102** 184801
[17] Pukhov A and Meyer-ter-Vehn J 2002 *Appl. Phys.* B **74** 355
[18] Abramyan L A, Litvak A G, Mironov V A and Sergeev A M 1992 *JETP* **75** 978
[19] Geissler M, Tempea G, Scrinzi A, Schnurer M, Krausz F and Brabec T 1999 *Phys. Rev. Lett.* **83** 2930
[20] Sprangle P, Esarey E, Krall J and Joyce G 1992 *Phys. Rev. Lett.* **69** 2200
[21] Esarey E, Schroeder C B and Leemans W P 2009 *Rev. Mod. Phys.* **81** 1229

Spin effects in strong-field laser-electron interactions

S Ahrens[1], T-O Müller[2], S Villalba-Chávez[3], H Bauke[1] and C Müller[3]

[1] Max-Planck-Institut für Kernphysik, Saupfercheckweg 1, 69117 Heidelberg, Germany
[2] Physik Department, Technische Universität München, 85747 Garching, Germany
[3] Institut für Theoretische Physik I, Heinrich-Heine-Universität, 40225 Düsseldorf, Germany

E-mail: c.mueller@tp1.uni-duesseldorf.de

Abstract. The electron spin degree of freedom can play a significant role in relativistic scattering processes involving intense laser fields. In this contribution we discuss the influence of the electron spin on (i) Kapitza-Dirac scattering in an x-ray laser field of high intensity, (ii) photo-induced electron-positron pair production in a strong laser wave and (iii) multiphoton electron-positron pair production on an atomic nucleus. We show that in all cases under consideration the electron spin can have a characteristic impact on the process properties and their total probabilities. To this end, spin-resolved calculations based on the Dirac equation in the presence of an intense laser field are performed. The predictions from Dirac theory are also compared with the corresponding results from the Klein-Gordon equation.

1. Introduction

Quantum mechanical interaction processes between electrons and high-intensity laser fields in a relativistic regime have encountered a sustained interest in recent years (see [1, 2] for reviews). A particular question in this regard is to which extent the electronic spin degree of freedom may influence the electron-laser interaction. When a free electron is exposed to the electromagnetic field of a plane laser wave, the electronic spin vector performs a precessive motion. However, after the laser pulse has passed and the interaction is over, the electron will be in the same spin state as it was before the interaction started [3].

Distinct electron spin transitions can occur when the electron participates in a scattering process while being subject to the laser field. Then, the electronic spin state before and after the interaction with the field may change. Corresponding spin transitions have theoretically been revealed, for instance, in multiphoton Compton scattering [4, 5, 6], laser-assisted Mott scattering [7, 8], and photo-induced production of electron-positron pairs [9, 10]. Spin effects in strong laser fields were also investigated with regard to bound-electron dynamics [11], two-photon ionization [12] and high-harmonic generation [13]. Moreover, characteristic differences between fermionic and bosonic particles were found in pair production in an oscillating electric field [14] and in recent investigations of the Klein paradox [15, 16].

In this contribution, further strong-field laser-electron interaction processes are studied with respect to the relevance of spin effects. First, we consider Kapitza-Dirac scattering of electrons on a standing laser wave of high intensity and frequency (Section 2). Our analysis relies on the Dirac equation which is solved by numerical integration. In a certain regime of interaction parameters, characteristic electron spin dynamics involving Rabi oscillations and complete spin-flip transitions is identified. Differences between the Kapitza-Dirac scattering of spin-half and

Journal of Physics: Conference Series **414** (2013) 012012 doi:10.1088/1742-6596/414/1/012012

spin-zero particles are emphasized. Afterwards, we turn to spin-sensitive calculations of electron-positron pair creation in very intense laser fields. By using the polarization operator in the presence of a plane-wave laser field, we examine the production of pairs induced by a high-energy photon travelling through a plane laser wave of circular polarization (Section 3). A comparison between the total production rates of spin-half versus spin-zero particles is drawn, this way complementing the earlier studies in [9, 10]. Finally, electron-positron pair production in combined laser and nuclear Coulomb fields is studied within an S-matrix formalism using Dirac-Volkov states and the spin projection operator (Section 4). In a nonperturbative regime of interaction, the degree of longitudinal polarization of one of the produced particles is obtained and its dependence on the particle momentum is discussed. We finish with some concluding remarks (Section 5).

Relativistic units with $\hbar = c = 4\pi\varepsilon_0 = 1$ are used unless explicitly stated otherwise.

2. Spin dynamics in Kapitza-Dirac scattering

The quantum mechanical diffractive scattering of an electron beam from a standing wave of light, formed by two counterpropagating waves of equal frequency and intensity, is referred to as Kapitza-Dirac effect [17]. The process, which relies on the quantum wave nature of the electron, constitutes an analogue of the classical diffraction of light on a grating, but with the roles of light and matter interchanged. In its original version [18], the effect represents a combined absorption and emission process involving two photons: the electron absorbs one photon of momentum $\vec{k}$ from, say, the right-travelling laser beam and emits a photon of momentum $-\vec{k}$ into the counterpropagating laser beam. The momentum of the incident electron thus changes by $2\vec{k}$. Seminal theoretical work on the Kapitza-Dirac effect and its multiphoton generalization to intense laser fields has been carried out starting in the 1960s [19].

It is interesting to note that a clear experimental confirmation of the Kapitza-Dirac effect as originally proposed has been achieved only recently [20]. This observation has led to a newly revived interest of theoreticians in the effect. During the last decade, a variety of additional aspects of Kapitza-Dirac scattering has been studied such as the influence of the electronic wave packet size [21] and diffractive scattering in the case of counterpropagating laser beams of non-equal frequencies [22]. Besides, an analysis based on the Pauli equation showed that spin effects are very small at low laser intensities [23]. Both the experiment and the theoretical studies have relied on rather moderate, nonrelativistic laser intensities [24].

The recent advent of novel high-intensity light sources raises the question as to how the Kapitza-Dirac effect is modified in the relativistic regime where also spin effects may be expected to occur. To this end, a treatment of the process based on the Dirac equation is required. It has been provided recently [25]. One can show that – rather than in the original Kapitza-Dirac effect with two-photon exchange – pronounced spin effects may arise when the electron scatters from the standing wave with participation of an *odd* number of laser photons. Such an interaction channel can be accessed by a suitable choice of the electron and field parameters. The latter can be determined from a generalized Bragg condition which follows from the laws of energy and momentum conservation.

The relevant Dirac equation may be written as

$$i\frac{\partial}{\partial t}\Psi(\vec{x},t) = \left[\left(\hat{\vec{p}} + e\vec{A}(\vec{x},t)\right)\cdot\vec{\alpha} + \beta m\right]\Psi(\vec{x},t)\,, \qquad (1)$$

where $-e$ and m are the electron charge and mass, respectively, $\hat{\vec{p}} = -i\nabla$ is the momentum operator, and $\vec{A}(\vec{x},t)$ denotes the vector potential of the standing laser wave. Besides, $\vec{\alpha} = (\alpha_x, \alpha_y, \alpha_z)$ and β are the Dirac matrices. Due to the periodicity of the vector potential, the electronic wave function may be decomposed into a discrete set of plane waves, according to

$$\Psi(\vec{x},t) = \sum_{n,\zeta} c_n^\zeta(t)\psi_{n,\vec{p}}^\zeta(\vec{x}) , \tag{2}$$

with free Dirac states $\psi_{n,\vec{p}}^\zeta(\vec{x}) = u_{\vec{p}+n\vec{k}}^\zeta \mathrm{e}^{i(\vec{p}+n\vec{k})\cdot\vec{x}}$ of momentum $\vec{p}+n\vec{k}$ ($n = 0, \pm1, \pm2, \ldots$). The index $\zeta \in \{+\uparrow, +\downarrow, -\uparrow, -\downarrow\}$ labels the sign of the energy and the spin projection along the laser electric field vector. Insertion of the ansatz (2) into the Dirac equation (1) yields a system of coupled differential equations for the expansion coefficients $c_n^\zeta(t)$, which represent the transition amplitudes to the various states. The system of equations can be solved by numerical integration, starting from the initial condition $c_0^{+\uparrow}(0) = 1$ and all other $c_n^\zeta(0) = 0$. It describes an incident electron of momentum $\vec{p}$ and positive spin projection.

As an example, we consider Kapitza-Dirac scattering of an electron which is incident with a weakly relativistic momentum of 176 keV at a small angle of $\vartheta = 0.4°$ with respect to the laser beam axis. The laser frequency amounts to 3.1 keV and each laser beam has a peak intensity of $2.0 \times 10^{23}\,\mathrm{W/cm^2}$. Under these circumstances, the electron may be scattered with absorption of two photons from the right-travelling laser beam and emission of one photon into the left-travelling beam. The electron momentum thus changes by $3\vec{k}$.

Figure 1 shows the time dependence of the occupation probabilities of the initial state (with momentum $\vec{p}$) and the scattered state (with momentum $\vec{p}+3\vec{k}$). The latter probability consists of two contributions, corresponding to a spin-preserving or spin-flipping transition. In the figure, these contributions have been added to give the spin-summed probability $|c_3(T)|^2 := |c_3^{+\uparrow}(T)|^2 + |c_3^{+\downarrow}(T)|^2$ (see the black solid line) [26]. Note that the results shown in Fig. 1 represent occupation probabilities *outside* the laser field; that is, after each value of T the interaction between the electron and the field has been faded out in the numerical simulation.

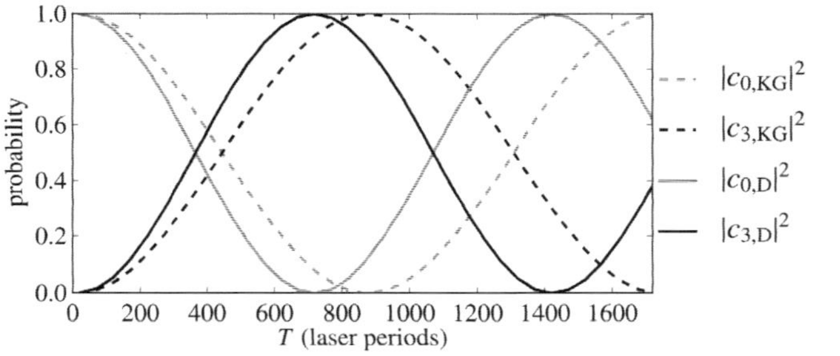

Figure 1. Rabi oscillations of the occupation probabilities $|c_0|^2$ and $|c_3|^2$ of the initial state and the scattered state, respectively, as a function of the interaction time T for a three-photon Kapitza-Dirac effect. The Rabi flopping dynamics is faster for a Dirac (D) particle than a Klein-Gordon (KG) particle.

The occupation probability oscillates back and forth between the two momentum modes $\vec{p}$ and $\vec{p}+3\vec{k}$, with a Rabi frequency of [25]

$$\Omega_{R,\mathrm{D}} = \Omega_0\sqrt{\frac{25}{2}\left(\frac{p_E}{k}\right)^2 + 1} . \tag{3}$$

Here, we introduced the abbreviation $\Omega_0 = e^3E^3/(24m^3k^2)$, with the laser electric field amplitude E, and the electron momentum component $p_E = |\vec{p}|\sin\vartheta$ along the electric field.

It is important to note that, if the electron was a spinless particle, the Rabi frequency would differ from Eq. (3). For a Klein-Gordon particle it would read

$$\Omega_{R,\mathrm{KG}} = \Omega_0\frac{5}{\sqrt{2}}\frac{p_E}{k} . \tag{4}$$

The difference may be understood by the following argument. In order to induce a transition involving three photons, it is necessary to have in the interaction Hamiltonian terms which are linear in the laser field. In the case of a spinless particle, the $\hat{\vec{p}} \cdot \vec{A}$ interaction term is of this kind. It leads to the linear momentum dependence of the Rabi frequency in Eq. (4). In the case of a spin-half particle there is in addition a term which corresponds to the magnetic $\vec{\sigma} \cdot \vec{B}$ interaction in the Pauli equation. This term, which is momentum-independent and capable of inducing spin-flip transitions, leads to the difference between Eqs. (3) and (4).

The different Rabi frequencies $\Omega_{R,\mathrm{D}}$ and $\Omega_{R,\mathrm{KG}}$ give rise to different diffraction patterns for fermionic versus bosonic particles, as Fig. 2 shows. The fact that the electron carries spin, enhances the Rabi frequency such that, after an interaction time corresponding to 520 laser cycles, the scattering probability is significantly larger than if the electron was spinless ($|c_{3,\mathrm{D}}(T)|^2 \approx 80\%$ versus $|c_{3,\mathrm{KG}}(T)|^2 \approx 60\%$). The left panel of Fig. 2 also shows the contributions to the scattered state stemming from the spin-preserving and the spin-flipping transition. Note that the scattering probability of a Klein-Gordon particle differs not only from the spin-summed scattering probability in the Dirac case but also from the scattering probability without spin flip. This is because the interaction time is rather large, $T \sim \Omega_R^{-1}$. In contrast, when $T \ll \Omega_R^{-1}$ so that $\sin^2(\Omega_R T/2) \approx (\Omega_R T/2)^2$, the scattering probability of a Klein-Gordon particle would coincide with the spin-preserving scattering probability of a Dirac particle.

Hence, under certain conditions, the electronic spin degree of freedom can have a substantial impact on observable features of the three-photon Kapitza-Dirac effect.

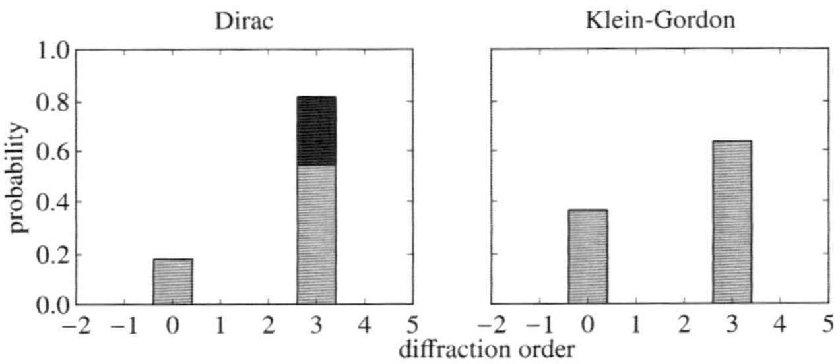

Figure 2. Diffraction probability after an interaction time of 520 laser cycles ($T \approx 0.7\,\mathrm{fs}$) for Dirac and Klein-Gordon particles. In the Dirac case, the light (dark) gray bars represent the spin-up (spin-down) probabilities. The other interaction parameters are given in the text.

3. Spin dependence of photo-induced pair production

Electron-positron pair creation can occur in a variety of physical contexts. In this section we focus ourselves on a generalized nonlinear version of the Breit-Wheeler reaction. In the generalized process an incident high-energy photon decays into a pair with simultaneous absorption of photons from a strong laser field. It has been studied first in the early 1960s [27, 28] and was observed experimentally in 1997 [29]. Spin-resolved calculations were provided in [9, 10]. For very recent theoretical investigations of the process, we refer the reader to [30, 31].

A comparative analysis between the photo-production rates $\mathcal{R}^{(0)}$ and $\mathcal{R}^{(\frac{1}{2})}$ of a pair of spin-zero and spin-$\frac{1}{2}$ particles, respectively, also provides evidence for the role played by the internal spin degrees of freedom [32]. The respective calculations of the production rates can be based either on the quantum mechanical transition amplitude between Volkov states or, alternatively, on a consideration of the unitarity condition of the dispersion S-matrix. This latter procedure provides the optical theorem, whose application to the case under consideration allows us to establish a relation between the photo-production rate and the imaginary part of the vacuum polarization tensor $\Pi^{(0,\frac{1}{2})}_{\mu\nu}$. With the incident γ-photon energy ω_γ, it reads

$$\mathcal{R}^{(0,\frac{1}{2})}=\sum_{\ell=1,2}\frac{\epsilon^{\mu}_{\ell}\epsilon^{*\nu}_{\ell}}{2\omega_\gamma}\mathrm{Im}\ \Pi^{(0,\frac{1}{2})}_{\mu\nu}=-\frac{1}{\omega_\gamma}\mathrm{Im}\left[\mathrm{Tr}\ \Pi^{(0,\frac{1}{2})}_{\mu\nu}\right]. \tag{5}$$

It is worth noting at this point that the rates $\mathcal{R}^{(0,\frac{1}{2})}$ refer to the average over the polarization vectors ϵ^{μ}_{ℓ} of the incident high-energy photon beam whose intensity is much smaller than the intensity associated with the field of the strong wave. Observe, in addition, that the second step in the expression above comes out as a result of using the completeness relation $g^{\mu\nu}=-\sum_\ell \epsilon^{\mu}_{\ell}\epsilon^{*\nu}_{\ell}$. Because of this fact, the photo-production rate of a pair of scalar (or fermionic) particles is finally expressed in terms of the trace (Tr) of the corresponding polarization tensor. The latter quantity can be obtained from the general expression of $\Pi_{\mu\nu}$ in the field of a circularly polarized monochromatic plane-wave which was derived in [33]. As a consequence,

$$\mathcal{R}^{(0)}(\xi,\lambda)=-\frac{\alpha m^2}{2\pi\omega_\gamma}\mathrm{Im}\int_0^1 dv\int_0^\infty\frac{d\rho}{\rho}e^{-\frac{2i\eta}{1-v^2}}\left\{1-e^{iy}+2\xi^2\sin^2(\rho)\right\}, \tag{6}$$

$$\mathcal{R}^{(\frac{1}{2})}(\xi,\lambda)=\frac{\alpha m^2}{\pi\omega_\gamma}\mathrm{Im}\int_0^1 dv\int_0^\infty\frac{d\rho}{\rho}e^{-\frac{2i\eta}{1-v^2}}\left\{1-e^{iy}-2\xi^2\frac{1+v^2}{1-v^2}\sin^2(\rho)\right\}, \tag{7}$$

where α is the finestructure constant and the abbreviations $\eta=(\rho/\lambda)\left[1+\xi^2\left(1-\sin^2(\rho)/\rho^2\right)\right]$ and $y=\frac{2\xi^2}{1-v^2}\frac{\rho}{\lambda}\left(1-\sin^2(\rho)/\rho^2\right)$ have been introduced, with $\lambda=k_\gamma\cdot k/(2m^2)$. Here, $k_\gamma=(\omega_\gamma,\vec{k}_\gamma)$ and $k=(\omega,\vec{k})$ are the four-momenta of the γ-photon and the laser photons, respectively. The remaining parameter of the theory is

$$\xi=\frac{e|\vec{a}|}{m}, \tag{8}$$

with the amplitude $|\vec{a}|$ of the laser vector potential which is taken in radiation gauge.

Unfortunately, an analytical evaluation of $\mathcal{R}^{(0,\frac{1}{2})}$ is quite difficult to perform in full glory. However, the dependence on the parameters λ and ξ allows us to obtain closed-form expressions for the rates in various asymptotic regimes of interest. Let us start by considering the case in which $\xi<1$ and $\lambda\gg 1+\xi^2$. Then, the scalar and fermionic pair production rates are given by

$$\begin{aligned}\mathcal{R}^{(0)}&\simeq\frac{\alpha m^2\xi^2}{4\omega_\gamma}\left[\left(1+\frac{n_*}{1+\xi^2}\right)\sqrt{1-n_*}-\frac{n_*}{1+\xi^2}\left(1-\frac{1}{2}n_*\right)\ln\left(\frac{1+\sqrt{1-n_*}}{1-\sqrt{1-n_*}}\right)\right]\theta(1-n_*),\\ \mathcal{R}^{(\frac{1}{2})}&\simeq\frac{\alpha m^2\xi^2}{2\omega_\gamma}\left[\left(1+\frac{n_*}{1+\xi^2}-\frac{n_*^2}{2(1+\xi^2)}\right)\ln\left(\frac{1+\sqrt{1-n_*}}{1-\sqrt{1-n_*}}\right)-\left(1+\frac{n_*}{1+\xi^2}\right)\sqrt{1-n_*}\right]\theta(1-n_*).\end{aligned} \tag{9}$$

Some comments are in order. First of all, $n_*=2m^2(1+\xi^2)/k_\gamma\cdot k$ denotes the minimal number of laser photons needed to overcome the energy threshold in the production process. Note that the presence of the unit step function $\theta(1-n_*)$ indicates that the reaction takes place with the absorption of one photon from the strong wave. We emphasize that the previous expressions apply to a wider region with respect to the parameter ξ and include, as a particular case, the first Born approximation ($\xi\ll 1$). The explicit expressions associated with this limit can be read off from (9) by replacing $n_*\to 1/\lambda$ and setting $\xi=0$ within the squared brackets.

The behaviours of $\mathcal{R}^{(0)}$ and $\mathcal{R}^{(\frac{1}{2})}$ under the asymptotic condition studied so far are displayed in Fig. 3 as a function of λ. Several curves are shown corresponding to different values of ξ. We point out that, for a common value of ξ, the rate based on the Dirac theory is generally found to be larger than the corresponding results for Klein-Gordon particles, i.e., $\mathcal{R}^{(\frac{1}{2})}>\mathcal{R}^{(0)}$.

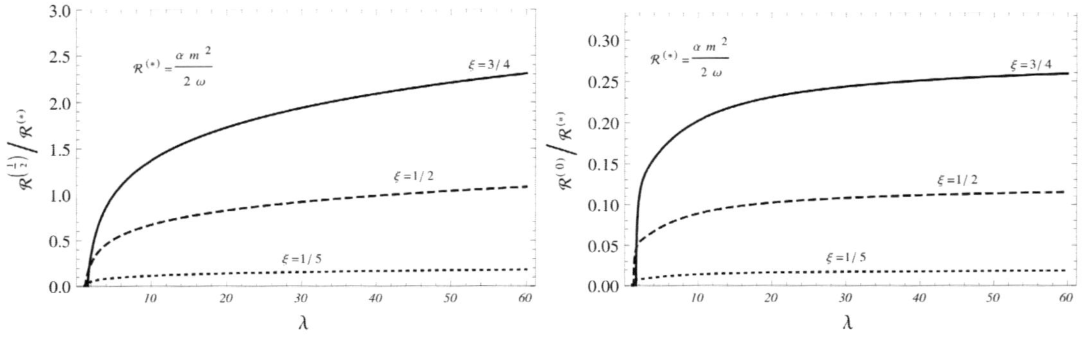

Figure 3. Dependence of the photo-production rates of a pair of fermionic particles (left panel) and scalar particles (right panel) on the parameter λ for different values of $\xi < 1$.

Consideration of the asymptotic regime $\xi \gg 1$ is also of interest. This corresponds to the situation in which a large number of photons of the strong wave are absorbed (multiphoton reaction). In this case, the fermionic and scalar rates are parametrized by a unique quantity $\zeta = \lambda\xi$ which is independent of the frequency of the strong wave (quasistatic limit):

$$\mathcal{R}^{(0)} \simeq \frac{\alpha m^2}{12\sqrt{3}\pi\omega_\gamma}\int_1^\infty \frac{du}{u\sqrt{u(u-1)}} K_{\frac{2}{3}}\left(\frac{4u}{3\zeta}\right)(4u-1), \quad (10)$$

$$\mathcal{R}^{(\frac{1}{2})} \simeq \frac{\alpha m^2}{6\sqrt{3}\pi\omega_\gamma}\int_1^\infty \frac{du}{u\sqrt{u(u-1)}} K_{\frac{2}{3}}\left(\frac{4u}{3\zeta}\right)(8u+1) \quad (11)$$

where $K_{\frac{2}{3}}(x)$ denotes the Macdonald function. Using the behaviour of this function for small and large arguments, respectively, one obtains

$$\mathcal{R}^{(0)} \simeq \frac{\alpha m^2 \zeta^{\frac{2}{3}}}{6\sqrt{3}\pi\omega_\gamma}\left(\frac{3}{2}\right)^{\frac{2}{3}}\frac{\Gamma^2\left(\frac{2}{3}\right)}{\Gamma\left(\frac{13}{6}\right)}, \qquad \mathcal{R}^{(\frac{1}{2})} \simeq \frac{5\alpha m^2 \zeta^{\frac{2}{3}}}{6\sqrt{3}\pi\omega_\gamma}\left(\frac{3}{2}\right)^{\frac{2}{3}}\frac{\Gamma^2\left(\frac{2}{3}\right)}{\Gamma\left(\frac{13}{6}\right)} \quad \text{for} \quad \zeta \gg 1 \quad (12)$$

$$\mathcal{R}^{(0)} \simeq \frac{\alpha m^2 \zeta}{16\omega_\gamma}\sqrt{\frac{3}{2}}e^{-\frac{4}{3\zeta}}, \qquad \mathcal{R}^{(\frac{1}{2})} \simeq \frac{3\alpha m^2 \zeta}{8\omega_\gamma}\sqrt{\frac{3}{2}}e^{-\frac{4}{3\zeta}} \quad \text{for} \quad \zeta \ll 1 \quad (13)$$

with $\Gamma(x)$ denoting the Gamma function. It follows from the expression above that $\mathcal{R}^{(\frac{1}{2})} = 5\mathcal{R}^{(0)}$ for $\zeta \gg 1$, whereas $\mathcal{R}^{(\frac{1}{2})} = 6\mathcal{R}^{(0)}$ for $\zeta \ll 1$. We point out that the production rates $\mathcal{R}^{(\frac{1}{2})}$ for fermion pairs may also be found in the books [34].

The general conclusion associated with these results is that the spin degree of freedom is clearly manifest in the total pair production rates. In particular, the lack of spin in the case of scalar particles suppresses the pair creation rate as compared to the predictions from Dirac theory. Thereby, the spin dependence is non-trivial because it differs from a simple factor of four which would account for the number of possible configurations of two spin-$\frac{1}{2}$ particles.

4. Spin-resolved multiphoton pair production on a nucleus

Electron-positron pairs can also be produced in the combined fields of a strong laser beam and an atomic nucleus [1, 2] (for some very recent papers see [35, 36]). When the nucleus is counterpropagating the laser beam with a high Lorentz factor γ, then the relativistic Doppler shift of the laser photon energy and laser field strength can be exploited. The process may be considered as a generalization of the well-known Bethe-Heitler effect to the multiphoton case.

Recently, spin effects in the multiphoton Bethe-Heitler process have been examined [37, 38, 39]. These studies were motivated by the question whether high degrees of spin polarization can be obtained when particle pairs are produced via the absorption of many photons. So far, the asymptotic regimes with $\xi \ll 1$ and $\xi \gg 1$ have been considered [see Eq. (8)]. Here we study the intermediate regime characterized by a laser intensity parameter of order unity ($\xi \sim 1$).

Within the strong-field approximation, the amplitude for multiphoton Bethe-Heitler pair production may be written as

$$\mathcal{M}_{p_+s_+,p_-s_-} = -ie\int d^4x\,[\Psi^{(-)}_{p_-,s_-}]^\dagger V(r)\Psi^{(+)}_{p_+,s_+}. \tag{14}$$

The electron and positron are described by Dirac-Volkov states $\Psi^{(\pm)}_{p_\pm,s_\pm}$ which are labeled by the free four-momenta $p_\pm = (E_\pm, \vec{p}_\pm)$ and spin quantum numbers $s_\pm$ outside the field. The laser field is assumed to be a circularly polarized plane wave. Below we shall consider the spin basis of helicity states, describing right- and left-handed particles, respectively. $V(r) = Ze/r$ denotes the Coulomb potential of the nucleus of charge number Z in its rest frame; it is taken into account within the lowest order of perturbation theory.

From the amplitude (14) one can derive the pair production rate in the usual manner. First, by expanding the periodic parts in the integrand of Eq. (14) into a Fourier series, one can decompose the amplitude into the contributions stemming from absorption of a certain photon number n, according to $\mathcal{M} = \sum_n \mathcal{M}^{(n)}$. By taking the absolute square of the amplitude, integrating over the degrees of freedom of the produced electron, and summing over n one obtains the double-differential rate $d^2\mathcal{R}^{\mathrm{R,L}}/dE_+d\vartheta_+$ for detecting a positron with a certain emission angle ϑ_+ and energy E_+ and either positive $s_+ = s_{\mathrm{R}}$ or negative $s_+ = s_{\mathrm{L}}$ helicity [38]. The asymmetry ratio between the rates for measuring right-handed and left-handed positrons,

$$P_\parallel(E_+, \vartheta_+) = \frac{d^2\mathcal{R}_{\mathrm{R}} - d^2\mathcal{R}_{\mathrm{L}}}{d^2\mathcal{R}_{\mathrm{R}} + d^2\mathcal{R}_{\mathrm{L}}}, \tag{15}$$

determines the degree of longitudinal polarization. We point out that, within the present approach, the degree of longitudinal polarization for the electron will exhibit the same energy and angle dependence.

Figure 4 shows the degree of longitudinal polarization of the positrons produced via the multiphoton Bethe-Heitler effect by an intense laser field with $\xi = 0.5$. The results refer to the nuclear rest frame. In this frame, the laser photons have an energy of $\omega_{\mathrm{nuc}} = 0.4\,\mathrm{MeV}$ such

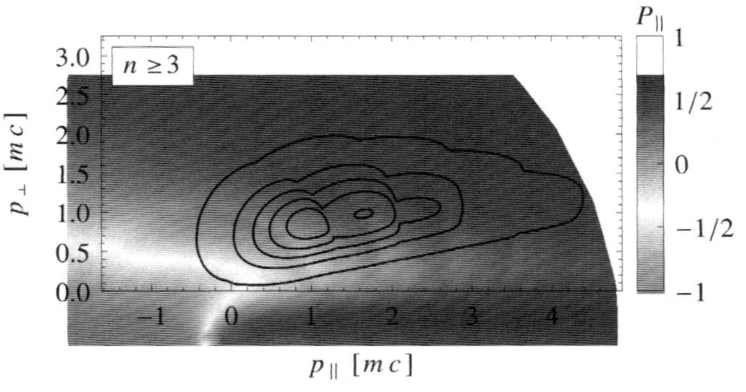

Figure 4. Color-coded degree of longitudinal polarization [see Eq. (15)] as a function of the transverse and parallel momentum components of the positron with respect to the laser beam (including positron energies up to $5mc^2$). Regions of negative (positive) degrees of polarization are encoded red (blue). Contours of the double-differential production rate are plotted as black lines.

Journal of Physics: Conference Series **414** (2013) 012012 doi:10.1088/1742-6596/414/1/012012

that electron-positron pairs can be produced by absorption of three or more photons. As can be seen from the figure, most of the momentum configurations support high positive degrees of polarization. Negative degrees of polarization only appear for particles that are emitted roughly antiparallel to the laser beam with relatively low energies. However, for practical purposes it is relevant to correlate the degree of polarization obtainable for a certain positron momentum vector with the corresponding rate of production.

Therefore, in addition to the color-coded degree of polarization, Fig. 4 also contains contour lines indicating the absolute value of the double-differential production rate. They are plotted at every fifth part of the maximum rate. Note that the contributions of the various photon orders are clearly visible in the contour lines. For the present parameters, the dominating photon orders are $n = 4, \ldots, 8$ whose relative contributions to the total production rate are 23%, 34%, 23%, 12% and 5%, respectively, whereas the relative contribution of the lowest energetically allowed order $n = 3$ is very small ($\sim 3.5 \times 10^{-4}$) due to a channel-closing effect. The majority of particles is produced inside the two innermost contour lines, where only moderate degrees of polarization are achieved, as the color-coding illustrates.

The parameter regime just discussed in the nuclear rest frame can be realized, for instance, by using a high Lorentz factor of the nucleus of $\gamma = 2000$ in the laboratory frame and an extreme ultraviolet laser photon energy of 100 eV. With an invariant intensity parameter of $\xi = 0.5$ this corresponds to a lab-frame laser intensity of $I \approx 4.4 \times 10^{21}$ W/cm^2.

Figure 5 shows the energy differential production rate in the laboratory frame in scaled units (black solid line). Note that, in contrast to the results from the nuclear rest frame, the production rates stemming from different photon orders cannot be resolved, which is a result of the Lorentz transformation. The maximum of the spin-summed production rates occurs at a positron energy of around $0.6\gamma mc^2$. As the red dashed line in Fig. 5 shows, this is where the degree of longitudinal polarization is just the lowest. For those energies where nonnegligible rates are attained, the average degree of longitudinal polarization [40] is bounded to values between about ± 0.2.

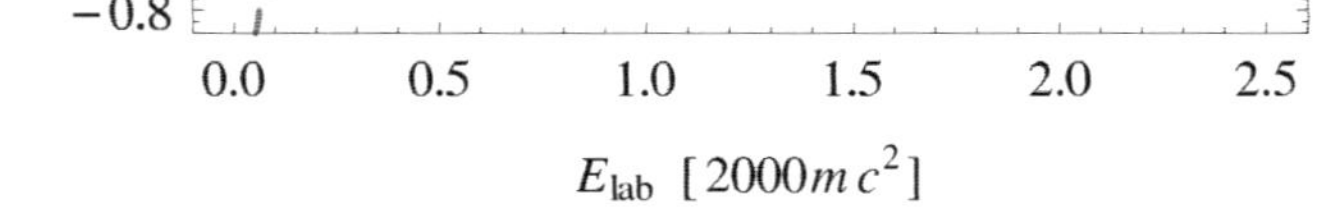

Figure 5. Scaled production rate, differential with respect to the positron energy (black solid line), together with the corresponding degree of longitudinal polarization (red dashed line) in the laboratory frame. The parameters of the laser-nucleus collision are $\omega_{\rm lab} = 100$ eV, $\xi = 0.5$ (i.e. $I \approx 4.4 \times 10^{21}$ W/cm^2) and $\gamma = 2000$. At the maximum of the differential rate, the degree of polarization is very small.

In conclusion, only relatively low degrees of longitudinal polarization can be obtained via the multiphoton Bethe-Heitler process in the intermediate intensity regime. This is in agreement with the previous results for $\xi \gg 1$ and $\xi \ll 1$ in [37, 38]. It is also in accordance with the result of [39] where it was shown that, in the quasistatic regime, the spin polarization vector of the positron is directed, to leading order, along the transverse momentum component. Hence, although the pairs are produced by absorption of several laser photons which all carry the same helicity, the transfer of helicity is less efficient than in the traditional Bethe-Heitler process involving one photon of high energy [41].

5. Conclusion

We studied the influence of the electron spin on three different fundamental quantum processes which can occur in the presence of strong laser fields. It was shown that, as one may have expected, the electron spin plays an important role in pair production processes such as the multiphoton Breit-Wheeler and Bethe-Heitler reactions. Besides, under certain conditions, the electron spin can also have a characteristic impact on Kapitza-Dirac scattering from a standing wave of light where it manifests itself in the Rabi oscillation period and the diffraction pattern.

We point out that with the present contribution our previous articles [25, 32, 37, 38, 39] have been extended by examining spin effects in electron-positron pair creation by the multiphoton Bethe-Heitler effect in an intermediate coupling regime ($\xi \sim 1$), which has not been considered before. Besides, we explicitly showed the Rabi oscillation dynamics of a Klein-Gordon particle scattering from a standing laser wave and the photo-production rates of Dirac versus Klein-Gordon particles at $\xi < 1$.

As a general result we note that, in all three cases under consideration, the process probabilities for Dirac particles were found to be higher than for Klein-Gordon particles.

Acknowledgments

S. V.-C. gratefully acknowledges support from the Alexander von Humboldt Foundation.

References

[1] Ehlotzky F, Krajewska K and Kamiński J Z 2009 *Rep. Prog. Phys.* **72** 046401
[2] Di Piazza A, Müller C, Hatsagortsyan K Z and Keitel C H 2012 *Rev. Mod. Phys.* **84** 1177
[3] A very readable consideration of free-electron spin dynamics in a strong laser field may be found in: Walser M W, Urbach D J, Hatsagortsyan K Z, Hu S X and Keitel C H 2002 *Phys. Rev.* A **65** 043410
[4] Panek P, Kamiński J Z and Ehlotzky F 2002 *Phys. Rev.* A **65** 022712
[5] Ivanov D Y, Kotkin G L and Serbo V G 2004 *Eur. Phys. J.* C **36** 127
[6] Boca M, Dinu V and Florescu V 2012 *Nucl. Instrum. Meth. Phys. Res.* B **279** 12
[7] Szymanowski C, Taïeb R and Maquet A 1998 *Laser Phys.* **8** 102
[8] Panek P, Kamiński J Z and Ehlotzky F 2002 *Phys. Rev.* A **65** 033408
[9] Tsai Y S 1993 *Phys. Rev.* D **48** 96
[10] Ivanov D Y, Kotkin G L and Serbo V G 2005 *Eur. Phys. J.* C **40** 27
[11] Hu S X and Keitel C H 1999 *Phys. Rev. Lett.* **83** 4709
[12] Faisal F H M and Bhattacharyya S 2004 *Phys. Rev. Lett.* **93** 053002
[13] Walser M and Keitel C 2001 *Opt. Commun.* **199** 447
[14] Popov V S 1972 *Sov. Phys. JETP* **34** 709
[15] Krekora P, Su Q and Grobe R 2004 *Phys. Rev. Lett.* **92** 040406
[16] Wagner R E, Ware M R, Su Q and Grobe R 2010 *Phys. Rev.* A **81** 024101
[17] Batelaan H 2007 *Rev. Mod. Phys.* **79** 929
[18] Kapitza P L and Dirac P A M 1933 *Proc. Cambridge Philos. Soc.* **29** 297
[19] Fedorov M V 1967 *Sov. Phys. JETP* **25** 952
[20] Freimund D L, Aflatooni K and Batelaan H 2001 *Nature* (London) **413** 142
[21] Efremov M A and Fedorov M V 2000 *J. Phys.* B **33** 4535
[22] Smirnova O, Freimund D L, Batelaan H and Ivanov M 2004 *Phys. Rev. Lett.* **92** 223601
[23] Rosenberg L 2004 *Phys. Rev.* A **70** 023401
[24] Relativistic treatments of the Kapitza-Dirac effect based on the Klein-Gordon equation were given in: Haroutunian V M and Avetissian H K 1975 *Phys. Lett.* **51A** 320; Fedorov M V and McIver J K 1980 *Opt. Commun.* **32** 179
[25] Ahrens S, Bauke H, Keitel C H and Müller C 2012 *Phys. Rev. Lett.* **109** 043601
[26] The amplitudes $c_3^\zeta(T)$ with $\zeta \in \{-\uparrow, -\downarrow\}$ corresponding to negative electron energies are practically zero.
[27] Reiss H R 1962 *J. Math. Phys.* **3** 59
[28] Narozhny N B, Nikishov A I and Ritus V I 1965 *Sov. Phys. JETP* **20** 622
[29] Burke D *et al.* 1997 *Phys. Rev. Lett.* **79** 1626
[30] Titov A I, Takabe H, Kämpfer B and Hosaka A 2012 *Phys. Rev. Lett.* **108** 240406
[31] Krajewska K and Kamiński J Z 2012 *Phys. Rev.* A **86** 052104
[32] Villalba-Chávez S and Müller C 2012 *Phys. Lett.* B, in print (arXiv:1208.3595)

[33] Baier V N, Milstein A I and Strakhovenko V M 1976 *Sov. Phys. JETP* **42**, 961
[34] Baier V N, Katkov V M and Strakhovenko V M 1998 *Electromagnetic processes at high energies in oriented single crystals* (Singapore: World Scientific);
Greiner W and Reinhardt J 2009 *Quantum Electrodynamics* (Berlin, Heidelberg: Springer)
[35] Krajewska K and Kamiński J Z 2012 *Phys. Rev.* A **85** 043404
[36] Di Piazza A and Milstein A I 2012 *Phys. Lett.* B **717** 224
[37] Müller T O and Müller C 2011 *Phys. Lett.* B **696** 201
[38] Müller T O and Müller C 2012 *Phys. Rev.* A **86** 022109
[39] Di Piazza A, Milstein A I and Müller C 2010 *Phys. Rev.* A **82** 062110
[40] The longitudinal degree of polarization shown in Fig. 5 involves an average over the positron emission angle which varies within very narrow limits in the laboratory frame due to the large value of the Lorentz factor.
[41] Olsen H and Maximon L C 1959 *Phys. Rev.* **114** 887

Journal of Physics: Conference Series **414** (2013) 012013 doi:10.1088/1742-6596/414/1/012013

New mechanism for the generation of electron-positron pairs in Laser-Matter Interaction: Resonantly Enhanced Pair Production in a molecular system

François Fillion-Gourdeau[1,2,4], **Emmanuel Lorin**[2,1], **André D. Bandrauk**[3,1]

[1] Centre de Recherches Mathématiques, Université de Montréal, Montréal, Canada, H3T 1J4
[2] School of Mathematics and Statistics, Carleton University, Ottawa, Canada, K1S 5B6
[3] Laboratoire de chimie théorique, Faculté des Sciences, Université de Sherbrooke, Sherbrooke, Canada, J1K 2R1
[4] Fields Institute, University of Toronto, Toronto, Canada, M5T 3J1

E-mail: filliong@CRM.UMontreal.ca

Abstract. A new mechanism for pair production from the interaction of a laser with two nuclei is presented. The latter takes advantage of the Stark effect in diatomic molecules and the presence of molecular resonances in the negative and positive energy continua. Both move in the complex energy plane as the interatomic distance and the electric field strength are varied. We demonstrate that there is an enhancement of pair production at the crossing of these resonances. This mechanism is studied in a very simple one-dimensional model where the nuclei are modeled by delta function potential wells and the laser by a constant electric field. The position of resonances is evaluated by using the Weyl-Titchmarch-Kodaira theory, which allows to treat singular boundary value problems and to compute the spectral density. The rate of producing pairs is also computed. It is shown that this process yields a positron production rate which is approximately an order of magnitude higher than in the single nucleus case and a few orders of magnitudes higher than Schwinger's tunneling result in a static field.

1. Introduction

Pair production from variations of Schwinger's mechanism is presently a very active topic of research because it is now possible to reach laser intensities exceeding $I \sim 10^{23}$ W/cm^2 [1]. Although the corresponding field strength ($E = (2I/c)^{1/2}$) is still far from the "critical" Schwinger field $E_S \sim 10^{18}$ V/m, at which spontaneous emission of pairs from a static electric field becomes important, it may still be possible to observe electron-positron production by using other electric field configurations. For instance, different mechanisms have been studied such as counterpropagating lasers [2, 3, 4, 5], counterpropagating lasers with space dependence [6], the interaction of laser field with heavy nuclei [7, 8, 9, 10] and the combination of rapidly and slowly varying fields [11, 12]. The effect of the temporal laser pulse shape has also been investigated [13, 14, 15]. Most of these ideas take advantage of some of the following effects:

- Time-dependence of the laser field, which effectively shift energy levels by $\hbar\omega$ (where ω is the laser frequency).

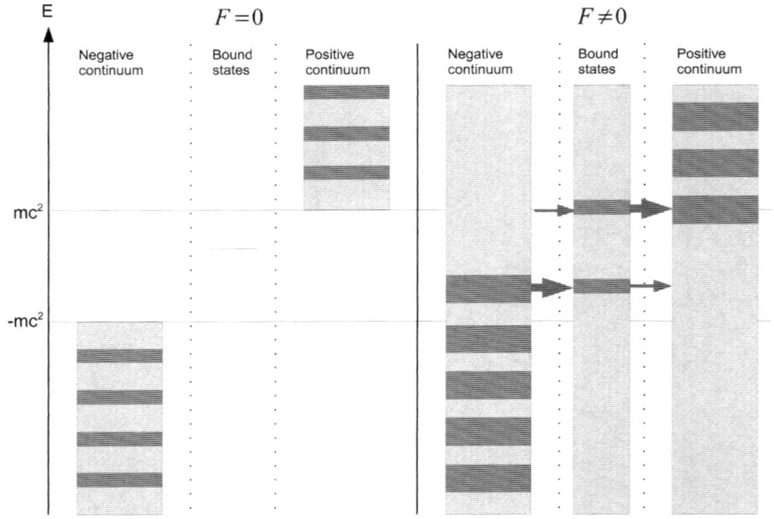

Figure 1: Description of the REPP mechanism [18]. The dark gray regions are the position of resonances (high density of states) while the light gray regions correspond to accessible energies which have a lower density of state.

- The presence of a nucleus, which have a ground state at lower energy level than in vacuum.
- Volume effects: in a laser field, pairs are produced in a bulk. Thus the small production rate (per unit volume) can be compensated by a large enough volume.

The mechanism we are proposing borrows ideas from a well-known mechanism in molecular high intensity laser physics, namely Charge Resonance Enhanced Ionization (CREI), which describes the non-perturbative ionization of diatomic molecules [16, 17]. Our main idea is to use the interaction of a laser field with a fully stripped diatomic molecule (two-center system) [18]. In this case, it is well-known that for large interatomic distance R and in the presence of an electric field of strength F, the ground state and the first excited states are shifted in the complex energy plane: the ground and first excited states are Stark shifted by $\Delta E \sim \mp RF/2$, respectively, and they gain an imaginary part because in the field, they become unstable resonances by decaying into continuum energy states. More interesting is the fact that in such system, there are also resonances in the negative energy sea which are related to the backscattering of negative energy states on the molecular potential. These resonances are Stark shifted to higher energy by the electric field [19]. When the ground state resonance crosses one of the resonances coming from the negative energy sea, it is possible to have a transition between the negative and positive energy states resulting in an enhancement of pair production. We call this the Resonantly Enhanced Pair Production (REPP) mechanism [18] and it is depicted in Fig. 1.

In this proceeding, the REPP is studied in a very simple one-dimensional model: the two nuclei are modeled by delta function potential wells while the laser field is treated in the adiabatic regime. In this limit, we can consider a static electric field corresponding to the extrema of the field oscillations of an intense laser pulse [17, 20]. Although this model is very simple, it contains the main physical features allowing to investigate the REPP mechanism. Other 1-D models have been considered in the literature [15, 21, 14, 22].

This article is organized as follows. In Section 2, we present the model. In Section 3, we evaluate the position of resonances using the Weyl-Titchmarsh-Kodaira (WTK) theory. The pair production rate is computed in Section 4 and numerical results are in Section 5. We conclude in Section 6. Also, units in which $\hbar = c = m = 1$ (where m is the electron mass) and $e = \sqrt{\alpha}$ ($\alpha = e^2/\hbar c$ is the fine structure constant) are utilized in numerical calculations, in comparison to atomic units where $e = \hbar = m = 1$ and $c = \alpha^{-1}$.

2. Simple diatomic model

The model considered in this work is a time-independent one-dimensional model, which allows us to compute analytical solutions. Our analysis is based on the Dirac equation as it is the basic ingredient to evaluate pair production from the "vacuum". In the case considered, the Dirac equation is given by

$$E\psi(x) = \left[-ic\alpha\partial_x + \beta mc^2 + V(x) + A_0(x)\right]\psi(x), \quad x \in \mathbb{R}, \tag{1}$$

where in 1-D, ψ is a bi-spinor, V is the scalar potential of the nuclei, $A_0(x)$ is the scalar potential of the laser field, m is the electron mass and E is the energy. In the first part of this work (in Section 3), where the position of resonances are evaluated in the WTK theory, it is convenient to work in a representation where the Dirac matrices are given by

$$\text{Representation 1: } \alpha = -\sigma_y \text{ and } \beta = -\sigma_z. \tag{2}$$

This is chosen to recover the WTK theory for systems of equations [23, 24]. In the second part of this work (in Section 4), where pair production is evaluated, it is actually more suitable to use

$$\text{Representation 2: } \alpha = \sigma_z \text{ and } \beta = \sigma_x, \tag{3}$$

as it yields simpler equations. Both representations obey the appropriate anticommutation relations.

The nuclei are modeled by the two center potential

$$V(x) = -g\left[\delta(x-R) + \delta(x+R)\right], \quad x \in \mathbb{R}, \tag{4}$$

where $2R$ is the interatomic distance and g is the strength of the Dirac delta potential wells (physically, this parameter is very similar to the electric charge of the nucleus). As in non-relativistic quantum mechanics, this type of potential can be characterized by boundary conditions at the potential well position. However, the mathematical derivation is more intricate due to the singularities because they involve the product of distributions as the wave function has a jump discontinuity ($\psi(x) \sim H(x+R) + H(x-R)$, where H is the Heaviside function). Thus, the Dirac equation has terms of the form $\delta(x \pm R)H(x \pm R)$ which are not well-defined mathematically. Nevertheless, there exists a mathematical framework which includes products into the definition of distributions: the Colombeau theory of generalized functions [25, 26]. Using this theory, along with conjugation invariance and properties of self-adjoint extensions, it is possible to single out one boundary condition. In representation 1, it is given by [19]

$$\psi_1(\pm R^+) = \left[1 + \frac{g^2}{4c^2}\right]^{-1}\left[\left(1 - \frac{g^2}{4c^2}\right)\psi_1(\pm R^-) - \frac{g}{c}\psi_2(\pm R^-)\right], \tag{5}$$

$$\psi_2(\pm R^+) = \left[1 + \frac{g^2}{4c^2}\right]^{-1}\left[\left(1 - \frac{g^2}{4c^2}\right)\psi_2(\pm R^-) + \frac{g}{c}\psi_1(\pm R^-)\right], \tag{6}$$

where we defined $\psi(\pm R^+) = \lim_{\epsilon\to 0}\psi(\pm R + \epsilon)$ and $\psi(\pm R^-) = \lim_{\epsilon\to 0}\psi(\pm R - \epsilon)$.

3. WTK theory and the position of resonances

The WTK method was developed to provide a generalization of eigenvalue expansion to singular operator[1] and infinite domain, for 1-D second order equations [27]. This was also generalized to the 1-D Dirac equation in [23, 24]. This method is based on the WTK theorem which states that there exists a square integrable solution when the imaginary part of the energy $\mathrm{Im}(E) > 0$. Within this framework, it is also possible to evaluate the spectral density $\rho(E)$ which contain all the information on the spectrum of our operator: bound states appear as poles of ρ while continua occurs when $\rho(E) \neq 0$. Resonances, which will play a very important role in the following, are poles of ρ in the complex energy plane. As these give the most relevant information from the physical point of view, we will focus on this quantity in this work.

Within the model presented in the previous section, we consider the two center problem in a static electric field which allows to describe molecular CREI in intense laser fields [16, 17]:

$$A_0(x) = -Fx, \tag{7}$$

where F is the field strength. It can be shown in the present model that the position of poles in the complex plane is obtained [19] by solving numerically the equation

$$\begin{aligned} & \left[iG_u^+U_1^+[a,y(R)] - G_u^-U_1^-[a,y(R)]\right] \\ & \times\left[iH_v^-U_2^+[a,y(-R)] - H_v^+U_2^-[a,y(-R)]\right] \\ - & \left[iH_u^-U_2^+[a,y(-R)] - H_u^+U_2^-[a,y(-R)]\right] \\ & \times\left[iG_v^+U_1^+[a,y(R)] - G_v^-U_1^-[a,y(R)]\right] = 0, \end{aligned} \tag{8}$$

which corresponds to the spectral density denominator. Here, we have $y(x) = e^{-i\frac{\pi}{4}}\sqrt{\frac{2c}{F}}\left(\frac{E+Fx}{c}\right)$, $a := i\frac{m^2c^3}{2F} - \frac{1}{2}$ and $U_{1,2}^{\pm} := U_{1,2} \pm \tilde{U}_{1,2}$ while the other quantities are defined in Appendix A.

The numerical results for the real part of the pole position is shown in the Section 5 in Fig. 3, for $g = 0.8$ and $F = 0.2$.

4. Pair production rate

The pair production rate from the vacuum in a strong background static field of amplitude F is related to its corresponding Dirac equation. This physical quantity can be evaluated using the following considerations. First, we assume that the static electric field has a finite extent and is non-zero in the region $[-L, L]$. In this case, it is possible to define "asymptotic states" at $x = \mp\infty$: in these regions, the particles are free and there is a natural separation between the negative and positive energy solutions. This allows us to evaluate the number of pairs produced from a solution of the Dirac equation. This first assumption is required because the external field is static: in the time-dependent case, we would require the vanishing of the field at $t = \pm\infty$. Then, it has been argued that in the static case, the average number of pairs $\langle n \rangle$ is given by [28, 29, 30, 31, 32]

$$\frac{d\langle n \rangle}{dtdE} = \frac{1}{2\pi}|A(E)|^2, \tag{9}$$

where A is the coefficient of the positive energy solution propagating towards $x = +\infty$, at the right of the potential (see Fig. 2).

[1] Here, by singular, we mean that one term of the differential equation diverges on the boundary of the domain

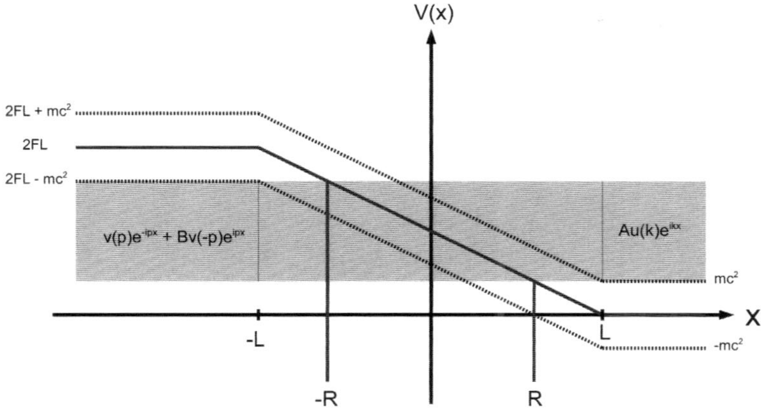

Figure 2: Simple model to study REPP with appropriate asymptotic particle amplitudes [18]. The electric field has a finite extent in space. In blue is the Klein region where it is possible to have a transition from a negative to positive energy states.

The calculation of pair production reduces therefore to a transmission-reflection problem where the incident, reflected and transmitted waves are given respectively by:

$$\psi_{\text{inc.}}(x) = v(p)e^{ip(E)x}, \tag{10}$$

$$\psi_{\text{ref.}}(x) = Bv(-p)e^{-ip(E)x}, \tag{11}$$

$$\psi_{\text{trans.}}(x) = Au(k)e^{ik(E)x}. \tag{12}$$

Here, u, v are the positive and negative energy free spinors in representation 2: their explicit expression will be given below.

The electric potential is divided in three spatial regions as (see Fig. 2)

$$A_0(x) = \begin{cases} 2FL & \text{for } x \in (-\infty, -L] \\ -F(x-L) & \text{for } x \in (-L, L) \\ 0 & \text{for } x \in [L, \infty) \end{cases} \tag{13}$$

where F is the field strength (we are working in a gauge where the vector potential is $A_x = 0$ and the electric field is related to the potential as $E_x = -\partial_x A_0(x)$) and $2L$ is the length over which the electric field is constant. Outside the interval $[-L, L]$, the electric field vanishes.

For the other scalar potential, three cases are considered:

(i) No nucleus: $V(x) = 0$

(ii) Single nucleus: $V(x) = -g\delta(x)$

(iii) Two nuclei: $V(x) = -g\delta(x - R) - g\delta(x + R)$

Then, Eqs. (5) and (6) are used to match the wave function at $x = 0, \pm R$, for case 2 and 3 respectively. The Dirac equation can be solved analytically everywhere, for $x \in \mathbb{R}$, in terms of parabolic cylinder functions.

The last missing ingredients to compute pair production are the negative and positive energy free spinors. There is a well-known procedure to compute these quantities where one seeks plane wave solutions. This yields

$$u(k) = \frac{1}{\sqrt{2E}} \begin{bmatrix} \sqrt{E + ck(E)} \\ \sqrt{E - ck(E)} \end{bmatrix}, \tag{14}$$

$$v(p) = \frac{1}{\sqrt{2(E-2FL)}} \begin{bmatrix} \sqrt{(E-2FL) + cp(E)} \\ -\sqrt{(E-2FL) - cp(E)} \end{bmatrix}, \tag{15}$$

where $k(E) = \frac{1}{c}\sqrt{E^2 - m^2c^4}$ and $p(E) = \frac{1}{c}\sqrt{(E-2FL)^2 - m^2c^4}$.

Now, imposing the continuity of the wave function at the region boundaries (at $x = \pm L$) and using Eqs. (5) and (6) to include the potential wells, we obtain the following conditions for the three cases [18]:

(i) No nucleus:

$$v(p)e^{ipL} + Bv(-p)e^{-ipL} = a_1U_a(-L) + a_2U_b(-L),$$
$$a_1U_a(L) + a_2U_b(L) = Au(k)e^{ipL}.$$

(ii) Single nucleus:

$$v(p)e^{ipL} + Bv(-p)e^{-ipL} = a_1U_a(-L) + a_2U_b(-L),$$
$$a_1U_a(0) + a_2U_b(0) = G^{-1}\left[b_1U_a(0) + b_2U_b(0)\right],$$
$$b_1U_a(L) + b_2U_b(L) = Au(k)e^{ipL}.$$

(iii) Two nuclei:

$$v(p)e^{ipL} + Bv(-p)e^{-ipL} = a_1U_a(-L) + a_2U_b(-L),$$
$$a_1U_a(-R) + a_2U_b(-R) = G^{-1}\left[b_1U_a(-R) + b_2U_b(-R)\right],$$
$$b_1U_a(R) + b_2U_b(R) = G^{-1}\left[c_1U_a(R) + c_2U_b(R)\right],$$
$$c_1U_a(L) + c_2U_b(L) = Au(k)e^{ipL}.$$

Here, we defined:

$$U_{a,1}(x) \equiv U(\gamma, y(x)), \tag{16}$$

$$U_{b,1}(x) \equiv U(-\gamma, -iy(x)), \tag{17}$$

$$U_{a,2}(x) \equiv mc\sqrt{\frac{c}{2F}}e^{i\frac{3\pi}{4}}U(\gamma+1, y(x)), \tag{18}$$

$$U_{b,2}(x) \equiv \frac{1}{mc}\sqrt{\frac{2F}{c}}e^{-i\frac{\pi}{4}}U(-\gamma-1, y(x)), \tag{19}$$

where $\gamma = i\frac{m^2c^3}{2F} - \frac{1}{2}$, which is related to the probability of producing one pair in a static field as $P_S \sim e^{i\pi(\gamma+1/2)}$ [33], and $U(\gamma, z)$ is the parabolic cylinder function. The system of equations obtained for each case can be used to solve for the integration constants $A, B, a_{1,2}, b_{1,2}, c_{1,2}$ and thus, to evaluate pair production via Eq. (9). The numerical results are presented in the following.

5. Results

In this section, numerical results are presented. First, in Fig. 3, we show the position of resonances, obtained from the WTK theory, and the particle spectrum, obtained from the quantum field theory calculation. We fix the electric field to $F = 0.2$ ($\times 10^{18}$ V/m) and the

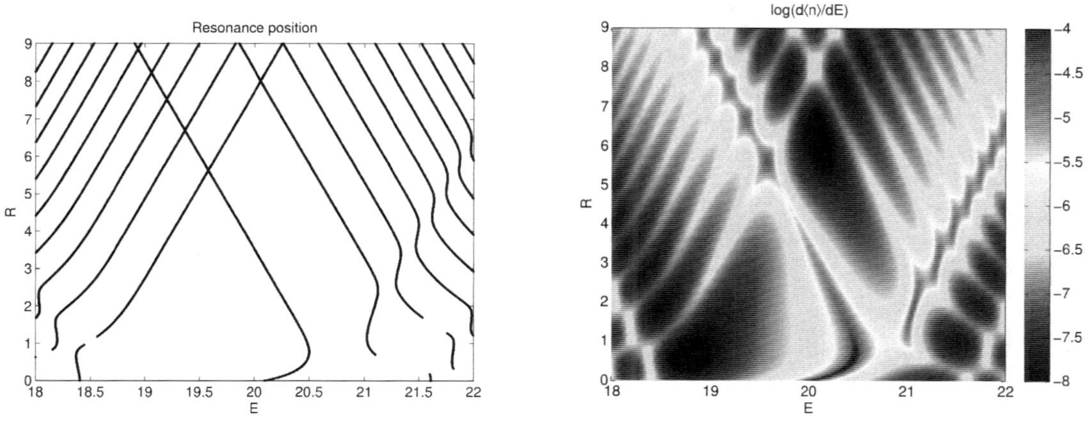

(a) (b)

Figure 3: In (a), we show the position of resonances and in (b), the particle spectrum $d\langle n\rangle/dEdt$ [18], as a function of interatomic distance. The parameters are chosen as $g = 0.8$ (corresponding to U^{91+}), $F = 0.2$ and $L = 100$.

potential wells strength to $g = 0.8$. This value of g reproduces the ground state energy of the $1s$ orbital of the U^{91+} atom. By looking at these two pictures, it is clear that the spectrum $d\langle n\rangle/dtdE$ is enhanced when the resonances are crossing (when two lines are crossing in (a)). For instance, there is a peak in the spectrum at $E \approx 19.5$ and $R \approx 5.5$, where the ground state resonance crosses with a resonance coming from the negative energy states.

This phenomenon actually enhances the number of pair produced, which is shown in Fig. 4, for $F = 0.2$ and $F = 0.09$. The results are compared to the single and no nucleus cases. We can see a peak in $\langle n\rangle$ at the interatomic distance where the resonances are crossing (for instance, at $R \approx 5.5$ for $F = 0.2$). As the field strength is reduced, the total yield is exponentially suppressed, as in Schwinger's result, but the relative enhancement is increased and can reach up to an order of magnitude (and maybe higher at lower field strength). We note also that the presence of resonances at small R also enhances pair production because in that regime, the molecular system becomes similar to an atomic system with a net nuclear charge of $Z_{\text{eff}} \approx 2Z$. As the ground state resonance is then closer to the negative energy sea, it is easier to have a transition between the positive and negative energy states, which enhances pair production. This phenomenon could be observed in a relativistic heavy ion collision experiment where the collision region is immersed in an external electric field.

6. Conclusion

In this work, the spectral density and the position of resonances were computed for a simple model by using the WTK theory. The pair production rate was also evaluated. It was shown that there is an enhancement of positron production when the resonances from the negative energy states cross with the resonances of the positive energy states. This is the mechanism that enhances pair production at large interatomic distance and is called the REPP phenomenon.

It is interesting to estimate the intensity at which REPP starts to be important in a real molecular two center system. For the sake of this discussion, we adopt the atomic units system and we consider two uranium nuclei separated by a distance of $R = 10$ a.u.. In this more realistic case, the nuclei should be described by a Coulomb-like potential which has an infinite number of bound states in the mass gap. This is different from the potential considered in this

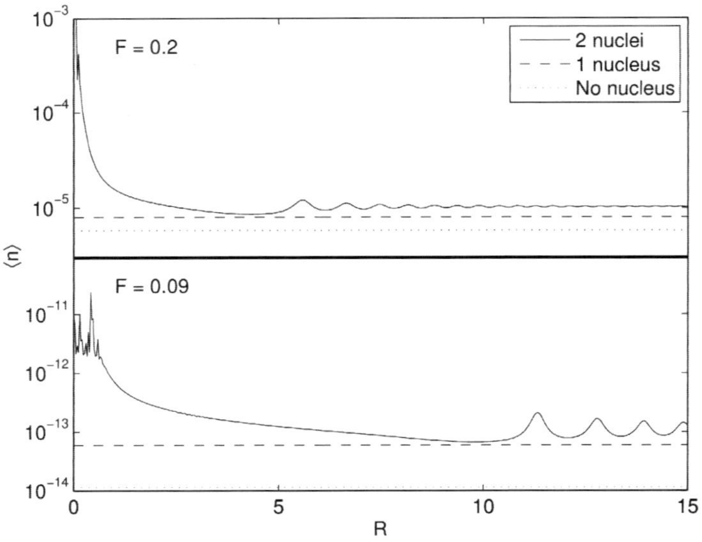

Figure 4: Total rate $d\langle n\rangle/dE$ as a function of interatomic distance, for $F = 0.2$ and $F = 0.09$ [18].

study which can only support two bound states. Nevertheless, REPP will proceed in a similar way as in our model because the behavior of the ground state resonance is qualitatively the same in both approaches. Of course, there can be a transition between the negative energy resonances and any of the bound state resonances, but the main contribution will come from the ground state, as we have shown in our analysis. Also, the magnitude of the enhancement depends on the potential shape and on the stability of the states subjected to the electric field. For this reason, we expect that peaks in the pair production rate for a Coulomb potential will be broader than the point-like interaction considered in this study. The energy of the ground and first excited states in this case is quasi-degenerate and is given by the ground state energy of the single nucleus case, that is $E_{\text{ground}} \approx E_{\text{exc}} \sim 13908.$ a.u.. One approximate condition for REPP to occur is that the ground state reaches $-mc^2 \approx -18769.$ a.u., given that its Stark shift is approximately $\Delta E \sim -RF/2$. We find that this occurs when $F \approx 6536.$ a.u. $= 3.36 \times 10^{15}$ V/m. This value of electric field strength corresponds to a laser intensity of

$$I_{\text{REPP}} \approx 1.5 \times 10^{24} \text{W/cm}^2, \tag{20}$$

as compared to atomic units of intensity, $I_0 = 3.54 \times 10^{16}\text{W/cm}^2$ and field $F_0 = 5.0 \times 10^{11}\text{V/m}$. This value of intensity is relatively close to the highest intensities reached in recent experiments [1]. It is also similar to recent estimations of the intensity at which pair production would occur in counterpropagating lasers [34] and in head-on collisions of a laser (with both high and low frequency components) with protons [10]. The present model suggests that REPP is a process that could be used to detect the production of pairs from the QED vacuum and to probe the vacuum properties in the nonperturbative relativistic regime of laser-matter interaction.

Acknowledgments
The authors would like to thank S. Selstø and S. Chelkowski for numerous discussions on REPP and other topics.

Appendix A. Definition of functions

Here and in the following , we define

$$\begin{aligned} U_1(a,y) &:= U(a,y), && \text{(A.1)}\\ U_2(a,y) &:= U(-a,-iy), && \text{(A.2)}\\ \tilde{U}_1(a,y) &:= mc\sqrt{\frac{c}{2F}}e^{i\frac{3\pi}{4}}U(a+1,y), && \text{(A.3)}\\ \tilde{U}_2(a,y) &:= \frac{1}{mc}\sqrt{\frac{2F}{c}}e^{-i\frac{\pi}{4}}U(-a-1,y), && \text{(A.4)} \end{aligned}$$

where U is a parabolic cylinder function. Then we have

$$\begin{aligned} u_1(x) &= c_1U_1[a,y(x)]+c_1\tilde{U}_1[a,y(x)]+c_2U_2[a,y(x)]+c_2\tilde{U}_2[a,y(x)], && \text{(A.5)}\\ iu_2(x) &= c_1U_1[a,y(x)]-c_1\tilde{U}_1[a,y(x)]+c_2U_2[a,y(x)]-c_2\tilde{U}_2[a,y(x)], && \text{(A.6)}\\ v_1(x) &= c_1'U_1[a,y(x)]+c_1'\tilde{U}_1[a,y(x)]+c_2'U_2[a,y(x)]+c_2'\tilde{U}_2[a,y(x)], && \text{(A.7)}\\ iv_2(x) &= c_1'U_1[a,y(x)]-c_1'\tilde{U}_1[a,y(x)]+c_2'U_2[a,y(x)]-c_2'\tilde{U}_2[a,y(x)], && \text{(A.8)} \end{aligned}$$

Letting $y_0 := e^{-i\frac{\pi}{4}}\sqrt{\frac{2c}{F}}\left(\frac{E}{c}\right)$ be the value where $x=0$, the integration constants are

$$\begin{aligned} c_1 &= -\frac{1}{2}\frac{U_2(a,y_0)-\tilde{U}_2(a,y_0)}{\tilde{U}_2(a,y_0)U_1(a,y_0)-U_2(a,y_0)\tilde{U}_1(a,y_0)}, && \text{(A.9)}\\ c_2 &= \frac{1}{2}\frac{U_1(a,y_0)-\tilde{U}_1(a,y_0)}{\tilde{U}_2(a,y_0)U_1(a,y_0)-U_2(a,y_0)\tilde{U}_1(a,y_0)}, && \text{(A.10)} \end{aligned}$$

and

$$\begin{aligned} c_1' &= \frac{i}{2}\frac{U_2(a,y_0)+\tilde{U}_2(a,y_0)}{\tilde{U}_2(a,y_0)U_1(a,y_0)-U_2(a,y_0)\tilde{U}_1(a,y_0)}, && \text{(A.11)}\\ c_2' &= -\frac{i}{2}\frac{U_1(a,y_0)+\tilde{U}_1(a,y_0)}{\tilde{U}_2(a,y_0)U_1(a,y_0)-U_2(a,y_0)\tilde{U}_1(a,y_0)}, && \text{(A.12)} \end{aligned}$$

Finally, we have the constants:

$$\begin{aligned} G_u^{\pm} &:= \left[1+\frac{g^2}{4c^2}\right]^{-1}\left[\left(1-\frac{g^2}{4c^2}\right)u_{2,1}(R)\pm\frac{g}{c}u_{1,2}(R)\right], && \text{(A.13)}\\ G_v^{\pm} &:= \left[1+\frac{g^2}{4c^2}\right]^{-1}\left[\left(1-\frac{g^2}{4c^2}\right)v_{2,1}(R)\pm\frac{g}{c}v_{1,2}(R)\right], && \text{(A.14)}\\ H_u^{\pm} &:= \left[1+\frac{g^2}{4c^2}\right]^{-1}\left[\left(1-\frac{g^2}{4c^2}\right)u_{1,2}(-R)\pm\frac{g}{c}u_{2,1}(-R)\right], && \text{(A.15)}\\ H_v^{\pm} &:= \left[1+\frac{g^2}{4c^2}\right]^{-1}\left[\left(1-\frac{g^2}{4c^2}\right)v_{1,2}(-R)\pm\frac{g}{c}v_{2,1}(-R)\right]. && \text{(A.16)} \end{aligned}$$

References

[1] Mourou G A, Tajima T and Bulanov S V 2006 *Rev. Mod. Phys.* **78**(2) 309–371
[2] Brezin E and Itzykson C 1970 *Phys. Rev. D* **2**(7) 1191–1199
[3] Marinov M S and Popov V S 1977 *Fortschritte der Physik* **25** 373–400 ISSN 1521-3979
[4] Narozhnyi N and Nikishov A 1974 *JETP* **65**(3) 427

[5] Avetissian H K, Avetissian A K, Mkrtchian G F and Sedrakian K V 2002 *Phys. Rev. E* **66**(1) 016502
[6] Ruf M, Mocken G R, Müller C, Hatsagortsyan K Z and Keitel C H 2009 *Phys. Rev. Lett.* **102**(8) 080402
[7] Müller C, Voitkiv A B and Grün N 2003 *Phys. Rev. A* **67**(6) 063407
[8] Di Piazza A, Hatsagortsyan K Z and Keitel C H 2008 *Phys. Rev. Lett.* **100**(1) 010403
[9] Müller C, Voitkiv A B and Grün N 2003 *Phys. Rev. Lett.* **91**(22) 223601
[10] Di Piazza A, Lötstedt E, Milstein A I and Keitel C H 2009 *Phys. Rev. Lett.* **103**(17) 170403
[11] Jiang M, Su W, Lv Z Q, Lu X, Li Y J, Grobe R and Su Q 2012 *Phys. Rev. A* **85**(3) 033408
[12] Dunne G V, Gies H and Schützhold R 2009 *Phys. Rev. D* **80**(11) 111301
[13] Di Piazza A 2004 *Phys. Rev. D* **70**(5) 053013
[14] Dumlu C K and Dunne G V 2010 *Phys. Rev. Lett.* **104**(25) 250402
[15] Dumlu C K and Dunne G V 2011 *Phys. Rev. D* **83**(6) 065028
[16] Zuo T and Bandrauk A D 1995 *Phys. Rev. A* **52**(4) R2511–R2514
[17] Bandrauk A D and Légaré F 2012 *Progress in Ultrafast Intense Laser Science VIII* (*Springer Series in Chemical Physics* vol 103) ed Yamanouchi K, Nisoli M, Hill W T, Castleman A W, Toennies J P, Yamanouchi K and Zinth W (Springer Berlin Heidelberg) pp 29–46 ISBN 978-3-642-28726-8
[18] Fillion-Gourdeau F, Lorin E and Bandrauk A D 2012 *Phys. Rev. Lett.* In press
[19] Fillion-Gourdeau F, Lorin E and Bandrauk A D 2012 *Journal of Physics A: Mathematical and Theoretical* **45** 215304
[20] Milosevic N, Krainov V P and Brabec T 2002 *Phys. Rev. Lett.* **89**(19) 193001
[21] Akkermans E and Dunne G V 2012 *Phys. Rev. Lett.* **108**(3) 030401
[22] Cheng T, Ware M R, Su Q and Grobe R 2009 *Phys. Rev. A* **80**(6) 062105
[23] Titchmarsh E C 1961 *The Quarterly Journal of Mathematics* **12** 227–240
[24] Titchmarsh E C 1961 *Proceedings of the London Mathematical Society* **s3-11** 169–192
[25] Colombeau J F 1983 *Journal of Mathematical Analysis and Applications* **94** 96 – 115
[26] Colombeau J F 1990 *Bulletin of the American Mathematical Society (New Series)* **23** 251 – 268
[27] Titchmarsh E C 1962 *Eigenfunction expansions associated with second-order differential equations, 2nd ed.* (Oxford University Press)
[28] Damour T 1977 *1st Marcel Grossmann Meeting on General Relativity* ed Ruffini R pp 459–482
[29] Wang R C and Wong C Y 1988 *Phys. Rev. D* **38**(1) 348–359
[30] Hansen A and Ravndal F 1981 *Physica Scripta* **23** 1036
[31] Kim S P, Lee H K and Yoon Y 2010 *Phys. Rev. D* **82**(2) 025015
[32] Chervyakov A and Kleinert H 2009 *Phys. Rev. D* **80**(6) 065010
[33] Schwinger J 1951 *Phys. Rev.* **82**(5) 664–679
[34] Bell A R and Kirk J G 2008 *Phys. Rev. Lett.* **101**(20) 200403

Tailoring isolated attosecond pulses using quantum path interferences

Carlos Hernández-García and Luis Plaja
Grupo de Investigación en Óptica Extrema (GIOE), Universidad de Salamanca, Salamanca, E37002, Spain
E-mail: carloshergar@usal.es

Abstract. By performing calculations on the generation of isolated attosecond pulses by intense and few-cycle laser fields, we demonstrate that the pulse temporal width is sensitive to the angle of emission. We show that this effect results from the interference of the short and long electronic trajectory contributions, responsible for high-order harmonic generation. In particular, we find that the shortest pulses are emitted off-axis, which are substantially shorter than those emitted on-axis, where single trajectory contributions dominate.

1. Introduction

There is an astonishing simplicity in some processes of intense laser-matter interactions. At high field amplitudes, there is a large probability for the electron to tunnel out of the atom. Afterwards, the field interaction dominates and the electrons evolve almost freely in the continuum. High-order harmonics are emitted when these fast electrons are driven back to rescatter the ion, converting their kinetic energy into high-frequency radiation, in a process known as high-order harmonic generation (HHG) [1, 2]. The simplicity of this process is mapped into the quantum mechanical wavefunction, described by a small set of semiclassical orbits, or Feynmans paths [3]. The energy conservation demands that each orbit contributes to a single harmonic, with a phase given by its action. However, as different paths may have the same rescattering energy, the final harmonic signal is the result of the interference of the degenerated quantum path contributions.

One of the most exciting perspectives of HHG by intense lasers is the possibility of synthesizing XUV pulses of sub-femtosecond duration [4, 5, 6]. An attosecond pulse train (1 as $= 10^{-18}$ s) is obtained by the selection of the higher frequency part of the harmonic spectrum which conform the plateau region. For the correct synthesis, the spectrum should approach two conditions: on one side its structure should approach that of a frequency comb, in which the harmonic intensities are similar; on the other side, the relative phase between the harmonics should be nearly constant (phase locking) [4]. Fortunately, these two conditions are approached in the typical harmonic spectra generated during the interaction of intense fields with matter.

However, the HHG mechanism itself prevents the harmonics to be emitted in the Fourier limit. Since the different electronic trajectories rescatter at different times [7], a chirp is imprinted in the emitted radiation. The emission from the short trajectories is positively chirped (the lower harmonics are emitted before the higher), while the emission from the long trajectories exhibits

Journal of Physics: Conference Series **414** (2013) 012014 doi:10.1088/1742-6596/414/1/012014

a negative chirp (the higher harmonics are emitted before the lower) [8, 9]. This chirp is referred in the literature as *attochirp*.

The first experimental evidence of an attosecond pulse train was obtained by selecting five consecutive harmonics generated in argon, obtaining 250 as pulses [10]. Almost simultaneously, isolated pulses with durations of 650 as were generated by spectrally filtering a few cut-off harmonics produced by an ultrashort laser pulse [11]. At present, after postcompression of the attosecond pulses, isolated pulses of temporal durations < 100 attoseconds have been measured experimentally [12, 13]. Recent works proposed the synthesis of X-ray waveforms in the zeptosecond regime [14], using mid-infrared laser sources to drive ultra-high order harmonics generation [15].

In this contribution we explore the relevance of macroscopic phase-matching conditions for the final phase-locking of the harmonics. In particular, we show that at certain angles of detection, where the short and the long trajectories are relevant, the interference of these contributions can lead to a final attosecond burst with smaller duration than the corresponding to a single trajectory. This effect was already observed in [16] for longer pulses, and the interference between short and long paths was identified to occur between paths from two consecutive ionization maxima. Here we explore this effect over isolated attosecond pulses driven by few-cycle pulses.

2. Methods and setup

Our computations are carried on using the extended strong field approximation (SFA+) [17] for the single atom response, combined with a discrete dipole approach for computing the harmonic propagation [18]. Our approach is based in the integral solution of the wave equation, instead of resorting to its numerical integration [19]. Every point in the target is treated as source of an elementary wave, which is then propagated to the detector. The final field at the detector is, thus, the coherent addition of these elementary contributions. We take into account propagation effects in the fundamental field such as the free charges and neutrals. More information on this method and the SFA+ model used to compute the single-atom harmonic generation may be found in [20, 16].

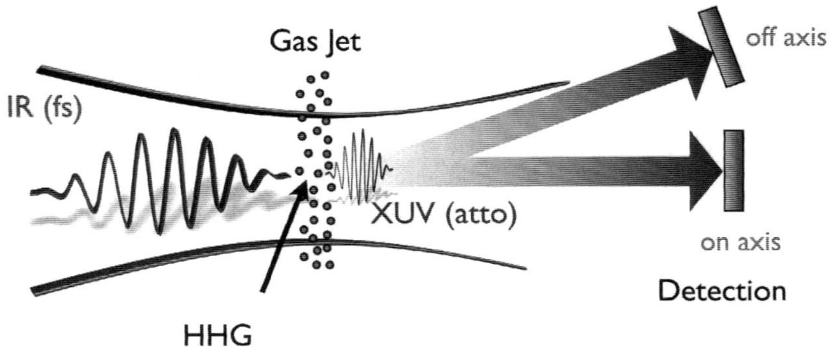

Figure 1. Scheme of the off-axis detection geometry used to demonstrate the angular chirping compensation of the attosecond bursts. A Gaussian beam propagating along the z direction is focused into an argon gas jet. The gas jet profile is constant in the x direction, while Gaussian in y and z directions, with 500 μm full width at half maximum (FWHM), and peak density 10^{18} atoms/cm^3. For the angular resolved detection, we consider a window of 0.5 mrad in the far field profile.

We have performed our computations assuming a fundamental field modeled by a $\sin^2$ envelope at 800 nm, with intensity at focus $\simeq 2.45 \times 10^{14}$ W/cm^2 and polarized along the

x direction. The beam is assumed Gaussian, propagating along the z direction. The beam waist at focus is 22.5 μm and the confocal parameter results $b = 3.98$ mm. The target is modeled as an argon gas jet flowing perpendicularly to the Gaussian beam. The gas jet profile is constant in the x direction, while Gaussian in y and z directions, with 500 μm full width at half maximum (FWHM), and peak density 10^{18} atoms/cm^3. For the angular resolved detection, we consider a window of 0.5 mrad in the far field profile, that would correspond experimentally to an annular gate. Figure 1 illustrates the scheme of this setup.

It is possible to obtain an isolated attosecond pulse using few-cycle driving pulses [6, 11], since high order harmonics are then generated in a single rescattering event. For that purpose we have performed our computations assuming a fundamental few-cycle laser pulse of 1.4 cycles (3.8 fs) FWHM. Other techniques have been applied for isolating a single attosecond pulse [5, 21, 22], whereas the extraction of the different attosecond pulses from the train has also been reported recently [23].

3. Compensating the attochirp of isolated pulses

We plot in Fig. 2 the he attosecond pulses obtained with our numerical simulation after Fourier transforming the detected harmonic spectra. The low energy part of the harmonic spectra has been filtered with an Al plate 100 μm width. We have considered two different situations: target placed before the laser focus (left column) and after the focus (right column). The first row in Fig. 2 –plots (a) and (b)– corresponds to the attosecond pulses detected on-axis, while the second row shows the off-axis detection –plots (c) and (d)–, where the spatial window was centered at 3.3 mrad. First, we observe that the CEP choice and and the parameters of the propagation led to an isolated attosecond pulse when the gas jet is placed before the focus, whereas a two-pulse train structure after the focus. This behavior can be modified by changing the CEP of the incident laser pulse.

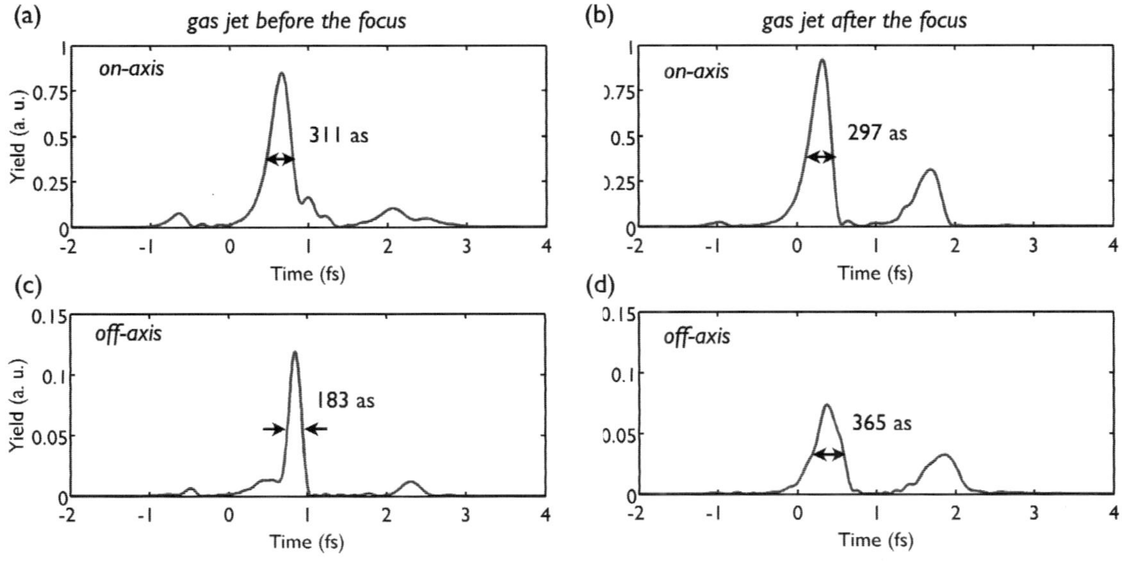

Figure 2. Temporal distribution of the attosecond pulses when the high-order harmonics are detected in a window (a) on-axis and (b) centered at 3.3 mrad, when the gas jet is placed 1 mm before the focus position. Plots (c) and (d) represent the same situation as **c** and **d**, but when the gas jet is placed 1 mm after the focus position. The driving laser pulse is 1.4 cycles (3.8 fs) FWHM $\sin^2$, 800 nm, with intensity at focus $\simeq 2.45 \times 10^{14}$ W/cm^2.

More interesting is the comparison between the attosecond pulse width at different detection angles. From plots (a) and (c) we can observe how the attosecond pulse width has been reduced from 311 as on-axis to 183 as off-axis. This reduction has a physical explanation in the interference between the radiation bursts associated with the rescattering of short and long paths, which is more acute off-axis, where both trajectories have comparable weights, rather than on-axis, where short trajectories dominate. For the same reason, this effect is more acute when the gas jet is placed after the focus. In that situation, short trajectories are phase-matched on-axis, whereas as we increase the detection angle, long trajectories become dominant. In an intermediate situation, both quantum paths can have similar weights and interfere. In the temporal domain, this interference has been identified to occur between the short and long trajectories originated respectively, at two consecutive ionization maxima [16]. Counterintuitively, even with few-cycle pulses, the weight of quantum paths from two consecutive ionization maxima in each half-cycle is high enough to interfere with similar weight.

On the other hand, if we place the gas jet after the focus, phase-matching favors only short trajectories at any angle. As a consequence, in plots (b) and (d) we do not observe any major change in the attosecond pulse width when increasing the detection angle.

Let us now study in detail how the isolated attosecond pulse changes with the detection angle for a gas jet placed before the focus. In Fig. we present the isolated attosecond pulse and a time-frequency analysis of the spectra when detected (a-b) on-axis, and at (c-d) 0.7, (e-f) 1.3, (g-h) 2.0, (i-j) 2.6, (k-l) 3.3 and (m-n) 4.0 mrad off-axis. If we look at the time-frequency analysis on-axis –plot (b)– we observe that the electron's recollision energy shows a roughly linear dependence on time, in which the lower harmonics are emitted before the higher ones, giving rise to a positive chirp typical from the harmonic spectrum emitted by short trajectories. However, as we increase the detection angle, we observe how the time-frequency structure changes leading to a vertical structure when detected above 3.3 mrad –plots (l) and (n)–. In those cases, the destructive interference between paths tends to confine the harmonic radiation into a narrower window, reducing drastically the positive chirp associated to short trajectories alone.

In the left column of Fig. 3 we observe how one can tailor the attosecond pulse waveform by changing the detection angle. Although the efficiency of the signal decreases when for off-axis detection, the pulse width is reduced to nearly half of the width obtained on-axis.

4. Conclusions

We have proposed a route to directly compensate the chirp of isolated attosecond pulses by detecting them at a certain angle from the propagation axis. Performing a time-frequency analysis, we have identified the interference between different quantum paths as the underlying mechanism for this compensation. We have presented the detection angle as a parameter for controlling the macroscopic signatures of these single-atom processes.

Acknowledgments

We acknowledge support from Junta de Castilla y León (Consejería de Educación and Fondo Social Europeo), Spanish MINECO (FIS2009-09522, Consolider Program SAUUL CSD2007-00013), and Centro de Láseres Pulsados, CLPU.

References

[1] Corkum P B 1993 *Phys. Rev. Lett.* **71**, 1994
[2] Schafer K, Yang B, DiMauro L F and Kulander K C 1993 *Phys. Rev. Lett.* **70**, 1599
[3] Salières P, Carré B, Le Déroff L, Grasbon F, Paulus G, Walther H, Kopold R, Becker W, Milosevic D, Sanpera A and Lewenstein M 2001 *Science*, **292**, 902
[4] Farkas G and Toth C 1992 *Phys. Lett. A* **168**, 447
[5] Corkum P B, Burnett N H and Ivanov M Y 1994 *Opt. Lett.* **19**, 1870
[6] Christov I P, Murnane M M and Kapteyn H C 1997 *Phys. Rev. Lett* **78**, 1251

Journal of Physics: Conference Series **414** (2013) 012014 doi:10.1088/1742-6596/414/1/012014

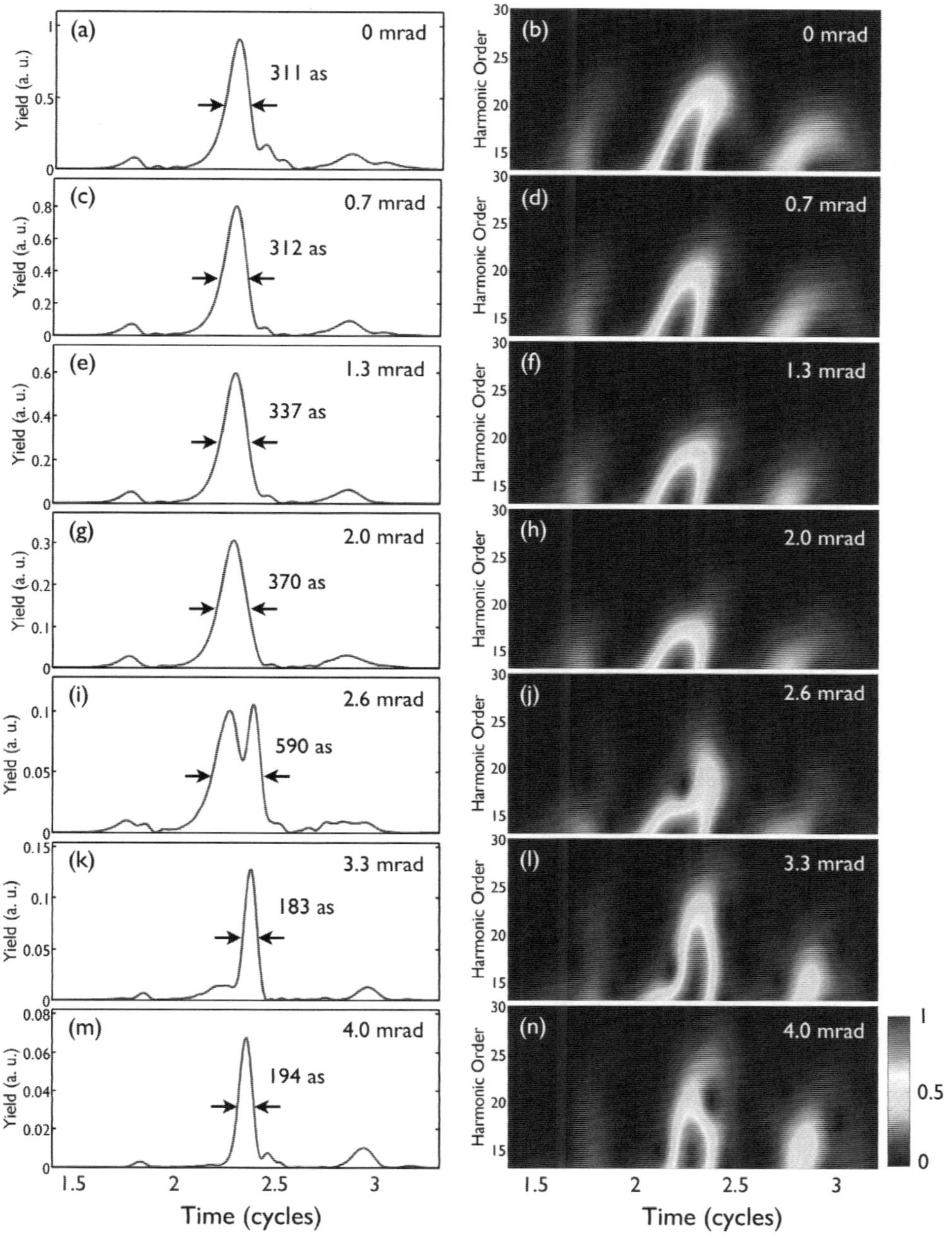

Figure 3. Temporal distribution of the attosecond train of pulses (left column) and time-frequency structure (right column) when the high-order harmonics are detected in a window (a-b) on-axis in the first row, and centered at (c-d) 0.7, (e-f) 1.3, (g-h) 2.0, (i-j) 2.6, (k-l) 3.3 and (m-n) 4.0 mrad off-axis in the following rows. The gas jet is placed 1 mm before the focus position, and the laser pulse is 1.4 cycles (3.8 fs) FWHM $\sin^2$, 800 nm, with intensity at focus $\simeq 2.45 \times 10^{14}$ W/cm^2

[7] Antoine P, L'Huillier A and Lewenstein M, 1996 *Phys. Rev. Lett.* **77,** 1234

[8] Mairesse Y, de Bohan A, Frasinski L J, Merdji H, Dinu L C, Monchicourt P, Breger P, Kovacev M, Taïeb R, Carré B, Muller H G, Agostini P and Saliéres P 2003 *Science,* **302,** 1540

[9] Varjú K, Mairesse Y, Carré B, Gaarde M B, Johnsson P, Kazamias S, López-Martens R, Mauritsson J, Schafer K J, Balcou Ph, LHuillier A and Saliéres P 2005 *J. of Mod. Opt.*, **52**, 379

[10] P. M. Paul, E. S. Toma, P. Breger, G. Mullot, F. Augé, Ph. Balcou, H. G. Muller and P. Agostini *Science* **292,** 1689 (2001).

[11] Hentschel M, Kienberger R, Spielmann C, Reider G A, Milosevic N, Heinzmann U, Drescher M and Krausz F 2001 *Nature* **414**, 509

[12] Goulielmakis E, Schultze M, Hofstetter M, Yakovlev V S, Gagnon J, Uiberacker M, Aquila A L, Gullikson E M, Attwood D T, Kienberger R, Krausz F and Kleineberg U 2008 *Science,* **320** 1614

[13] Zhao K, Zhang Q, Chini M, Wu Y, Wang X and Chang Z 2012 *Opt. Lett.* **37**, 389

[14] Hernández-García C, Pérez-Hernández J A, Popmintchev T, Murnane M M, Kapteyn H C, Jaron-Becker A, Becker A and Plaja L 2012 Zeptosecond keV X-ray waveforms driven by mid-infrared laser pulses, *submitted*

[15] Popmintchev T, Chen M-C, Popmintchev D, Arpin P, Brown S, Ališauskas S, Andriukaitis G, Balčiunas T, Mücke O, Pugzlys A, Baltuška A, Shim B, Schrauth S E, Gaeta A, Hernández-García C, Plaja L, Becker A, Jaron-Becker A, Murnane M M and Kapteyn H C 2012 *Science* **336**, 1287

[16] Hernández-García C and Plaja L 2012 *Journal of Physics B: At. Mol. Opt. Phys.* **45**, 074021

[17] Pérez-Hernández J A, Roso L and Plaja L 2009 *Opt. Exp.* **17**, 9891

[18] Hernández-García C, Pérez-Hernández J A, Ramos J, Conejero Jarque E, Roso L and Plaja L, 2010 *Phys. Rev. A* **82**, 033432

[19] L'Huillier A, Li X F and Lompré L A 1990 *J. Opt. Soc. Am. B*, **7**, 527

[20] Pérez-Hernández J A, Hernández-García C, Ramos J, Conejero E, Plaja L and Roso L 2011, New Methods For Computing High-OrderHarmonic Generation and Propagation (Progress in Ultrafast Intense Laser Science VII), (pp 145-162), Springer chapter 7 pp 145-162
Kuhn T 1998 Density matrix theory of coherent ultrafast dynamics Theory of Transport Properties of Semiconductor Nanostructures (Electronic Materials vol 4) ed E Scholl (London: Chapman and Hall) chapter 6 pp 173214

[21] Sola I J, Mével E, Elouga L, Constant E, Strelkov V, Poletto L, Villoresi P, Benedetti E, Caumes J-F, Stagira S, Vozzi C, Sansone G and Nisoli M, 2006 *Nature Phys.* **2**, 319

[22] Sansone G, Benedetti E, Calegari F, Vozzi C, Avaldi L, Flammini R, Poletto L, Villoresi P, Altucci C, Velotta R, Stagira S, De Silvestri S and Nisoli M 2006 *Science* **314**, 443

[23] Vincenti H and Quéré F 2012 *Phys. Rev. Lett.* **108**, 113904

Population trapping in Xe atoms

A Azarm[1], S M Sharifi[1], A Sridharan[1], S Hosseini[1], Q Q Wang[1], A M Popov[2,3], O V Tikhonova[2,3], E A Volkova[2] and S L Chin[1]
[1] Department of Physics, Engineering Physics and Optics & Center for Optics, Photonics and Laser (COPL), Université Laval, Québec City, Québec, G1V 0A6 Canada
[2] D V Skobeltsyn Institute of Nuclear Physics, Moscow State University, 119234, Moscow, Russia
[3] Physics Department, Moscow State University, 119991, Moscow, Russia

E-mail: alexander.m.popov@gmail.com

Abstract. We study the interference stabilization (population trapping) of Xe atoms using a fs Ti – Sapphire laser both experimentally and theoretically. The investigation is performed for two pulses of different duration. The signature of population trapping arising from the dynamic multiphoton resonance of the initial state and a group of Rydberg states of the atom is found to exist. The results obtained can be considered as the manifestation that population trapping is indeed a universal phenomena.

1. Introduction

The idea of interference stabilization (IS) of Rydberg atomic states in a strong laser field was proposed by Fedorov and co-workers [1] in 1988. In accordance with [1,2], the physical reason for the stabilization is the interference of the amplitudes of transitions to the continuum from the excited Rydberg states that are coherently repopulated by the Λ -type Raman transitions in the strong laser field. The suppression of the ionization in this case is a result of the overlapping of the ionization widths of all these Rydberg states and the destructive interference of the amplitudes of the transitions to the continuum. The threshold of the interference stabilization phenomenon can be estimated as

$$\varepsilon_0/\omega^{5/3} > 1 \qquad (1)$$

where ω is the laser frequency and ε_0 is the amplitude of the laser's electric field strength [2]. For the Ti:Sapphire laser ($\hbar\omega = 1.55$ eV) condition (1) is fulfilled for laser intensities larger than $I > I^* \approx 10^{13}$ W/cm^2. The existence of the interference stabilization was first experimentally demonstrated in the most unambiguous way in [3-5].

The possibility of IS for the atom in the ground atomic state which is coupled with the continuum by the multiphoton transition was proposed in [6] and studied theoretically in detail in [7]. The key element of such interpretation is the multiphoton resonance of the ground state and a set of high-lying Rydberg states AC Stark shift taking into account, and then the IS of population trapped in excited Rydberg states. Recent data on population trapping in Rydberg states obtained by the numerical solution of the time – dependent Schroedinger equation (TDSE) for Hydrogen and model Silver atoms are discussed in [8-10].

Independently, the same idea of population trapping in Rydberg states was suggested by S.L. Chin et al [11,12] for interpretation of experimental results on the strong – field ionization yields in atoms of noble gases and molecules, respectively. It was found that in a certain intensity range below the saturation intensity, the ion yield would decrease subtly before increasing again. This was explained by the dynamic resonance of the ponderomotively shifted Rydberg states resulting in population trapping. In [13] the population trapping in high – lying Rydberg states was demonstrated to exist also in another molecular system. Paper [14] gives a review of experimental work of S.L. Chin et al. on this subject in which real application to facilitate the remote detection of THz pulses through filamentation was discussed.

The trapping of population in highly excited Rydberg states was also observed in recent experiments with He atoms irradiated by Ti-Sa femtosecond pulse [15]. The authors of this paper proposed the model of the frustrated tunnel ionization. In accordance with this theoretical model the capture to highly excited states take place as a result of the recombination process on parent ion during the turn-off time of the laser pulse caused by the Coulomb interaction with the atomic core. Such a temporal dynamics is in contradiction with the model based on IS phenomenon where the population of excited Rydberg states is found to be pronounced already at the first ramp of the laser pulse as soon as the field amplitude becomes close to its peak value [10].

In this work we study both experimentally and theoretically the ionization dynamics features of Xe atom in the strong pulse of the Ti-Sapphire laser. The step structure of ionization yield versus laser intensity is interpreted as a manifestation of the population trapping in excited states and confirmation of interference stabilization phenomenon. The effect of the pulse duration and spectral bandwidth on population trapping are analyzed in the frames of IS model.

2. Experimental Setup

A schematic diagram of the experimental setup is shown in figure 1. Briefly, a commercial ~0.2 TW femtosecond Ti-sapphire laser system was used. It consists of an oscillator (Spectra Physics Tsunami followed by a chirped pulse amplification (CPA) module (Spectra Physics Spitfire). In the CPA section, a stretcher is used to stretch the pulses from 30 fs transform limited to around 200 ps positively chirped pulses. These were then amplified by a regenerative amplifier at 1 kHz repetition

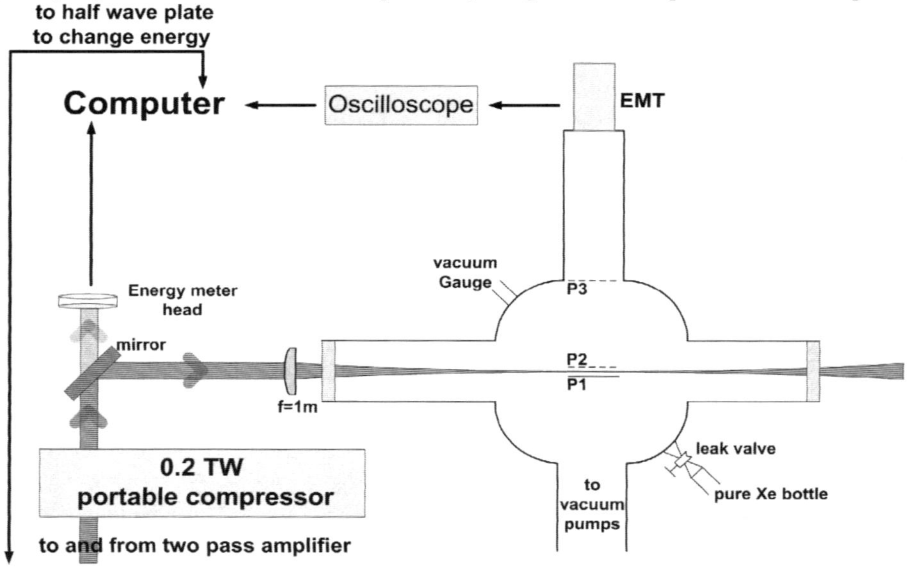

Figure 1. Sketch of the experimental setup.

rate. A pulse picker was used to select a ten hertz beam from the 1 kHz beam. The 10 Hz beam was sent into a two pass Ti-sapphire amplifier. A portable parallel grating compressor is used to give a transform – limited pulse with pulse duration of about 45 fs at FWHM (full width at half-maximum). The pulse duration is measured using a second – order single – shot autocorrelator (SSA, Positive Light). The output beam has a maximum energy of 20 mJ/pulse with a slight elliptic shape beam pattern with diameter of about 4.2 (vertically) and 4.7 (horizontally) mm (Full width at half maximum of the fluence). A half – wave plate and a polarizer are used inside the two pass amplifier to change the output energy of the linearly polarized pulses. They are installed before the gain medium.

To change the pulse duration and to compare two cases with the same pulse duration and different bandwidth, we used a narrow bandwidth band pass filter. Since after the amplification stage the power of the pulse can damage any optical component, we put the band pass filter before the two pass amplifier stage. Based on uncertainty principle and considering Gaussian pulse shape the transform limited pulse duration of 175 and 45 fs has the bandwidth of 5.3 and 20.9 nm respectively with central wavelength of 800 nm. In both cases, the beam was focused into an ultrahigh – vacuum (UHV) chamber by a 1 m plano–convex lens. The background pressure of the UHV chamber was 3×10^{-9} Torr. The UHV chamber was evacuated by a turbo molecular pump connected to a TriScroll pump (oil free rotary pump). The Xe pressure in the interaction chamber was controlled by a precision leak valve (see figure 1) and ranged from 8×10^{-8} to 7×10^{-4} Torr. Ion curves were produced by combining a series of intensity scans, each having a different fill pressure in the interaction chamber. Ion species were collected using a time – of – flight mass spectrometer with a 60 cm long drift tube. The laser was focused through the centre between two plates P1 and P2 (with an opening covered by a metallic mesh) spaced by 5 mm (see figure 1). The voltages applied to P1 and P2 were 4 and 3.5 kV, respectively. The opening of the TOF spectrometer, P3, was 25 mm covered by a metallic mesh.

3. Results and Discussions

3.1. Experimental results

First, let us discuss the experimental data for the ionization yield of Xe atoms (which is proportional to

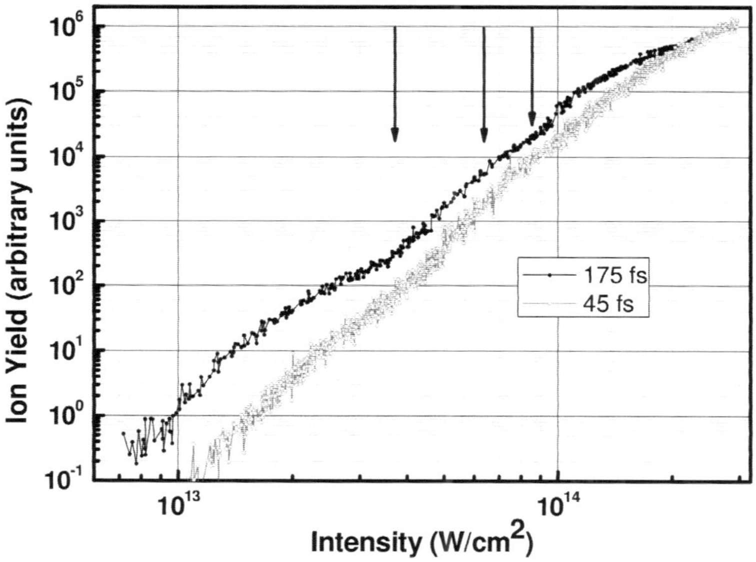

Figure 2. The dependence of the ion yield of Xe atom on the intensity for "long" (black) 175 fs pulse, and "short" (red) 45 fs pulse. Arrows indicate the intensities mentioned in the text.

the average probability W_i) as a function of laser intensity I. The results obtained in our experiment for "short" (45 fs) and "long" (175 fs) laser pulses with two different bandwidths are presented in figure 2. Both pulses are transform limited pulses. The main specific feature of the dependence obtained for the 175 fs pulse is its step – like structure. The steps on the dependence $W_i(I)$ are well pronounced for the intensity values around 3.7×10^{13}, 6.2×10^{13} and 8.3×10^{13}W/cm^2. As it was mentioned above, in [11, 12] it was supposed that the reason for such step – like dependence is the resonance character of the population trapping in a number of high-lying Rydberg states that are resistant to the ionization if the laser field strength exceeds the critical value (1). In comparison with the case of the "long" laser pulse such step – like structure is much less pronounced for the curve corresponding to the "shorter" 45 fs laser pulse. It means that the resonant character of the excitation process might have been smeared out due to the broader bandwidth in this case. Also there is a shift of the steps to higher intensities As a result, the ratio of the ion yields for "long" and "short" pulses $w_i^{(long)} / w_i^{(short)}$ demonstrates a non – monotonous behavior of the dependence on laser intensity (see figure 3). In the intensity range from 3×10^{13} W/cm^2 up to 10^{14} W/cm^2 below the saturation of ionization process, the ratio of ionization probabilities $w_i^{(long)} / w_i^{(short)}$ oscillates near the value of 3.8 corresponding to the ratio of pulse durations for "long" and "short" pulses. For the intensities higher than 10^{14} W/cm^2 the saturation of ionization process for "long" pulse appears to exist and the ratio of the yields $w_i^{(long)} / w_i^{(short)}$ has the tendency to decrease. On the other hand for intensities lower than 3×10^{13} W/cm^2 the ionization probability is dramatically enhanced for the case of the "long" pulse; the ratio $w_i^{(long)} / w_i^{(short)}$ at the intensity $\approx2\times10^{13}$ W/cm^2 reaches almost one order of magnitude higher than the level at the highest intensity. The physical reason for this dramatic enhancement of ionization will be discussed later.

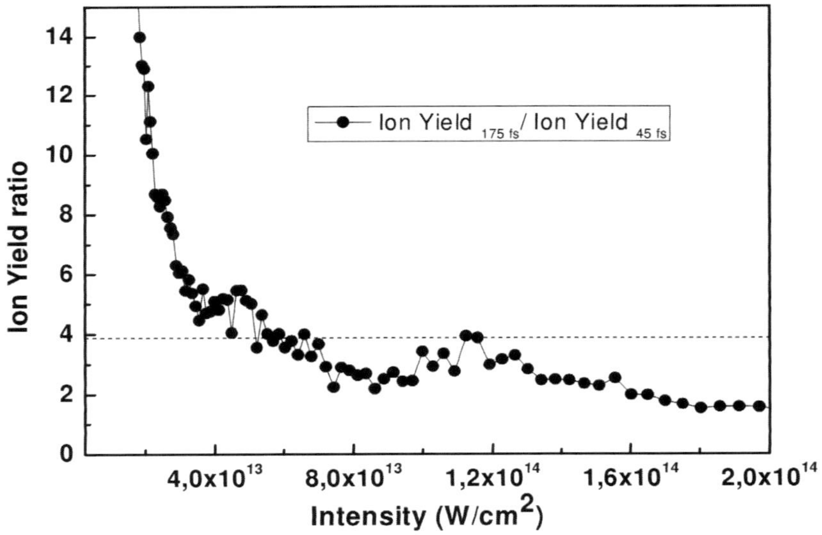

Figure 3. The ratio of ion yields $w_i^{(long)} / w_i^{(short)}$ for "long" (175 fs) and "short" (45 fs) laser pulses. Dash line indicates the ratio of pulse durations 3.8.

3.2. Numerical simulation

To confirm the effect of population trapping in high – lying Rydberg states we performed the numerical simulation of the dynamics of ionization of Xe atom for laser parameters similar to the experimental ones. Our simulation was based on the solution of the TDSE (time dependent Schroedinger equation) for the model single electron Xe – like atom. In our calculations the potential for the model Xe atom was chosen in the form

$$V(r) = -\frac{e^2}{\sqrt{r^2 + \alpha^2}} \tag{2}$$

with smoothing parameter $\alpha = 0.0943$ A. For such a value of α the ionization potential of the ground state is equal to $I_i = 12.13$ eV and coincides with the ionization potential of the real Xe atom. Two lowest s and p excited states have the energies of -3.21 eV and -3.37 eV respectively, which are close enough to the position of the real levels. This model is rather simple, but it provides both the exact value of ionization potential and existence of Rydberg series which is quite important for the qualitative comparison of the experimental and numerical data.

The laser pulse with $\lambda = 800$ nm ($\hbar\omega = 1.55$ eV) had a smoothed envelope with sine – squared ramps of duration t_f equal to integer number of optical cycles:

$$\varepsilon_0(t) = \varepsilon_0 f(t)\cos(\omega t) \tag{3}$$

where $f(t) = \sin^2(\pi t/2t_f)$ is the pulse envelope, so that the total pulse duration is $2t_f$ and pulse duration at FWHM is $\tau = t_f$. For example, for $t_f = 20T$ ($T = 2\pi/\omega$ is the optical cycle) we have $\tau = 53$ fs. All the details of the numerical procedure can be found in [16].

The main result of our numerical simulations is presented in figure 4 and demonstrates the non –

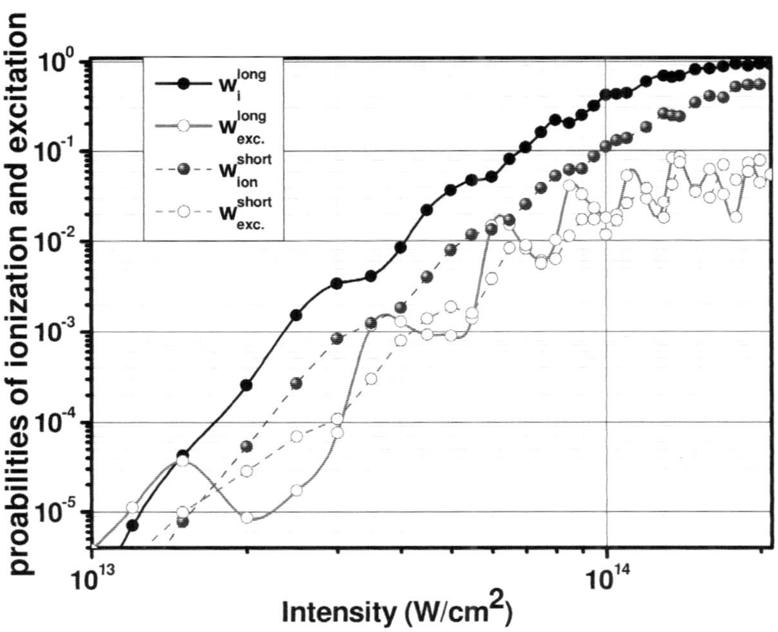

Figure 4. Probabilities of ionization (full circles) and excitation (open circles) per pulse in dependence of intensity of Ti:Sapphire laser pulse with $t_f = 20T$ (solid curves) and $t_f = 5T$ fs (dash curves) for a Xe atom obtained in numerical simulation.

monotonous dependence of the ionization and excitation probabilities of the Xenon atom versus laser intensity for two different pulse durations ($t_f = 20T$ and $t_f = 5T$). These two pulses are shorter in comparison with our experimental pulses of 45 and 175 fs. Nevertheless the comparison is possible and the ratio of durations R=4 is similar to the experimental ratio R=175/45~3.8. So, qualitative comparison is possible. The obtained dependence for both pulses demonstrates non – monotonous structure with each excitation maximum being accompanied by the minimum of ionization probability. The maxima of excitation probability correspond to the trapping of the population in high – lying Rydberg states with binding energies in a range of 0.1 – 0.4 eV and are similar to that were found in simulations [8,9]. These maxima are separated by approximately the same value of laser intensity equal to $\Delta I \approx 2.5 \times 10^{13}$ W/cm^2, which exactly corresponds to the ponderomotive shift of the continuum boundary $U_p = e^2\varepsilon_0^2/4m\omega^2$ equal to photon energy $\hbar\omega$. The maxima in excitation probability are better pronounced for longer pulse with $t_f = 20T$ and observed at laser intensities of $3{,}7\times10^{13}$, $6{,}2\times10^{13}$ and $8{,}6\times10^{13}$ W/cm^2. These values correspond to the "steps" on the ionization curve and we would like to stress that these values of intensity are in rather good coincidence with the positions of the "steps" on the ionization curve obtained experimentally (see figure 2). For shorter laser pulse with $t_f = 5T$ the dependence of excitation probability versus intensity is smoother, but it is also characterized by a set of minima and maxima. Their positions are shifted a little in comparison with the case of long ($t_f = 20T$) pulse. As a result the steps on the ionization curve for longer pulse duration are better pronounced and also shifted a little in comparison with their positions for shorter pulse. This shift in position of maxima in the dependence of population trapped in Rydberg states versus intensity manifests itself in the non – monotonous character of the ratio of ionization yields $w_i^{(long)} / w_i^{(short)}$ calculated for pulses with $t_f = 20T$ and $t_f = 5T$. This dependence is presented at figure 5 and demonstrates the features similar to that obtained in experiments discussed above.

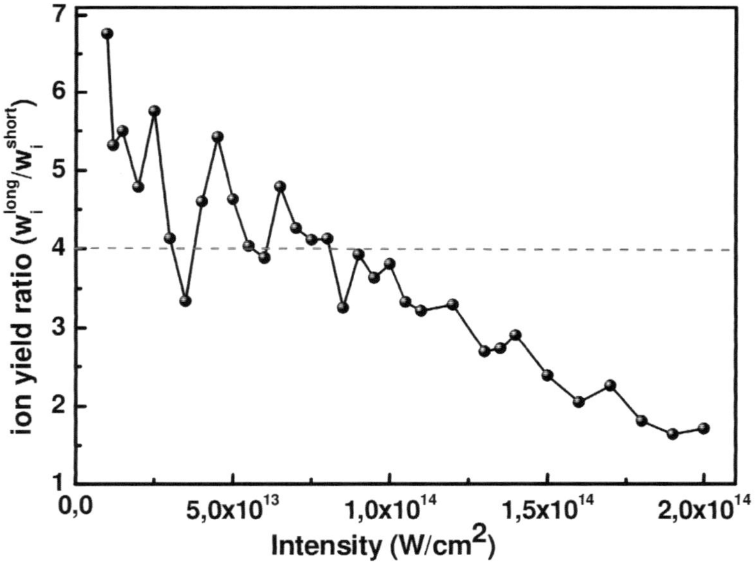

Figure 5. The ratio of ion yields $w_i^{(long)} / w_i^{(short)}$ calculated for "long" (20T) and "short" (5T) laser pulses. Dash line indicates the ratio of the pulse durations.

Indeed, the curve oscillates near the value of 4 corresponding to the ratio of pulse durations for "long" and "short" pulses, has the tendency to decrease in the range of high intensities ($I \geq 9\times10^{13}$ W/cm^2). Also the significant increment of the ratio is seen for low intensity limit ($I \geq 1-3\times10^{13}$ W/cm^2). Moreover, the position of minima on the dependence $w_i^{(long)} / w_i^{(short)}$ on laser intensity are in excellent coincidence with the maxima of excitation probability for long pulse that observed on the curve (see figure 4). Hence, the observed oscillation structure of the ratio of ionization yields can be treated as the manifestation of population trapping in high – lying Rydberg states in our experiments.

To provide more insight into the process of ionization and population trapping in high – lying Rydberg states, let us analyze the energy spectra of photoelectrons obtained for a 20T pulse for intensities corresponding to the maximum and minimum of the excitation on the curve in Fig. 4. These spectra represent a number of ATI peaks separated by $\hbar\omega$ from each other. Due to the ponderomotive shift of the continuum boundary the energy of each ATI peak decreases with increment of laser intensity. The "channel closing" effect [17] and the change of the multiphoton order of the ionization process should take place as soon as the peak laser intensity is increased by 2.5×10^{13} W/cm^2. Such a behavior of ATI peaks is clearly seen at figure 6, where the dependence of positions of several first ATI peaks on the peak laser intensity is presented.

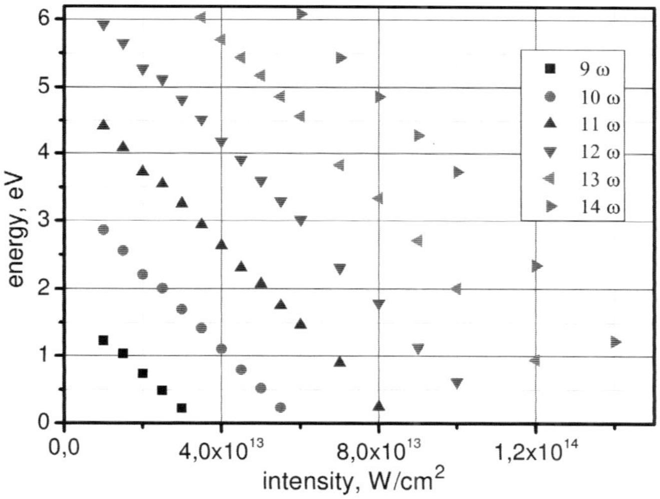

Figure 6. Energies of several first ATI peaks of the photoelectron spectra versus the laser intensity, *n* is the number of absorbed photons.

As can be seen from figure 6 the intensities 3.7×10^{13}, 6.2×10^{13} and 8.6×10^{13} W/cm^2 providing approximately the maximum of excitation probability correspond to the case just after closing of the 9-th, 10-th and 11-th channels of multiphoton ionization when one more photon is needed to be absorbed for the ionization. In this situation the energy of the first ATI peak is about 1.5 eV. The absorption of 9, 10 or 11 photons for the above mentioned intensities lead to the transition to the high – lying Rydberg states near the continuum boundary that will be efficiently populated. The reason of the observed population trapping consists in the repopulation of the resonantly coupled Rydberg band by Raman – type transitions via the continuum that are very efficient in a strong field and results in the significant suppression of the ionization process (see [8-10] for details).

In contrast, for intermediate values of laser intensity the multiphoton transition to the continuum provides the first ATI peak at the energy ≈ 0.75 eV and multiphoton excitation gives the possibility to populate only relatively low – lying states with binding energy around 1 eV and lower principal quantum numbers. Since for this situation the density of Rydberg states is significantly lower and their resistance to ionization is smaller in comparison with the high – lying states near the boundary of the continuum, the excitation probability is found to be suppressed in this case.

In order to compare results of numerical simulations with real experimental observations the integration over the focal volume should be taken into account. In assumption of cylindrical form of the focal volume and the Gaussian radial profile with halfwidth r_0

$$I(r) = I_0 \exp(-r^2/r_0^2) \quad (4)$$

we obtain

$$W_{i(excit)} = \frac{2}{r_0^2}\int_0^{\infty} w_{i(excit)}(I(r))rdr \quad (6)$$

The averaged over the focal spot numerical data for ionization and excitation probabilities from figure 4 are presented at figure 7. As can be seen, the existing maxima and minima of ionization/excitation probabilities are partly smoothed and the obtained curves for ionization yield are quite similar to the experimental data (see figure 2).

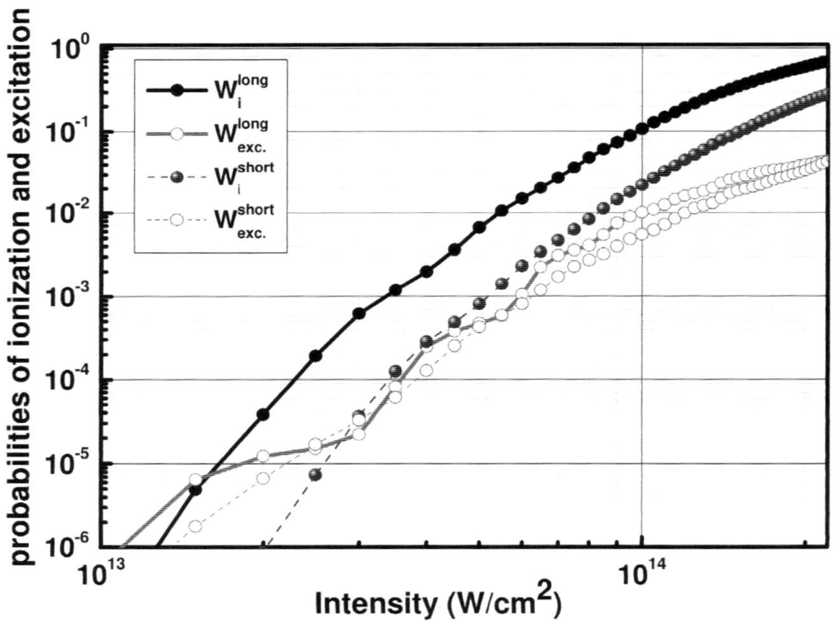

Figure 7. Probabilities of ionization (full circles) and excitation (open circles) averaged over focal spot distribution of intensity in dependence on intensity of Ti:Sa laser pulse with $t_f = 20T$ (solid curves) and $t_f = 5T$ fs (dash curves) for a Xe atom obtained in numerical simulation.

To conclude the discussion of both experimental and numerical results we would like to stress that the population trapped is really an important phenomenon in physics of strong laser field interaction with atoms. To support this statement we present the ratio of trapped population to the ionization yield versus laser intensity for both "short" and "long" pulses averaged over the Gaussian distribution of the intensity in the focal spot (see figure 8). Both curves have the nonmonotonous structure which can be

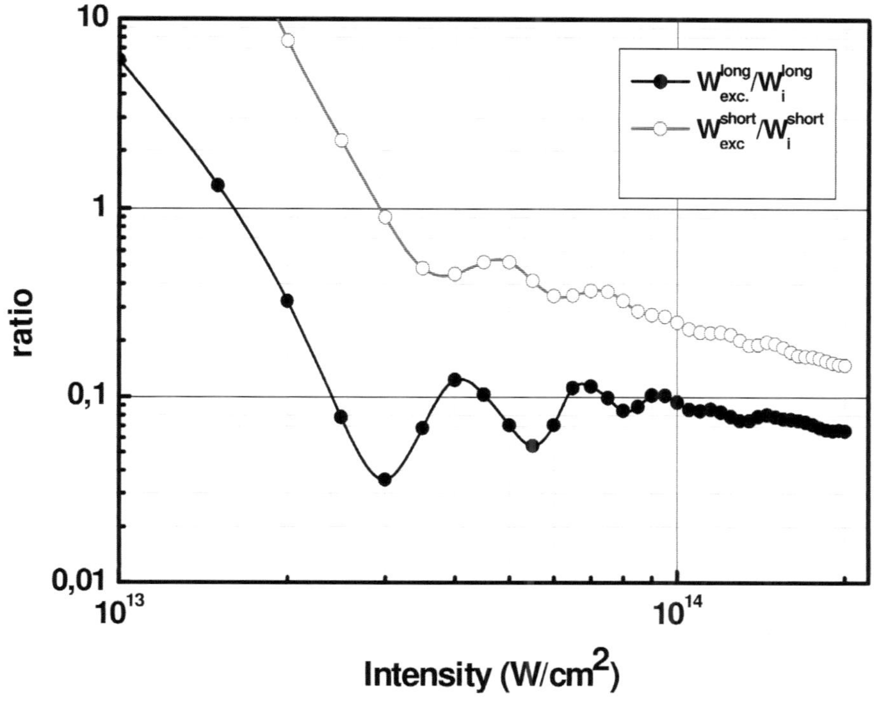

Figure 8. The ratio of trapped populations to ionization yields for "long" (full circles) and "short" (open circles) laser pulses averaged over focal spot versus the intensity of Ti:Sapphire laser pulse for a Xe atom obtained in numerical simulation.

considered as the manifestation of the multiphoton resonances between the ground and highly excited Rydberg states near the boundary of the continuum. The ratio w_{exc} / w_i is larger for "short" pulse. That is due to the wider bandwidth of the short pulse and more effective interaction of the ground state with highly excited states via the multiphoton transition. On the other hand, the resonance features of the curve are less pronounced in comparison with the case of the "long" pulse. It should be emphasized that it is in weak fields ($I \approx 1 - 2\times 10^{13}$ W/cm^2) the probability of excitation is even larger than the probability of ionization, and the ratio w_{exc} / w_i exceeds the value of unity. Unfortunately for such intensities the level of both signals is too low and can be difficultly studied experimentally or numerically. Nevertheless even for "long" pulse with rather high intensity the ratio of the population trapped in the set of Rydberg states to the ion yield oscillates at a level of 10%.

4. Conclusion

So far, population trapping in atoms in a strong laser field was observed through measuring the ion yield as a function of intensity. The qualitative agreement was found to exist between the experimental data and results of numerical simulation based on TDSE solution. The obtained dependences are very similar to the experimental data. The specific features of the ionization yields (step structure) as well as non – monotonous dependence of the ratio of ionization yields for long and short pulses $w_i^{(long)} / w_i^{(short)}$ vs laser intensity can be considered as the manifestation of the population trapping in high-lying Rydberg states. The observed phenomena were interpreted as the confirmation of interference stabilization phenomenon, which is indeed a universal process during strong laser field interaction with atoms and molecules.

Acknowledgments
This work was partially supported by NSERC, Canada Research Chairs, CIPI, and FQRNT of Quebec, Ministry of Science and Education of Russian Federation, Russian Foundation for Basic Research (project 12-02-00064). Technical support and assistants of Mr. M Martin in the Ultrafast Intense Laser Science Lab. at Laval University, where the experiment was carried out is highly appreciated. The simulations were performed on supercomputer SKIF-MSU "Chebyshev" in Moscow State University.

References

[1] Fedorov M and Movsesian A M 1988 *J. Phys. B* **21** L155
[2] Fedorov M V 1988 *Atomic and free electrons in a strong light fields* (Singapore: World Scientific)
[3] Jones R R, Schumacher D W and Bucksbaum P H 1993 *Phys. Rev. A* **47** R49
[4] De Boer M P, Hoogenraad J H, Vrijen R B, Noordam L D and Muller H G 1993 *Phys. Rev. Lett.* **71** 3263
[5] Hoogenraad J H, Vrijen R B and Noordam L D 1994 *Phys. Rev. A* **50** 4133
[6] Dubrovskii Yu V, Ivanov M Yu and Fedorov M V 1991 *Sov. Phys. JETP* **99** 228
[7] Popov A M, Tikhonova O V and Volkova E A 2000 *Laser Phys.* **10** 898
[8] Popov A M, Tikhonova O V and Volkova E A 2010 *Laser Phys.* **20** 1028
[9] Volkova E A, Popov A M and Tikhonova O V 2011 *JETP* **113** 394
[10] Fedorov M V, Poluektov N P, Popov A M, Tikhonova O V, Kharin V Yu and Volkova E A 2012 *IEEE J. of Selected Topics in Quantum Electronics* **18** 42
[11] Talebpour A, Chien C Y and Chin S L 1996 *J. Phys. B* **29** 5725
[12] Talebpour A, Liang Y and Chin S L 1996 *J. Phys. B* **29** 3435
[13] Azarm A, Ramakrishna S, Talebpour A, Hosseini S, Teranishi Y, Xu H L, Kamali Y, Bernhardt J, Lin S H, Seideman T and Chin S L 2010 *J. Phys. B* **43** 235602
[14] Chin S L, Azarm A, Xu H L, Wang T J, Sharifi M, Talebpour A 2012 Experiments in population trapping in atoms and molecules by an intense short laser pulse in *Progress in Ultrafast Intense Laser Science,* vol. 8 (PUILS 8), ed Kaoru Yamanouchi, Mauro Nisoli and Wendell T. Hill, III, editors. Springer Chapter 4, p.79-95
[15] Nubbemeyer T, Gorling K, Saenz A, Eichmann U and Sandner W 2008 *Phys. Rev. Lett.* **101** 233001
[16] Popov A M, Tikhonova O V and Volkova E A 2010 *Laser Phys.* **21** 1593
[17] Muller H G, van Linden Van Den Heuvell H B, Agostini P, Petite G, Antonetti A, Franco M, and Migus A 1988 *Phys. Rev. Lett.* **60** 565

Pair production in a plane wave due to a thermal background

B. King
Ludwig-Maximilians-Universität, Theresienstraße 37, 80333 München, Germany.

A. Di Piazza
Max-Planck-Institut für Kernphysik, Saupfercheckweg 1 D-69117 Heidelberg, Germany.

E-mail: dipiazza@mpi-hd.mpg.de

Abstract. Ever since Schwinger's publication [J. Schwinger, Phys. Rev. **82**, 664 (1951)], the maxim that there can be no pair creation from vacuum in a plane wave has been often cited. We put forward an analysis showing that in any real situation, where thermal effects are present, in a single plane-wave field, even in the limit of zero frequency (a constant crossed field), pair creation can indeed occur. Interestingly, we find that the pair-production rate depends non-perturbatively on both the temperature and the amplitude of the constant crossed field.

1. Introduction

One immediate consequence of Dirac's relativistic equation for the electron is that, given sufficient energy density, massive particles can be generated by electromagnetic radiation, in the form of particle-antiparticle pairs. Early results by Sauter [1], Heisenberg and Euler [2] and Weisskopf [3] were followed by an approach based on effective field theory by Schwinger [4] who derived the rate of pair creation per unit volume in a uniform and constant electric field of strength E. This rate has a non-perturbative dependency on the combination E/E_{cr}, where E_{cr} is the so-called "critical" electric field $E_{cr} = m^2c^3/e\hbar = 1.3 \times 10^{16}$ Vcm^{-1}, equivalent to a critical peak intensity of $I_{\text{cr}} = 4.6 \times 10^{29}$ Wcm^{-2}, where e is the positron charge and m is its mass. More than a decade later, an intensive campaign was begun to investigate quantum electrodynamics in intense external fields, given the impetus by key initial results such as the rate of photon decay into electron-positron pairs [5, 6], which also depends non-perturbatively on the external field through the parameter $\chi = e\hbar\sqrt{|(k_\mu F^{\mu\nu})^2|}/m^3c^4$, with photon four-vector k_μ and background field $F^{\mu\nu}$. Several factors have led to a heightened and sustained interest in better understanding and devising ways of detecting such phenomena. This is in part due to the planning of upcoming laser facilities, such as ELI (Extreme Light Infrastructure) [7], XCELS (eXawatt Center for Extreme Light Studies) [8] and HiPER (High Power laser Energy Research) [9], where it is intended that electromagnetic fields will be generated orders of magnitude larger than the current record of around 10^{22} Wcm^{-2} [10], and even up to a percent of the critical value. It is also due in part to recent theoretical advances, both analytical [11, 12, 13, 14, 15] and from numerical simulation [16, 17, 18], that strongly indicate that pair creation could be observed at orders of magnitude well below the critical field (a recent review of strong-field QED effects can be found in [19]).

Journal of Physics: Conference Series **414** (2013) 012016 doi:10.1088/1742-6596/414/1/012016

Pair creation is predicted to occur in various contexts. For example, possible terrestrial laboratory experiments in which two intense electromagnetic plane waves propagate against each other with [20] and without [21, 22, 23] the presence of a nucleus modelled by a Coulomb field; in an astrophysical context, such as the decay of photons in the strong magnetic field of magnetars [24]; and also in constant electric fields catalysed with a thermal background, having already been calculated in different formalisms [25, 26, 27, 28]. However, Schwinger famously derived a "no-go theorem" for pair creation, stating in [4], that "there are no nonlinear vacuum phenomena for a single plane wave, of arbitrary strength and spectral composition". First, this is forbidden as in a plane wave, photons propagate parallel to one another and so cannot physically interact with one another. Second, the only two unique electromagnetic relativistic invariants vanish in a single plane wave.

In the current paper, we do not challenge that a single plane wave induces no pair creation in vacuum, rather we explain how, when one takes into account the inevitable presence of some background heat radiation, pair creation can indeed proceed in single plane waves in all realisable scenarios, a point also demonstrated in [29]. Specifically, this will be shown to be the case for a constant crossed field, which can be reached by taking the zero-frequency limit of an infinite plane-wave. It will turn out that the "thermal" rate for pair creation by a photon gas at a given temperature T in a constant crossed field of strength E will be non-perturbative in both variables. In addition to the qualitative result and the interesting appearance of a thermodynamic variable in the pair-creation exponent, these results could be of relevance when describing heavy-ion collisions, relativistic plasmas and pair production processes near neutron stars.

2. Thermal photon gas in an external plane wave

The case we are considering is an intense external electromagnetic plane wave in thermal equilibrium with a photon gas at temperature T. The intensity can be quantified with the classical non-linearity parameter, often simply referred to as the "intensity-parameter" $\xi = (m/\omega_l)(E/E_{\text{cr}}) \gg 1$, where ω_l is the angular frequency of the photons of the external field, and we have already set here and subsequently $\hbar = c = k_{\text{B}} = 1$ (k_{B} is the Boltzmann constant). The non-linearity parameter also qualifies the type of pair-creation. $\xi \ll 1$ refers to a multi-photon process, whereas $\xi \gg 1$ refers to a tunneling process, which is characteristically frequency-independent, tending to this limit with increasing ξ [30]. This corresponds to the so-called "constant crossed field" when applied to a plane wave, for which the field is crossed ($\boldsymbol{E} \cdot \boldsymbol{B} = 0$, $\boldsymbol{E} \cdot \boldsymbol{E} = \boldsymbol{B} \cdot \boldsymbol{B}$) and constant. As modern laser systems have demonstrated values of ξ as large as around 10^2 [10], we are motivated to focus our study on constant crossed fields interacting with the thermal photon gas.

The process of lowest-order pair-creation can be calculated from the polarisation operator, shown in Fig. 1. The optical theorem relates the imaginary part of the forward scattering amplitude to the process acquired by cutting a symmetric diagram in half. In Fig. 1, we show two ways that this can be achieved, which we have calculated agree with one another to leading order. One can either calculate the scattering of free photons in the external field and sum over a thermal Bose-Einstein distribution (the left-hand diagram), or one can calculate the polarisation of the virtual pairs due to an external field when they exchange a thermal photon (the right-hand diagram). A vacuum polarised by a field $F^{\mu\nu}(x)$, with four-co-ordinate x, which varies slowly over the reduced Compton wavelength $\lambda_C = 1/m$, can in general be investigated via the effective Lagrangian formulation of QED [31]. Schwinger's comment referred to this object being identically zero for all plane-wave backgrounds. In a purely plane wave and by extension, a constant crossed field, the only two unique relativistic invariants that one can construct: $-F^{\mu\nu}F_{\mu\nu} = 2(E^2 - B^2)$ and $-F^{\mu\nu}F^*_{\mu\nu} = 4(\boldsymbol{E} \cdot \boldsymbol{B})$, where $F^*_{\mu\nu} = \epsilon_{\mu\nu\rho\sigma}F^{\rho\sigma}/2$ and $\epsilon_{\mu\nu\rho\sigma}$ is the rank-four Levi-Civita symbol (antisymmetric in all indices and $\epsilon^{0123} = 1$), both

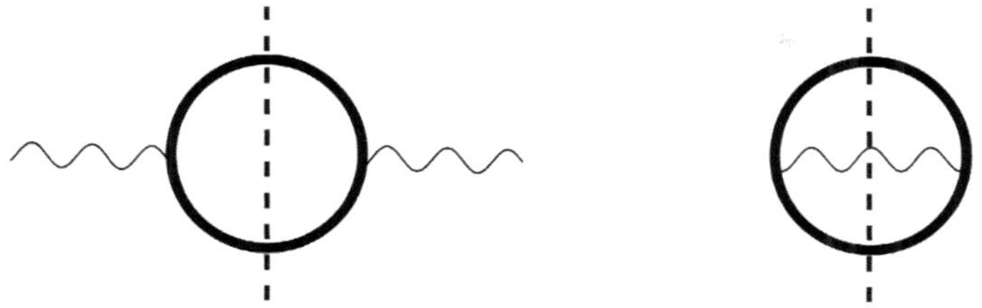

Figure 1. Two different approaches to calculating pair-creation by thermal photons (wavy lines) in a plane wave (in which the virtual electron-positron loop is dressed). Using the optical theorem to split the diagrams in half, summing pair creation by a single photon over a thermal ensemble (left) gives the same leading-order rate as splitting the two-loop polarisation operator with a thermalised internal photon.

vanish identically. Suppose we specialise the discussion to constant fields $F^{\mu\nu}(x) = F^{\mu\nu}$. Then when a thermal bath of photons is present, whose constituent photons have on-shell wavevectors $k^{\mu} = (\omega, \boldsymbol{k})$, one can form a third unique relativistic invariant $(k_{\mu}F^{\mu\nu})^2$, which is in general non-zero. In addition to the classical non-linearity parameter ξ, which diverges for a constant crossed field, QED phenomena in the strong constant fields can also be described with the quantum non-linearity parameter $\chi = e\sqrt{|(k_{\mu}F^{\mu\nu})^2|}/m^3$, directly related to this third, non-zero, relativistic invariant. χ will therefore be the microscopic quantity defining the macroscopic interaction between external field and thermal bath.

We limit ourselves to the most accessible parameter range for experiments, namely $\chi \ll 1$. If the thermal photons have an energy approximately equal to the temperature, then this condition on χ corresponds to $\delta = (T/m)\chi_E = (T/m)(E/E_{\text{cr}}) \ll 1$. Throughout our analysis, we make two key assumptions: i) processes involving larger numbers of thermal photons can be neglected (i.e. the perturbative expansion in thermal photons is valid) and ii) pair creation due to thermal photon-photon inelastic scattering is much less likely than due to thermal photons inelastically scattering with the external-field. The first assumption is valid if $|e|T/m \ll 1$ (in [32] it was shown that a perturbative expansion for the interaction between the photon and electron-positron field is valid when $\alpha\chi^{2/3} \ll 1$ where $\alpha = e^2$ is the fine-structure constant). This limit also allows us to assume the formation region of each pair is much smaller than the average region between pair-generation events, so that an incoherent sum over the thermal ensemble is a good approximation. The validity of the second condition has been already investigated in [29], which we mention here, is fulfilled when $\sqrt{\delta} \gg T/m$.

3. Thermal pair creation rate via sum over ensemble

Before proceeding with the calculations, we note that, although we demand a thermal equilibrium for our result to be exact, the approach of summing over the thermal ensemble can be used to approximate pair-creation rates for physically-relevant scenarios. One such example is the irradiation of solid-state targets such as thin foils by ultra-intense lasers. In a recent experiment, prolific positron creation was observed [33, 34] by irradiating target foils of various proton numbers using an intense laser for which $\xi \gg 1$, allowing processes to be locally approximated with the constant crossed field rates. It was shown that hot photons with energies up to the pair-creation threshold and beyond were generated, allowing a distribution with a high "effective" temperature to also be considered, albeit out of equilibrium. Integration of the pair creation rate over this ensemble could then be used as an approximation to the non-equilibrium rate. It should be mentioned however, that an accurate calculation for pair creation rates in a laser-irradiated relativistic plasma must take into account the many competing channels such as pair production from photon-electron collisions and from the Bethe-Heitler process (photon-ion collisions), which can play a substantial role in the plasma's evolution.

We now turn to the main calculation of this article, pair creation in the interaction of a constant crossed field with a pure thermal gas. In light of the above discussion, we stress that since our analysis assumes a thermal equilibrium, as soon as the first pair is created, the vacuum fermion state no longer exists and the result is no longer exact. Instead, one would have to perform a calculation that took into account effects such as Pauli blocking. Therefore, the word "rate" should more precisely refer to the probability per unit volume per unit time for pair creation to begin, rather than the number of pairs produced per unit volume per unit time. Proceeding via the optical theorem, which for a single photon interacting with a virtual pair in the presence of the external field, relates the rate, $R_a(k, \chi_E)$ to the polarisation operator $\Pi_{\mu\nu}(k, \chi_E)$ via $R_a(k, \chi_E) = \text{Im}[l_a^\mu \Pi_{\mu\nu}(k, \chi_E) l_a^{*\nu}]/\omega$, for normalised photon four-polarisation l_a^μ ($\sum_a l_a^\mu l_a^{*\nu} \to -g^{\mu\nu}$), $a = \{1, 2\}$. For a constant crossed field background (see e.g. [35] for a derivation), the single-photon rate is then given by:

$$R_a(k, \chi_E) = -\frac{2\alpha m^2}{3\omega} \int_4^\infty dv \frac{2v + 1 + 3(-1)^a}{v\sqrt{v(v-4)}} \frac{\text{Ai}'(z)}{z}, \tag{1}$$

where $z = (v/\chi)^{2/3}$, $\text{Ai}(\cdot)$ is the Airy function of the first kind (defined as in [36], with normalisation $1/\pi$).

Since we are interested in cases where $\delta \ll 1$, we can assume that $\chi \ll 1$ is satisfied by the vast majority of photons in the distribution. This allows us to expand Eq. (1), which will be further justified in the discussion, to give:

$$R_a(k, \chi_E) \sim \alpha m \frac{3 + (-1)^a}{16} \sqrt{\frac{3}{2}} \frac{m}{\omega} \chi e^{-\frac{8}{3\chi}}. \tag{2}$$

By summing Eq. (2) over the Bose-Einstein distribution, one can then obtain the probability of a density of pairs being created in the interaction of thermal photons and external field $n_{\text{th}}(T, \chi_E) = dN_{\text{th}}/dV dt$, well-approximated by an integration:

$$n_{\text{th}}(T, \chi_E) = \sum_a \int \frac{d^3k}{(2\pi)^3} \frac{1}{e^{\omega/T} - 1} R_a(k, \chi_E). \tag{3}$$

Inserting the asymptotic rate Eq. (2) into Eq. (3), with the substitutions $\phi = \omega/T$ and $y = 1 - \cos\theta$, with $\cos\theta = \boldsymbol{k} \cdot \boldsymbol{\varkappa}/\omega$, one can see that at small χ, only large values of ϕ contribute to the integral, giving eventually:

$$n_{\text{th}}(T, \chi_E) \sim \frac{3\sqrt{3}T^3 \alpha m \chi_E}{32\sqrt{2}\pi^2} \int_0^\infty d\phi \int_0^2 dy\, \phi^2 y \frac{e^{-\phi}}{e^{8/3\phi y \delta}}. \tag{4}$$

By using the following integral identities, the remaining integrations can be analytically performed,

$$\int_0^\infty dz\, z^2 e^{-z - a^2/z} = 2a^3 K_3(2a), \tag{5}$$

$$\int_0^2 dy\, y^{-1/2} K_3(b/\sqrt{y}) = 4b^{-1} K_2(b/\sqrt{2}), \tag{6}$$

where $K_n(\cdot)$ is the modified Bessel function of the second kind, giving:

$$n_{\text{th}}(T, \chi_E) \sim \frac{\alpha T^2 m^2}{\pi^2} \sqrt{\frac{3}{2}} K_2\left(\sqrt{\frac{16}{3\delta}}\right). \tag{7}$$

The analysis shows that this result is valid at $\sqrt{\delta} \ll 1$ (see [29]), therefore, by using the expansion of the modified Bessel function for large arguments:

$$K_2(a/\sqrt{\delta}) \sim \sqrt{\frac{\pi\sqrt{\delta}}{2a}}\mathrm{e}^{-\frac{a}{\sqrt{\delta}}}, \tag{8}$$

for $a, \delta > 0$, one then acquires [29]:

$$n_{\mathrm{th}}(T, \chi_E) \sim \frac{3^{3/4}\alpha}{4\pi^{3/2}}\, m^4 \left(\frac{T}{m}\right)^2 \delta^{1/4}\mathrm{e}^{-\frac{4}{\sqrt{3\delta}}}, \; \delta = \frac{T}{m}\chi_E. \tag{9}$$

This result for the density of created pairs demonstrates an interesting dependency on environmental variables. It is non-perturbative in both the external field and the temperature, although only lowest-order thermal photons were included. The non-perturbative dependency in the pre-exponent and exponent is also quite different to other results for thermal pair creation rates. For example, the thermal part of the rate of pair creation at finite temperature in a constant electric field, for the relevant case of $T/m, \chi_E \ll 1$, has been shown to be [31, 25] $n_E(T, \chi_E) = (\alpha\pi^2/90)m^4(T/m)^4\chi_E^{-2}\exp(-\pi/\chi_E)$. For a constant electric field, the temperature enters only in a perturbative sense, in the pre-exponent. In contrast, here it occurs on the same footing as the external field – indeed the exponent is the geometric average of the purely thermal pair-creation exponent ($2m/T$, see [29]) and the external-field part of the photon-seeded pair-creation exponent ($8/3\chi_E$). One reason why in a constant crossed field, the temperature could be expected to play such an important role is that when the temperature is zero, the pair creation rate is identically zero. The logarithm of the rate $\log_{10} n_{\mathrm{th}}(T, \chi_E)$ calculated from Eq. (9) for a space-time four-volume $\lambda_C^4 = m^{-4} = 7.4 \times 10^{-53}$ cm^3 s is plotted in Fig. 2. Above the dashed line, the expected number of pairs created in an optical strong laser beam of typical four-volume $\Omega_l = \tau_l V_l$ with duration $\tau_l = 10$ fs and spatial extent $V_l = \pi \times (0.8\,\mu\mathrm{m})^2 \times c\tau_l = 6 \times 10^{-12}$ cm^3 (where we temporarily reinstate the speed of light, c). The parameters used in the plot have been chosen so that the background process of purely thermal pair creation is negligible compared to the external-field process [29]. We can justify using the expression for single-photon pair creation asymptotically small in χ, by comparing these results with a numerical integration of the full single-photon pair creation rate over the thermal ensemble (i.e. inserting Eq. (1) in Eq. (3)). In Fig. 3 we see that the relative difference in using the asymptotic rate remains less than ten percent for values of $\delta = 0.03$. We conclude that Eq. (9) is a good approximation for parameters of interest. That this deviation is a function only of δ can be seen by performing the variable substitution $\phi = \omega/T$ in Eq. (3), leading to the combination $|\Delta n_{\mathrm{th}}(T, \chi_E)|/n_{\mathrm{th}}(T, \chi_E)$ depending only on δ.

4. Thermal pair creation rate via effective action

As displayed in Fig. 1, there are two different approaches one can use to acquire the thermal pair creation rate. Here we concentrate on the second diagram - use of the optical theorem on the two-loop polarisation operator with thermalised internal photons. This is equivalent to averaging the one-loop polarisation tensor over a thermal photon bath, and brings additional insight into the result Eq. (9). Since our calculations are valid in thermal equilibrium, we can use the Matsubara formalism. If we assume that only the photon propagator need be thermalised (this will be justified later), the Matsubara rate n_{M} is given by:

$$n_{\mathrm{M}}(T, \chi_E) = -T\,\mathrm{Im}\int \frac{d^3k}{(2\pi)^3} \sum_{n=-\infty}^{\infty} \frac{1}{k_n^2}\, g_{\mu\nu}\Pi^{\mu\nu}(k_n, \chi_E), \tag{10}$$

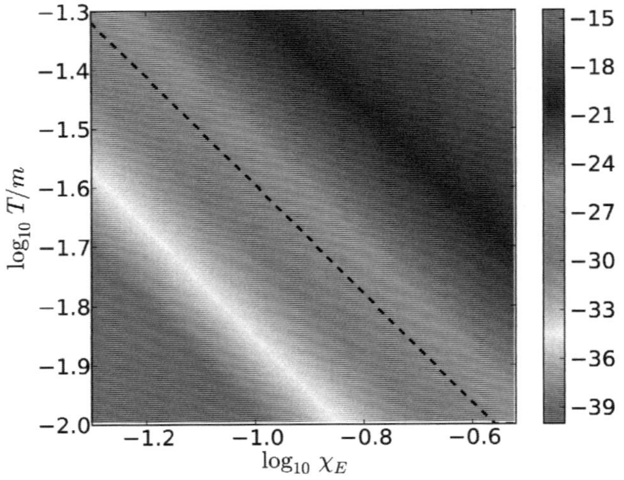

Figure 2. In the plot is the logarithm of the expected number of pairs created in the fundamental space-time volume of dimension reduced Compton wavelength $\lambda_C^4 = 7.4 \times 10^{-53}$ cm^3 s when a thermal photon gas is polarised by a constant crossed field. The number of created pairs is calculated using Eq. (9). The parameters in the plot have been chosen so that the background process of zero-field pure thermal pair creation is negligible compared that due to the external field. The expected number of pairs created by this process in a typical strong optical laser four-volume $\Omega_l = 6 \times 10^{-26}$ cm^{3}s, rises above one for regions above the dashed line.

where $k_n^\mu = (2\pi i n T, \boldsymbol{k})$ and $\Pi^{\mu\nu}(k, \chi_E)$ is the polarization operator in a constant crossed field [37]. If one transforms the sum into a complex integral using an integral relation (see Eq. (4.11) in [38]), then it can be shown that the thermal average rate in the previous section Eq. (3) and the Matsubara rate Eq. (10) exactly coincide. We can use the two-loop effective in the Matsubara formalism for a general constant-field background [25, 31], which can be written as a sum over an integer n of a triple propertime integral. If one takes the crossed-field limit and performs a saddle point expansion of the propertime integrals, Eq. (9) is recovered when the condition $\gamma_T = T^2/m^2\delta \ll 1$ is fulfilled. This restriction on γ_T agrees with assumptions made in the previous section.

For increasing temperature, one should in principle also take into account the thermal fluctuations of the fermion loop. However, it has been shown [31, 39, 40] that for the temperatures range of interest $T \ll m$, the thermal correction to the electrons and positrons is exponentially suppressed by a factor $\sim \exp(-m/T)$. This suppression justifies our reasoning throughout this paper that thermalising only the photon should be a good approximation when $T \lesssim m$. Moreover, our results add to the general trend, that the thermal contribution toward pair creation generally exceeds that of the zero-temperature contribution, in the limit of weak fields (see also [31]). Here it is particularly evident, as the zero-temperature pair creation rate in a constant crossed field is identically zero.

5. Non-dissipative interaction

The interaction between the external field and thermal photon gas is mediated by the polarised virtual pairs of the vacuum, depicted in the polarisation tensor in Fig. 1. In addition to the dissipative process of pair creation, this interaction also provides the non-dissipative process of

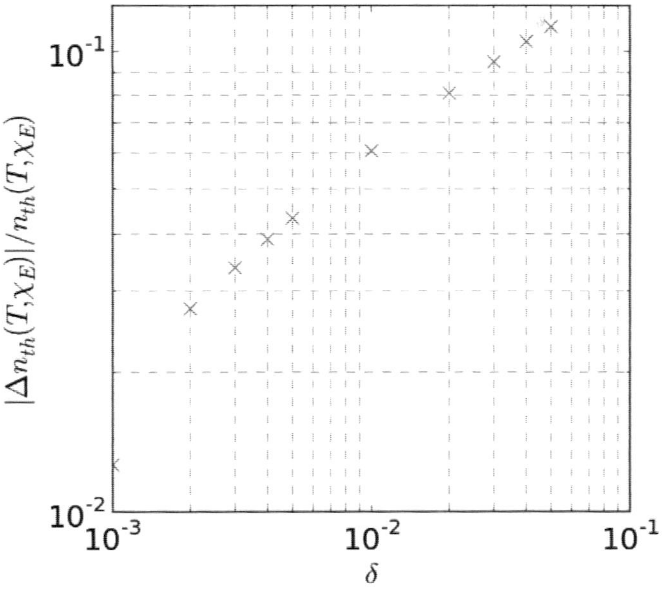

Figure 3. This log-log-plot compares the accuracy of the thermal rate calculated using the asymptotic single-photon pair creation formula for $\chi \ll 1$ compared to using the exact single-photon formula (using Eq. (9) and Eq. (1) in Eq. (3) respectively). The relative difference between the two approaches, $|\Delta n_{\rm th}(T,\chi_E)|/n_{\rm th}(T,\chi_E)$ is plotted as a function of its only dependent variable, δ.

thermal photon scattering in the external field. One can derive [37] an altered dispersion relation for a thermal photon with on-shell four-momentum $k_0^\mu = (\omega_0, \boldsymbol{k}_0)$ in vacuum:

$$\omega_{1,2}(\chi) = \omega_0 - \alpha\frac{11 \pm 3}{180\pi}\frac{m^2}{\omega_0}\chi_0^2, \tag{11}$$

with $\chi_0 = e\sqrt{|(k_{0,\mu}F^{\mu\nu})^2|}/m^3 \ll 1$. Here we make the following observation. Common thermodynamical quantities can be derived from the partition function Z:

$$Z = \mathrm{Tr}\ \mathrm{e}^{-\hat{H}/T} \tag{12}$$

and when the Hamiltonian operator $\hat{H}$ is expanded in thermal photon energy eigenstates, which are modified in the plane-wave background, thermodynamical quantities such as the photon gas pressure, would seem to also, as a consequence, be modified. This hints a new possibility for searching for vacuum polarisation effects - by reading the signatures in thermodynamic changes due to microscopic strong-field QED effects. Here, as the change in refractive index is perturbative in the external field, so are the modifications in such thermodynamical quantities. More details can be found in [41].

6. Discussion and conclusion

We have shown how, when one acknowledges the inevitable presence of photons from a thermal and by extension non-thermal background, in all realistic physical scenarios, pair creation can ensue even in a constant crossed field and hence a single plane wave. This does not contradict

Schwinger's maxim, but demonstrates the contrast of what is predicted to occur in an experiment compared to ideal, theoretical, conditions. Moreover, the probability for this new process of thermal pair creation in a constant crossed field has an interesting and non-trivial form. It is non-perturbative in the external field strength as is the case for photon-seeded pair creation in a constant crossed field, but also non-perturbative in temperature, which enters on the same footing as the external field. Contrasted with non-perturbative pair-creation rates in just an external field, or just a finite-temperature background, the rate when both are present is related to their geometric average. In addition, we have noted that the presence of the external field and the elastic scattering of thermal photons it invokes, will alter the value of thermodynamical variables such as the equilibrium pressure of a photon gas.

References

[1] Sauter F 1931 *Z. Phys.* **69** 742
[2] Heisenberg W and Euler H 1936 *Z. Phys.* **98** 714
[3] Weisskopf V 1936 *Kgl. Danske Videnskab. Selskab, Mat.-fys. Medd.* **14** 6
[4] Schwinger J 1951 *Phys. Rev.* **82** 664–679
[5] Brown L S and Kibble T W B 1964 *Phys. Rep.* **133** A705–A719
[6] Nikishov A I and Ritus V I 1964 *Sov. Phys. JETP* **19** 529–541
[7] 2012 Extreme Light Infrastructure `http://www.extreme-light-infrastructure.eu`
[8] 2012 eXawatt Center for Extreme Light Studies `http://www.xcels.iapras.ru`
[9] 2012 High Power Energy Research `http://www.hiperlaser.org`
[10] Yanovsky V *et al.* 2008 *Opt. Express* **16** 2109
[11] Narozhny N B *et al.* 2004 *Phys. Lett. A* **330** 1–6
[12] Bell A R and Kirk J G 2008 *Phys. Rev. Lett.* **101** 200403
[13] Fedotov A M *et al.* 2010 *Phys. Rev. Lett.* **105** 080402
[14] Sokolov I V *et al.* 2010 *Phys. Rev. Lett.* **105** 195005
[15] Bulanov S S *et al.* 2010 *Phys. Rev. Lett.* **105** 220407
[16] Nerush E N *et al.* 2011 *Phys. Rev. Lett.* **106** 035001
[17] Elkina N V *et al.* 2011 *Phys. Rev. ST Accel. Beams* **14**(5) 054401
[18] Ridgers C P *et al.* 2012 *Phys. Rev. Lett.* **108**(16) 165006
[19] Di Piazza A *et al.* 2012 *Rev. Mod. Phys.* **84** 1177–1228
[20] Di Piazza A *et al.* 2009 *Phys. Rev. Lett.* **103** 170403
[21] Brezin E and Itzykson C 1970 *Phys. Rev. D* **2** 1191
[22] Popov V S 1971 *Sov. Phys. JETP* **13** 185
[23] Dunne G V, Gies H and Schützhold R 2009 *Phys. Rev. D* **80** 111301
[24] Thompson C 2008 *Astrophys. J.* **2** 1258–1281
[25] Gies H 2000 *Phys. Rev. D* **61** 085021
[26] Gavrilov S P and Gitman D M 2008 *Phys. Rev. D* **78** 045017
[27] Kim S P, Lee H K and Yoon Y 2010 *Phys. Rev. D* **82** 025016
[28] Monin A K and Zayakin A V 2008 *JETP Lett.* **87** 709
[29] King B, Gies H and Di Piazza A 2012 *Phys. Rev. D* **86**
[30] Ritus V I 1985 *J. Russ. Laser Res.* **6** 497–617
[31] Dittrich W and Gies H 2000 *Probing the Quantum Vacuum* (Berlin: Springer-Verlag)
[32] Narozhny N B 1980 *Phys. Rev. D* **21**(4) 1176–1183 URL `http://link.aps.org/doi/10.1103/PhysRevD.21.1176`
[33] Chen H *et al.* 2009 *Phys. Rev. Lett.* **102** 105001
[34] Chen H *et al.* 2009 *Phys. Plasmas* **16** 122702
[35] Baĭer V N, Mil'shteĭn A I and Strakhovenko V M 1976 *Sov. Phys. JETP* **42** 961–965
[36] Olver F W J 1997 *Asymptotics and Special Functions* (A K Peters Ltd., 63 South Avenue, Natick, MA 01760: AKP Classics)
[37] Ritus V I 1971 *Ann. Phys.* **69** 555–582
[38] Quirós M 2007 *Acta Phys. Pol. B* **38** 3661–3703
[39] Elmfors P and Skagerstam B S 1995 *Phys. Lett. B* **348** 141–148
[40] Gies H 1999 *Phys. Rev. D* **60** 105002

[41] King B 2010 *Vacuum Polarisation Effects in Intense Laser Fields* Ph.D. thesis University of Heidelberg `http://www.ub.uni-heidelberg.de/archiv/10846`

Further Development of the Laser Tweezers Technique for Biomedical Applications

K Afanasiev, A Korobtsov, S Kotova, N Losevsky, A Mayorova, V Patlan, V Volostnikov

P.N. Lebedev Physical Institute of Russian Academy of Sciences, Samara Branch, 221, Novo-Sadovaya Str., Samara 443011, Russia

E-mail: kotova@fian.smr.ru

Abstract. The results of the experimental work aimed at a functional enhancement of the existing laser tweezers system by introducing a pixel-addressable liquid-crystal spatial light modulator (Holoeye HEO-1080P) are presented. The use of the modulator allows us to generate light fields of complicated structure including ones with the vortex component and control the objects positions in real time. The special software is developed to form an array of optical traps with the number of elements up to thirty two with the capability of individual or subgroup control. The method of a spatial separation of the modulator aperture is implemented. The ability to control the lateral power distribution of the light field as well as the value of its orbital moment brings new possibilities of a precise manipulation of microobjects including biological ones.

1. Introduction

Since the time of the laser tweezers introduction they are extensively used in biology to manipulate different biological objects. There are several trends in the studies on the application of the optical manipulation in biology and medicine carried out by several research groups (see e.g. reviews [1, 2] and the references therein). A large number of studies are focused on a quantitative evaluation of fundamental properties of various biological objects for example, nucleic acids, proteins, isolated organelles and whole cells (RBCs, white blood cells, etc.) with the use of optical traps. Changes of biological objects characteristics under the effects of various internal and external factors, as well as during some diseases are also studied (e.g. [3-6]). As a rule, the standard Gaussian single-beam traps or dual-traps are used for these studies.

Another line of the researches is represented by the works on further enhancements of the trapping technique itself, particularly maintaining 3D capture spatial stability [7, 8], possibility of the real-time process control, employment of the beams with dedicated complex configurations and demonstration of the applicability of the method to alive objects such as E-coli [9], Saccharomyces cerevisiae [10] and other bacteria, RBCs [8, 11, 12], white blood cells.

In our previous studies [13, 14] we have demonstrated that the functional capabilities of the laser manipulators could be significantly enhanced owing to the use of beams with predetermined distributions of the intensity and angular momentum specifically so called spiral beams. Beams with the nonzero angular momentum were generated by using the phase and amplitude–phase masks. Such beams made it possible to rotate weakly absorbing particles and move them along a given trajectory.

Experimental results on the microobject motion along various trajectories (circle, triangle contour, self-intersecting curve, and Archimedes spiral) were presented.

Liquid crystals spatial light modulators (LS SLM) essentially expand the possibilities of the laser tweezers and are extensively used for optical trapping (e.g. [15, 16, 17]). In this study we further extend the LC SLM-based tweezing technique. In particular, we have explored the formation of optical traps and their arrays of different types as well as vortex fields by means of the LC SLM. Optical fields of the predetermined spatial structure are promising for the ordering microbjects into various spatial configurations, objects sorting and for the use as the optical shield. Vortical traps open up new possibilities for the control of microobjects motion, rotation and deformation. The advantages of the LC SLM use for the formation of vortex traps are also demonstrated. Though our experiments were carried out with the latex spheres, we believe that the mechanical effects demonstrated here will be useful for biomedical technology applications as well.

2. The experimental setup

Figure 1 shows a typical scheme of the setup for the laser manipulations with the built-in spatial light modulator.

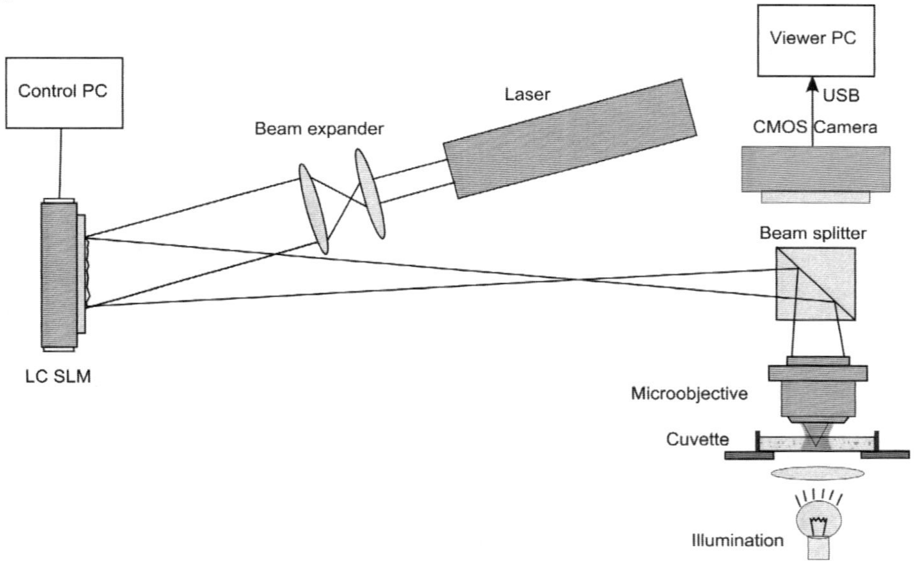

Figure 1. The scheme of experimental setup.

A diode-pumped frequency-doubled solid-state laser was used in our experiments as the light source. The laser wavelength was 0.53 μm and maximal output power was 500 mW. Along with the main radiation, a frequency component with λ=1.06 μm was also present in the output spectrum. CMOS photodetectors are highly sensitive to this wavelength which may lead to strong flare lighting. To optically isolate the detector from a parasitic radiation, the green filter was inserted in the path of the beam. The filter was placed behind the beam expander to prevent the formation of the thermal lens in the body of the filter, which could occur due to the IR-radiation absorption by the filter. The expanded beam with a quasihomogeneous intensity distribution was directed on the spatial light modulator LC SLM HOLO EYE 1080 with the number of controlled elements 1920×1080 and ability to specify 256 phase gradation within the range from 0 to 2π.

To provide the adequate operation of the modulator, the optical scheme was arranged in such a way that the incident laser radiation angles differed from the right angles by less than 7^0. The required phase distribution of the modulator was achieved with the aid of a separate computer via the DVI port of the video-card. The light field formed with the LC SLM was directed into the objective of the modernized microscope with a beam splitter. A dielectric mirror was used as the beam splitter, with the reflectance of 70% at the incident angles of 45^0 for radiation at λ=0.53 μm. The mirror was made using 1 mm thick glass substrate. During the experiments on manipulation, the immersion micro-objective with a numerical aperture of 1.2 was used. Calibrated latex microspheres of various diameters suspended in water served as the objects for the laser trapping and manipulation.

The sequence of microscopic images for the observation and analysis was captured and recorded with the 1.3 Mpixel CMOS camera (DCIM130) connected to the PC through the USB interface.

3. The formation of optical traps with the LC SLM and experiments on microobjects dynamic laser manipulation

The special software is developed to form an array of optical traps with the number of elements up to thirty two with the possibility of individual or subgroup control. The method of a spatial separation of the modulator aperture is implemented. The advantages of this method are the high diffraction efficiency and the complete independence of every element control. For each individual domain it is possible to specify its wedge, spherical and cylindrical components as well as a component corresponding to the axicon. Thus the traps in the shape of points, circles, line segments with individually controllable spatial characteristics or other arbitrary distributions are formed in the plane of manipulation. The examples of the phase distributions and the corresponding intensities in the plane of manipulation for 8 independent traps are presented in Figure 2. Three point traps, three traps in the form of rings and two traps in the form of the line segment (both vertical and horizontal) are shown here.

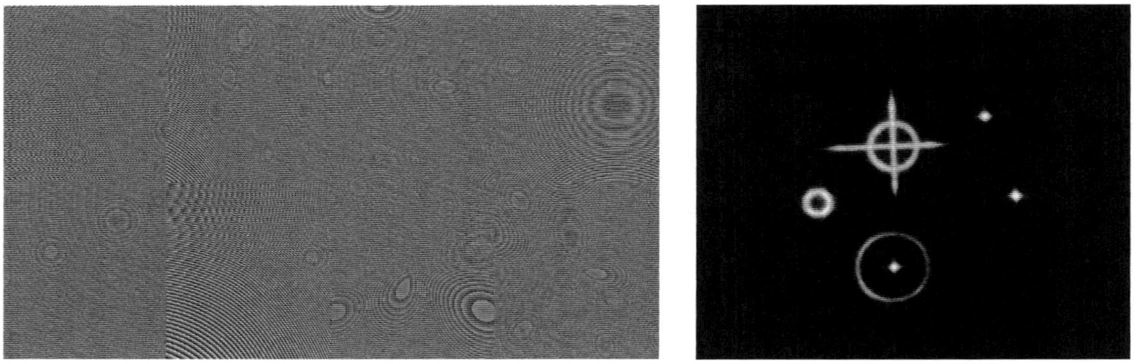

a b

Figure 2. The formation of optical traps. The phase distributions (a) and the corresponding far field intensities (b).

It is possible to form single traps or an array of up to 32 traps of specified forms (points, circles or line segments). Thus the trapping of up to 32 objects can simultaneously be realized with the capabilities of an independent dynamic control of both a separate trap and of the traps group in whole or in part. It should be noted that the LC SLM are extensively used for the formation and dynamical motion of the group of points traps [15-17] and as a rule, the so-called holographic optical traps are formed with the LC SLM. In this study we demonstrate the capabilities of the manipulation with the complicated configuration traps, for example in the form of a circle and line segment (shown below) as well as the traps on the base of vortex fields (presented in Section 4).

The experiments on the dynamic laser manipulation of microobjects were carried out with the use of the experimental setup with the built-in LC SLM presented in Figure 1. The laser trapping of the 1.2 μm-diameter latex particles was realized with the use of the ring-shaped optical trap that subsequently changed its shape from a ring to an ellipse. The arrangement of captured particles is changed accordingly, as it is illustrated in Figure 3. One of the possible applications for this manipulation is the deformation of biological objects, cells for instance. To achieve this effect, one should form the ellipse of a diameter comparable with the diameter of the cell.

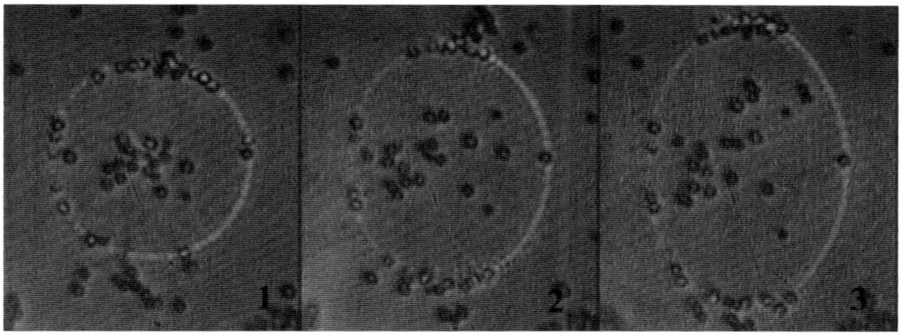

Figure 3. The change of the trap form from ring to ellipse (Media 1). The laser power is 100 mW.

Figure 4 demonstrates the experiment by means of the optical trap in the form of a line segment. The latex spheres of 1.2 μm-diameter are used as laser manipulated objects. The line segment operates as the optical shield for particles since a pure area is formed in the fluid flow. The line segment optical trap can be applied for the separation of objects by their size due to the strong dependences of line segment trapping power on the objects size. The ratio of the trapping powers for 3.2 μm spheres and 1.2 μm spheres was evaluated. This ratio was 4.2 for the trap in the form of line segment and 1.4 for the point trap.

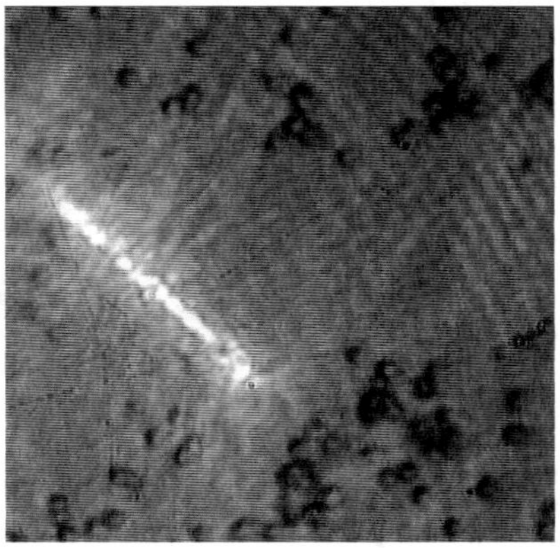

Figure 4. The laser manipulation with the use of the optical trap in the form of a line segment (Media 2). The laser power is 100 mW.

4. Experiments on manipulation with vortex fields

The vortex fields have non-zero orbital momentum and can be formed in the shape of curves. Owing to this they can be used to rotate the absorbing particles around the beam axis, to move them along a fixed trace and to apply non-uniform deformations. The experiments on a continuous rotation and control of the angular position of microobjects by means of the fields characterized by the non-zero angular momentum are carried out in many laboratories (e.g. see reviews [18, 19]). The advantageous feature of our experiments is the possibility to move the objects along the trajectories of widely various forms.

Our work is based on the use of the so-called spiral beams. These beams were for the first time discovered and developed by E. Abramochkin and V. Volostnikov [20, 21]. Basing on the spiral beams optics and Gerchberg-Saxton iterative algorithm, we have developed a new way of the phase masks computation [13]. We used the phase distribution of the corresponding spiral beam as the zero

Journal of Physics: Conference Series **414** (2013) 012017 doi:10.1088/1742-6596/414/1/012017

approximation and therefore the vortex light fields could be formed by means of only the phase mask with the high efficiency of the energy conversion (about 80%). The use of only the phase masks is really important for the qualitative dynamic manipulation. The light fields in the shape of the boundary of a triangular, square, Archimedes spiral and other contours were formed with the use of diffractive optical elements on the dichromate gelatin. And the motion of micron-size objects along these trajectories was experimentally demonstrated. The motion was achieved without any mechanical shifts in the system of the laser tweezers, only by means of the transmission of the light beam momentum to the particles. Note, that the spatial distribution of the transverse component of the momentum is determined by spatial phase structure of the light field.

The use of LC SLM HOLO EYE 1080 in the system of the laser tweezers opens up the new capabilities and advantages for the dynamical laser manipulation. The experiments with vortex fields formed by means of the LC SLM were carried out.

The frames illustrating the particles movement along two concentric rings formed with the vortex axicons are shown in Figure 5. The phase distribution needed for the forming of the respective intensity distribution is presented in Figure 6. One can see that the particles movement along the ring-shape trajectory is realized due to the transmission of the light beam momentum to the particles, and this is consistent with the phase distribution of the formed vortex fields (Figure 6.). Since the topological charges of the inner and exterior rings have the opposite signs (in other words, the vortex twists of the light fields are oriented in opposite directions for the inner and exterior rings), the particles move in opposite directions along the inner and exterior rings. Going back to Figure 3 you can see that the particles were trapped within the maximal intensity region, while the movement along the trajectory was not observed during that experiment since the used light field was not vortex.

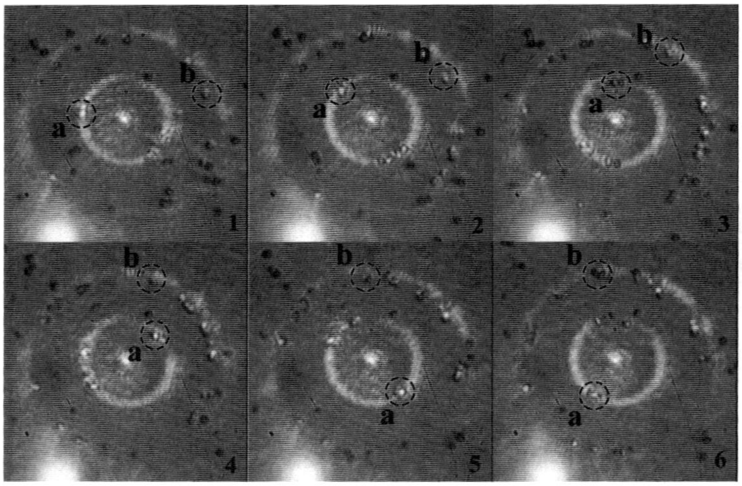

Figure 5. The particles movement along the inner and exterior rings of the vortex optical traps according to the phase distribution (for example, the particle "a" moves clockwise and the particle "b" moves counterclockwise) (Media 3).

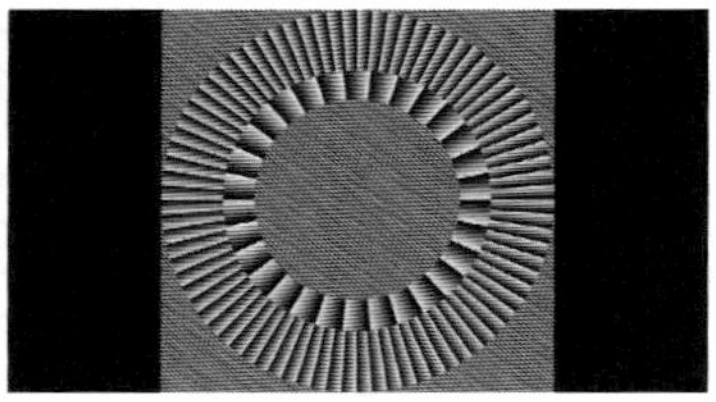

Figure 6. The phase distribution.

The motion of microobjects of various diameters (0.46 μm, 1.2 μm, 3.2 μm) along the Archimedes spiral was carried out. The light field was formed by means of the LC SLM. The laser power was 140 mW, Archimedes spiral's radius was 8.2 μm and light line width (for half the intensity value) was about 1 μm. The results of the angular velocity motion (ω) dependence (at the second spiral turn) on the particles size (D) are presented in Table 1.

Table 1. The dependence of the angular velocity of the motion (ω) on the particles size (D)

D, μm	ω, degree /second
0.46	25±3
1.2	37±4
3.2	27±2

It is seen from the table that the maximal angular velocity of the motion was observed for the spheres of 1.2 μm size. This fact could be explained by the following. For the particles of the diameter less than the beam width the amount of the light beam angular momentum transmitted to the particles increases as square of the particle diameter (D^2) due to the increase of the sectional area of microspheres. At the same time the force of the viscous drag increases proportionally to the diameter (D). Thus the increase of the microspheres diameter to the size close to the spiral turn width results in the increase of the microspheres angular velocity. For the case of the particles with the diameter more than the width of the light line forming the spiral, a different dependence is observed. The particle sectional area intercepting the light field increases approximately as the first degree of the particle diameter (D). Therefore the particles of this size and more have to move with the equal velocities. A certain decrease of the velocity with the increase of their diameters observed within the given size range could be explained by the increase of the friction force on the substrate. This force is proportional to the mass of the particle i.e. to the third degree of the diameter, D^3. Besides, the microobjects oscillations in the transverse plane were observed during their motion along the trajectory. It is the result of the Brownian motion effect due to the thermal motion of liquid molecules, since the particles are very small and the experiments were carried out at the room temperature. The influence of the Brownian motion is especially appreciable for the particles of the submicron size.

It should be noted that generally the rate of movement depends on the geometrical parameters of particles, their refraction indexes and the environment, as well as the laser power and light field spatial structure. The comparison of the experiments with the use of the dichromate gelatin masks and those with the use of the LC SLM, demonstrates that in the case of the light fields formation by means of the LC SLM it is possible to reach the increase of the particle velocity (about by 3 times) due to the adaptable capabilities to form the light field thoroughly.

The advantages of the use of multielement modulators in laser manipulation are the capability of the fast load of the phase distribution on LC SLM array. It is resulted in the abilities of the fast change of the motion direction and dynamical control of the trajectory form. In Figure 7 the change of the motion direction of the 1.2 μm diameter latex spheres with the load of the phase distribution with direct and invert phase on LC SLM array is presented. Also the change of the form trajectory of the particle's motion from Archimedes spiral to triangle form was realized.

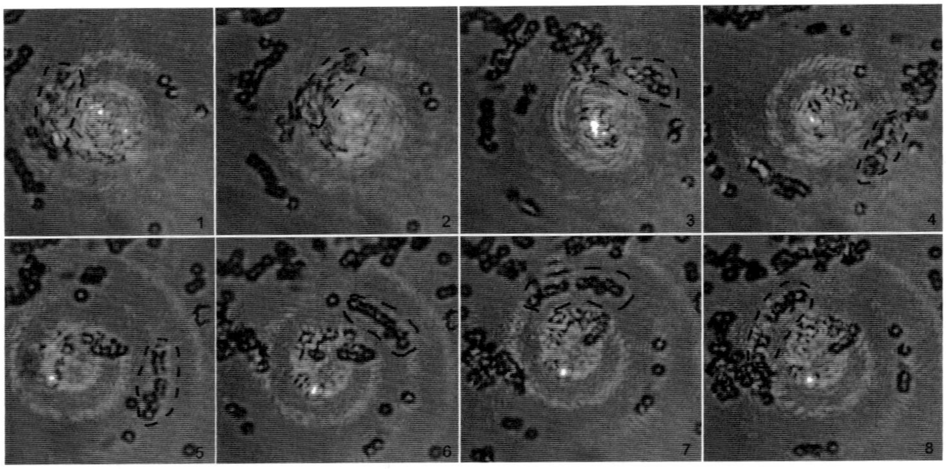

Figure 7. Fast change of the motion direction of the 1.2 μm diameter latex spheres (Media 4). In frames 1-4 the particles move clockwise while in frames 5-8 the particles move counterclockwise wise.

5. Conclusion

Several methods for generating various traps and dynamic control of the manipulation process are presented in this study. The use of the vortex fields in the form of curves makes it possible to move the micro-objects along these curves. The light field formation with the help of the phase LC SLM permits to produce a universal set-up capable of a dynamic change of the traps types, including a single-point trap, point traps arrays, vortex traps in the form of curves. The results obtained can be used for a further development of new technologies in biophotonics and micromechanics.

The work was supported by the Special Federal Program "Scientific and scientific-pedagogical personnel of the innovative Russia" (2009–2013), (National Contract № 14.740.11.0063).

6. References

[1] Zhang H and Liu K-K 2008 *J. R. Soc. Interface* **5** 671
[2] Banerjee A and Chowdhary S 2011 *J. Biomedical Optics* 16 051302-1
[3] Bambardekar K, Dharmadhikari A, Dharmadhikari A, Mathur D and Sharma Sh 2008 *J. Biomedical Optics*, **13**, 064021-1
[4] Fontes A, Barjas Castro M, Brandao M, Fernandes H, Thomaz A, Huruta R, Pozzo L, Barbosa L, Costa F, Saad S and Cesar C 2011 *Journal of Optics* **13** 044012
[5] Rancourt-Grenier S, Wei M-T, Bai J-J, Chiou A, Bariel P, Duval P-L and Sheng Y 2010 *Opt. Express* **18** 10462
[6] Korobtsov A, Kotova S, Losevsky N, Mayorova A, Patlan V, Timchenko E, Lysov N and Zarubina E 2012 *Laser Physics* **22** (7) 1265
[7] Kumar R, Shakher C and Mehta D 2010 *Laser Physics* **20** 1514
[8] Sheu F-W, Lan T-K, Lin Y-Ch, Chen S and Ay C 2010 *Opt. Express* **18** 14724
[9] Li X and Sun X 2010 *Laser Physics* **20** 1774
[10] Skidanov R, Rykov M, Iannacchione G and Krivoshlykov S 2012 *Computer optics* **36** 377-386
[11] Spyratou E, Makropoulou M and Serafetinides A, 2011 *Proc. of SPIE-OSA Biomedical Optics* **8092** 80920T
[12] Gu M, Kuriakose S and Gan X 2007 *Opt. Express* **15** 1369
[13] Abramochkin E, Afanasiev K, Korobtsov A, Kotova S, Losevsky N, Mayorova A, Rakhmatulin M and Volostnikov V 2006 *Laser Physics* **16** (5) 842-848

[14] Abramochkin E, Afanasiev K, Volostnikov V, Korobtsov A, Kotova S, Losevsky N, Mayorova A and Razueva E *Bulletin of the Russian Academy of Sciences: Physics* 2008 **72** (1) 68–70
[15] Bowman R, Gibson G and Padgett M 2010 *Opt. Express* **18** 11785–790
[16] Jesacher A, Maurer C, Bernet S, Schwaighofer A and Ritsch-Marte M 2008 *Opt. Express* **16.** 4479–86
[17] Chapin C, Germain V and Dufresne E 2006 *Opt. Express* **14.** 13095–100
[18] Dholakia K and Čižmár T 2011 *Nature Potonics* **5** 335
[19] Padjett M and Bowman R 2011 *Nature Potonics.* **5** 343
[20] Abramochkin E and Volostnikov V 1993 *Optics Comm.***102** 336-350
[21] Abramochkin E and Volostnikov V 1996 *Optics Comm.* **125** 302-323

Journal of Physics: Conference Series 414 (2013) 012018 doi:10.1088/1742-6596/414/1/012018

Noninvasive control of dental calculus removal: qualification of two fluorescence methods

S. Gonchukov[1], A. Sukhinina[1], D. Bakhmutov[2], and T. Biryukova[2]

[2]National Nuclear Research University MEPhI
[1]Moscow State Medical Dental University

Abstract

The main condition of periodontitis prevention is the full calculus removal from the teeth surface. This procedure should be fulfilled without harming adjacent unaffected tooth tissues. Nevertheless the problem of sensitive and precise estimating of tooth-calculus interface exists and potential risk of hard tissue damage remains. In this work it was shown that fluorescence diagnostics during calculus removal can be successfully used for precise noninvasive detection of calculus-tooth interface. In so doing the simple implementation of this method free from the necessity of spectrometer using can be employed. Such a simple implementation of calculus detection set-up can be aggregated with the devices of calculus removing.

1. Introduction

The periodontitis is one of the most frequent dental diseases. Its development is accompanied with the calculus growth on the teeth surface. Dental calculus contains microorganisms, which vital activity results in the destruction of the periodontal ligament and even loss of the teeth. Calculus removal is the main condition of periodontitis prevention. Its full removal is the principal aim of the treatment. On the other hand such a medical procedure must not result in tooth tissue harming. The problem is accentuated in the case of subgingival calculus removal as it covers the dental root in periodontal chamber [1-2]. The desire of complete calculus removal leads to fatal hard tissue damage because the procedure of calculus removal must be stopped precisely at calculus–tooth interface. Residual calculus detection by tactile, visual, radiograph, ultrasonic, optical coherence tomography, light scattering, IR and THz imaging are not always precise, convenient and cheap [2–7].

In view of its potential for dental diagnostics, fluorescence spectroscopy in optical range has been the object of research for the last years [8-23]. Such a diagnostics is sensitive and noninvasive. However the opportunities of calculus–tooth interface detection using fluorescence technique have not been investigated yet. The aim of this work is to determine the optimal spectral bands of excitation and registration of the fluorescence necessary for precise detection of calculus–tooth interface. We also aim to optimize the simple implementation of the method free from necessity of spectrometer using.

2. Materials and methods

10 human teeth extracted due to the reason of periodontitis of considerable degree were submitted to the in vitro experiments. The patients were from 40 to 50 years old. The fluorescence was stimulated by cw He-Ne (633 nm) laser and LED (369 nm) irradiation. The power density of irradiation did not exceed 100 mW/cm^2. The fibre-optic spectrometer LESA-5 on the basis of diffraction grating and linear array detector was used for the spectral analysis of fluorescence. The laser and LED beams were transported through a fibre bundle to the tip of the handpiece within a central fibre. Six surrounding fibres were arranged around this central fibre what allowed the fluorescent light emitted from the irradiated spot to be collected. The fibre tip was situated close to the investigated locus on the tooth surface. Probing area was equal to 200 μm. The wide-band cut-off glass filters KS-18 and ZhS-10 were used to suppress diffused exciting light and surrounding light.

The simple implementation of calculus detection set-up free from necessity of spectrometer using was proposed in [24]. The investigated tooth was illuminated by laser or LED directly or with the help of a fibre. Emitted signal from irradiated spot was collected by the fibre with the tip close to the investigated locus on the tooth surface and detected by photomultiplier. The measurements were fulfilled in direct-current and alternating-current regimes. In the second case exciting light was modulated by the obturator at the frequency of 500 Hz and the emitted signal was recorded in the frequency band 50 Hz. In such kind of implementation the correct choice of type and thickness of wide-band cut-off glass filter in front of a photomultiplier is important as the diffused exciting light must be completely suppressed, while the Stokes components of fluorescence pass a filter without essential losses.

The device SMA-30 was used for layer by layer dental calculus removal using rotating sapphire fraizes. Calculus thickness was determined with the help of optical coherence tomography with longitudinal spatial resolution at a level of 10 μm [25].

3. Results

As it was shown recently [24] the maximum difference between fluorescence intensity from dental calculus and tooth root takes place in two spectral ranges: 620 – 645 nm and 340 – 370 nm. This fact can be used as a basis for calculus–tooth interface detection.

Spectra of fluorescence excited at wavelengths of 369 nm and 633 nm are shown in Fig. 1. It is seen from Fig. 1, that the fluorescence intensity of calculus exceeds its value for dental tissue at the excitation by red light, while at the ultraviolet excitation this situation is opposite because of different optical properties of fluorophores and structures of irradiated objects in these spectral domains. Typical tomogram of tooth with calculus, one part of which was removed by sapphire fraize, is presented in Fig. 2.

Experimental dependencies of fluorescence intensity as a function of dental calculus thickness are shown in Fig. 3. The simple implementation of calculus detection set-up free from necessity of spectrometer was used for these measurements. In comparison with spectral measurements on the basis of spectrometer LESA-5 in this case the difference between the fluorescence signals from calculus and from the root is more essential as these signals are detected using wide-band filters but not narrow band (10 nm) of spectrometer.

For convenience the fluorescence intensity from the calculus with initial thickness d_0 = 120 μm is designated by a number 1 for both excitation wavelengths. As can be seen, intensity of fluorescence excited at wavelength of 633 nm decreases approximately on one and a half order. In this case the calculus fluorescence is the principal contribution to fluorescence signal. Reduction of this signal is well described by exponent $\exp[-440(d_0 - d)]$, considering the decrease of exciting and fluorescence light in calculus [26]. The behavior of fluorescence intensity excited at wavelength of 369 nm differs from the one described above. In such a case the fluorescence intensity increases four orders. As a result the difference between the fluorescence intensities excited at wavelength of 633 nm and of 369 nm in the vicinity of calculus-tooth interface is more than five orders.

At the calculus thickness 60 μm and more the detected signal is determined by both the calculus fluorescence and the root fluorescence. At $d < 60$ μm the fluorescence from the root predominates. Therefore the detected signal quickly increases while approaching calculus-tooth interface.

Thus, the fluorescence excitation in ultraviolet has the obvious advantage in comparison with the excitation by red light. Indeed, the fluorescence intensity excited at wavelengths of 369 nm is more than five orders the fluorescence intensity excited at wavelengths of 633 nm in the vicinity of calculus-tooth interface. A gain in fluorescence signal near interface can be used for the development of the system controlling the calculus removing operation. By estimate, using 10% change of fluorescent signal could guarantee positioning near $d = 0$ with the error at a level 2 μm. Such a quality of calculus removing has never been achieved.

4. Conclusion

It is shown that the fluorescence excited at near ultraviolet is the most suitable to precise calculus-tooth interface detection. The simple set-up free from the necessity of spectrometer using can be successfully used for these measurements. This set-up can aggregate with the devices of dental calculus removing. It should also be noted that compact, economical and low cost LED can be used for the fluorescence excitation.

References

1. J. Eberhard, H. Ehleres, W. Falk, Y. Acil, H-K. Albers, and S. Jepsen. J. Clin. Periodontol, **30**, 511 (2003)
2. G. Meisner, B. Oehme, J. Strackeljan, and T. Kocher. J. Clin. Periodontol, **33**, 195 (2006)
3. P.R. Sherman, L.H. Hutchens, L.G. Jewson, J.M. Moriarty, G.W. Greco, and W.T. McFall, J. Periodontol. **61**, 3 (1990)
4. J.S. Rees, M. Addy, and J. Hughes, J. Clin. Periodontol. **26**, 106 (1999)
5. A. Tugnait, V. Clerehugh, and P.N. Hirschmann, J. Dent. **28**, 219 (2000)
6. T. Kocher, J. Strackeljan, and D. Behr, J. Dent. Res. **79**, 829 (2000)
7. Y.C. Sim, I. Maeng, and J.-H. Son, Curr. Appl. Phys. **9**, 946 (2009)
8. F. Krause, A. Braun, and M. Frentzen, Laser Med. Sci. **18**, 32 (2003)
9. E. Kurihara, T. Koseki, K. Gohara, T. Nishihara, T. Ansai, and T. Takehara, J. Periodont. Res. **39**, 59 (2004)
10. W. Buchalla, A´ .M. Lennon, and T. Attin, J. Periodont. Res. **39**, 327 (2004)
11. F. Schwarz, K. Bieling, S. Venghaus, A. Sculean, S. Jepsen, and J. Becker, J. Clin. Periodontol. **33**, 200 (2006)
12. Y.L. Qin, X.L. Luan, L.J. Bi, Z. L¨u, Y.Q. Sheng, G. Somesfalean, C.N. Zhou, and Z.G. Zhang, J. Photochem. Photobiol. **B 87**, 88 (2007)
13. D.E. Clarke, J. Vet. Dent. **16**, 165 (1999)
14. W.C. Dolowy, M.L. Brandes, M. Gouterman, J.D. Parker, and J. Lind, J. Vet. Dent. **12,** 105 (1995)
15. D. Bakhmutov, S. Gonchukov, O. Kharchenko, O. Nikiforova, and Yu. Vdovin, Laser Phys. Lett. **1**, 565 (2004)
16. D. Bakhmutov, S. Gonchukov, O. Kharchenko, O. Voytenok, and B. Zubov, Laser Phys. Lett. **5**, 375 (2008)
17. D. Bakhmutov, S. Gonchukov, and A. Sukhinina, Laser Phys. Lett. **7**, 384 (2010)
18. M.L. Sinyaeva, Ad.A. Mamedov, S.Yu. Vasilchenko, A.I. Volkova, and V.B. Loschenov, Laser Phys. **14**, 1132 (2004)
19. Q.G. Chen, B. Lin, Z.B. Chen, H.H. Zhu, and H. Chen, Laser Phys. Lett. **7**, 752 (2010)
20. S. Gonchukov, A. Sukhinina, and D. Bakhmutov, Kratkie Soobshchenia Po Fizike P.N. Lebedev Physics Institute of the Russian Academy of Sciences, issue **6**, 20 (2010), in Russian
21. D. Bakhmutov, S. Gonchukov, A. Sukhinina, and O. Yanushevich, Stomatologiya **89**, 27 (2010), in Russian
22. D. Bakhmutov, S. Gonchukov, A. Sukhinina, and O. Yanushevich, Ortodontiya **1**, 22 (2010), in Russian

Journal of Physics: Conference Series **414** (2013) 012018 doi:10.1088/1742-6596/414/1/012018

23. D. Bakhmutov, S. Gonchukov, and A. Sukhinina. Fluorescence spectroscopy of dental calculus. Laser Phys. Lett. **7**, 384 (2010)
24. S. Gonchukov, T. Biryukova, A. Sukhinina, and Yu. Vdovin. Laser Phys. Letters, **7,** 812 (2010)
25. Handbook on Optical Coherence Tomography (Fismatlit, Moscow, 2007), in Russian
26. Angewandte Lasermedizin. Laser- und Medizin-Technologie, Berlin, 1997

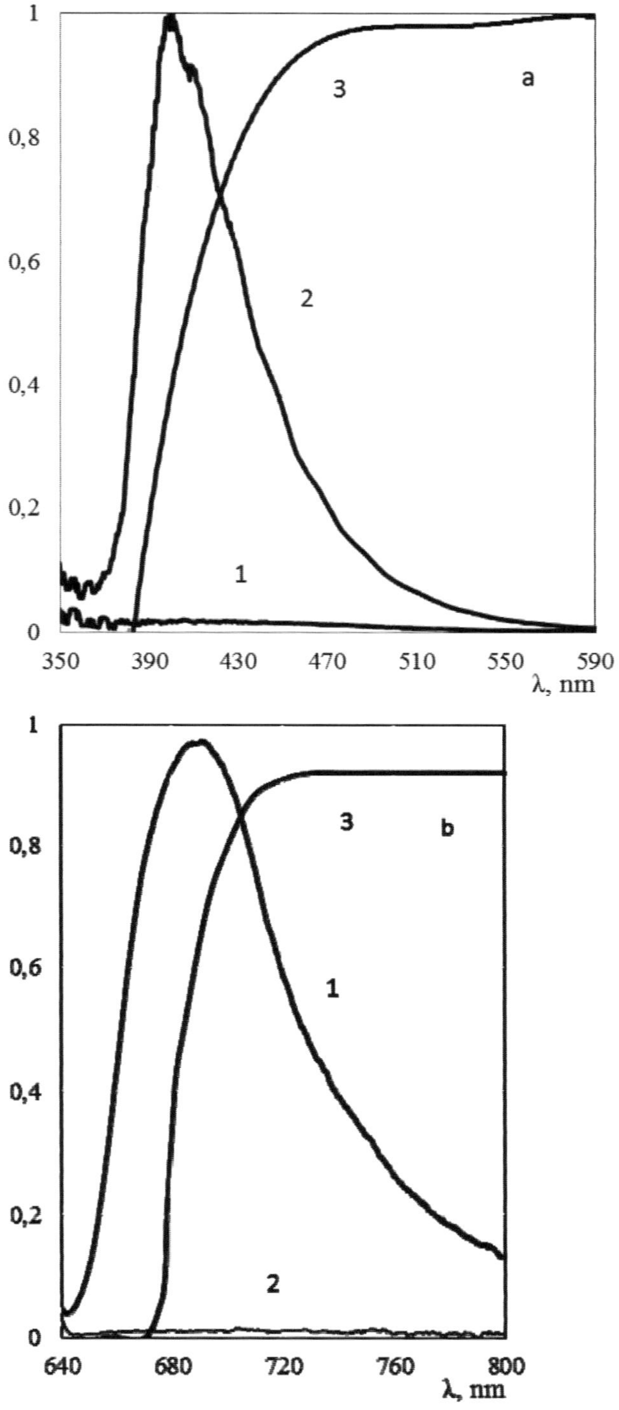

Fig. 1. Spectra of the fluorescence excited at wavelengths of 369 nm (a) and 633 nm (b): 1 – calculus, 2 – root, 3 – transmittances of ZhS-10 (a) and KS-18 (b) wide-band cut-off filters.

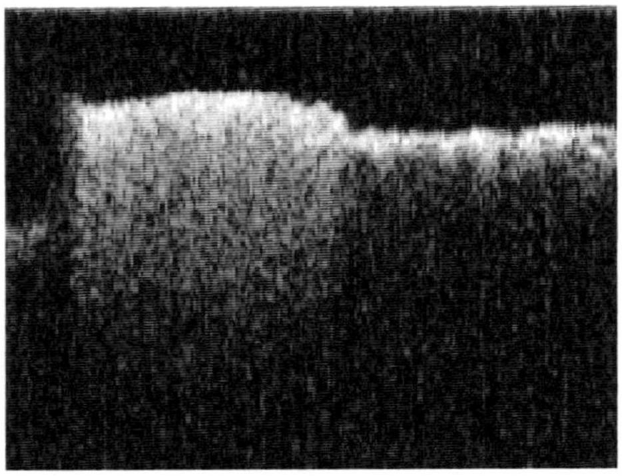

Fig. 2. Typical OCT image of the tooth with fractionally removed (right-hand site of image) calculus. Vertical dimension of the image is equal to 1 mm.

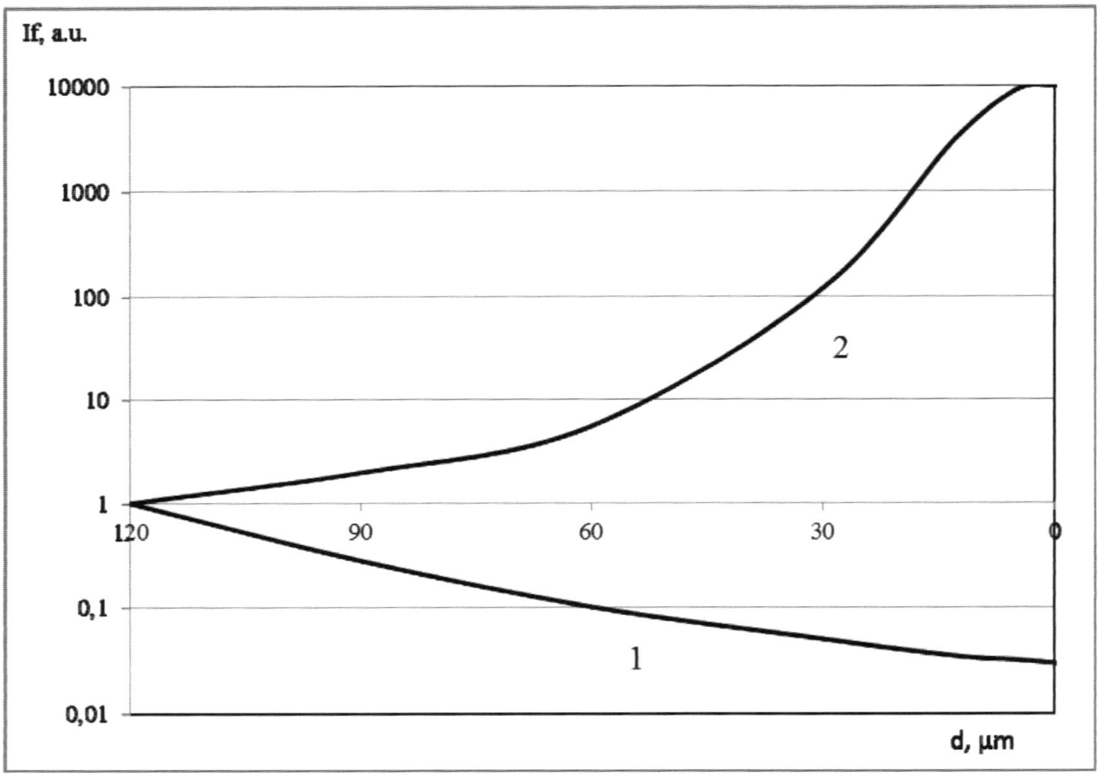

Fig. 3. Experimental behaviour of fluorescence intensity excited at wavelengths of 633 nm (1) and 369 nm (2) as a function of layer by layer removed calculus.

Magnetic nanoparticles as a seed layer for growing ZnO nanowires for optical applications

M. S. AlSalhi[a,b], M. Atif[a,b,f], Anees A. Ansari[c], K. Khun[d], Z. H. Ibupoto[d], M. Willander[d,e]

[a]Physics and Astronomy Department, College of Science, King Saud University, Riyadh, Saudi Arabia
[b]Research Chair for Laser Diagnosis of Cancer, King Saud University, Riyadh, Saudi Arabia
[c]King Abdullah Institute for Nanotechnology, King Saud University
[d]Physical Electronic and Nanotechnology Division, Department of Science and Technology, Campus Norrköping, Linköping University, Sweden
[e]Visiting Professor College of Science, King Saud University, Riyadh, Saudi Arabia
[f]National Institute of Laser and Optronics, Nilore, Islamabad, Pakistan
Corresponding Author: Dr M Atif, atifhull@gmail.com

Abstract

In the present work, cerium oxide CeO_2 nanoparticles were synthesised by sol-gel method and used for the growth of ZnO nanorods. The synthesised nanoparticles were studied by x-ray diffraction technique [XRD]. Furthermore, these nanoparticles were used as seed layer for the growth of ZnO nanorods by following the hydrothermal growth method. The structural study of ZnO nanorods was carried out by using field emission scanning electron microscopy [FESEM], and x-ray diffraction [XRD] techniques. This study demonstrated that the grown ZnO nanorods are well align, uniform, good in crystal quality and possess diameter of less than 200 nm. Energy dispersive x-rays [EDX] revealed that the ZnO nanorods are only composed of zinc, cerium as seed atom and oxygen atoms and no any other impurity in the grown nanorods. Moreover, photoluminescence [PL] approach was applied for the optical characterisation and it was observed that the near-band-edge emission [NBE] was same to that of zinc acetate seed layer, however the green emission and orange/red emission peaks were slightly raised due to possible higher level of defects in the cerium oxide seeded ZnO nanorods. This study provides an alternative approach for the synthesis of controlled ZnO nanorods using cerium oxide nanoparticles as seed nucleation layer which in reverse describe the application of these nanoparticles as well as due to controlled morphology of ZnO nanorods the performance of nanodevices based on ZnO can be increased using these particles as seed.

Keywords: sol-gel method, ZnO nanorods, field emission scanning electron microscopy, photoluminescence, x-ray diffraction

1. Introduction

Recently, the researchers are paying more attention towards manganese oxide nanomaterials because of their tremendous applications in the research areas of catalysis, electrochemistry, molecular adsorption, magnetite, and batteries [1-4]. CeO_2 is more highlighted oxide member of rare earth oxides and frequently have been used in the glass-polishing materials, as ultraviolet absorbent and automotive exhaust promoter [5-9]. Currently, cerium oxide nanoparticles have been getting more attention due to their extra ordinary luminescence, magnetic and electronic properties. Because of these well-known properties, CeO_2 nanoparticles significantly behave in different way than the bulk materials [10-13]. Neverthelessness, during the practical use the material behaviour is highly dependent on the properties exhibited by the constituent CeO_2 nanoparticles. Due to fact of marvellous properties exhibited by the nano sized particles of CeO_2, hence several methods have been proposed for the synthesis of cerium oxide nanoparticles including homogeneous precipitation [14], hydrothermal preparation [15-17] and different decomposition methods [18-19]. Among different solution based chemical synthesis methods, the sol gel method has superiority due to following facts such as homogeneity, better stoichiometry, almost free of any impurity phase-powders at low temperature and ease in the formation of dense monoliths, thin films or nanoparticles.

ZnO is becoming more popular among researchers since last two decades due its marked physical properties including wide direct bandgap (3.37 eV), high exciton binding energy (60 meV) which make it highly useful for the optoelectronic applications [20]. ZnO has been doped either intrinsically or extrinsically with different dopants particularly for achieving the desired electrical properties such as metallic, semiconducting and insulating characteristics [21, 22]. ZnO exists in various nanostructures, specially one dimensional nanostructure such as nanorods, nanowires, nanoflowers, nanotubes and nanobelts are widely used due to significantly improved performance for the piezoelectric and sensing devices [23, 24], and also involved in the fabrication of UV lasers, dye sensitized solar cells and photo catalysis [25].

ZnO nanorods have been grown by several growth methods such as epitaxial Electrodeposition, catalyst-assisted vapor-liquid-solid, thermal evaporation, metal organic chemical vapor deposition, radio frequency magnetron sputtering and solution based methods [26-30]. However, low temperature growth is significantly followed for the growth of ZnO nanorods due to several advantages of this method such as high deposition rate, ease in handling of dopant impurity and the most favourable priority of this method is its low temperature for the growth of ZnO nanorods [25]. The perpendicularly oriented ZnO nanorods to substrate have been remained in attention for the improvement in the performance of light emitting and field emission devices [31-33]. This can be achieved by the gas phase based growth techniques, however these techniques needs highly sophisticated tools, single crystalline substrates of high price for the well aligned growth, high working temperature of 450 -900 °C, and also suffers some other problems such as weak uniformity of grown nanostructures and minimum amount of grown material on the substrate [34].

In the hydrothermal method, ZnO crystals layers have been considered important for the growth of ZnO nanorods along with calcination at higher than 300 °C for the crystallization of ZnO crystal seed layers using oxidized zinc foil in the absence of seed layer [35]. Wu et al. has shown the well oriented ZnO nanorods on zinc oil substrate at high temperature, and Naoyuki Ueno et al. has grown ZnO nanorods on silicon substrate coated with seed layer of organic–inorganic composite using low temperature aqueous chemical growth technique [36]. Wu et al. approach can only be applied on zinc foil plate and it is also accompanied by calcination and Naoyuki Ueno et al. used silicon substrate which is expensive substrate relative that of glass and gold coated substrates [35-36].

In this study, we have grown well aligned, highly dense and uniform ZnO nanorods on the glass and gold coated substrates using and CeO_2 magnetic nanoparticles as seed layer without zinc acetate seed solution by following the hydrothermal growth technique.

2. Materials and experimental section

2.1 Materials

Zinc nitrate hexahydrate, zinc acetate dihydrate, hexamethylenetetramine [HMT], Cerium (III) nitrate [$(Ce(NO_3)_3{\bullet}6H_2O)$], ammonium hydroxide [$NH_4OH$], hydrochloric acid

[HCl] were purchased from sigma Aldrich Sweden. All other chemicals used were of analytical grade.

2.2. Synthesis of cerium oxide and manganese oxide nanoparticles

For the synthesis of CeO_2 nanoparticles $(NH_4)_2Ce(NO_3)_6$ and NH_4OH were used as starting precursor without any further purification. 1.0 g Ammonium cerium nitrate $[(NH_4)_2Ce(NO_3)_6]$ or Cerium (III) nitrate $[(Ce(NO_3)_3 \cdot 6H_2O)]$ was dissolved in 20 ml distilled water after that 5 ml ammonium solution (1M) was added into the solution mixture with the constant stirring at 70 °C for 30-45 minutes. A grey colour precipitate was occurred by the addition of ammonia solution. Ammonia solution was added in excess by drop-wise until a pH of 9.0 was achieved. The resulting mixture was left to stir continuously at 70 °C for 3 h to complete the reaction. After complete stirring, the pale yellow precipitates of $Ce(OH)_4$ were appeared, then centrifuged and washed several times with deionized water to remove the excess ammonium ion and nitrate from the mixture. The prepared yellow precipitates were dried in air at 100 °C temperature and afterwards annealed at 300 °C for 2 h.

2.3. Preparation of seed solution of prepared magnetic nanoparticles for the growth of ZnO nanorods

The preparation of homogeneous and completely dissolved solution of magnetic metal oxide nanoparticles as a seed layer is critical issue and in this work we prepared seed solution of manganese oxide and cerium oxide into 0.1 M solution of hydro chloric acid. The concentration of each metal oxide nanoparticle was 85 mg /5 ml as seed layer for the growth of ZnO nanorods after the optimization.

2.4. The fabrication of ZnO nanorods on the glass and gold coated substrates

Following steps have been used for the growth of ZnO nanorods on the glass substrate and gold coated glass substrate. Firstly, the glass substrates were washed with isopropanol in ultra-sonic bath for 10 minutes and cleaned by deionized water at room temperature. In order to have coated layer of gold on the glass substrates the evaporator Satis (725) was used for the deposition of 10 nm thickness titanium as adhesive layer, then followed by 100 nm thickness of gold. Then the cleaned substrates were spin coated three times with seed layer of magnetic nanoparticles at 2500 r.p.m and annealed at 120 °C for 20 minutes. Afterwards, the seeded substrates were fixed in Teflon sample holder and dipped in equimolar solution of zinc nitrate hexahydrate and HMT, and then samples were left in oven at constant temperature of 93 ^{o}C for 5 to 7 hours. After the completion of growth duration samples were taken out from oven, washed with deionized water and dried with nitrogen gas. The grown ZnO nanostructures were further characterised by field emission electron microscopy, x-ray diffraction technique, and transmission electron microscopy.

3. Results and discussion

3.1. The XRD characterisation of synthesised magnetic nanoparticles and grown ZnO nanorods with seed layer of these magnetic nanoparticles

Fig. 1 (a) shows the X-ray powder diffraction pattern of the sol-gel prepared CeO_2 nanoparticles. The results of XRD pattern indicate that the CeO_2 nanoparticles are well-crystallized and patterns are in good agreement with a cubic structure. The principle reflection peaks of CeO_2 in diffraction pattern are measured, which correspond to the (111), (200), (220) and (311) planes. These reflection peaks can be indexed to the pure cubic fluorite structure of CeO_2. The intensities and positions of the diffraction plane which are perfectly similar to the JCPDS card and no differences between them and reported data. The broadened peaks in the diffracto-gram confirm the nanocrystalline nature of the prepared CeO_2 nanoparticles. According to the Scherrer equation, the strongest peak (111) at $2\theta = 28.4^{o}$ and the peak (220) at $2\theta = 47.17^{o}$ were used to calculate the average crystallite size of CeO_2 nanoparticles, determined to be around 3-4 nm.

Fig. 1 (b) shows the X-ray powder diffraction pattern of grown ZnO nanorods on the gold coated glass substrate using cerium oxide CeO_2 nanoparticles as a seed layer. It can be seen from the diffracto gram that peak at 002 planes is quite dominant over other appeared peaks which clearly demonstrates the well orientation of ZnO nanorods in c-axis and the growth patterned of ZnO nanorods could be perpendicular to the gold coated substrate. Fig. 1(c) shows the energy dispersive x-ray study of the cerium oxide seeded ZnO nanorods and it can be seen from the figure that the ZnO nanorods are composed of Zn and O atoms and the cerium atoms are also present on the surface which acted as catalyst for the proper oriented growth of ZnO nanorods.

3.2. The field emission scanning electron microscopy (FESEM study of magnetic particles seeded ZnO nanorods

Fig. 2 shows the FESEM image of ZnO nanorods grown on the gold coated glass substrate using CeO_2 nanoparticles as seed layer and this study has shown that ZnO nanorods were very well aligned, highly dense and uniform on the substrate. The diameter of the present nanorods is about 100 nm to 20 nm with length of 1μm. The diameter of ZnO nanorods depends on the concentration of precursors and duration of the growth process. The length is function of time and it increases with time. The FESEM and XRD results are consistent with one another. The diameter of ZnO nanorods is about 100 nm to 200 nm and length of 1μm. The length and diameter depend on the conditions of reaction bath. The FESEM characterisation repeated the same results as XRD has shown for the magnetic nanoparticles seeded ZnO nanorods and both analyses are fully agreed to one another.

2.3. Optical characterisation of cerium oxide CeO_2 seeded ZnO nanorods

The aim of this investigation was to determine the imperfection of crystal structure of nanomaterial, so for this analysis photoluminescence (PL) technique is potential analytical tool. It is well known that ZnO materials with high defect densities such polycrystalline thin films or powders, usually exhibit strong PL bands of defect origin within the visible spectral region. On other hand, these emissions are suppressed in single crystalline epitaxial films or bulk ZnO, where near-band-edge (NBE) emission due to free exciton recombination with in the near-ultra-violet spectral region dominates. In case of ZnO nanostructures, such as ZnO nanorods due to their high surface to volume ratios which increases the importance of recombination via surface states. Moreover, it is found that presence of boundaries and pilling of nanorods might lead to the enrichment of the defect states. The presence of these defects in nanostructure in higher intensity of defect – related visible emissions as compare to that of the band edge emission.

In our case, we observed the increased in the green emission peak around 530 nm and a red or orange emission around 673 nm for CeO2 seeded ZnO nanorods as compare to zinc acetated seeded ZnO nanorods as shown in figure 5 (a-b). The increased green and red or orange peaks could be assigned to the oxygen vacancies, interstitial zinc ions and impurities [37-39]. However, the near–band-edge emission (NBE) was found to be same in the both cases.

3. Conclusion

In this study, cerium oxide magnetic nanoparticles were synthesised and applied as a seed layer for the growth of ZnO nanorods using low temperature growth technique. The cerium oxide nanoparticles were characterised by [XRD]. In addition to this, the structural characterisation of grown ZnO nanorods was carried out using field emission scanning electron microscopy, [EDX] and [XRD] techniques. These techniques revealed that the grown ZnO nanorods are well aligned and perpendicular to the substrate and the obtained results for each technique are consistent to one another. Moreover, optical characterisation was performed for the understanding of photocatlytic effect shown by the cerium oxide nanoparticles for ZnO nanorods. It was observed that the green emission peak and orange/red emission peaks were slightly increased as compare to the zinc acetated seeded ZnO nanorods. The present study has shown that the cerium oxide nanoparticles can be used as seed layer for controlling the alignment of ZnO nanorods which in result enhances the working performance of nanodevices based on the ZnO nanostructures.

Acknowledgement

This project was supported by King Saud University, Deanship of Scientific Research, College of Science Research Center.

Journal of Physics: Conference Series **414** (2013) 012019 doi:10.1088/1742-6596/414/1/012019

References

1. A. E. Fischer, K. A. Pettigrew, D. R. Rolison, R. M. Stroud, J. W. Long, Nano Lett. **7**, 281(2007).
2. Z. Yang, Y. Zhang, W. Zhang, X. Wang, Y. Qian, X. Wen, X. Yang, J. Solid State Chem. **179**, 679 (2006).
3. J. B. Fei, Y. Cui, X. H. Yan, W. Qi, Y. Yang, K.W. Wang, Q. He, J.B. Li, Adv. Mater. 20, 452 (2008).
4. Z. Y. Yuan, Z. L. Zhang, G. H. Du, T. Z. Ren, B. L. Su, Chem. Phys. Lett. **378**, 349 (2003).
5. D.S. Bae, B. Lim, B. I. Kim, K.S. Han, Mater. Lett. **56**, 610 (2002).
6. T. Tsuzuki, P. McCormick, J. Am. Ceram. Soc. **84**, 1453 (2001).
7. S. Colussi, C. de Leitenburg, G. Dolcetti, A. Trovarelli, J. Alloys Compd. **374,** 387 (2004).
8. M. Boaro, A. Trovarelli, J.H. Hwang, T.O. Mason, Solid. State Ion. **147**, 85 (2002).
9. L. F. Liotta, G. Pantaleo, A. Macaluso, G. Marc, S. Gialanella, G. Deganello, J. Sol–Gel Sci. Technol. **28**, 119 (2003).
10. M. Boaro, M. Vicario, C. de Leitenburg, G. Dolcetti, A. Trovarelli, Catal. Today **77**, 407 (2003).
11. T. Masui, M. Yamamoto, T. Sakata, H. Mori, G. Adachi, J. Mater. Chem. **10,** 353 (2000).
12. T. Hoshino,Y.Kurata,Y. Terasaki, K. Susa, J. Non-Cryst. Solids **283,** 129 (2001).
13. P. Jasinski, T. Suzuki, H.U. Anderson, Sens. Actuators B **95**, 73 (2003).
14. E. Muccillo, R. Rocha, S. Todokoro, J. Rey, R. Muccillo, M. Steil, J. Electroceram. **13**, 609 (2004).
15. M. Kamruddin, P.K. Ajikumar, R. Nithya, A.K. Tyagi, B. Raj, Scr. Mater. **50,** 417 (2004).
16. N.C. Wu, E.W. Shi, Y.Q. Zheng, W.J. Li. Am. Ceram. Soc. **85**, 2462 (2002).
17. S. Lakhwani, M.N. Rahaman. Mater. Res. **14**, 1455 (1999).
18. H. Wang, J. J. Zhu, J.M. Zhu, X.H. Liao, S. Xu,T. Ding, H.Y. Chen, Phys. Chem. Chem. Phys. **4**, 3794 (2002).
19. Q. Zhang, Z. Yang, B. Ding, Mat. Sci. Forum Vols. **610-613,** 233 (2009).
20. L. Pavesi, D. L. Negro, C. Mazzoleni, G. Franzo, d F. Priolo, Nature, **408**, 440 (2000).

21. C. H. Ahn, W. S. Han, B. H. Kong, H. K. Cho, Nanotech. **20**, 015601 (2009).
22. Y. Zeng, T. Zhang, L. Wang, R. Wang, W. Fu, H. Yang, J. Phys. Chem. C **113**, 3442 (2009).
23. K. Keis, J. Lindgren, S. E. Lindquist, A. Hagfel dt, Langm. **16**, 4688 (2000).
24. J. B. Baxter, E. S. Aydil, Appl. Phys. Lett. **86**, 053114 (2005).
25. T. H. Fang, S. H. Kang,J. Phys. D: Appl. Phys. **41**, 245303 (2008).
26. D. J. Park, D. C. Kim, J. Y. Lee, H. K. Cho, Nanotech. **18**, 395605 (2007).
27. J. J. Yu, S. C. Liu, Adv. Mater. **14**, 215 (2002).
28. D. C. Kim, B. H. Kong, H. K. Cho, J. Y. Lee, D. J. Park, Nanotech. **18**, 015603 (2007).
29. G. Z. Shen, Y. Bando, C. J. Lee, J. Phys. Chem. B **109**, 10578 (2005).
30. Y. B. Li, Y. Bando, D. Golberg, Appl. Phys. Lett. **84**, 3603 (2004).
31. O. Lupan, T. Pauporté, B. Viana, Adv. Mater. **22**, 3298 (2010).
32. X. Fang, J. Li, D. Zhao, D. Shen, B. Li, X. Wang, J. Phys. Chem. C **113**, 21208 (2009).
33. C. H. Chen, S. J. Chang, S. P. Chang, M. J. Li, I. C. Chen, T. J. Hsueh, C. L. Hsu, Appl. Phys. Lett. **95**, 223101 (2009).
34. V. Strikant, D.R. Clark, J. Mater. Res. **12**, 1425 (1997).
35. X. Wu, H. Bai, C. Li, G. Lu, G. Shi, Chem. Commun. Camb. **15**, 1665 (2006).
36. N. Ueno, K. Nakanishi, T. Ohta, Y. Egashira, N. Nishiyama, Thin Solid Films **520**, 4291 (2012).
37. L. Wang, et al., Appl. Phys. Lett. **86**, 024108 (2005).
38. D.C. Agarwal, et al., J. Phys. D: Appl. Phys. **42**, 035310 (2009).
39. E. De La Rosa, et al., J. Phys. Chem. C **111**, 8489 (2007).

Journal of Physics: Conference Series **414** (2013) 012019 doi:10.1088/1742-6596/414/1/012019

Figure 1(a-c) : **a. The XRD pattern for a. CeO_2 nanoparticle b. ZnO growth with CeO_2 seed, c. The EDX pattern for ZnO nanorods grown with CeO2.**

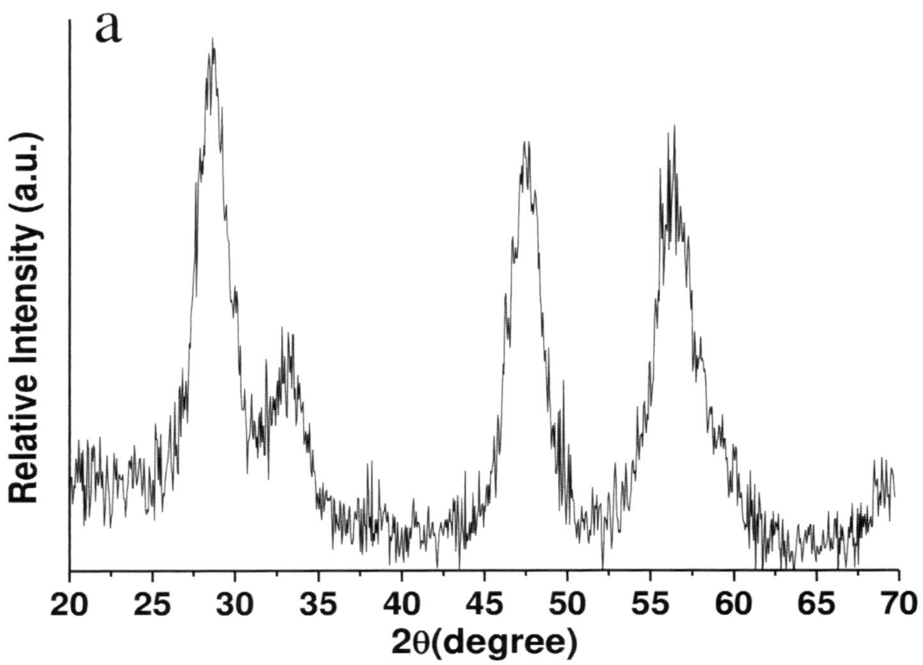

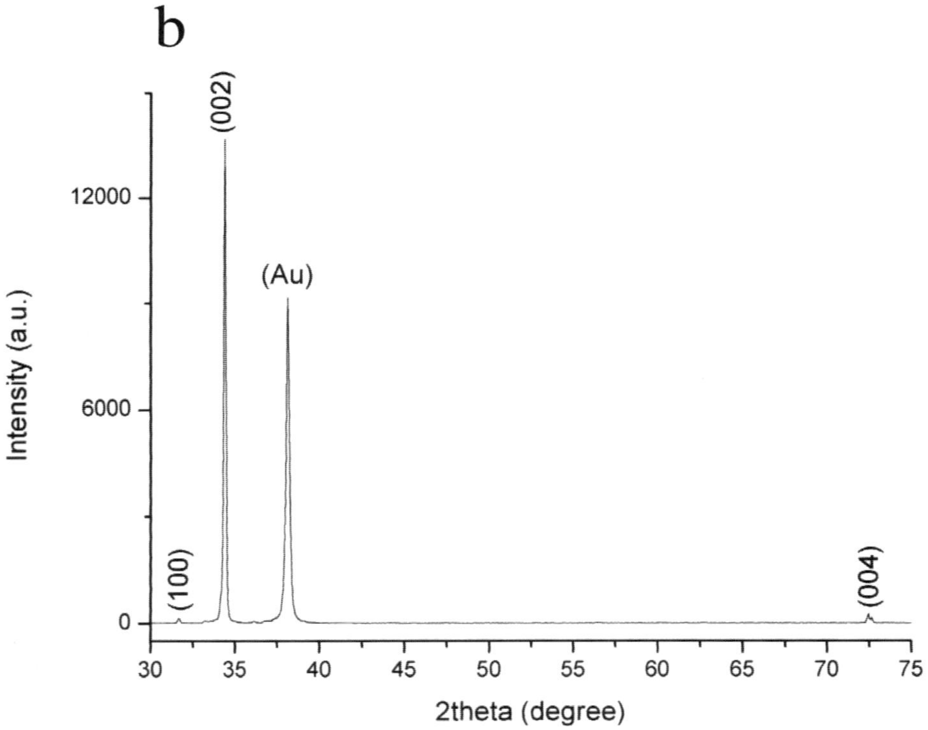

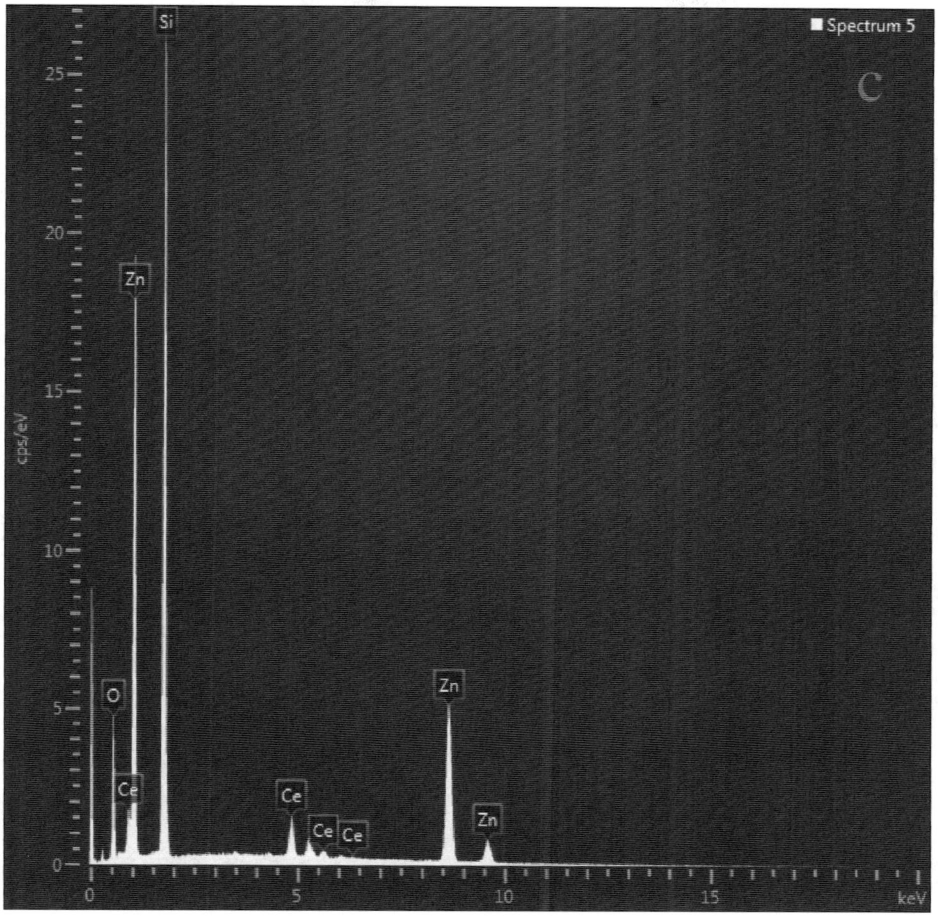
Spectrum 5
C
Si
Zn
O
Ce
Ce
Ce
Ce
Zn
Zn
cps/eV
0
5
10
15
20
25
0
5
10
15
keV

Journal of Physics: Conference Series **414** (2013) 012019 doi:10.1088/1742-6596/414/1/012019

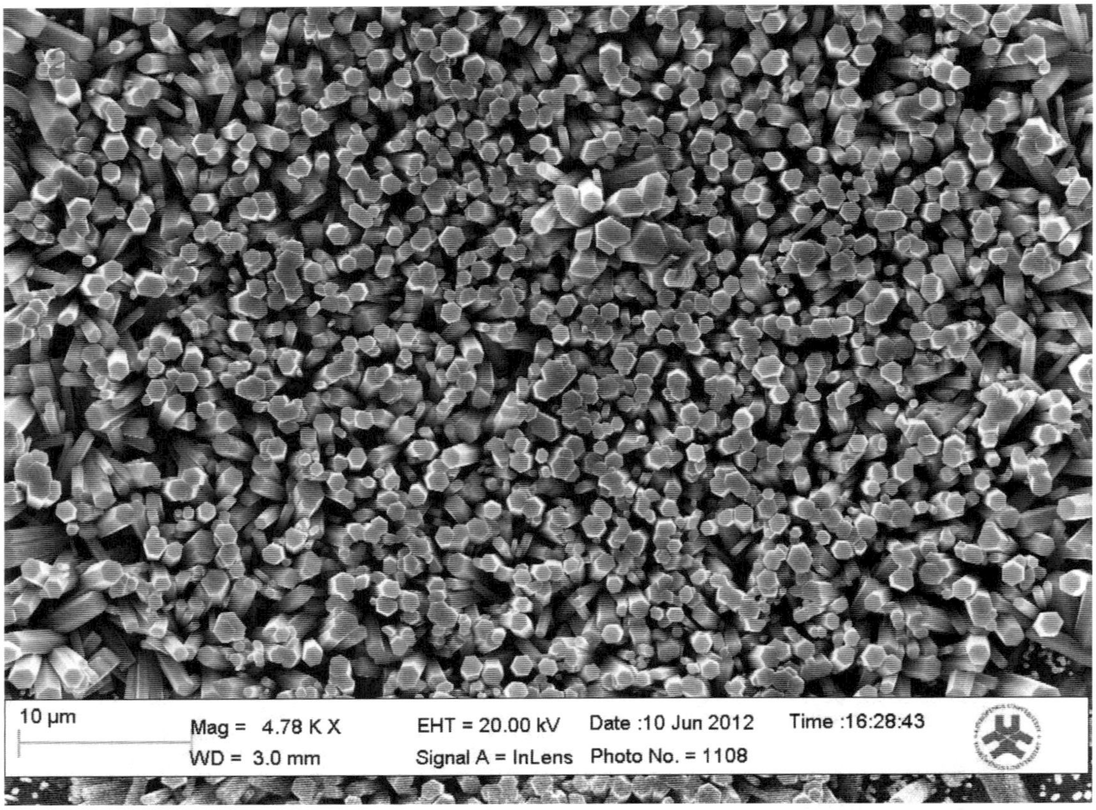

Figure 2 : The FESEM image for ZnO nanorods grown with CeO2 seed.

Journal of Physics: Conference Series **414** (2013) 012019 doi:10.1088/1742-6596/414/1/012019

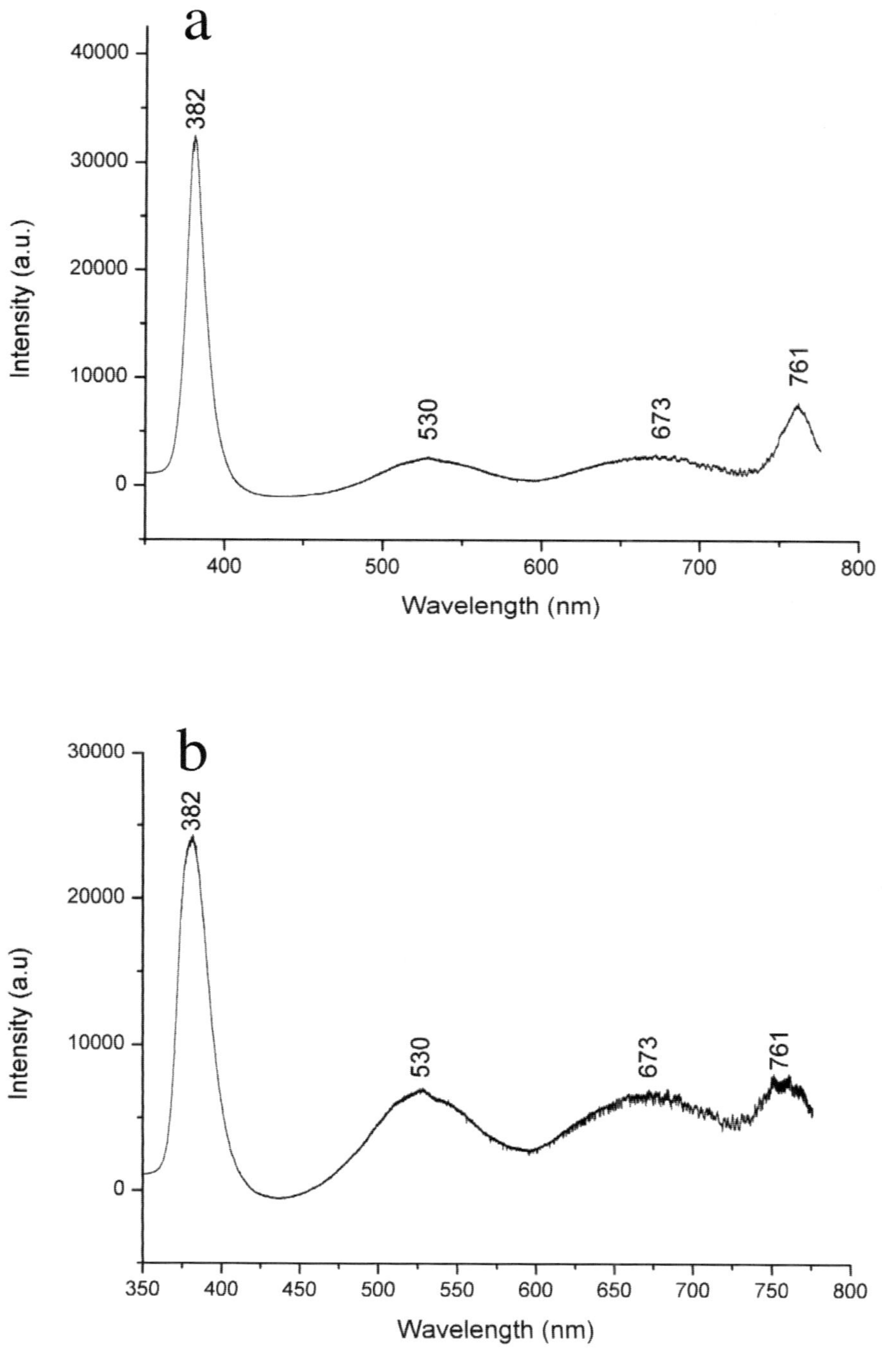

Figure 3 : The PL study of ZnO nanorods a. growth with Zinc acetate seed, and b. growth CeO2 seed.

Journal of Physics: Conference Series **414** (2013) 012020 doi:10.1088/1742-6596/414/1/012020

Experimental verification of the wave model of the genetic coding

T.Yu. Moguilnaya[a], V.I.Tomilin[b] A. A Agibalov[a]

[a] CJSC "SERVET-M", 2 8th Radial Str., Moscow, Russia, 115404, tel/fax +7(495) 984-7639

[b] "MATI"-Russian state technological university, 3 Orshanskaya Str., Moscow, Russia, 121552, tel/fax +7(499) 141-95-03

INTRODUCTION

Close Corporation SERVET is a multifunctional commercial enterprise, which develops new technologies and produces equipment in the fields of housing and public utilities, environmental protection and health care, nature management and mining operations.

During past two years our company has been involved in research and development programs in the sphere of medical diagnostics and treatment, manufacturing new drugs and medical equipment, intended for timely detecting infectious diseases and identifying qualitative and quantitative characteristics of infectious agents both in a human organism and in other environments (water, soil, foodstuff, premises, etc).

Having completed a long research work containing different aspects of physics, virology and microbiology, we came to an obvious conclusion, that the physio-chemical properties (characteristics) of various pathogens (viruses, bacteria etc.) are strictly specific for each stage of their development in an organism or environment. Our experiments stem from the notion that a DNA generates a certain radiation, a so-called wave genetic code, which is unique for every type of DNA. So it is assumed that a DNA type can be identified by its wave genetic code. The ideas and theoretical models for the wave genetic code are known well enough [1] but they lack an experimental verification.

The research results enabled us to work out an operating model for a unique medical device for complex noninvasive and aliquot-free express diagnostics (i.e. without blood test and/or laboratory sampling of human tissue) for viruses and infectious disease agents. The model is based on the comparison of the electromagnetic radiation of a human with the electronic analogues of disease agents (virus and bacterium characteristics) stored into the device database. The device is called Intest Diagnostics Unit (for more details, see [2]).

For the moment we have managed to do the following:

1) to obtain the licenses of the Russian Federation №2190350 Method of Noninvasive Express Diagnostics (Appendix 2) and №2190349 Device for Noninvasive Express Diagnostics (Appendix 3);

2) to start the licensing procedure in the EU and the USA;

3) to develop a set of design documents needed for a large-scale production of Intest Diagnostics Units;

4) to develop a special software for visualizing and analysing the qualitative and quantitative diagnostic results;

5) to produce four production sample devices of Intest Diagnostic Unit;

6) to start registering Intest Diagnostics Unit in Health Ministry of the Russian Federation;

7) to develop the technology for creating and populating the database for Intest Diagnostic Unit (for storing characteristics of viruses, bacteria, fungi and parasites) – at present the database contains the description of 31 infectious disease agents including the characteristics of the agents of such dangerous diseases as viral hepatitis B and C (genotype 1B), human immunodeficiency, tuberculosis, etc (for the list of the infectious agents, detected with Intest Diagnostic Unit, see Appendix 4);

8) to carry out a medical testing of Intest Diagnostic Unit in Hematological Research Centre, Russian Academy of Medical Sciences – a total of 151 persons were examined for infectious agents with Intest Diagnostics Unit, while the results were checked by the standard laboratory methods (immune-enzyme analysis and polymerase chain reaction). On comparing the results, the detection reliability with Intest Diagnostic Unit was found to be 95%. (Note that Test Report was signed by the commission headed by Director of the Centre, an academician of Russian Academy of Sciences and of Russian Academy of Medical Sciences). Moreover, the medical testing revealed several extra advantages of Intest Diagnostic Unit compared to the current laboratory diagnostics:

- Fast diagnostics (2-5 min vs. 1-3 days);
- Lightweight, portable and free-running device;
- Suitability for mass examination;
- Ensuring a quick on-spot rejection of donors on account of infection risk.

The experiments, presented in this contribution, confirm the ideas of "the wave genetic code" and also demonstrate that the DNA molecules can keep memory of an external electromagnetic field.

We studied the spectra of stimulated Brillouin scattering (SBS), and the spectra of superluminescence of the pathogens themselves (hepatitis C, rabbit pox, salmonella, etc.), and their isolated DNA in different colloidal solutions (water, saline solution, alcohol). Also, the radiation from the objects under virus testing was analyzed by MNOS (metal -nitride-oxide-semiconductor) transistors, which compared this radiation and the own radiation of the viruses under study. Checking the unique characteristics of the field was based also on spectral analysis of the modulated laser light, and analysing the MNOS structures with electron-optical moire, followed by subsequent mathematical processing.

TESTED MODEL

The following theoretical propositions were checked:

- the possibility of creating the condition for the excitation of the basic (allocated) collective Freilich mode due to the coherent resonant interaction of the electromagnetic radiation and Freilich-oscillator;
- the possibility of modulating laser in the infrared range, that is, the possibility of "writing down" the information about the wave genetic code on laser radiation.

It was found that the information about the wave genetic code remains intact after the conversion of the modulated signal into another frequency range.

EXCITATION OF FROLISH MODES

The possibility of creating the conditions for the excitation of the basic (allocated) collective Freilich mode due to the coherent resonant interaction of the electromagnetic radiation with Freilich-oscillator has been investigated. Within the limits of concepts of laser physics it is a matter of creating the inverse population between the quantum levels of the allocated oscillatory mode. Hence, it boils down to obtaining superfluorescence and laser generation with the use as working bodies the molecules of DNA, RNA, protein as well as such structures as ribosomes, polyribosomes and chromosomes.

The Freilich-mode [1] is modelled as a two-level quantum system (where level 1 is the basic condition, 2 - the top one), this system is raised by the resonant peak-modulated electric field (For. 1):

$$E(t) = E_o g(t) \cos wt \qquad \text{(For. 1)}$$

where E_o - an intensity amplitude of the field, g(t) – a factor of modulation,

$w = w_{21}$ (w_{21} - the frequency of the transition from the excited level to the ground level).

The feasibility of such modes and their dependence on the modulation of radiation were checked. As seen from equation (7), the spectrum of pulsations of diagonal matrix elements besides Rabi frequency includes also Stokes and anti-Stokes combinational frequencies (For. 2):

$$\nu_\Omega = \Omega_0 \pm n\nu (n = 1,2 \ldots) \qquad \text{(For. 2)}$$

Suppose the following condition is satisfied (For. 3) for a definite n:

$$\Omega_0 = n\nu;$$
$$q_{nl} = \frac{\Omega_0}{\nu} = n \qquad \text{(For. 3)}$$

Then, as appears from (7), the constant component of the probabilities disappears, so that the inversion of level population densities looks like (For. 4):

$$\Delta\rho = \langle\rho_{22}\rangle - \langle\rho_{11}\rangle = -J_n(n\gamma) \qquad \text{(For. 4)}$$

The effect of inversion ($\Delta\rho > 0$) is realised the following condition (For. 5):

$$J_n(n\gamma) < 0 \qquad \text{(For. 5)}$$

If the modulation depth parameter (γ) is in the range, where values of Bessel function (J_n) are negative, the mode of system overexcitation is realised (containing the information about biomacromolecular and permolecular structures).

Thus, by selecting a modulation mode it is possible to cause Freilich mode, i.e the appearance of Stokes and anti-Stokes combinational frequencies.

So, it suffices to check out the Stokes and anti-Stokes excitation in a solution containing DNA, along with the registration of the intensity distribution of the laser modulation that causes excitation in the solution.

DESCRIPTION OF EXPERIMENTS

The experiment was carried out as follows. Biological objects under study (table 1) were dissolved in a physiological saline at a concentration of a pathogen from 1 to 107 oml/(1-7 lgtds50). The radiation was passed through the air at three different wavelengths. Semiconductor lasers with wavelengths of 670 nm, 870 nm and 1013 nm and a high-stable power were used as a source of radiation to stimulate fluorescence.

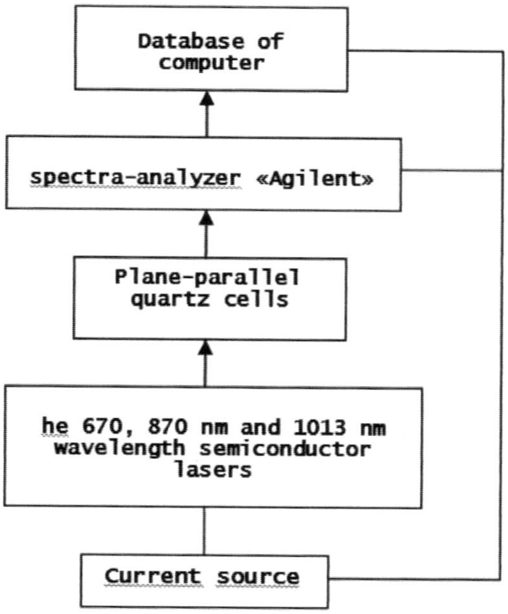

Fig. 1. The principal diagram of the device.

The radiation of the lasers was consistently fed to the spectrum analyzer and further into the computer memory (Fig. 1). A high degree of averaging was chosen for the stability of the results. Each result was recorded twice with an interval of at least five minutes. Optical waveguides were used to deliver the output radiation to the environment. The spectral resolution equaled 0,5 nm. A plane-parallel cuvette with viruses was irradiated in series at wavelengths of 1017, 810, 670 nm. However, to choose the most informative wavelength and reduce the possibility of second sort error at processing the results, we recorded the spectra of the second harmonicas for 810 nm and 670 nm lasers, namely, 1620 nm and 1340 nm, accordingly. The radiation, delivered to the cell, was not focused and had the density below the Brillouin scattering threshold.

At the beginning of each series of experiments, in addition to the spectra, which carry information, there were recorded also the spectra of laser radiation and that of radiation passing through a clean cuvette. The results were stored in a computer memory, while the subsequent processing of the spectra was carried out using a specialized software package for the statistical processing of experimental data.

An analyzer of optical spectrum Agilent 86140B (USA) was used to register spectra and fluorescence intensities. The sensitivity of the device was 69 dBm. The measurements were taken in a transparent quartz cuvette with a plane-parallel surfaces in red and infrared ranges. The optical path length of the cuvette was 3 mm, while its temperature was stabilized at +37°C throughout the experiments. The superluminescence was registered in a range from 600 nm to 1600 nm, yet a more careful attention was given to the spectral area from 780 nm to 840 nm, because the previous experiments and literary data testify to the greatest fluorescence intensity of carbon nanotubes in this range of optical spectrum. The intensity of fluorescence was changed 1 time per min.

RESULTS AND DISCUSSION

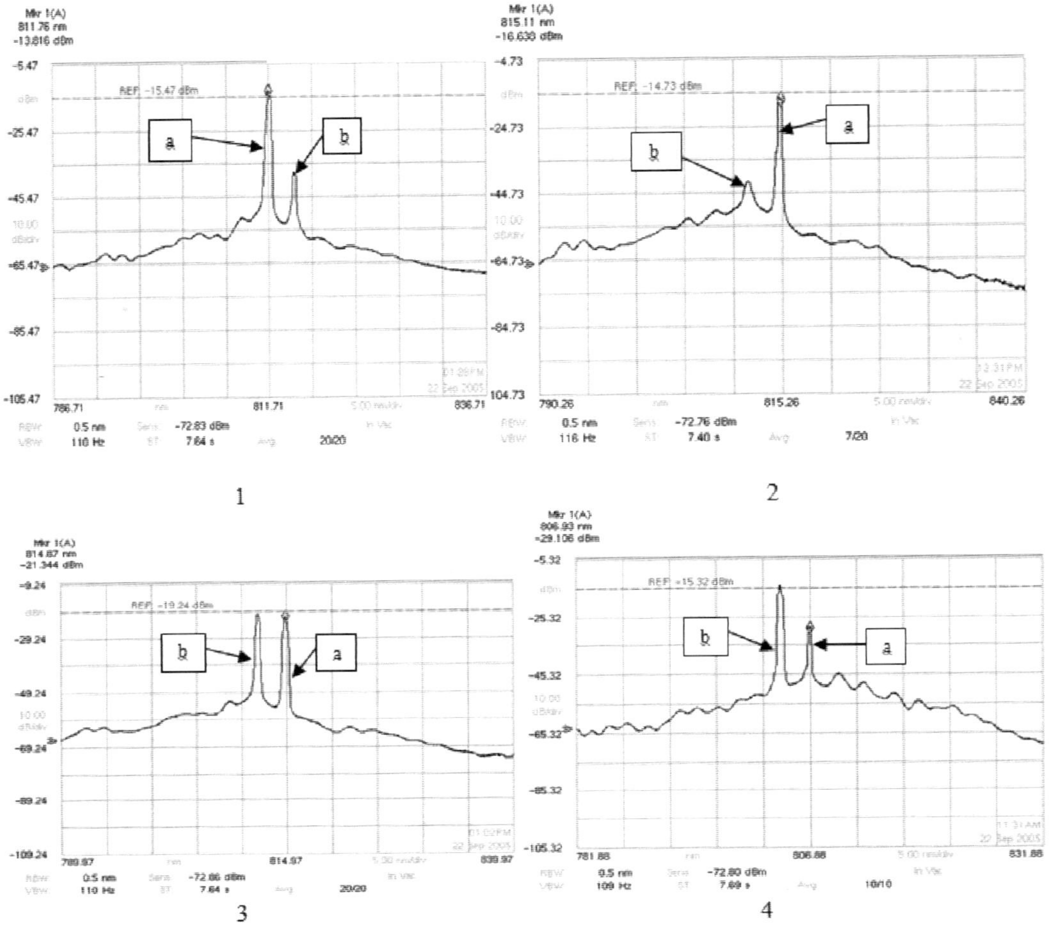

Fig. 2. Dependence of Stokes and anti-Stokes components on laser modulation for rabbit pox viruses with active agents of 10^1 OOE/ml: a - laser spectrum, b - rabbit pox virus.

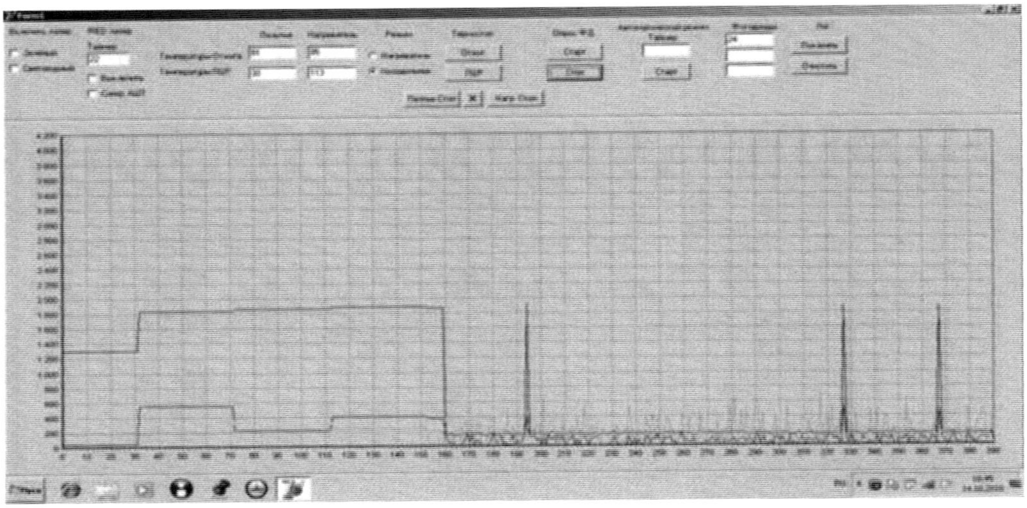

Fig. 3. Dependence of the intensity-modulated laser beam on time.

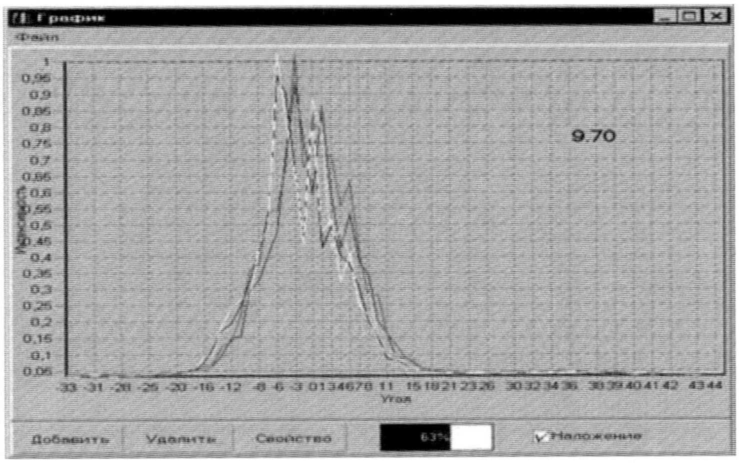

Fig. 4. The angular distributions of Stokes and anti-Stokes component.

Fig. 2 shows the dependence of Stokes and anti-Stokes components on the laser modulation for rabbit pox viruses with active agents 10^1 OOE/ml.

- 1 no modulation - corresponds to spectral distribution 1 Fig 2;
- 2 modulation corresponds to spectral distribution 2 Fig 2;
- 3 modulation corresponds to spectral distribution 3 Fig 2;
- 4 modulation corresponds to spectral distribution 4 Fig 2.

On increasing the modulation depth, the peak shifted from the Stokes region to the anti-Stokes region, while the peak intensity increased, even at a constant virus concentration. In case 4, the peak intensity of anti-Stokes component exceeded the peak intensity, related to the laser source, but retained its logarithmic dependence on concentration [2]. If the radiation was fed into the solution not along the waveguide, but through the air, there appeared an intense laser pumping mode corresponding to the object luminescence. The results were presented in our paper [2], with all the experimental dependencies in good agreement with the dependence by formula 8. Fig. 4 displays the intensity distribution of the angles of Stokes and anti-Stokes components. The fact that the polarization of radiation depends weakly on modulation is mainly defined by the type of the object, as shown in our previous studies.

CHECKING INDIVIDUAL DNA CHARACTERISTICS

Individual characteristics of DNA were checked in two ways. In the first procedure the reference radiation of the laser source was recorded and then the direct scattering spectra were analysed. The experimental setup is the same as that in the diagram of Fig. 1. The problem was solved by the method of electron-optical moire with subsequent mathematical processing.

The scattering spectra of 31 cytopathogenic viruses and bacteria were studied, see Table 1.

Table 1. The list of infectious agents investigated with Intest diagnostic complex.

	VIRUSES		BACTERIA
1	Herpes simplex –Virus	17	Mycobacterium tuberculosis avium
2	Herpes Zoster-Virus	18	Mycobacterium tuberculosis bovis
3	Herpes genitaler - Virus	19	Mycoplasma hominis
4	Epstein-Barr-Virus	20	Neisseria gonorrhoeae
5	Cytomegalovirus	21	Neisseria meningitidis
6	Hepatitis A -Virus	22	Peptostreptococcus anaerobius
7	Hepatitis B -Virus	23	Proteus mirabilis
8	Hepatitis C-Virus (Genotype 1B)	24	Staphylococcus aureus
9	Delta-hepatitis-Virus	25	Streptococcus ß-hemolytischer
10	AIDS-Virus	26	Streptococcus pneumoniae
	BACTERIA	27	Streptococcus viridans
11	Campylobacter jejuni		FUNGUS and PARASITES
12	Chlamydia psittaci	28	Candida albicans
13	Chlamydia trachomatis	29	Lamblia intestinalis
14	Enterococcus	30	Plasmodium malariae
15	Helicobacter pylori	31	Trichomonas vaginalis
16	Mycobacterium tuberculosis hominis		

DESCRIPTION OF EXPERIMENTS 1

The experiment was conducted in the same manner as in the previous case (see Fig. 1), except that radiation was fed to the cell along the waveguide, not through the air. The power of the laser source was doubled. In this case, Stokes and anti-Stokes components of stimulated Brillouin scattering is determined by internal factors.

RESULTS AND DISCUSSION 1

Fig. 5 presents the intensity distribution of a laser with a wavelength of λ=0.81 mcm. The modulation rate was varied by changing the time of a filling factor from 0.0001 sec till 1 sec and the radiation intensity of a laser source. The spectra of several other objects are given here.

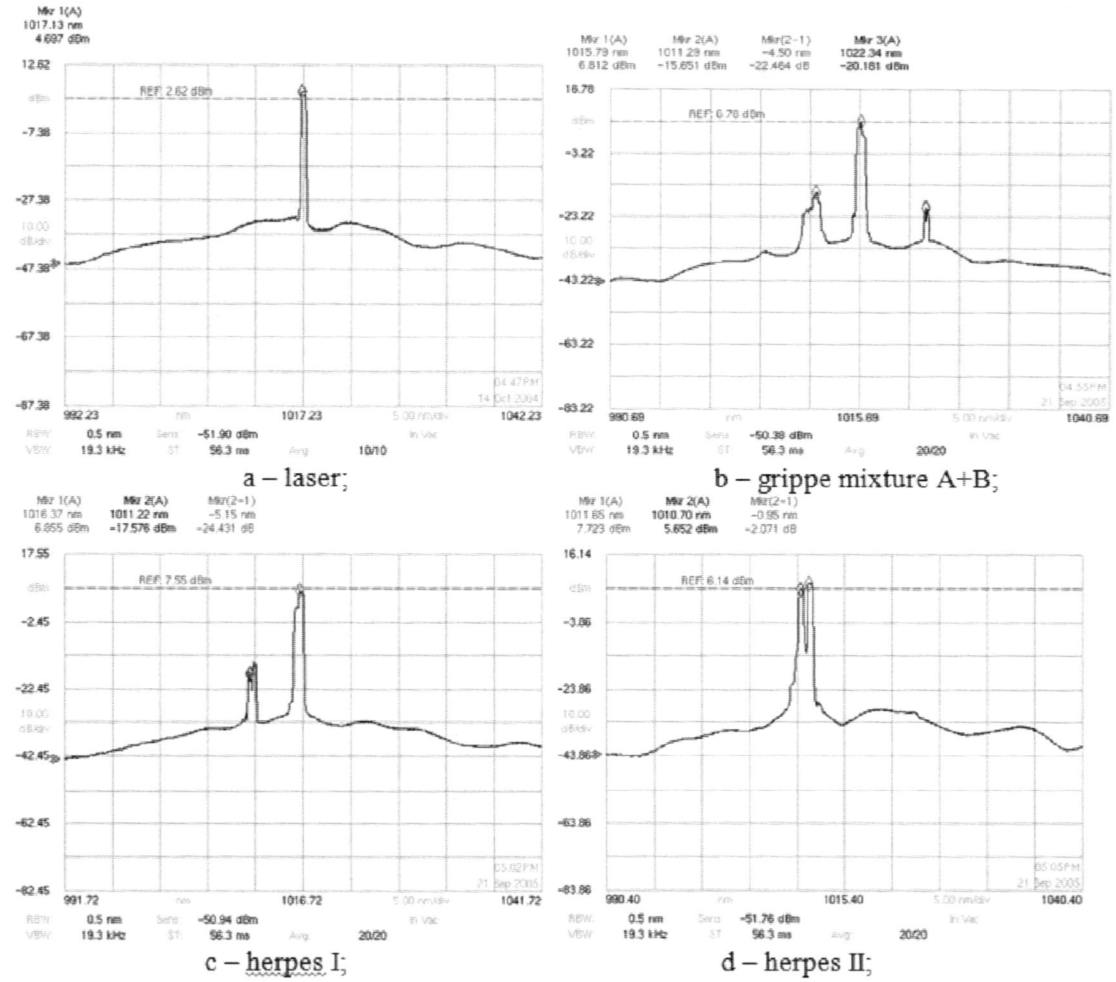

a – laser; b – grippe mixture A+B; c – herpes I; d – herpes II;

Fig. 5. The intensity distribution of a laser.

DESCRIPTION OF EXPERIMENTS 2

The second method is as follows:

For the elementary design type (with one source of radiation) the principal scheme of data recording (see Fig. 2) represents a device, which works as follows. From a monochromatic radiation source 1 radiation arrives at cell 2, containing the base medium and the object under study (viruses, microorganisms). Then, radiation arrives at unit 3 for recording and reading optical information (for example, for recording by means of a two-layer plate - copper, cupric oxide). After transforming optical information into an electrical signal, it arrives at unit 4 for information processing and storage, consisting of a microprocessor and integrated circuits.

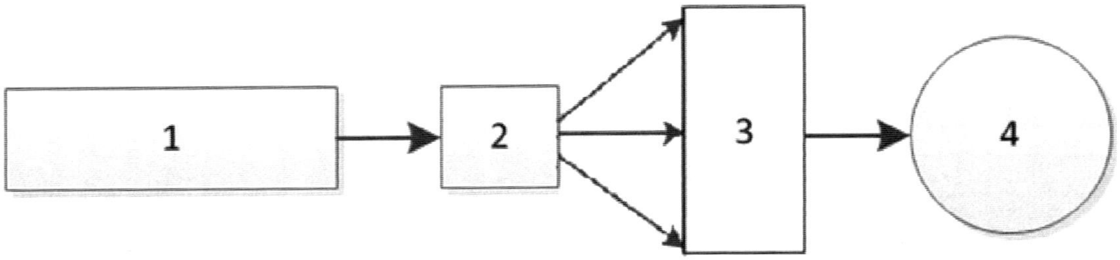

Fig. 6. Principal scheme of a device for recording data.

The recording cell contains at least two elements, an insulator and a conductor (silver or aluminium can be used as a conducting material). The cell of a multilayered design can also be used. According to available data, the cell keeps its the information structure for no less than 3-5 years at a room temperature without exposure to external electromagnetic fields, for example, an ultraviolet radiation. A power source supplying both alternating and direct currents is used in the circuit. A recording cell database and a switchboard for connecting the cells can also be built in a single unit of control and analysis.

After exposing an object to electromagnetic radiation of a weak power (electric current), the wave pack of a spectral structure, which corresponds to the features of spectral characteristics of all the objects, is stimulated in this object. Thus, the current in the circuit repeats the features of spectral characteristics of the object. In the case of a direct current in the circuit, one of the electrodes is active (measuring), another one is a base electrode. In the case of an alternating current, each of the electrodes performs alternately the specified functions, according to an instant direction of the current. The power supply is desirable to choose in the form of a source of adjustable basic voltage. It allows to set the level of a direct current or an alternating current, which passes through the investigated object, so that to ensure the best information response.

The physical mechanism of diagnostics of Intest Diagnostic Unit is based on comparing the electromagnetic field characteristics of a biological object (a person or an animal) and the records of electromagnetic field characteristics of the superfluorescence (generation) in samples prepared in advance, which consist of the base solution and a virus. As is known, the virus consists basically of proteins and amino-acids possessing unique spectral characteristics.

If a biological agent is present in an investigated object or a chemical substance corresponds to one of the connected cells, the dynamic characteristics will cause an increasing current due to the signal interference. After digitizing the current, it is processed by the algorithms of the unit. On the basis of conformity between the dynamic characteristic parameters of the structure and the cell along with a change in the value of the measured current, one can determine the presence of the chemical substance or the biological agent corresponding to the connected cell in the investigated object.

The accuracy of the method is proportional to the area of a measured object, which is due to decreasing the useful signal with its reduction, while the noise remains constant and reduces the accuracy of recognising viruses.

The results of recording were checked by visualising and analysing the power-information fields of the biological objects, which were registered as a power-information field pack in two-shutter field-effect transistors. Shooting was done by an electro-phonograph EG-100A with an accelerating voltage of 40 kV, using a rectangular grid with 100x100 mcm cells and systems of a plate-ball type (dia 0.5 m), to which the potential of the field analogue of a virus or bacterium was applied. The moire picture turned out to be at double exposure on a photo-plate due to the lack of external disturbance of the applied field. The results of shooting are presented in Fig. 3.

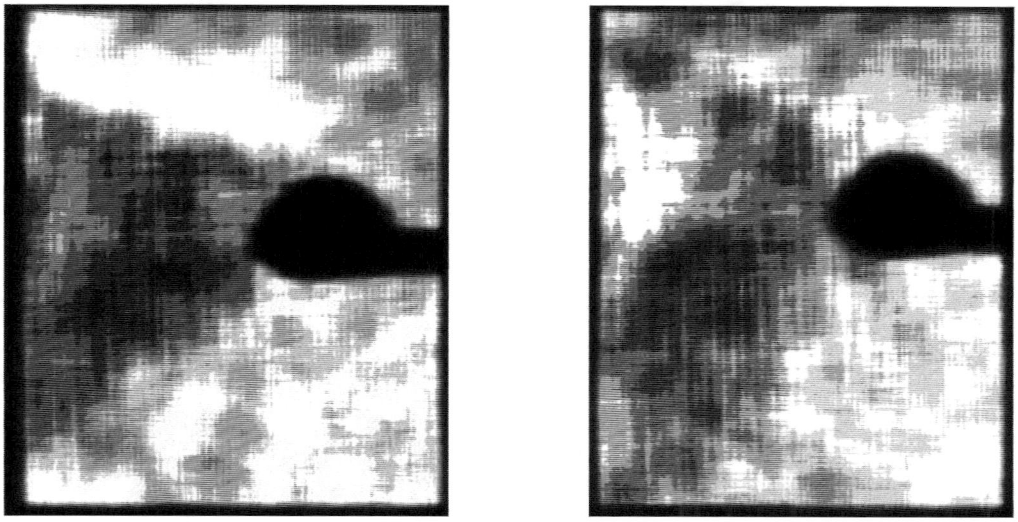

a – cholera; **b – staphylococcus;**

Fig. 7. Results of mathematical processing moire pictures of the biological object fields: a - cholera; b - staphylococcus.

The pictures were scanned, digitized and then mathematically processed by means of specially selected function (in the form of "sombrero") and "the drop" method with the change of scale and area by steps. The results of processing moire pictures by means of the wavelet-analysis can be presented as pictures or schedules, given in Fig. 4. The obtained pictures represent an intermediate stage of mathematical processing. The degree of darkness in them is defined by the value of a field-energy in each point. Then further step-by-step processing the obtained pictures is done, which results in the schedules of energy or degree of darkness dependent on the area.

The results of mathematical processing show that the dependence inherent in a given object corresponds to its field analogue with sufficient reproducibility in both qualitative and quantitative terms. Thus it is shown that the method of electron-optical moiré allows registering the electromagnetic parameters of the information fields of objects of various natures.

The information about a wave code remains intact after converting the corresponding modulated signal into another frequency range.

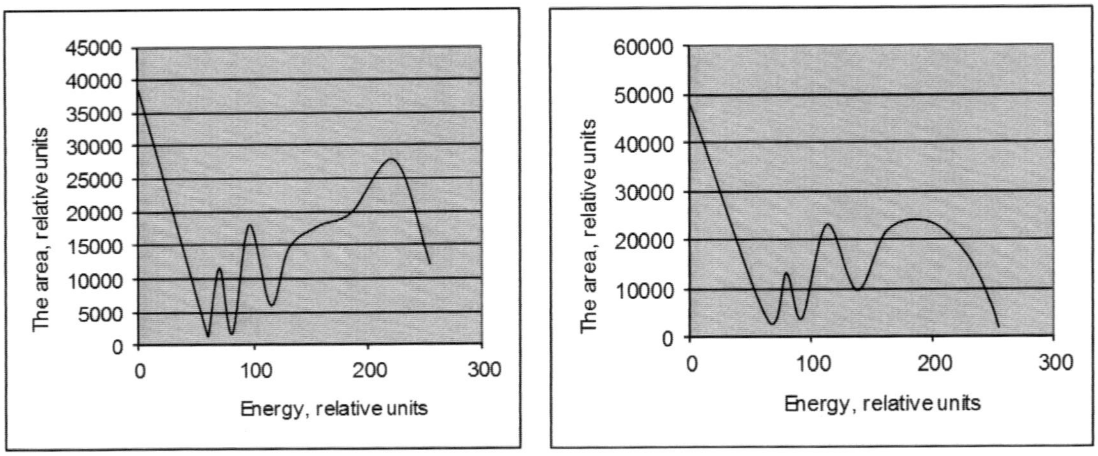

a – cholera; b – staphylococcus;

Fig. 8. Results of wavelet-analysis of moiré pictures of the biological object fields.

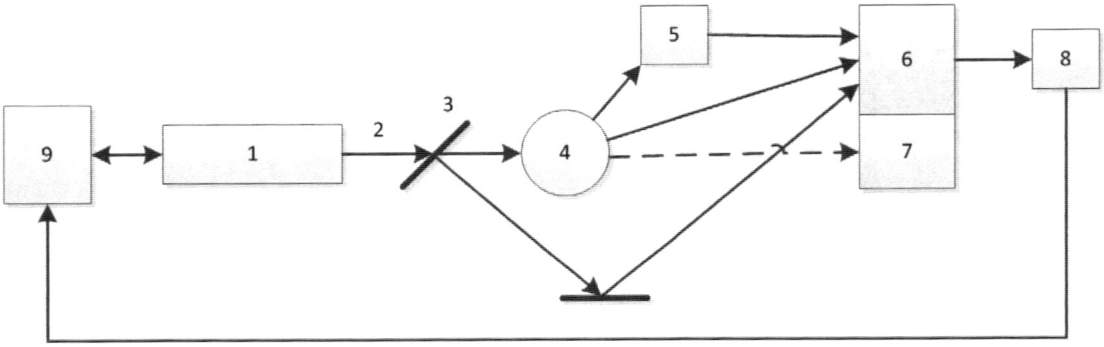

Fig. 9. A circuit for comparing an excited test signal and the standardised signals stored in a computer memory.

From monochromatic light source 1 radiation passes to cell 4, containing the base medium and an object under study (viruses, microorganisms). Then the radiation passes into unit 6 for reading and recording optical information. After transforming optical information into an electrical signal, it passes into unit 8 for information processing and storage, consisting of a microprocessor and integrated circuits. A more complex design version of the device (Fig.8) works as follows. Radiation from three lasers with various wavelengths in a visible and near infrared range is consistently directed from unit 1 (for example, through a fiber waveguide 2) into unit 3 for measuring light. Then a part of radiation is directed onto cell 4 containing the base medium (water, spirit, salt solutions) and the biological agent (virus, bacterium) under study. After passing cell 4, the radiation is fed simultaneously into spectrum analyzer 5 and unit 6 for recording and reading the optical information, then a re-recorded carrier containing the established information is fed to the rotary device 7, allowing to change an angle between a source and a receiver of optical information. Unit 6 for recording and reading optical information (for example, a device containing a registering layer on phase transitions or a holographic device) is needed for a long-term storage of the database for numerous researched object spectra and for their exact reproduction while rewriting and duplicating .After analysing the readings of spectrum analyzer 5, the narrower range of the spectrum and the angle between the source of radiation in unit 2 and the receiver in unit 6 are selected for recording so that the most typical features were represented. A more detailed description of the device is given in [2].

Significant changes of the current in the registration circuit were noted if spectral characteristics of the radiation from an object under testing were fully identical to the spectral characteristics of own radiation from one of the identified virus. If the spectral characteristics of the radiation from an object could not be identified with any of the identified viruses, i.e. the object did not contain the virus; the fluctuations of the current were within the measuring errors. The experimental distribution of the circuit current and the spectral distributions of the radiation from DNA under study were compared with the theoretical distributions, suggested in [1]. The theory and the experiment appeared to be in good qualitative agreement.

The current distribution in the circuit for the case of matching the pathogen in the cuvette and the pathogen in the PC memory (Fig. 10).

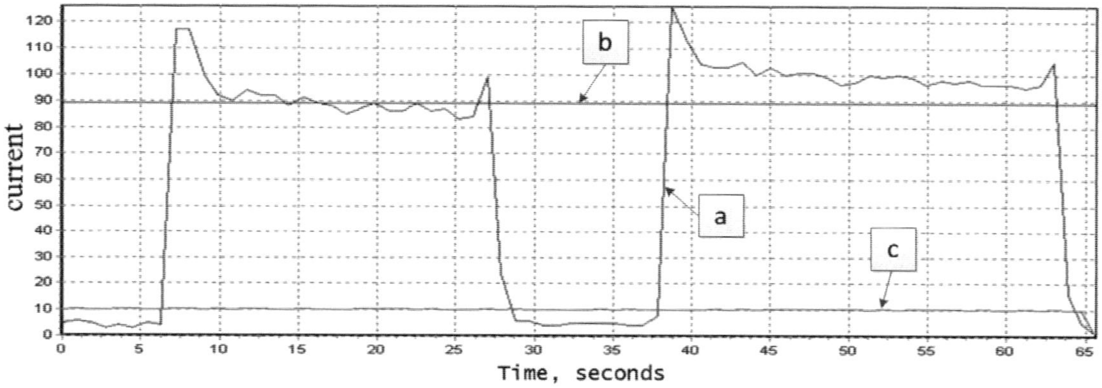

Fig. 10. The current distribution in the circuit if the pathogen in the cuvette does not match the pathogen in the PC memory.

CONCLUSIONS

The scattered laser radiation in a solution of an object containing DNA may cause an excitation Frohlich mode, with the intensity proportional to the concentration of a dissolved object.

Spectral distributions of various objects containing DNA are unique.

The information encrypted in DNA can be recorded by the laser.

The spectral distribution of the DNA luminescence taken together with the transmission spectra of the laser can be the basis for a series of devices for the diagnostics of pathogenic organisms.

The series of experiments allowed us to identify the method underlying the design principles of a stand for monitoring the Intest Diagnostic Unit efficiency.

REFERENCES

1. P.P. Gariaev, Wave Genetic Code, Moscow, 1997, 108 p.

2. T.Yu. Moguilnaya*[a], G.I. Andreev[a], A.A. Agibalov[a], A.G. Botikov[b], E.V. Kosenkov[b], E.A. Saguitova[a], Instrumentation for Noninvasive Express Diagnostics of Bacteriophages and Viruses by Optical Method, (Smart Medical and Biomedical Sensor Technology, Proceeding of SPIE 28-29 October 2003, Providence, Rhode Island, USA, Volume 5621, pp 18-29).

Effect of CO_2 pulsed laser irradiation on improving the biocompatibility of a polyethersulfone film

S. Jelvani[1*], H. Pazokian[1], S. Moradi Farisar[2]

[1] *Laser and Optics Research School, Tehran, Iran*

[2] *Departments of Physics, Khaje Nasir Toosi University of Technology, Tehran, Iran*

Abstract

In this paper a 200 ns pulsed TEA CO_2 laser is used for treatment of polyethersulfone (PES) films surface. The laser induced structures and chemical compositions on the surface upon irradiation are studied. The hydrophilicity and biocompatibility of the irradiated surfaces are examined by contact angle and platelet adhesion measurements, respectively. The optimum number of pulses and fluence for improving the surface biocompatibility are found.

Key words: pulsed CO_2 laser, polyethersulfone film, contact angle, platelet adhesion

***Corresponding author: sjelvani@aeoi.org.ir**

1. Introduction

Polymers are widely used as biomaterials in medical applications. In order to enhance the biocompatibility feature, polymers surface should be modified with proper surface treatment techniques. A number of surface modification methods have been employed to achieve such ends. These include chemical treatment, plasma treatment, electron beam irradiation and laser irradiation [1-7]. Lasers can offer the user an exceedingly high degree of process controllability and flexibility.

Polyethersulfone is one of the most favorable polymers for hemodialysis membranes and thus improving its biocompatibility is essential if it is to be applied in blood contact works. Recently, some researchers have used laser irradiation for polyethersulfone film surface modification [8-12]. Pazokian et al have carried out laser treatment using excimer lasers at 193, 248 and 308 nm. They have shown that increasing the hydrophilicity of polyethersulfon (PES) surface leads to increasing biocompatibility in contact with blood. On the other hand the roughness is the most important parameter in cell culture on the surface and also, the biocompatibility is depended on the irradiation fluences [13-16].

In this work the effect of CO_2 pulsed laser irradiation on improving the blood compatibility and hydrophilicity of the PES film surface is studied.

2. Materials and methods

2.1. Sample preparation

The samples (PES,Ultrason E6020, Mw=58000, flakes) were supplied by BASF Co. PES was compressed at 250 °C and 30 Mpa ,using a laboratory press (Mini Pestpress,10 poyosciti), into pieces of about 80-400 μm in thickness and 10 mm× 20 mm in dimension. The surfaces of all samples were ultrasonically washed in bath with methanol before irradiation.

2.2. Laser treatment

The samples were irradiated using a line tunable pulsed TEA CO_2 laser at the wavelength of 9.58 μm. The laser was operated at the laser fluence of 150-400 mJ/cm^2 and repetition rate of 1-5 Hz.

2.3. Sample Characterization

2.3.1. Contact Angle Measurement

Hydrophilicity was evaluated by measuring the static water contact angle. Static contact angles were measured using the sessile drop method by a contact angle goniometer (Kruss G10).

All water drop contact angles are the mean value of five measurements on the surface ± standard error ($\theta \pm 2^\circ$).

2.3.2. Platelet Adhesion Measurements

Human blood was collected into the 250 ml blood bag. The blood was centrifuged to obtain platelet rich plasma (PRP). The PRP concentration was determined by a Cobas blood counter and used to normalize all the experiments. The PRP was placed on PES films of 1cm^2 area and kept for 1 h at 37 °C. Then the films were taken out and dip-rinsed twice with phosphate buffer saline solution (PBS) in order to remove the platelets that were not attached to the film. The number of adhered platelets was determined by lactate dehydrogenase (LDH) method [17]. The films were put in 5 ml of PBS containing Triton- X100 for 1 h at room temperature to lyse the adhered platelets.

The LDH activity of lysate was measured with an enzymatic method in which the adhered platelets were counted using a calibration curve of platelet counts. The change in ultraviolet absorption at 340 nm was measured using a spectrophotometer. The platelet adhesion test was repeated three times using different PRP and the mean value was obtained.

3. Results and Discussion

3.1. Contact-Angle Changes

Contact angle measurement is a simple method for investigation of the hydrophilicity of a surface [18].

Table1. Shows the contact angles measured on the irradiated surfaces with 2000 pulses at the repetition rate of 4 Hz and different laser fluences.

Table 1.Contact angle of water droplet on untreated and laser-treated surfaces with different laser fluences with 2000 pulses

Fluence	0 mJ/cm^2	170 mJ/cm^2	200 mJ/cm^2	225 mJ/cm^2	250 mJ/cm^2	300 mJ/cm^2
Contact angle(degree)	79	67.53	68.90	63.20	72.60	74.70

As it is indicated in this Table, the surface wettability tends to increase with increasing the laser fluence; the contact angle reaches a minimum value and then increases with further increases in the fluence. The fluence at which the contact angle becomes minimum (F_{min}) is subthreshold fluence (the threshold fluence is found experimentally to be ~ 250 mJ/cm^2). Irradiation with fluences above F_{min} always causes a decrease in surface wettability (increasing contact angle) due to the cleavage of chemical bonds (Fig. 1). As well know, Oxygen containing group on polymer surfaces are responsible for the change of surface hydrophilicity [17-19], then by controlling the laser fluence , the surface could be modified as polar or non-polar .

F=200 mJ/cm^2 F=170 mJ/cm^2 F=0 mJ/cm ^{2}

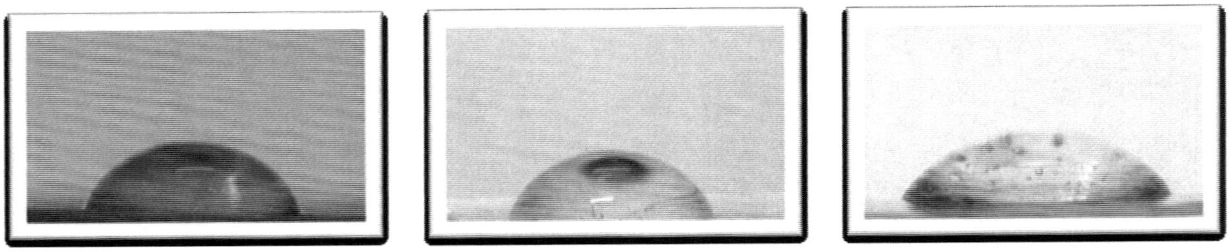

F=300 mJ/cm^2 F=250 mJ/cm^2 F=225 mJ/cm^2

Fig.1-The images of the water drop on PES surface at different laser fluences and 2000 pulses.

Figure 2 shows the contact angle on the surface as a function of pulse numbers at the fluence of 225 mJ/cm^2 and repetition rate of 4 Hz.

As it is shown in this figure, the contact angle initially decreases with increasing number of pulses up to 5000 pulses and then increases with further pulses.

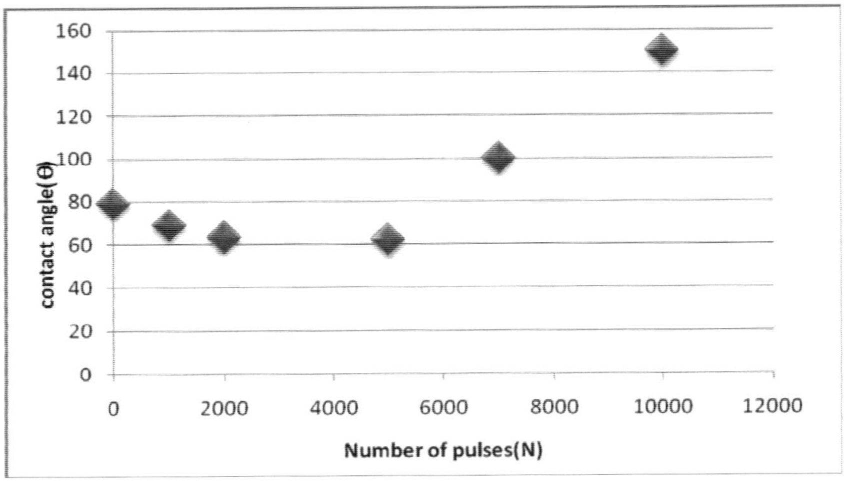

Fig.2-Dependence of the contact angle of the pulsed TEA CO_2 laser irradiated PES films as a function of pulses number at the laser fluence of 225 mJ/cm^2

It seems likely that the - oxidized groups would be decomposed with further pulses (more than 5000). The images of the water drop on PES surfaces at different number of pulses are shown in figure3.

3.2. Platelet Adhesion

To test the thrombogenecity of PES surface the LDH method was used. This method provides a quantitative determination of the number of platelets adhering to the surface and providing procoagulant sites [19,20].

Figure 4 shows platelet adhesion as a function of pulse numbers. As can be seen in the figure, the optimal laser pulses required to reduce the platelet adhesion is 1000 pulses at a fluence of 225 mJ/cm^2.

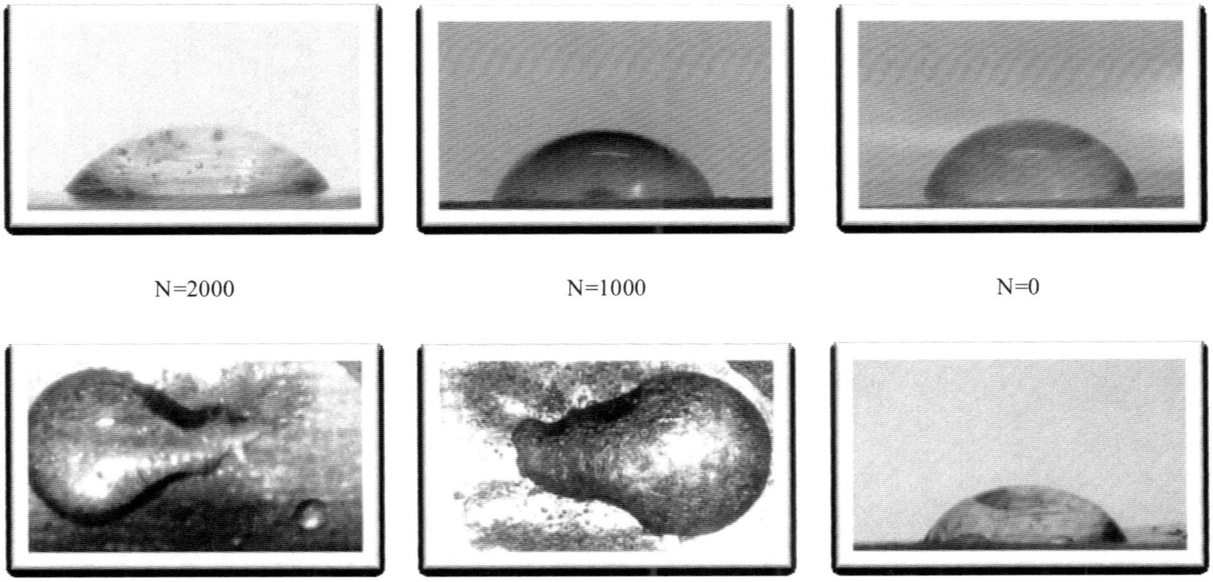

Fig.3-The images of the water drop on PES surface irradiated with different laser pulses number and the fluence of 225 mJ/cm^2.

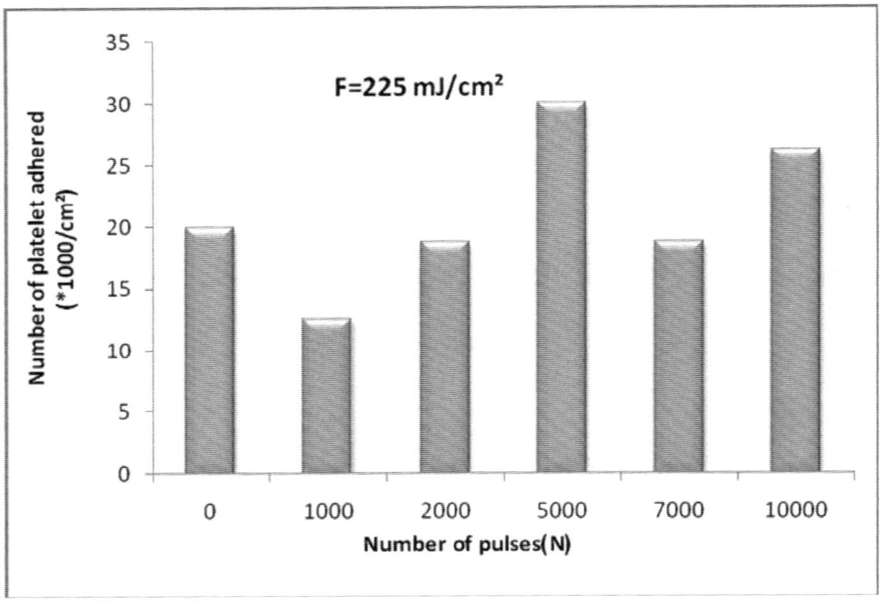

Fig.4- Number of platelet adhered to the surface as a function of number of laser pulses at fluence of 225 mJ/cm^2.

Laser treatment of polymers introduces specific chemical groups onto the surface that not only increase the wettability on the surface but also have a specific effect on the interaction of platelets with surface.

The radicals formed on the CO_2 laser irradiated PES surface are modified into corresponding peroxides in contact with air.

4. Conclusions

In this work, the effect of pulse numbers and the fluence of a pulsed TEA CO_2 laser irradiation on the biocompatibility of PES were investigated. The results showed that, irradiation below the ablation threshold (at fluence 225 mJ/cm^2 and 5000 pulses) increases the surface wettability. The obtained results also indicated that, at the fluence of 225 mJ/cm^2 there is an optimum for number of pulses to reduce platelet adhesion on the surface. Therefore by controlling the laser fluence and number of pulses, the surface could be modified for manners of polar groups or non-polar groups. It concluded that wettability and platelet adhesion are related to surface chemistry.

References:

[1] F.Garbassi, M.Morra, E.Occhiello, Polymer surface: from physics to technology, Wiley,New York(1994).

[2] J.R.Hallaham, A.T.Bell, Techniques and application of plasma chemistry, Wiley,New York(1974).

[3] J.F.Rabek, Int.polym.sci.Technol,Vol.12(1985).

[4] Astm, Selecting Generic biological test methods for material and devices,F748-95.

[5] S.A.Jenekhe, S.J.Tibbetts, J.Polym.sci.phys.Ed.,Vol.201, pp.26(1988).

[6] L.E.S.Brink, S.J.G.Elabers, T.Robbertsen, P.Both, J.Membr .Sci,Vol.76,pp.281(1993).

[7] H.Mirzedeh, A.A.Katbab, R.P.Burford, Radiat.phys.chem.,Vol.41, pp.507 (1993).

[8] H. Pazokian, S. Jelvani, J. Barzin, M. Mollabashi, Sh. Abolhosseini, Optics Communications" 284(2011).

[9] H. Pazokian, M. Mollabashi, S. Jelvani, J. Barzin, S. Abolhoseini , Optical Engineering Vol.50(8) (2011).

[10] H.Pazokian, A.Selimis, E. Stratakis, M. Mollabashi, J. Barzin, S. Jelvani, Applied Surface Science, 258(2011).

[11] N.Hiroyuki,N.Masashi,N.Shozaburo,Y.Akira,Appl.phys.lett,vol55,no.5.pp.31 (1989).

[12] N.Hiroyuki,Sh.Masashi,Y.Akira,Appl.phys.lett,vol57,No.22.pp.26 (1990).

[13] H. Pazokian, A. Selimis, E. Stratakis, M. Mollabashi, J. Barzin, S. Jelvani, DOI: 10.1007/s00339-012-7142-9

[14] H. Pazokian, S. Jelvani, M. Mollabashi, J. Barzin, G.zizabadi Farahani App. Sur. Sci **257** 6186–6190(2011).

[15] H. Pazokian, A. Selimis, Jalal Barzin, Saeid Jelvani, Mahmoud Mollabashi, C. Fotakis, E. Stratakis ",Journal of Micromechanics and Microengineering,22(2012).

[16] H. Pazokian, J. Barzin , M. Mollabashi, S. Jelvani, , S. Abolhoseini,Laser Physics,22(5) (2012).

[17] H. Mirzadeh, M.Dadsetan, Radiatin Physics and Chemistry, 67 381-385(2003).

[18] M. Dadsetan, H. Mirzadeh, N. S. Sanjani, M. Daliri. Journal of Biomed Mater Res, 57,183, (2001).

[19] M. Dadsetan, H. Mirzadeh, N. S. Sanjani, P. Salehian, Journal of Biomed Mater Res, 54(4), 540-6(2001).

[20] J.F. Blanco, Q.T. Nguyen, P. Schaetzel, ,J .Membr. Sci, Vol. 186, pp. 267 (2001).

Laser surface modification of polyethersulfone films: effect of laser wavelength on biocompatibility

H. Pazokian[1], M. Mollabashi[2], S. Jelvani[1], J.Barzin[3]

[1] *Laser and Optics Research School, Tehran, Iran*

[2]*Department of Physics, Iran University of Science and Technology, Tehran, Iran*

[3]*Biomaterial Department, Iran Polymer and Petrochemical Institute, Tehran, Iran*

Abstract

In this paper laser ablation of polyethersulfone (PES) films regarding to the change in biocompatibility of the surface is investigated at 3 different wavelengths of 193nm (ArF), 248 nm (KrF) and 308 nm (XeCl). The optimum laser fluence and number of pulses for the improvement of the surface biocompatibility is found by examination of the surface behavior in contact with platelets and fibroblasts cells at 3 wavelengths. These biological modifications are explained by alteration of the surface morphology and chemistry following irradiation. The results show that the KrF laser is the best choice for treatment of PES in biological applications.

1. Introduction

Polymers are being used in a wide range of applications, particularly in medical device industry. Understanding the surface characteristics of the polymer is important since the surface properties of materials in contact with biological systems play a key role in determining the outcome of biological interactions [1, 2].

Lasers are used as a standard tool for improving the biocompatibility of the surface [3–6]. Different materials are irradiated with lasers at different experimental conditions depending on the desirable results for various applications.
Surface hydrophilicity of material is an important parameter in the medical applications. There have been remarkable advances in the understanding of the extraordinary wetting properties of rough surfaces which, contrary to the flat ones, can be tuned from superhydrophilic to superhydrophobic [7] opening up the possibility for many exciting applications [7–11]. Direct laser micro/nano structuring is a versatile method for controllable modification of materials' surfaces [12–15].

Surface biocompatibility is influenced by the surface morphology, chemistry and specially wettability modifications, which should be investigated to account for the biocompatibility alterations. In this scope, contact angle measurement is a simple method for measuring the hydrophilicity of a surface. Different combinations of pulse numbers and fluence at each wavelength give rise to various microstructures and chemical modifications.

In our last papers [16-21] we investigated the effect of KrF, XeCl and ArF laser irradiation on the surface chemical,morphological and wettability of the polyethersulfone surface, individually. In this paper a brief comparison review between different wavelengths is done.

2. Methods and experimental procedures

PES films were prepared as described elsewhere [16-21]. They were cut into pieces of 80–100 μm in thickness and 30mm×10mm in dimension and were washed with ethanol before irradiation.
PES samples were irradiated with ArF and XeCl lasers with pulse durations of ~25 ns. KrF laser was used at 3 pulse durations of 30 ns, 5 ps and 500 fs. The samples were irradiated with different number of pulses and fluences. Irradiation was done in ambient air and the central, fairly homogeneous area of the beam was selected for the observation of the chemical and morphological changes in the ns cases,while in the ultra-short irradiation a beam homogenizer was used in the experimental set up.
The laser-treated areas were characterized by field emission and scanning electron microscopy (FESEM& SEM).Chemical composition changes were investigated by Raman (λ = 473 nm, Nicolet Almega XR), attenuated total reflectance–Fourier transform infrared (ATR–FTIR, Thermo- Electron Nicolet 6700) spectroscopies. To characterize wettability,static contact angle measurements were performed by an automated tension meter, using the sessile drop method. A water droplet was gently positioned onto the surface using a microsyringe and images were captured to measure the angle formed at the liquid–solid interface. The mean value was calculated from at least five individual measurements. Successive measurements were reproducible within ±2°.

The number of adhered platelets was determined by LDH (lactate dehydrogenase) method [16, 17]; and cell culture studies were done using L929 mouse cells for investigation of the surface toxicity [16, 19].

3. Results and Discussions

3.1 Morphological changes

Journal of Physics: Conference Series **414** (2013) 012022 doi:10.1088/1742-6596/414/1/012022

3.1.1 KrF laser irradiation

For the ns laser sources at all of the wavelengths, the morphology comprised irregular micro and nanosized features at low fluences and microcones at moderate fluences. In the case of KrF micro ripples at low (below the ablation threshold) and micro cones and columns appear on the surface at moderate and high fluences respectively.The number of cones increases withthe number of pulses and their height with the irradiation fluence;until well-defined columns are formed on the surface at a fluence of 200 mJ/cm^2 [17].

On the other hand, the surface ablated with ps and fs pulses of KrF comprises microcones for all fluences used, while the roughness degree increases with the fluence. At the same fluence, the cone morphology and the roughness degree attained are more pronounced in the fs as compared to the ps case. Experiments performed at a different number of pulses for each pulseduration and fluence [17] indicate that the most probable mechanism behind microcones formation is the laser beam shielding by local inhomogeneities or surface instabilities induced by the laser irradiation process.

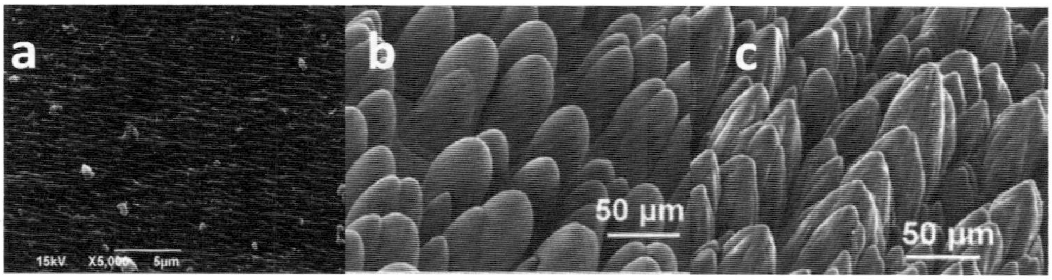

Fig.1 SEM micrographs of the sample irradiated with a) KrF laser with 7000 (ns) pulses at 15 mJ/cm^2, b)1000 (ns) pulses of 100 mJ/cm^2 and c)1000 (fs) pulses of 150 mJ/cm^2

3.1.2 ArF laser irradiation

A threshold fluence of about 13±3mJ/cm^2 was obtained for ArF laser ablation of PES [20]; then samples were irradiated at 7mJ/cm^2 (below the ablation threshold) and 21 mJ/cm^2 above the ablation threshold.Ablation with 193 nm results in nanostructures at low and microbeads (or cones) at moderate fluences that become decorated with nanometric secondary protrusions at higher fluences (fig.2). Such nanostructures may be formed due to the re-deposition of the ablated photofragments condensed on the surface between successive pulses, an effect which is often observed upon irradiation of polymers with ns pulses [22].

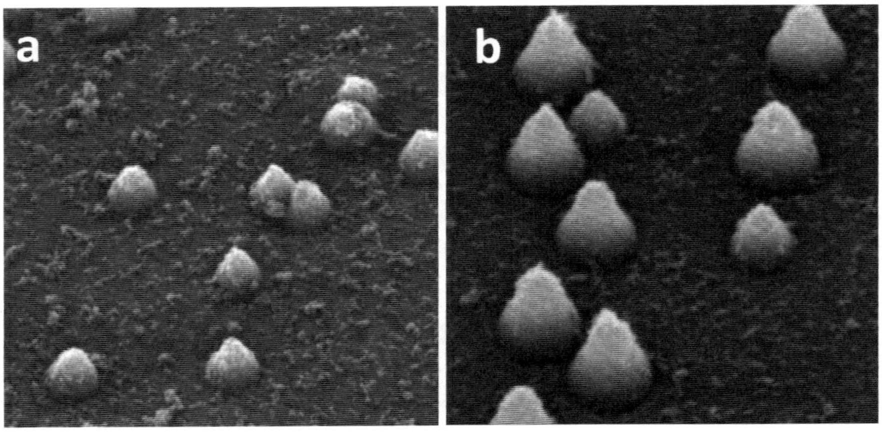

Fig.2 SEM micrographs of the samples irradiated with ArF laser with a) 60 pulses and b)100 pulses of 21 mJ/cm^2

3.1.3 XeCl laser irradiation

The laser ablation threshold of PES was experimentally measured to be about 60 mJ/cm^2 [21], and samples were irradiated at fluences of 25 and 50 mJ/cm^2 (below the ablation threshold), and 70 mJ/cm^2 in the range of the ablation threshold.
XeCl laser interaction of PES has a thermolytical nature rather than direct bond disassociation because of low absorption coefficient of PES (about 120 cm^{-1}) at308 nm. During XeCl laser ablation of PES, conical microstructures appear on the surface. The developed microstructures are laser fluence dependent as well as dose dependent, with the structures becoming more prominent with increasing pulse numbers. It is observed that at the fluence of 70 mJ/cm^2, the material removal from the surface begins to start with about 300 pulses of laser irradiation [21]. This isbecause of the weak absorption of the PES at 308 nm that causes ablation to occur first through incubation followed by subsequent ablation. During the incubation period, the material is converted either photochemically or thermally into a more highly absorbing material and the modified material then ablates as if it was strongly absorbing at that wavelength.

On the other hand, irradiation of the samples below the ablation (at fluences of 25 and 50 mJ/cm^2), leads to formation of a microscopically smooth surface because of melting and resolidification. The laser irradiated surface at 2 fluences regimes of below and above the ablation threshold is shown in Fig.3.

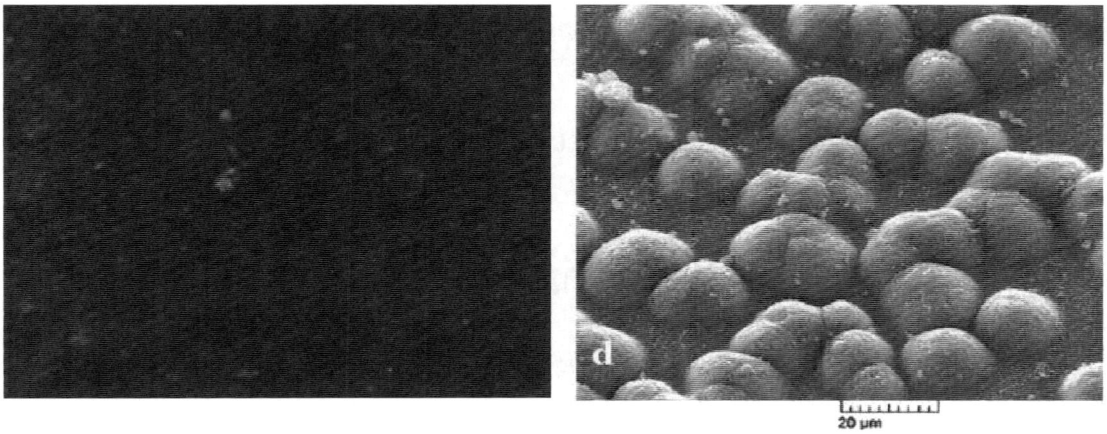

Fig.3 SEM micrographs of the samples irradiated with XeCl laser with a) 500 pulses of 50 mJ/cm2 and b)3000 pulses of 70mJ/cm^2

3.2 Hydrophilicity alteration versus Chemical modifications

At all the wavelengths the increasing wettability were observed in the lower fluence regime, i.e. for fluences below the ablation threshold. CA minima indicated is associated with a change of thermochemical composition of the surface following KrF laser irradiation, particularly to the release of sulfonic and disulfonic acids of the type R–SO_3H (R = H, phenols) as well as carboxylic acids R–CO_2H (R = phenols, methyl, H): all are well known as surfactants and formed by the exposure of polyarylsulfones to UV light [23]. The presence of disulfonic acid is indicated by the ATR–FTIR spectra of the respective samples [18]. Above the ablation the surface chemistry changes especially carbonization with ns laser pulses which are observed in Raman spectra of the irradiated and non-irradiated surface leads to decreasing the hydrophilicity[17, 18].

The functional groups induced on the surface by ArF laser irradiation were examined with ATR-FTIR spectroscopy [20]. The spectrum shows a new peak at 1746cm^{-1} assigned to (C- O group) appears with irradiation of the samples at the fluences of 7 and 21 mJ/cm^2. The peak intensity increases with increasing number of pulses at a fluence of 7mJ/cm^2. Laser irradiation up to3000 pulses produces radicals on the PES surface that are convertedto hydrophilic groups. Further irradiation leads to cleavageof chemical bonds and recombination of carbon radicals producing a surface with hydrophobic properties.

At a fluence of 21 mJ/cm^2, the peak is formed with lower pulse numbers (optimum numbers) and then disappears with 60 pulses. Material removal and cleavage of chemical bonds above the ablation threshold suppose to be responsible for the removal of radicals created on the surface with a slight increase in the number of pulses.The surface becomes very hydrophobic with increasing number of pulses. Inthe other word with increasing the number of pulses and the micro-cone formation, the presence of secondary nanofeatures onto microstructures may be responsible for the superhydrophobicity attained, since it is known that hierarchical roughness greatly enhances the wettability effects [3].

XeCl laser irradiation also can change surface chemistry or surface charge and leads to a change in water droplet contact angle onthe surface. Although here the chemical changes are not pronounced as in the case of KrF and ArF laser irradiation. The chemical modification depends on the laser fluence and number of pulses. The most optimal condition for improving the hydrophilicity of PES surface is500 pulses at a fluence of 25 mJ/cm^2 and it is believed to be a result of photo-oxidation on the irradiated surface. Further increase in number of pulses and fluence leads tocleavage of chemical bonds and recombination of carbon radicals and creation of hydrophobic properties on the surface [24].On the other hand, at higher fluences with increasing the roughness, and under certain surface conditions, the contact angle increases. The laser fluence and number of pulses for minimum contact angle have been presented in table 1.

Table1. The minimum contact angle and the optimum pulse number and fluence obtained at each wavelength

Wavelength (nm)	308 (XeCl)	248 (KrF)	193 (ArF)
Number of pulses	100	7000	3000
Fluence(mJ/cm^2)	70	15	7
Minimum Contact angle obtained	75°	16°	49°

3.3 Results for the biocompatibility tests

Two factors should be considered for cell adhesion on an irradiated surface: 1-chemical composition and 2-roughnessdue to structure formation or debris [16, 25]. According tothe contact angle measurements, it seems below the ablation threshold; chemical changes of the polymer surface including oxidization in ambient air and creation of polar groups lead to increased cell adhesion [19, 25].

In order to quantify how the L929 cells reacted to the different surface topographies, the cell culture studies were done on the irradiated and non-irradiated surfaces and the number of cells was measured for each of the samples. The cell numbers were counted at several positions on the substrate in order to determine the standard deviation. It was shown that [19] the number of cells adhered on the surface irradiated at a fluence of 15 mJ/cm^2 of nanosecond KrF laser is increased significantly. Formation of the structures on KrF laser treated surfaces above the ablation threshold has no significant effect on cell adhesion. The cell adhesion is not changed even with increasing the hydrophilicity of the surfaces in this regime. On the other hand below the ablation, despite of increasing the surface hydrophilicity in all 3 pulse durations of KrF due to the chemical changes, the cell adhesion only is increased with the presence of the periodic nano structures on the surface irradiated with ns pulses.

The data from the LDH method on ArF laser irradiated surfaces show that the extent of platelet adhesion on PES films surface can be controlled by a change in the fluence and pulse numbers. The optimum pulse numbers and the fluence required to reduce platelet adhesion is 3000 pulses and 7mJ/cm^2 respectively (fig.4). This result was expected since the platelet adhesion decreases with an increase in surface wettability. Measurements of platelets adhered in the samples irradiated above the ablation threshold, showed a dramatic increase inplatelet adhesion on the surface.

The platelet adhesion also decreases upon a XeCl laser treatment with various number of laser pulses below the ablation threshold. There is an optimum for number of pulses at each one of fluences to reduce platelet adhesion on the surface. The most optimal condition for decreasing platelet adhesion on the surface is a fluence of 25 mJ/cm^2 and 500 pulses (fig.4). On the other hand with irradiation of surface with 500 pulses at a fluence of 70 mJ/cm^2 the contact angle increaseswhile the number of adhered platelet decreases [16]. Results of the cell culture studies show that fewer cells areattached on untreated surface in comparison to the laser treated ones. Increasing the surface roughness at a fluence of the 70 mJ/cm^2increases the number of adhered cells to the surface.

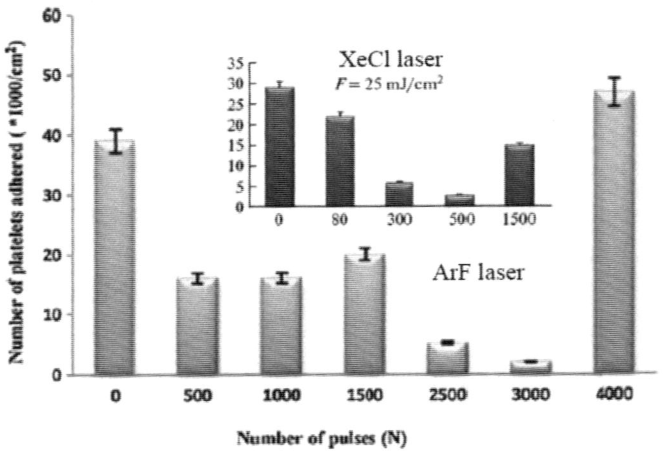

Fig. 4 the number of platelet adhered on the surface following XeCL and ArF laser

4. Conclusions

Laser irradiation of polymers can change the surface biocompability. The extent and types of modifications depends on the laser parameters and the polymer.The values for absorption coefficient of PES are about 140 and 6.1×10^{4} cm^{-1} and 1.8×10^{5} cm^{-1} at 308 and 248 nm and 193 nm respectively [11]. Regardless of the absorption of the surface at the laser wavelength, Irradiation of PES surface at the fluences below the ablation threshold, leads to increasing the surface hydrophilicity and biocompatibility especially for the decreasing the platelet adhesion in the blood contact works. It is observed that the wetting properties attained are mostly influenced by thephoto-induced chemical changes, being highly dependent on irradiation parameters.

The results obtained from the cell-behavior studieson the PES surface revealed that the surface morphology or roughness is more important than the chemistry in cell culture on the surface. Nanoscale hydrophilicripples appear on the surface following ns KrF laser irradiation significantly affect the adhesion of L929 cells on thesurface.

In conclusion it is found that although the XeCl and ArF laser irradiation improve the surface biocompatibility and wettability of the surface, the KrF laser with middle absorption coefficient for PES(respect to the XeCl and ArF laser irradiation) can be supposed as the best wavelength for the improving the surface hydrophilicity.

References

1. J. Gold, Euro. Cells and Mater 10, Suppl. 1, 2 (2005).
2. Evans Analytical Group, Analytical Solutions for Biotechnology (BN 1210, 2007).
3. A. Haruyama, J. Kato, A. Kameyama, Y. Hirai, andY. Oda, Laser Phys. 20, 881 (2010).
4. X. Jiang, H. M. Ge, J. J. Liu, and Q. S. Ren, LaserPhys. 21, 548 (2011).
5. D. Bauerle, J.D. Pedarnig, I. Vrejoiu, M. Peruzzi, D.G. Matei, D. Brodoceanu, Rom.
Rep. Phys. 57 (4) (2005) 935–952.
6. S. Thanawala, S.P. Khan, O. Palyvoda, D.G. Georgiev, I.A. Al-Homoudi, G. Newaz,G. Auner, Mater. Res. Soc. Symp. Proc. 872 (2005) J18.8.1.
7. E. Stratakis, A. Mateescu, M. Barberoglou, M. Vamvakaki, C. Fotakis, S. Anastasiadis, Chem. Commun. 46 4136 (2010)
8. L. Gao, T.J. McCarthy, X. Zhang, Wetting and superhydrophobicity*Langmuir***25** 14100 (Part of the 'Langmuir 25th Year: Special Issue) (2009)
9. M. Barberoglou, V. Zorba, A. Pagozidis, C. Fotakis, E. Stratakis, *Langmuir* **26** 13007 (2010)
10. E.L. Papadopoulou, M. Barberoglou, V. Zorba, A. Manousaki, A. Pagkozidis, E. Stratakis, C. Fotakis, *J. Phys. Chem.* C **113** 2891(2009)
11. G. Scheen, K. Ziouche, Z. Bougrioua, P. Godts, D. Leclercq, T. Lasri, *Langmuir* **27** 6490 (2011)
12. C.G. Khan Malek, Anal. Bioanal. Chem. 385 1351(2006)
13. E. Stratakis, A. Ranella, C. Fotakis, Biomicrofluidics 5013411(2011)
14. E. Stratakis, A. Ranella, M. Farsari, C. Fotakis, Progr. Quant. Electron. 33 127 (2009)
15. D. Bauerle, Laser Processing and Chemistry (Berlin:Springer)(2000)
16. H. Pazokian, J. Barzin, M. Mollabashi, S. Jelvani, S. Abolhosseini, Laser Phys. 22(5), 922 (2012)
17. H. Pazokian, A. Selimis, E. Stratakis, M. Mollabashi, J. Barzin, S.Jelvani, Appl. Surf. Sci. 258, 169 (2011)
18. H. Pazokian, A. Selimis, J. Barzin, S. Jelvani, M. Mollabashi, C.Fotakis, E. Stratakis, J. Micromech.Microeng. 22, 035001 (2012)
19. H. Pazokian, M. Mollabashi, A. Selimis, E. Stratakis, J. Barzin, S. Jelvani, ApplPhys A, DOI 10.1007/s00339-012-7142-9, (2012)
20. H. Pazokian, S. Jelvani, M. Mollabashi, J. Barzin, G. AzizabadiFarahani, Appl. Surf. Sci. **257**, 6186 (2011)
21. H. Pazokian, S. Jelvani, J. Barzin, M. Mollabashi, S. Abolhosseini, Opt. Commun.284 (2011) 363–367.
22. R.S. Taylor , K.E. Leopold , *J. Appl. Phys.* **64** 2815 (1988).

23. A. Rivaton, J.L. Gardette, *Polym. Degrad. Stab.* **66** 385 (1999).
24. E. Rebollar, G. Bounos, A. Selimis, M. Castillejo, S. Georgiou, Appl. Phys. A 92 1043 (2008)
25. W. Pfleging, R. Adamietz, H.J. Brückner, M. Bruns, A. Welle, Proc. SPIE 6459, 645907 (2007)

Journal of Physics: Conference Series **414** (2013) 012023 doi:10.1088/1742-6596/414/1/012023

Monitoring toxicants by stimulated Mandelshtam-Brillouin scattering (SMBS) in a turbulent water flow

T.Yu. Moguilnaya,[a] P.V. Bobkov,[b] V.A. Tomilin [b]
[a]JSC "SERVET-M", 2 8th Radial Str., Moscow, Russia,
mogilnay@mail.ru

[b] "MATI"-Russian state technological university, 3 Orshanskaya Str., Moscow, Russia,
Ist3@mail.ru

Abstract - In this article we considered a method of monitoring of toxicants in a turbulent water flow. The technique was based on SBS (stimulated Brillouin scattering). We have developed a mathematical model of light distribution in the medium with optical inhomogeneities. In our case quantity and form of inhomogeneities depended on time. Also experiments for check of adequacy of mathematical model were made. The results were used to refine the scheme of optical device for rapid analysis toxicants in drinking water.

1. INTRODUCTION

Currently there is an acute problem of controlling toxicants (pathogens, toxins, and other highly dangerous substances) in drinking water. However there are no devices that could determine the presence of toxins directly in a water flow with required accuracy. Dangerous doses of these components vary generally from fractions of a percent to nano-concentrations. The presence of these substances can be detected by modern methods of monitoring, developed recently in chemistry, chromatography and infrared spectroscopy only for stationary solutions. In a non-stationary case the stability of zero signal and a scale factor are beginning to depend not only on a detecting method but also on the design features of a sampling device and the degree of turbulence, leading to false positives and reduced system reliability. Moreover, a number of accurate methods of determining concentrations of some pathogens can be conducted only in specialized laboratories, while the measuring time ranges from several hours to several days. Thus, drinking water needs to be monitored continuously for the presence of pathogenic organisms.

For several years now the company "SERVET" has been conducting an experimental work on controlling the compositions of pathogenic microorganisms and nano-particles in aqueous solutions by the method of Mandelstam-Brillouin scattering (SMBS). Studies were carried out to determine the intensities of Stokes and anti-Stokes components as functions of the composition of a solution. It was shown that for most substances it remains a linear dependence on the optical density for both Stokes and anti-Stokes components up to 10% concentration in a solution. The method accuracy made it possible to determine silver concentrations up to 10^{-4}mg/l, and virus concentrations up to $10^{-8}$of infectious particles per 1 ml (-8 lg tds50) at high signal/noise.

Furthermore, being less dependent on flow parameters than other techniques, the method of stimulated Mandelstam-Brillouin scattering (SMBS) allows you to put an optical device directly into a pipeline and thus avoid the error correction for a sampling device. This advantage is used to control certain parameters in pipelines. Nevertheless, it was necessary to assess the impact of flow dynamics on the error signal occurrence, since errors might lead to false positives and reduced system reliability. A mathematical model of a turbulent flow in the pipe of a simple geometry was built to solve this problem, while calculations were carried out using software ANSYS. The model allows us to optimize various optical circuits in terms of minimizing the error of zero. Undoubtedly, assessing informative spectral characteristics and the parameters that affect the method sensitivity threshold is of great interest from a practical point of view.

2. THEORY

The non-stationary effect of SMBS in liquids has a certain peculiarity: it can be accompanied simultaneously by such effects as the stimulated Raman scattering, the stimulated Rayleigh scattering and the stimulated thermal scattering of light. The thresholds for these effects are usually higher than the threshold for SMBS. If there is a self-focusing, the threshold for Raman scattering and the SMBS threshold become close [1].

We observed a self-focusing of a laser beam in our experiments; therefore it was necessary to estimate theoretically the SMBS threshold to distinguish this effect from the others. The frequency of the sound generated by SMBS in the transient regime is defined by (1).

$$\Omega_{\text{мб}} = \omega_0 \frac{v}{c}[(n_1 - n_2)^2 + n_1 n_2 (2sin\,\frac{\vartheta}{2})^2]^{1/2}; \qquad (1)$$

where ϑ- a scattering angle; n_1, n_2 - the refraction indexes of the exciting and the scattered light waves.

If a generated sound enters a nonlinear mode, the effective sound absorption coefficient starts rising. It is defined by (2).

$$\alpha_{eff.} = \alpha R; \qquad (2)$$

where α - a normal linear sound absorption coefficient; R - an analogue of the Reynolds number defined by equation (3).

$$R = \frac{mp}{\eta_{eff.}\Omega_{BS}}; \qquad (3)$$

where p - a pressure amplitude of a sound wave; $\eta_{eff.}$ - the effective viscosity of a medium at a frequency Ω_{BS}.

If an acoustic non-linearity is observed, a hyper-sound of a basic frequency transforms into the harmonics of higher frequencies. The appearance of the second harmonic of a hyper-sound $f_{BS} \sim 10^{10} Hz$ was observed in water at a basic frequency of $f_{BS} \sim 5 \cdot 10^9 Hz$.

To verify the fact that we are dealing with the SMBS effect, rather than with a wing of Rayleigh scattering, we can estimate the ratios of the amplitudes of Stokes and anti-Stokes components to the amplitude of laser radiation. These values differ considerably from the data presented in [2]. Moreover, the intensities of these components, as was shown earlier, are logarithmic functions of concentrations of dissolved substances [1].

Paper [2] presents the evaluation of the SMBS threshold. This value is much more than the source radiation density of the laser used. It is known that the threshold depends on a self-focusing. As shown in [2], the dynamics of mediums is defined considerably by the process of hydrodynamic interaction of a liquid and the bubbles of gas, contained in it. Therefore to assess the parameters of changing the form and concentration of bubbles we developed a dynamic model. The model was based on Euler's equations, and the calculation results were used earlier while studying variations in concentration of air bubbles with the speed of movement [3].

For the analysis of this phenomenon it was necessary to estimate the amplitudes of falling, scattered and reflected waves. Also we needed to analyze a distribution diagram of the scattering intensity. A scattering medium is a colloidal solution consisting of individual molecules, dissolved objects and air bubbles. They build clusters in the form of an elongated polymer structure with electrically active ions. We assume that the bubbles are much larger in size than toxins, and their concentration exceeds the concentration of electrically charged impurities. The concentration of electrically neutral impurities is small compared to the concentration of air bubbles and objects under research. So, the model of scattering was built for a structure of solution shown in Fig. 1.

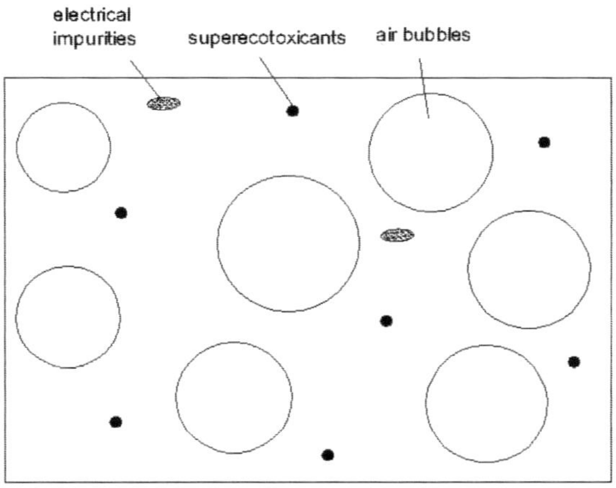

Fig.1. The structure of the solution.

The determination of the scattered field from a given distribution of inhomogeneities in a medium is as follows:

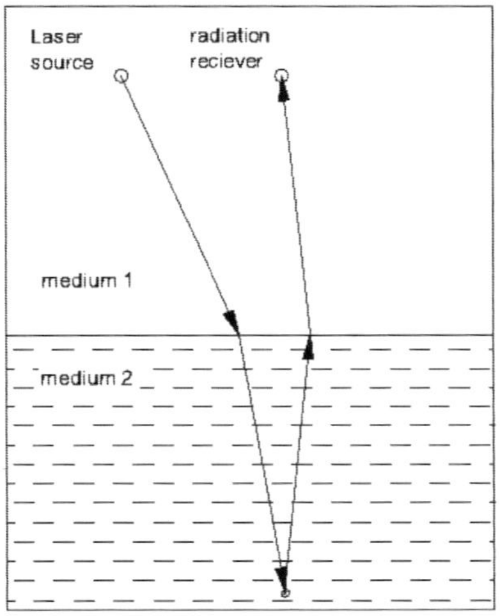

Fig.2. The trajectory of the signal.

After decomposing the spherical wave source into plane waves and taking into account the refraction at the interface, we can write the following expression for the field penetrating into a medium:

$$U(r_1) = \iint_{-\infty}^{+\infty} \frac{iP(k_0)exp(ik_0|C_1 - C_0| + ik_{0z}|Z_s - Z_0| + ik_{1z}|Z_1 - Z_s|)}{2(2\pi)^2 k_{0z}} d^2k_0; \qquad (4)$$

Where $P(k_0)$ - a refraction index for the transition from the upper wave (water) in the bottom half (cluster), the half-space, which is calculated using Fresnel formula (5).

$$k_{1z} = \sqrt{k_0^2 n^2 - k_0^2}; \qquad (5)$$

where $k_{1z}= k_{0z}n$;

k_0 - a wave vector in the upper half;

k_1 - a wave vector in the lower half;

c - a spatial vector in the horizontal plane;

h - a vertical coordinate of the surface relative to the chosen reference frame;

$r_0 = (c_0, z_0)$ - a coordinate of a point source on the surface of a medium;

$r_1 = (c_1, z_1)$ - a point in a medium.

Each plane wave is scattered by irregularities after penetration into the environment. Scattering by inhomogeneity can be calculated using a model of single scattering. In addition, each scattering heterogeneity becomes a source of spherical waves.

The scattered field at the receiver can be determined by the following equation:

$$E(r_0,k_0) = k_1^2 \iiint_{r_1} \Delta e(r_1)U(r_1) \times \iint_{-\infty}^{+\infty} \frac{iP(k_0)exp(ik_0|C_1-C_0|+ik_{0z}|Z_s-Z_0|+ik_{1z}|Z_1-Z_s|)}{2(2\pi)^2 k_{0z}} d^2k_0; \qquad (6)$$

where $P(k_0)$- a transmission coefficient of the interface for the transition from the lower half to the upper half.

Formula (7) is a solution of the direct problem.

$$f(c_F,k_0) = \iint_C E(c,k_0)M(c_F-c,k_0)dc \qquad (7)$$

where $f(c_F,k_0)$ - the field focused at a point of near-surface heterogeneity;

$M(k_0)$- a transmission coefficient of the interface for the transition from the lower wavelength (cluster) to the top half (water).

We consider the problem of the effect of SMBS in a standard setting. The condition for the effect of SMBS is as follows:

$$|E_2|^2 \geq \frac{8v_S^2 r^2 E_1 \alpha_s \alpha}{\gamma^2 k_1 k_s}; \qquad (8)$$

where E, r - the elastic constant and the density of a medium [3].

This evaluation shows that there is not enough power for the formation of waves on air bubbles. We observe the effect of SMBS only on dissolved components. Assume that the acoustic field is defined by equation (9).

$$U(r,t) = 0{,}5 \cdot U_s(r_s)exp[i(w_st-k_sr)] + m.k. \qquad (9)$$

where r_1, r_2, r_s- distances along directions k_1, k_2, k_s, so $r_i = \frac{(k_i \cdot r_i)}{k_i}$,

where $a_s = \frac{-h}{rv_s}$ - an acoustic damping constant.

With the emergence of an acoustic non-linearity, the hyper-sound of a basic frequency should transform into the harmonicas of higher frequencies, resulting in the strengthening of the second harmonica.

We can write the radiative transfer equation (10) to determine the field of dispersion in a medium with a finite value of concentration of scatterers. The scatterer is presented here as an ideal spherical particle.

$$I_{rel} = \frac{I}{I_0} = (r, gradS_i(r)) = -k \cdot S_i(r) + \frac{\sigma}{4\pi} \cdot \ \sum_{j=1}^{j=4} \Phi(f_{ij}(r,r_0), S_j(r_0))dr_0\,; \qquad (10)$$

where S_j - a relative component of the vector of Stokes parameters of the light field $f(c_F,k_0)$ in a medium for each element [3].

r - the direction of a light beam;

$k = \ (\sigma_i + \alpha_i)$ - a relative attenuation coefficient;

$\sigma_i = \sigma_b + \sigma_l + \sigma_{coll.}$- a relative scattering coefficient;

α - a relative absorption coefficient;

$f_{ij}(\varphi)$ - a relative scattering matrix;

φ - a scattering angle, i.e. an angle between the incident (r_o) and scattered (r) light beams;

Φ - a relative density of luminous flux;

$S_j(r_0)$ - a relative scattering on each element.

For the case of a semi-infinite layer the expression for reflection coefficients of multiple scattering in the first approximation becomes as follows:

$$ln \frac{R(\beta)}{R(\beta=0)} \approx -S(r,r_0)\eta \ \overline{\beta}; \qquad (11)$$

where q - an expansion coefficient of 1st degree, as shown in our earlier studies[3].

If a scattering medium with $\alpha = \alpha_0$, and $\sigma = \sigma_0$ has an impurity scattering or absorbing particles with $\alpha_1 = \alpha_2$,$\sigma_1 = \sigma_2$, then the rates of absorption of a dispersive medium is dependent linearly on concentration C, therefore:

$$\beta = \frac{\alpha_0 + c\alpha_1}{\sigma_0 + c\sigma_1} \qquad (12)$$

If an effective radius r of a scattering medium particle is large compared to wavelength λ, and if $\frac{4\pi\chi}{\lambda} \ll I$, where χ is an absorption coefficient for the substance of the particle, then:

$$\sigma = 2\pi r^2 N \qquad (13)$$

where N - a number of particles per unit volume.

Substituting respectively $\alpha = \pi r^2 kr \frac{4\pi\chi}{\lambda} N$, where k is a factor for the particle shape, $\beta = k\rho\chi, \rho = \frac{2\pi r}{\lambda}$, we obtain equation (14)

$$ln \frac{R(\beta)}{R(\beta=0)} \approx -S\eta \ \overline{k\rho x}; \qquad (14)$$

The relations obtained affect reflectivity only depending on $h_{ij}(r,r_0)$, while $S_{ij}(r,r_0)$ have to be determined experimentally. They can be obtained also by solving the transport equation. For our case, we got these relations in the approximation of spherical scatterers. This approximation was verified experimentally by us in [3].

As is known, the scattering function for a marker is determined by the formula calculated in the Rayleigh-Gans approximation:

$$I_1 = (\frac{2\pi}{\lambda})^4 \frac{V^2(\frac{n_1}{n_2}-1)^2}{8\pi^2 r^2} \cdot I_0(1 + cos^2\theta) \, G^2(2xsin\frac{\theta}{2}); \qquad (15)$$

where $G(U) = \overline{\frac{9\pi}{2U^3} I_{\frac{3}{2}}(U)}$, $x = \frac{2\pi dp}{\lambda} n_2$;

I_0, I_1 - intensities of the incident and scattered light;
λ - a wavelength of the incident light in a vacuum;
n - a refraction index for scattering particles;
V - a volume of a particle;
θ- a scattering angle;
r - a radius-vector from an observation point to a point of scattering;
$I_{\frac{3}{2}}(U)$- Bessel function.

$$I_2(\theta) = (\frac{2\pi}{\lambda})^2 \frac{a_f{}^4}{r^2} I_0 |A|^2 \qquad (16)$$

$$A(p,r) = \int_0^{\frac{\pi}{2}} (1 - e^{-ipsin\gamma}) I_0(2cos\, r) sin\, r\, dr \qquad (17)$$

The experimental verification of these relations is provided in work [3]. A general expression for the reflection index relation can be written as:

$$ln \frac{R(\beta)}{R(0)} \approx I_1 I_2 \eta \ \overline{\beta} \qquad (18)$$

We obtained a transcendental equation for R∞(0), which is related to the relative intensities in the previous experiments. The constant is determined only by the parameters of a device.

Studying scattering patterns for biological solutions is a difficult task. However, the following function can be accepted semi-empirically in many cases.

$$ln \frac{R(\beta)}{R(0)} = I_1 I_2 I_3 \ \overline{\beta_1 \beta_2 \beta_3} \qquad (19)$$

where I_i - relative intensities for the pollutants dispersed in a distilled water, in a turbulent flow and in a stationary aqueous solution, containing a toxin;

β_3 - a coefficient characterizing a scattering matrix, $\beta_3 = 1$ for a strictly parallel beam.

In the first approximation it holds:

$$\sigma = \alpha(1 - k_{pr}) \qquad (20)$$

On using a set of experimental data and averaging, a scattering coefficient can be determined in aqueous solution. Let us assume that we are dealing with a parallel beam, having the difference between the angle of incidence and the receiving angle of 180^0. Then we can assume that a = 1. If we have a scatter plot, then a scattering coefficient can be determined by equation (21).

$$\alpha(\varphi) = \frac{1-k(\varphi)}{1-k_2} \qquad (21)$$

where k_2 - a scattering coefficient at an angle of 180°;

$k(\varphi)$ - a scattering coefficient at an angle φ, which enters the expression for determining the intensity in the presence of lead salts in a medium.

This technique was tested in a typical medium for simulation - in milk. Milk is a colloidal solution of spherical particles of fat and protein. As a medium for simulation, it is used for the development of devices to study water in the seas and oceans. Thus, milk is known to be a good analogy of biological objects containing proteins and nucleic acids.

Formula (16) shows, that dispersion angles also can effect the determination of viruses, and therefore, the sensitivity range of the method. Thus, we paid particular attention to the calculations and experimental measurements of dispersion diagrams for the objects involved.

To establish any angular dependence we need to determine relative intensities. The following relation holds in the first approximation:

$$ln\ \frac{R(\beta)}{R(0)} = D(\frac{I}{I_0}) = ln\ (\frac{I}{I_0}) = I_1 I_2\ \ \overline{\beta_1 \beta_2}; \qquad (22)$$

then:

$$\overline{\beta_1(\varphi)\cdot\beta_2(\varphi)} = \frac{\frac{\Delta ln\ U_0}{\Delta\varphi}}{I_1 \cdot I_2}; \qquad (23)$$

From this relation we can deduce the following formula:

$$\overline{\beta_1(\varphi)\cdot\beta_2(\varphi)} = \sigma_1\sigma_2; \qquad (24)$$

I_1, $I2$ and $\frac{\Delta ln\ U_0}{\Delta\varphi}$ can be defined by formulas (16) and (17 .. 19).

Using only a single measurement for a differential scattering pattern results in a too big error. Therefore, ten scattering diagrams were calculated and used for deriving average values σ1 and σ2 in acetum.

For further developments we are interested in studying not only the total response of a scattering medium on toxins, but also in defining how exactly every dissolved component influences the scattering pattern. So,

$$\sigma_0 = \frac{\sigma_1\sigma_2}{\sigma_1^*}; \qquad (25)$$

where σ_1^* - an average value of a maximum diameter for pollutant dispersion.

After substituting semi-empirical relations, the equation was solved in Rytov approximation. The calculations were made using data of the diffraction phenomena at the output of the fiber from [3-5]. Fig.3 shows the dependence of the divergence angle of a radiation cone on a distance.

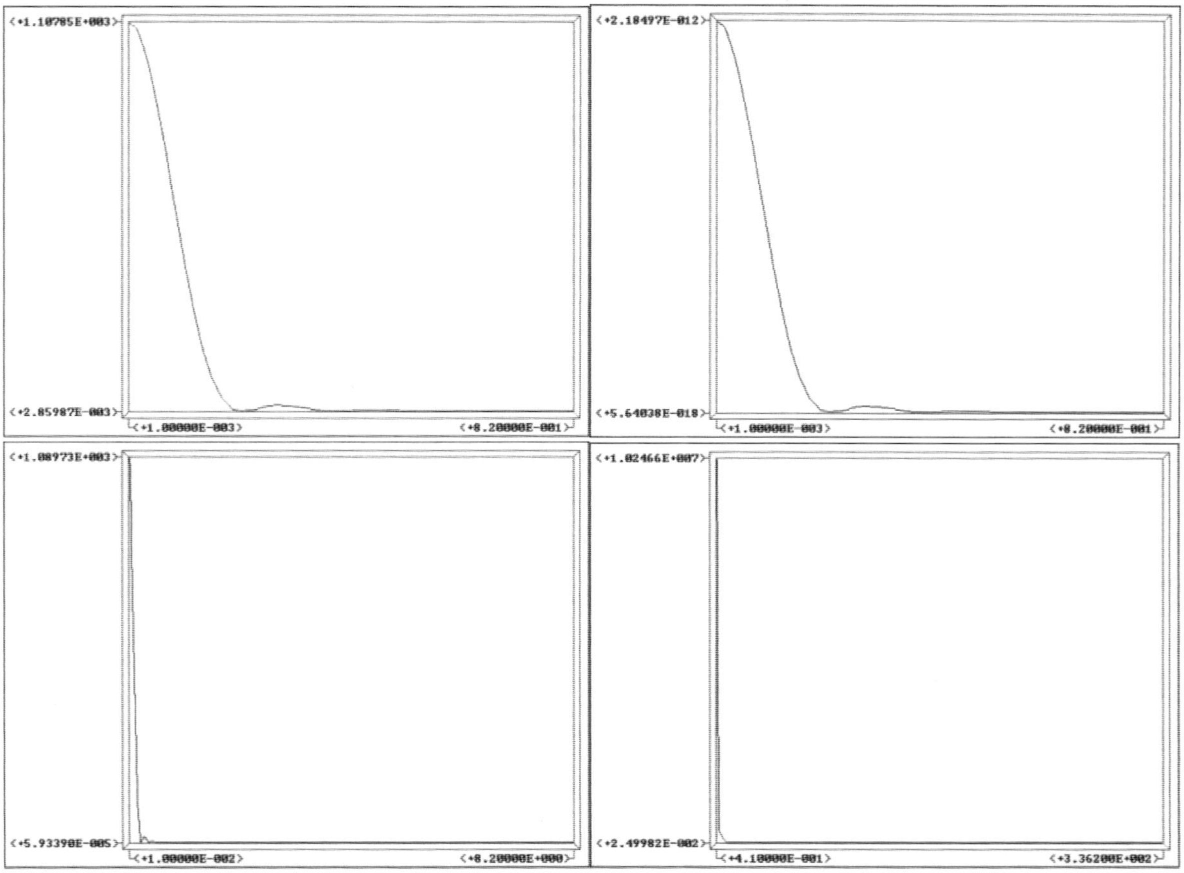

Fig.3. Dependence of the spectral dispersion on the amount of air bubbles.

Given a self-focusing associated with scattering from air bubbles, it should be noted that the threshold intensity for the solution is almost an order higher than the threshold set in [4]. Fig.4. shows the dependence of distribution of the relative intensity at the maximum on the concentration of air bubbles.

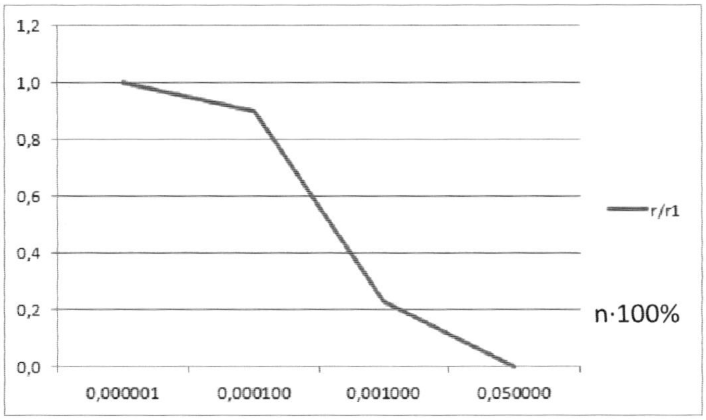

Fig.4. Dependence of beam radiuses relation on the concentration of air bubbles.

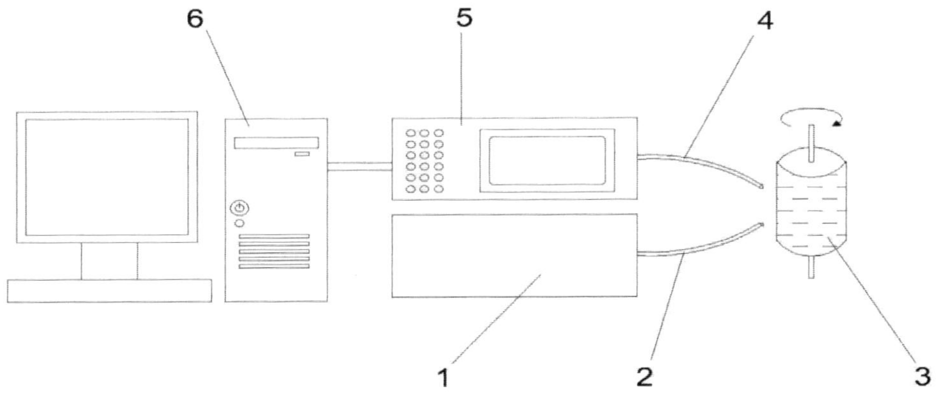

Fig.5. Experimental equipment.

3. EXPERIMENTAL SETUP

1 - a laser with a wavelength of λ = 830 nm (or 1010 nm), 2 and 4 - optical waveguides, 3 - a rotating cuvette, 5 - Agilent spectrum analyzer with a resolution of 0.05 nm, 6 - PC.

We employed semiconductor lasers with a wavelength of λ = 830 nm or 1010 nm and a maximum output power of 150 mW. The power of lasers could be changed gradually. Stokes and anti-stokes components were defined accurately at a working power of 40 mW. The whole setup was placed in a light-tight enclosure. A standard fiber with the following parameters was used: a numerical aperture of NA 0.15 micron and a core diameter of 105. The beam diameter was much smaller than the diameter of a cuvette.

4. DESCRIPTION OF EXPERIMENTS

Primers were selected to simulate dissolved particles. We used oligonucleotides FAM, HEX etc., because they have a spherical shape [3] and do not cause peaks at infrared luminescence, as shown in Table 1. The primers were diluted with water, added to a cuvette and pre-centrifuged to obtain an evenly distribution of a solution. The concentrations of primers ranged from 1 to 10 ^{7}oml / (7 lgtds50).

The laser radiation (1) with a wavelength of λ = 830 nm (or 1010 nm) was applied along a waveguide (2) to a rotating cuvette with water (3). Then the reflected scattered light was fed through a waveguide (4) into a high-precision spectrum analyzer Agilent (5) with a resolution of 0.05 nm. The obtained spectra were stored and processed by a PC (6). To get stable results we chose the averaging of a high degree. Each result was recorded twice with an interval of at least five minutes. A laser wavelength was chosen based on the table of optical properties of the primers used (Table 1).

Table1. The optical properties of primers.

Oligonucleotides	The maximum absorption	The maximum emission	absorption coefficient
FAM	494	519	65000
TET	521	536	99000
HEX	530	556	98000

Fig. 6-10 presents the following spectral distributions:

Fig. 6 shows the spectral distribution of the laser power irradiated to a stationary cell with distilled water.

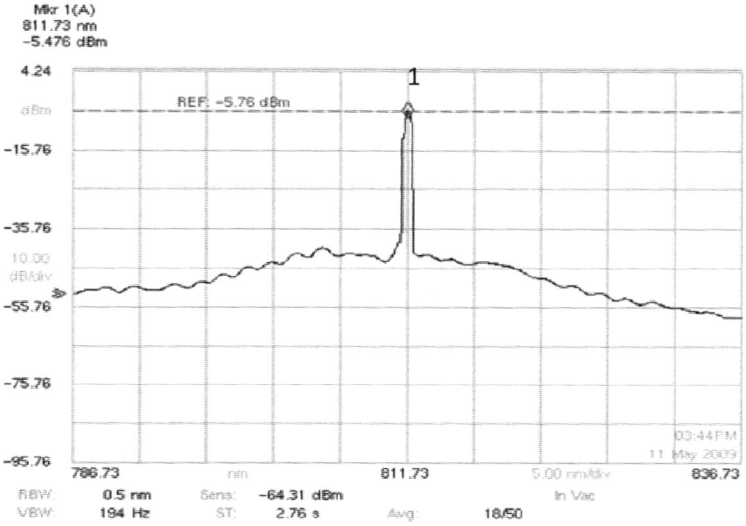

Fig.6. The inverse spectral scattering for a fixed cuvette with water.

The flow rate is zero in this case. There is only one peak here with the maximum at a wavelength of λ= 811.73 nm and an intensity equal to -5.476dBm. The peak corresponds to the laser field.

Based on our calculations, we chose an angular velocity of a cuvette. It was done carefully to exclude an additional Stokes component due to scattering from air bubbles as well as to avoid multiple scattering by the diffraction of inverse components. An angular velocity of a cuvette was determined using an analogue of Reynolds number.

Fig. 7 shows the scattering spectrum, observed at an angular velocity of a cuvette equal to 6 rot/min.

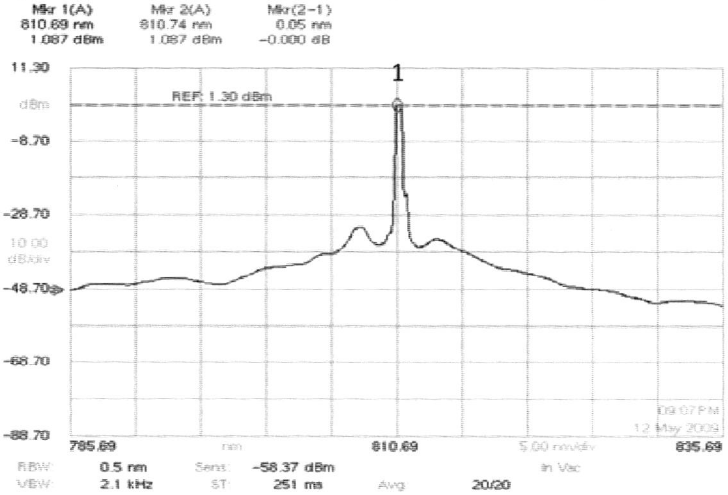

Fig.7. Spectral scattering at low flow rates.

Both Stokes and anti-Stokes components are weak. There is only one peak here with the maximum at a wavelength of λ= 810.69 nm and an intensity equal to -1.087 dBm. The peak corresponds to the laser field.

Fig. 8 shows the scattering spectrum, observed at an angular velocity of a cuvette equal to 30 rot/min.

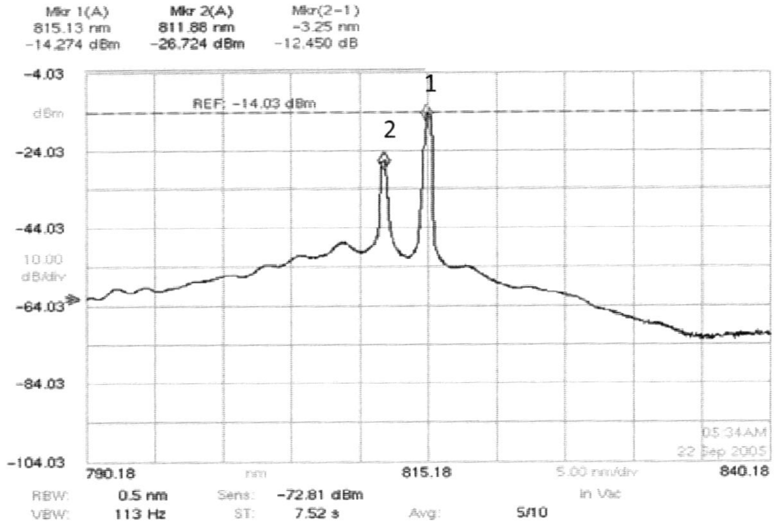

Fig.8. Spectral scattering at medium flow rates.

This case corresponds to the transient beam. We can see here that anti-Stokes component also increases with the velocity of the fluid. There are two peaks here: 1 - the maximum at a wavelength of λ=815,18 nm with an intensity equal to -14,274 dBm; 2 - the maximum at a wavelength of λ=811,88 nm with an intensity equal to -26,724 dBm. Peak 1 corresponds to the laser field, while peak 2 is associated with cavitation processes in the liquid.

Fig. 9 shows the scattering spectrum, observed at an angular velocity of a cuvette equal to 120 rot/min.

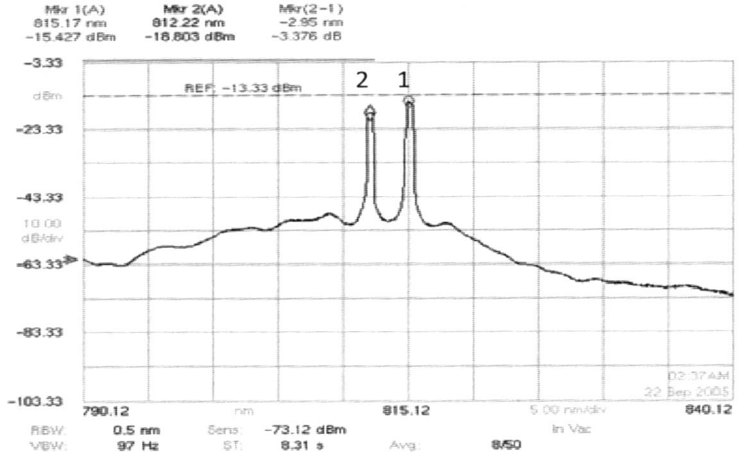

Fig. 9. Spectral scattering at high flow rates.

Again we can see that Anti-Stokes component increases with the velocity of the fluid. The second harmonic also increases significantly, as is shown in Fig. 11. There are two peaks here: 1 - the maximum at a wavelength of λ=815,17 nm with an intensity equal to -15,427dBm; 2 - the maximum at a wavelength of λ=812,22 nm with an intensity equal to -18,803 dBm. Peak 1 corresponds to the laser beam, peak 2 represents Anti-Stokes component. The intensity logarithm is proportional to the concentration of the marker.

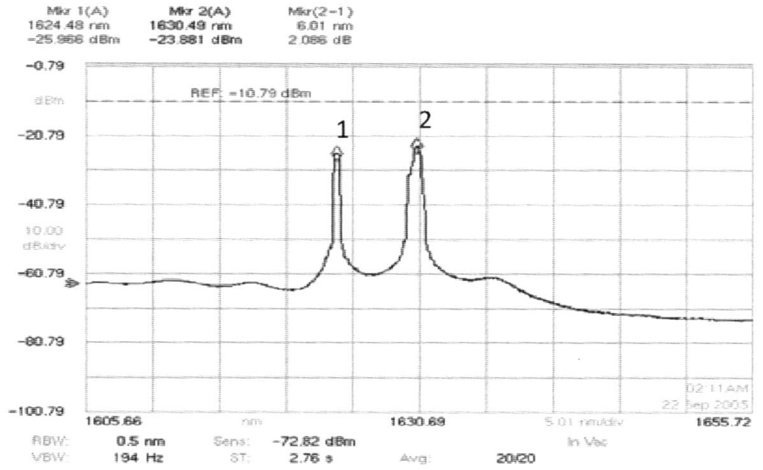

Fig.10. The intensity distribution of the second harmonic at high flow rates.

There are two peaks at a cuvette rotation speed equal to 120 rot/min.: 1 - the maximum at a wavelength of λ=1624,48 with an intensity equal to -25,966 dBm; 2 - the maximum at a wavelength of λ=1630,49 nm with an intensity equal to -23,881 dBm. The concentration of a marker is proportional to logarithm of the intensity for both of the peaks.

During the experiments we found that the intensities of both Stokes and anti-Stokes peaks increase with increasing an angular cuvette velocity. The complete analysis of the experimental data shows also that the second harmonic of the intensity increases significantly. However the intensity of the second harmonic is still much smaller than that of the first. Yet the second harmonic has a better signal/noise ratio, which makes possible to use it as an informative signal.

The experiments show that it was possible to observe only one of Stokes and anti-Stokes components up to a rotation speed about 60 rot/min. The results of both theoretical and experimental calculations were used to refine a scheme of an optical device for the rapid analysis of superecotoxicants in drinking water.

5. CONCLUSIONS

A mathematical model of the motion of a turbulent flow was developed to help us investigate the dynamics of air bubbles in water.

In a series of experiments for a flow of water with dissolved impurities, we found that the peaks that correspond to the second harmonic depend only on concentrations of dissolved substances. Analysing the spectral distributions of these peaks is more likely to determine the type of solute than analysing the Stokes and anti-Stokes peaks corresponding to the first harmonic.

We observed a self-focusing of a laser beam and found it to depend on the concentration of air bubbles.

Several peaks in the Stokes and anti-Stokes regions were observed with increasing flow rate. The intensity of the peaks was found to increase with the rising concentration of air bubbles.

The results were used to refine a scheme of an optical device for the rapid analysis of superecotoxicants in drinking water.

REFERENCES:

[1] Straunov V, and Faberlinski I 1969 SMBS and the stimulated thermal scattering of light *Advances in Physical Sciences* **1** 78

[2] Zverev G, and Martynov A 1989 Research of SMBS thresholds for different mediums on the waves of 0,35; 0,69 and 1,06 mcm. *Pis'ma GETF* **6** 11

[3] Moguilnaya T., Saguitova E, Prokhorov K, Grebenuk E, and Grebenuk N 2001 Laser instrumentation for the express diagnostics of impurities and toxins in liquid food, Photonic Detection and Intervention Technologies for Safe Food Proceeding of SPIE Vol. 4206 (2001) 256-265

[4] Fedotovski V, Vereshiagina T, Derbenev A, and Prokhorov Yu 1998 Theoretical and experimental research of the propagation features of low-frequency pressure waves in liquid mediums with air bubbles *FIZIKA* **1** 89-103

[5] Moguilnaya T, Bobkov P, Tomilin V, and Agibalov A Development of the device to control superecotoxicants in water supply systems 2012 *Pribori* **8** 23-31

Journal of Physics: Conference Series **414** (2013) 012024 doi:10.1088/1742-6596/414/1/012024

Potentiometric urea biosensor utilizing nanobiocomposite of chitosan-iron oxide magnetic nanoparticles

A. Ali,[1,2], M. S. AlSalhi[3,4], M. Atif[3,4,7], Anees A. Ansari[5], M. Q. Israr[1], J. R. Sadaf[1], E. Ahmed[2], O. Nur[1] and M. Willander[1,6]

[1]Department of Science and Technology, Campus Norrköping, Linköping University, SE-60174 Norrköping, Sweden.
[2]Department of Physics, Bahauddin Zakriya University, Multan, Pakistan.
[3]Physics and Astronomy Department, College of Science, King Saud University, Riyadh, Saudi Arabia
[4]Research Chair for Laser Diagnosis of Cancer, King Saud University, Riyadh, Saudi Arabia
[5]King Abdullah Institute for Nanotechnology, King Saud University, Riyadh-11451, P.O. Box-2455, Saudi Arabia
[6]Visiting Professor College of Science, King Saud University, Riyadh, Saudi Arabia
[7]National Institute of Laser and Optronics, Nilore, Islamabad, Pakistan

Corresponding Author: Dr M Atif, atifhull@gmail.com

Abstract

The iron oxide (Fe_3O_4) magnetic nanoparticles have been fabricated through a simple, cheap and reproducible approach. Scanning electron microscope, x-rays powder diffraction of the fabricated nanoparticles. Furthermore, the fabrication of potentiometric urea biosensor is carried out through drop casting the initially prepared isopropanol and chitosan solution, containing Fe_3O_4 nanoparticles, on the glass fiber filter with a diameter of 2 cm and a copper wire (of thickness ~500 μm) has been utilized to extract the voltage signal from the functionalized nanoparticles. The functionalization of surface of the Fe_3O_4 nanoparticles is obtained by the electrostatically immobilization of urease onto the nanobiocomposite of the chitosan- Fe_3O_4 in order to enhance the sensitivity, specificity, stability and reusability of urea biosensor. Electrochemical detection procedure has been adopted to measure the potentiometric response over the wide logarithmic concentration range of the 0.1 mM to 80 mM. The Fe_3O_4 nanoparticles based urea biosensor depicts good sensitivity with ~42 mV per decade at room temperature. Durability of the biosensor could be considerably enhanced by applying a thin layer of the nafion. In addition, the reasonably stable output response of the biosensor has been found to be around 12 sec.

Keywords: iron oxide (Fe3O4) magnetic nanoparticles, emission characteristics, potentiometric response, urea biosensor, chitosan solution

1. Introduction:

The growing demand for clinical diagnostics in relation to kidney and liver diseases has stipulated the development of new methods for rapid and accurate measurement of urea in samples like urine, serum and blood. Normal range of urea in urine and blood is 8-20 mg/dl. An increase in urea level in blood and urine causes renal failure, urinary tract obstruction, dehydration, shock, burns and gastrointestinal bleeding. However, reduced urea level may cause hepatic failure, nephritic syndrome and cachexia (low protein and high carbohydrate diets). In comparison to conventional methods of urea detection electrochemical biosensors have been considered to provide interesting alternatives due to their simplicity, low cost and high sensitivity [1, 2].

Since biosensors allow a wide range of transduction technology so they have advantage over ordinary electrochemical sensors. Among them, the urea biosensor has widely been distributed in nature, and its analysis is of considerable interest in clinical and agricultural chemistry [3, 4]. It is known to be an important maker for evaluating uremic toxin level.

Immobilization of Ur on a suitable matrix is an important step for the fabrication of electrochemical urea sensor. In this context, hybrid nanobiocomposites possess much attraction due to synergistic effect of two components in order to obtain improved biosensing characteristics. Metal oxide nanoparticles-chitosan (CH) based hybrid composites have attracted much interest for development of a desired biosensor (5-7). Metal oxide nanoparticles such as iron oxide (Fe_3O_4) (8-10), Zinc oxide (ZnO) (11-12), cerium oxide (CeO_2) (13-14) etc. have been reported as promising matrices for the immobilization of desired biomolecules.

Among various metal oxide nanoparticles, Fe_3O_4 nanoparticles have been found to be chemically and biologically inert, they can be coated with metal catalysts, enzymes, or antibodies to increase their functionalities for biosensor applications. Moreover, immobilization of bioactive molecules onto surface charged super paramagnetic Fe_3O_4 nanoparticles is of great interest since magnetic behavior of these bio conjugates may result in improved delivery and recovery of biomolecules for desired biosensing applications [15,16,17]. Fe_3O_4 nanoparticles have been proposed to provide a favorable micro environment for immobilization of desired enzymes directly onto electrode surface resulting in catalytic activity, electron transfer between medium and electrode or improved direct electron transfer of redox proteins [18,19,20].

Besides this existing problem of aggregation and rapid biodegradation of Fe_3O_4 nanoparticles on a given matrix containing biomolecules can perhaps be overcome by modifying these nanoparticles using CH by preparing hybrid nanobiocomposite.

Chitosan (CH) (N-deacetylated derivative of Chitin) is alinear copolymer of glucosamine and N-acetylglucosamine units and naturally found in the exoskeleton of crustaceans, fungal cell walls and in other biological materials. Among other reported properties CH has received attention for enzyme immobilization. For biosensor applications, recent attempts have been made for improving optical and electrical properties of CH by dispersing superparamagnetic Fe_3O_4 nanoparticles. (21, 22)

Ajeet Kaushik et al. (2009) have fabricated Fe_3O_4-CH hybrid nanobiocomposite for urea sensor [23]. The nano bio composite of Fe_3O_4 nanoparticles and chitosan have recently been reported for use in the detection of glucose, urea, mycotoxin, phenolic compounds and ferritin. [16, 24, 25,26]

We report results of studies relating to immobilization of Urs on to Fe_3O_4-CH nanobiocomposite deposited on glass fiber filter substrate for fabrication of urea sensor. The reported Fe_3O_4-CH nanobiocomposite have been fabricated by using simple two electrode method.

2. Materials and Methods:

Urease (E.C.3.5.1.5 from jack Bean 100 u/mg) and urea (Acsreagnt 99.9%) have been procured from Sigma Aldrich (USA). Phosphate buffer, 10mM solution (PBS) was prepared from Na_2HPO_4 and KH_2PO_4 (Sigma Aldrich) with Sodium Chlorate in 0.1315 mM and pH was adjusted to 7.4. A stock urea solution of 100mM was prepared in PBS. The low concentration standard solution of urea was also prepared before the measurements. For making Fe_3O_4-CH nanobiocomposite, following steps have been followed. Chitosan sol gel was prepared in 1% acetic acid and 1 M hydrochloric acid solutions and kept on stirring for

24 hours. The Fe_3O_4 magnetic nanoparticles were mixed with deionized water and stirred for 1 hour. The iron oxide magnetic nanoparticles were suspended in the sol gel of chitosan for making suspension. This suspension has been dispense on a copper wire (d=500um) mounted on a glass fiber filter. The iron oxide magnetic nanoparticles based biosensor electrode was developed by drop wise dispersion of suspension, based on sol gel with suspended iron oxide magnetic particles, on copper wire. Then the Fe_3O_4-CH nanobiocomposite based biosensing electrode was immobilized with urease using simple physical adsorption method. The urease solution was prepared in PBS of pH 7.4 with activity of 2mg/ml. The potentiometric measurement in different concentrations of urea was carried out using potentiometric method. The cell assembly was consisted of the following elements; the immobilized Fe_3O_4-CH nanobiocomposite based biosensing electrode was used as working electrode and Ag/AgCl as reference electrode. A pH meter (Model 215, Denver Instrument) was used to measure the potentiometric output voltage in the experiments.

3. Results and discussion

The representative XRD pattern showing all diffraction peaks of the iron oxide magnetic nanoparticle sample as shown in Figure 1 which can be readily indexed to the cubic spinel Fe_3O_4 having the lattice parameter α =8.072Å and reveals high consistency with the JCPDS card No. 76-1802. It can be clearly seen from a XRD diffraction pattern that synthesized iron oxide nanoparticles have high purity and good crystal quality. Moreover, broadening of the diffraction peaks confirms the small particle size and high crystalline nature of the material.

Schematic and sensing measurement in Figure (2) shows the experimental setup for the potentiometric biosensing mechanism consisting of a standard Ag/AgCl reference electrode and the immobilized Fe_3O_4/Cu working electrode. The chemical equation of urea hydrolysis illustrates the release of ammonia (NH_3) and CO_2 gases as given below:

$(NH_2)2CO + H_2O$..........................> $2NH_3 + CO_2$

When urease immobilized iron oxide magnetic nano particles mounted on glass fiber filter with Cu wire was tested into urea solution, then urease enzyme present on nano surface rapidly hydrolysis the urea into ammonia and carbon dioxide. The fact for the generation of output voltage is based on the reaction of ammonia with water through acid base chemistry resulting charged ions. The potentiometric response was carried out for different urea concentrations varying from 0.1 mM to 80 mM as shown in figure (3). As a result, highly quick output response was given by the proposed urease immobilized magnetic nanoparticles based sensor electrode for the detection of urea with sensitivity of 42 mV/decade.

The potential difference (EMF) between the urease/ Fe_3O_4-CH/Cu biosensor electrode and the reference electrode (Ag/AgCl), changes with the variation in the composition of the test electrolyte solution. According to the mechanism, the potential difference is accredited to the accumulation of ammonium ions on the surface of the urease/ Fe_3O_4-CH /Cu bioelectrode, where the potential of the reference electrode (Ag/AgCl) is illustrated to have a constant value of 222.34 mV at room temperature [27].

Moreover, the selectivity of the urease/ Fe_3O_4-CH /Cu bio sensor electrode was examined by adding different interferents, e.g. uric acid, glucose and sodium Pyruvate to the urea solution. It was also observed that the size of working electrode and volume of analyte solution has no effect on the output response of the proposed urea sensor electrode. The repeatability of urease/Fe_3O_4-CH/Cu biosensor electrode was carried out in the detected range of urea concentrations for three consecutive days and it was found that the sensor electrode had given almost similar response. The sensor electrode has demonstrated a fast time response less than 12 s for the detected concentrations of urea. The shelf life of ureases immobilized iron oxide magnetic particles was found to be three weeks and it was found that sensor electrode retained activity of 90% of initial activity, which reflects that the magnetic particles provide strong binding with the urease molecules.

4. Conclusion

A potentiometric urea biosensor based on Fe_3O_4-CH nanobiocomposite by immobilization of urease enzyme has been fabricated. The x-rays powder diffraction reveals purity and crystallinity of the iron oxide nanoparticle.

The potentiometric response of ~42 mV per decade at room temperature was observed over the wide logarithmic concentration range of the 0.1 mM to 80 mM .The presented biosensor shows negligible influence to interferents which reveals that this biosensor is useful for detection of urea in samples with some interferers. Throughout the experiment, the sensor showed good performance in sensitivity, stability, reproducibility and selectivity due to reasonable stable output response of the biosensor around 12 sec.

All these useful features can make the presented biosensor suitable for practical application in medical, food or other areas. Moreover, the presented biosensor is feasible in use for on-spot clinical diagnosis. Moreover, the fabrication method is found simple and easy and can be extended to immobilize other enzymes, and other bioactive molecules with low isoelectric points for multiple biosensor designs.

Acknowledgement

We are thankful to International Science Programme (ISP), Uppsala University, Sweden, Royal University of Phnom Penh (RUPP), Cambodia, and also the Deanship of Scientific Research at King Saud University through the research group project number: RGP- VPP-023, who financially supported this research work.

References

1. Rajesh, V. Bisht, W. Takashima, K. Kaneto, Biomater. **26**, 3683 (2005).
2. J. V. D. Melo, S. Cosnier, C. Mousty, C. Martelet, N. J. Renault, Anal. Chem. **74,** 4037 (2002).
3. Singhal, R. L., Gambhir, A., Pandey, M. K, Annapoorni, S., Malhotra, B. D., Biosens.Bioelectron. **17**, 697 (2002).
4. G. Zhao, J. J. Xu, H. Y. Chen, Electrochem. Commun. **8**, 148 (2006).
5. A. Maaref, H. Barhoumi, M. Rammah, C. Martelet, N. jaffrezic-Renault, C. Mousty, S. Cosnier, Sens. Actuators B, Chem. **123**, 671 (2007).
6. P. R. Solanki, A. Kaushik, A. A Anees, G. Sumana, B. D. Malhotra, Appl. Phys. Lett. **93,** 163903 (2008).
7. R. Khan, A. Kaushik, P.R. Solanki, A. A. Ansari, M. K. Pandey, B. D. Malhotra, Anal. Chimi. Acta **616**, 207 (2008).
8. L. M. Rossi, A.D. Quach, Z. Rosenzweig, Annal. Bioannal. Chem. **380**, 606 (2004).
9. G. K. Kouassi, J. Irudayaraj, G. McCarty, J. Nanobiotech. **3**, 1 (2005).

10. H. Wei, E. Wang, Anal. Chem. **80**, 2250 (2008).

11. S. P. Singh, S. Arya, P. Pandey, B. D. Malhotra, S. Saha, K. Sreenivas, V. Gupta, App. Phys. Lett. **91**, 063901 (2007).

12. A. Wei. X. W. Sun, J. X. Wang, Y. Lei, X. P. Cai, C. M. Li, Z. L. Dong, W. Huang, Appl. Phys. Lett. **89**, 123902 (2006).

13. A. A. Ansari, A. Kaushik, P. R. Solanki, B. D. malhotra, Electrochem. Commun. **10**, 1246 (2008).

14. A. A. Ansari, P. R. Solanki, B. D. malhotra, Appl. Phys. Lett. **92**, 263901 (2008).

15. L.M. Rossi, A. D. Quach, Z. Rosenzweig, Anal. Bioanal. Chem. **380**, 606 (2004).

16. A. Kaushik, R. Khan, P. R. Solanki, P. Pandey, J. Alam, S. Ahmad, B. D. malhotra, Biosens. Bioelectron. **24**, 676 (2008).

17. A. Kaushik, P. R. Solanki, A. A. Ansari, S. Ahmad, B. D. malhotra, Electrochem. Commun. **10**, 1364 (2008).

18. D. Cao, N. Hu, Biophys. Chem. **121**, 209 (2006).

19. G. K. Kouassi, J. Irudayaraj, G. J. McCarty, Nanobiotechnology, **31**, 1 (2005),

20. J. Gong, X. Lin, Microchem. J. **75**, 51 (2003).

21. A. Kaushik, R. Khan, P. R. Solanki, P. Pandey, J. Alam, S. Ahmad, B. D. malhotra, Biosens. Bioelectron. **24**, 676 (2008).

22. A. Kaushik, P. R. Solanki, A. A. Ansari, S. Ahmad, B. D. malhotra, Electrochem. Commun. **10**, 1364 (2008).

23. A. Kaushik, P. R. Solanki, A. A. Ansari, G. Sumana, S. Ahmad, Sensors and Actuators B **138,** 572 (2009).

24. A. Kaushik, P. R. Solanki, A. A. Ansari, S. Ahmad, B. D. malhotra, Electrochem. Commun. **10**, 1364 (2008).

25. S. F. Wang, Y. M. Tan, D. M. Zhao, G. D. Liu, Biosens. Bioelectron. **23**, 1781(2008).

26. S. F. Wang, Y. M. Tan, Anal. Bioanal. Chem. **387**, 703 (2007).

27. W. Stumm, J.J. Morgan, Aquatic Chemistry: An Introduction Emphasizing Chemical Equilibria in Natural Water Wiley, New York, 480 (1981).

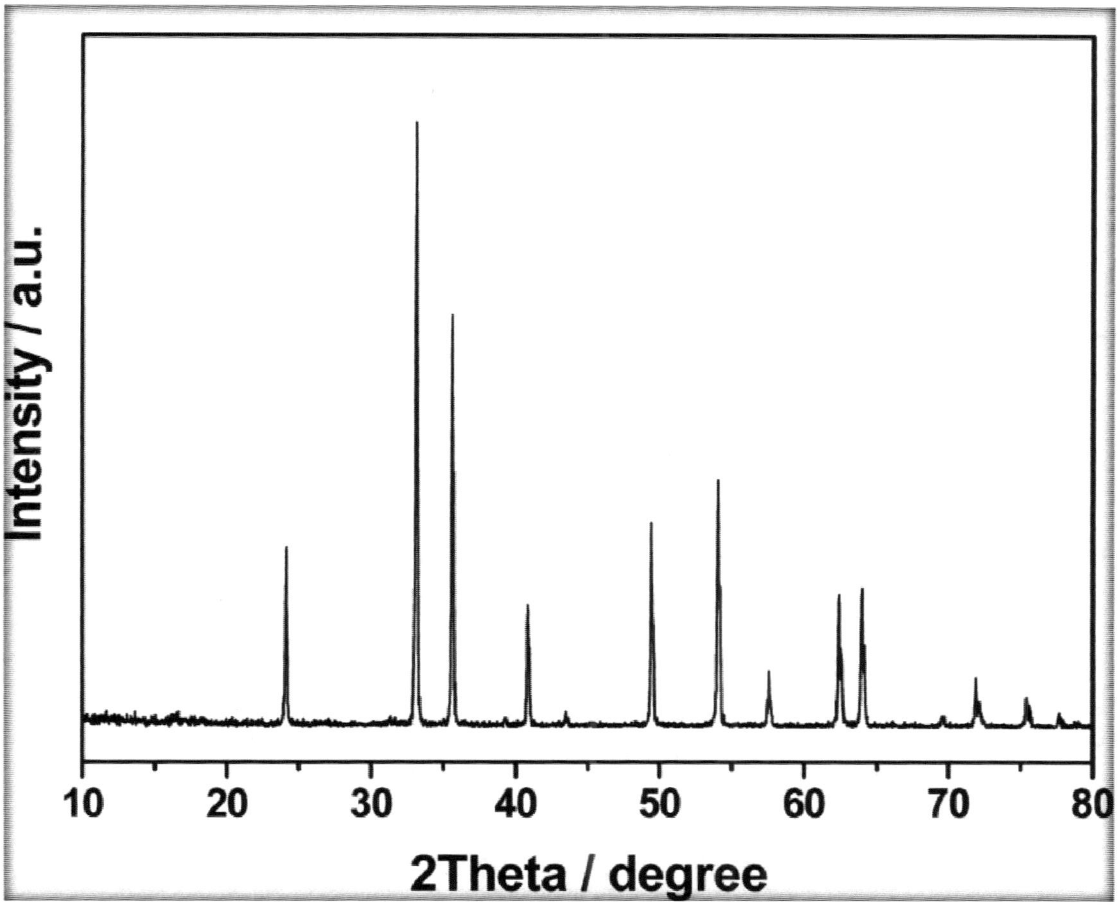

Figure 1: XRD pattern of iron oxide nanoparticles.

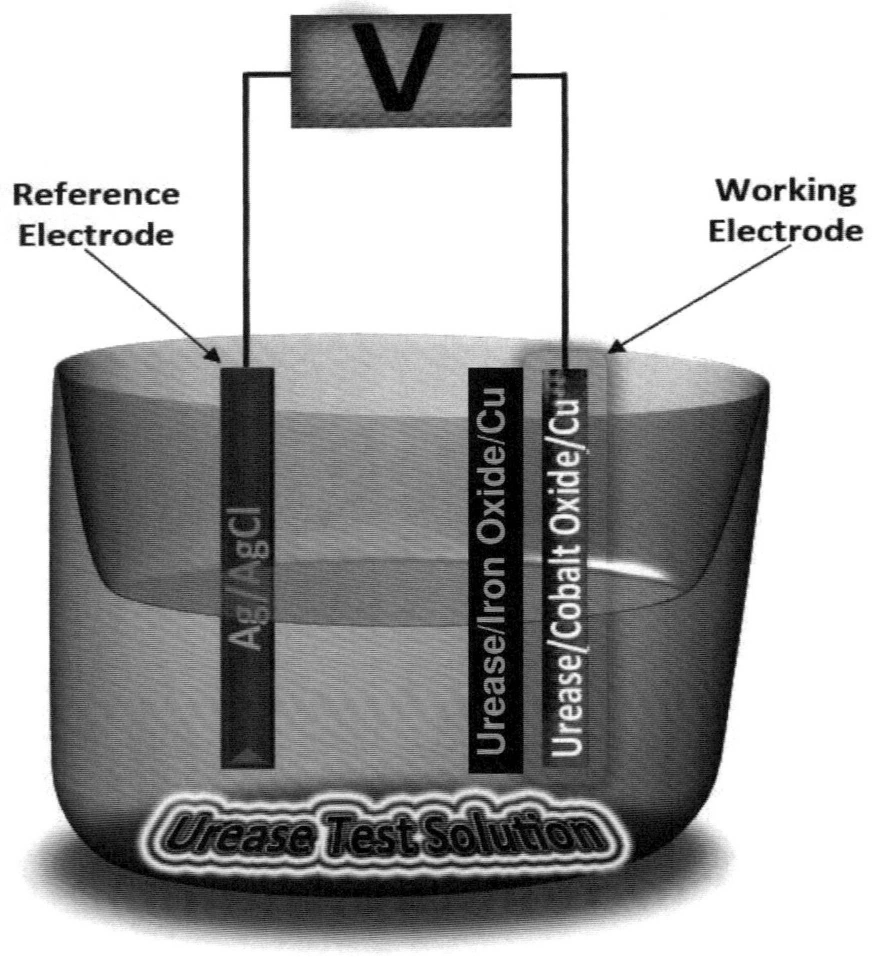

Figure 2: Experimental set-up for potentiometric urea measurements.

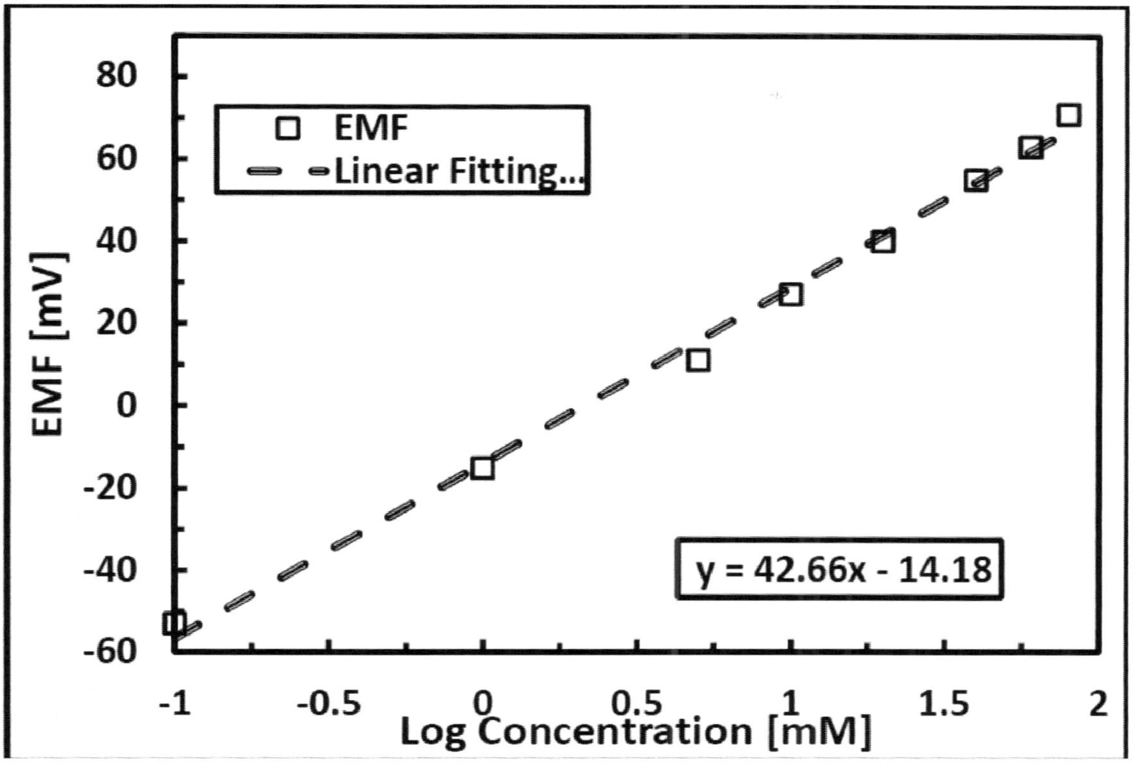

Figure3: Sensitivity response curve measured for logarithmic concentration range from 0.1 mM to 80 mM.

In Vitro Studies of Photosensitizer fluorescence changes on singlet oxygen mediated photobleaching

M. Atif

[a]Research Chair for Laser Diagnosis of Cancer, Physics and Astronomy Department, College of Science, King Saud University, Riyadh, Saudi Arabia

[b]National Institute of Laser and Optronics, Nilore, Islamabad, 45650, Pakistan

E-Mail: atifhull@gmail.com

Abstract:

The effects of photosensitizer concentration on the singlet oxygen mediated photobleaching have been investigated in keratinocytes. Cells having different photosensitizers concentration were identified on the basis of the fluorescence signal amplitude and their fluorescence decay then studied to see if there was a correlation with concentration. The experimental results shows that the photobleaching was more rapid at the higher concentration as compared to lower concentration which mean that singlet oxygen plays a vital role in the rapid m-tetrahydroxyphenylchlorin (mTHPC) induced fluorescence photobleaching causing the sensitizer to degrade.

Keywords: Photosensitizer concentration, singlet oxygen mediated photobleaching, keratinocytes, fluorescence signal amplitude and their fluorescence decay

1. Introduction

Singlet oxygen $^1O_2$ was first found in 1924 [1] and widely believed to be a more reactive form of oxygen, which oxidizes biological substrates [2]. It is not a free radical but an electronically excited form of oxygen. The generation of $^1O_2$ is carried out by photosensitization reactions [3]. An important characteristic of $^1O_2$ is that it readily combines chemically with a variety of biomolecules. Kautsky's investigations showed that singlet oxygen might be a reaction intermediate in dye-sensitized photo-oxygenation [4]. Several later observations support this idea and demonstrated that during detrimental oxidation of biomolecules, singlet oxygen is an important intermediate species. Moreover, the role of singlet oxygen during PDT is that it attacks the critical sites e.g., proteins, DNA, mitochondria and lipids. When the oxidative injury to cells exceeds a certain threshold level, the cell death occurs.

Zhao [5] addressed the impact of singlet oxygen as a higher energy state molecular oxygen species. It is considered the most active intermediates involved in chemical and biochemical reactions. In order to fully understand the reactivity of $^1O_2$, it is important to know about its physical and chemical nature. His work reviews the generation, quenching and detection of $^1O_2$. Moreover, the properties, possible chemical reactions and its biological role in photodynamic therapy (PDT) are also discussed. Fuchs et al, [6] described the current information on cutaneous oxygen physiology and oximetry, the role of oxygen and singlet oxygen in PDT.

The approach to modulate skin oxygen tension is also discussed. Their results reveal that effectiveness of PDT can be manipulated by modulating skin oxygen tension.

Georgakoudi et al, studied Photofrin bleaching via a singlet oxygen mediated mechanism [7]. Their results show that photobleaching is confined to regions of singlet oxygen abundance. Moreover, photosensitizer degradation results in decreased photodynamic oxygen consumption and as laser exposure precedes the supply of oxygen to cells at long distances from the source is due to diffusion. Due to this, it will gradually extend the distance from the oxygen supply over which significant singlet oxygen is produced. An important conclusion of this investigation is that sensitizer photobleaching is fundamentally linked to spatial distribution of photodynamic dose.

In the current study of normal keratinocytes, the observed sensitizer fluorescence changes using constant laser power are analyzed. The experimental results show that the availability of free oxygen may be limited even in the case of formalin fixed cells. This complex interaction between fluence rate and oxygen status during the fixed cell system in vivo determine the PDT dose delivered. This paper describes a performance study of the effects of photosensitizer concentration on singlet oxygen mediated photobleaching in keratinocytes.

2 Materials and methods

The cells used for this study are keratinocytes derived from human foreskin. The culture conditions and Foscan administration are the same as described in our previous reported studies [8-9].

2.1 Absorption spectrum of mTHPC at different concentration

Absorption spectrum of foscan has been recorded at different concentration using either a spectrophotometer or fluorimeter, respectively. On the basis of this absorption spectrum the absorption coefficient of mTHPC in solution is calculated using equation 1.2 of Beer-Lambert Law. Moreover, its maximum fluorescence intensity is also measured. Fig 1 (a) shows the plot of foscan at different concentration versus maximum fluorescence intensity while figure 1 (b) represents the absorption coefficient of mTHPC at different concentration against maximum fluorescence intensity.

2.2 Acquisition of fluorescence spectrum for Keratinocytes

A microscope with filtered light was used in order to locate and correctly position the individual cells. Moreover, laser was focused close to the nuclear region. Fluorescence spectra were collected from the area close to the nuclear region of the cells and analysed using Grams spectroscopy software (Grams/AI, Thermo Galactic, Salem, USA). The spectral region of investigation was in the range (635 – 675 nm) including the principal mTHPC emission peak (652 nm). A total of twenty spectrums were collected from each cell.This procedure was repeated for a set of 46 cells. Each spectrum was recorded over 2 seconds; sequential spectra were acquired in order to investigate the dynamics of mTHPC photobleaching.

a)

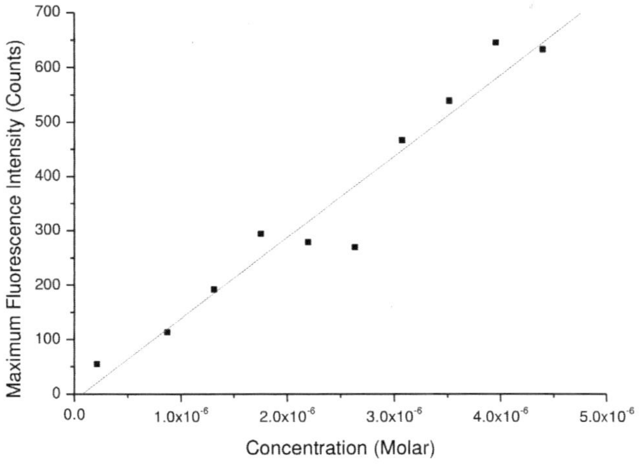

b)

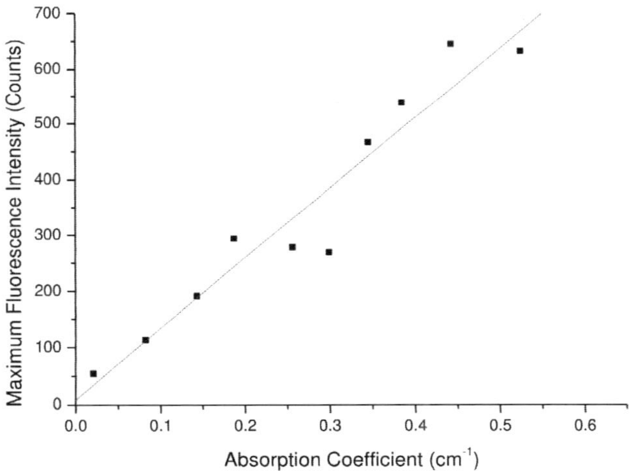

Fig 1 (a) mTHPC at different concentration versus maximum fluorescence intensity (b) represents the absorption coefficient of mTHPC at different concentration against maximum fluorescence intensity.

2.3 Continuous exposure

On irradiation of individual cells with a laser power of 0.0099 mW over a period of 40 seconds, twenty sequential spectra were recorded close to the nucleus of the cell. This procedure was repeated for a set of 46 cells and the mean of the 10 datum points at each exposure time calculated. These 46 cells are divided into six groups e.g., G1, G2, ...G6 etc on the basis of the fluorescence intensity observed. The different combinations of groups are investigated using constant laser power. This allowed comparison of the spectra within six groups of keratinocytes during a 40-second irradiation for photobleaching.

Different individual cells will take up the photosensitizer at different rates and there will be a subsequent distribution of the sensitizer into different parts of the cells and hence the fluorescence photobleaching.

Using the data acquired during the study an analysis was performed upon the shape of the fluorescence emission peak in order to assess possible perturbations due to self-absorption (inner filter effect). The analysis results reveal that the role of self-absorption is significantly small in determining the spectral profile of the fluorescence.

3 Results

3.1 Autofluorescence

Fluorescence spectra were also acquired from keratinocytes in the plateau phase, without mTHPC incubation. A low level (<2% of typical mTHPC signal) of autofluorescence was observed in the spectral regions of interest, and this was detected to bleach rapidly during the initial period of laser irradiation. Hence, autofluorescence has not been subtracted from the mTHPC spectra recorded. No fluorescence was found from regions outside the cells.

3.2 Continuous exposure at constant irradiance

mTHPC fluorescence spectra recorded from fixed keratinocytes show the progressive degradation of the 652 nm peak, following incubation with 0.044 mM mTHPC for 3 hours, during a continuous laser irradiation of 0.0099 mW for 40 seconds. The fluorescence intensity recorded from a set of 20 cells. When it is plotted against time, a significant variation in the magnitude of the fluorescence is found from different cells (normalized mean = 1 ± SD .23 around the mean for the first spectrum in each sequence).

Fig 2 shows the rate of photobleaching vs dose at constant irradiance for six groups of cells. From these results, we came to the conclusion that cells demonstrating high initial fluorescence (concentration) undergo bleaching at a faster rate then those with lower fluorescence.

A histogram of the fluorescence intensity for six groups of cells from keratinocytes is also shown in figure 3. Each group represents the mean 652nm emission intensity at constant laser power as indicated by the first spectrum of a bleaching sequence. These trends indicate that G1 has lower initial fluorescence implying lower concentration and this initial fluorescence goes on increasing till the last group with the increase in the concentration. The standard deviations in this histogram result from the large variation in fluorescence intensity for all groups and the ability of the cells to take up mTHPC at the incubation time selected.

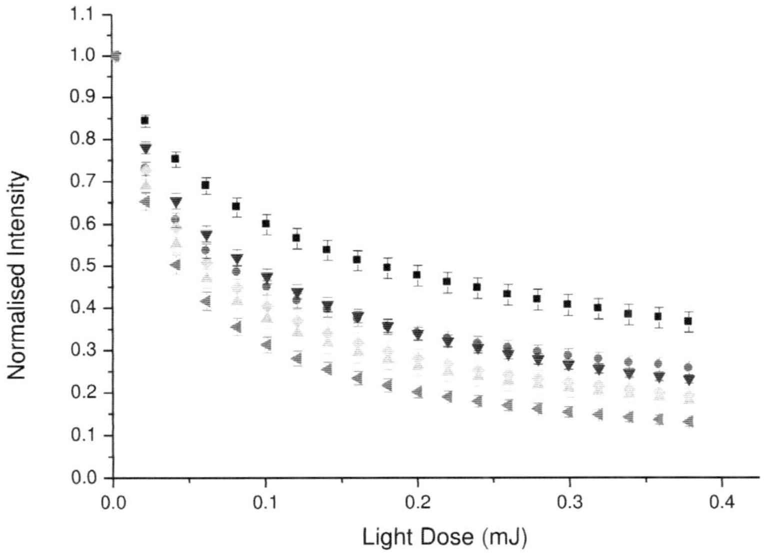

Figure 2: Photobleaching plot of mTHPC showing the normalised averaged data points from six groups (G_1 ■ G_2 ●G_3 ▲ G_4 ▼G_5 ◆ G_6◀) of Keratinocytes (0.044 mM mTHPC for 3 hours) exposed to constant laser power: 0.0099 mW.

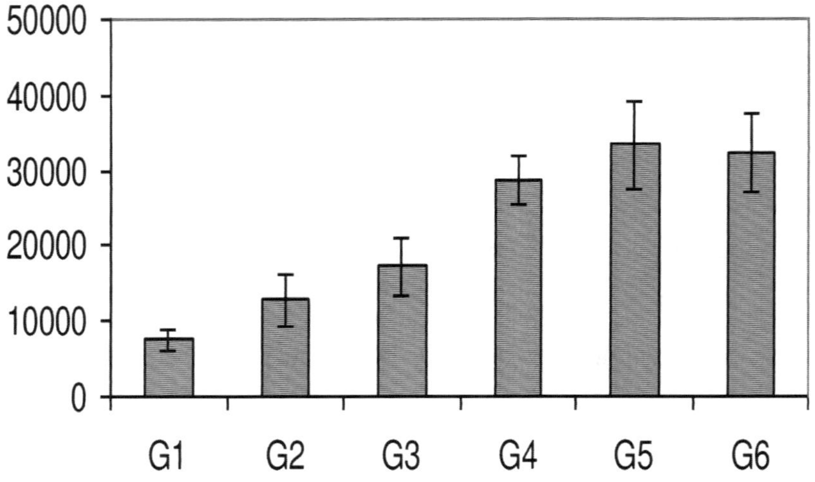

Figure 3: Initial mTHPC (652 nm) fluorescence intensity for keratinocytes among various groups

4. Discussion

The aim of this study is to analyze the effects of photosensitiser concentration in mTHPC induced photobleaching. Fuchs et. al. [6] demonstrated that within single cells there are large oxygen concentration gradients, especially close to the mitochondria. Similarly large variations are also observed at the tissue level in oxygen concentration over small distances particularly in skin, which possesses a complex morphological structure and angioarchitecture. Hence all measurements of oxygen concentration include averaging heterogenous oxygen concentrations.

Due to the observed sensitizer fluorescence changes, it has been classified into six groups according to the fluorescence intensity / counts as explained previously in Table 1. The rate of photobleaching can be explained in term of a bi-exponential decay, as a function of dose delivered. As the values of χ^2 are lower in terms of a bi-exponential decay rate, so the data from keratinocytes (3 hours mTHPC incubation) follow this bleaching function. This is similar with Moan et al. [10] works where a bi-exponential model was used to examine measurements of the loss of fluorescence intensity from PpIX during irradiation of WiDr cells. The main reason for using a bi-exponential decay rate is that first order decay kinetics reveals that photoproducts from one chromophore cannot cause the decay of a neighbouring chromophore, which is not true in the present case, since the decay rates increase with the increasing PpIX concentration which is in contrast to what one might expect for a pure second order process. An analysis of the described results indicate that in ALA-PCT PpIX is photodegraded mainly according to two or more first order processes i.e. a PpIX molecule is degraded by the photroproducts produced by itself, but that higher order processes definitely play significant roles.

In Fig 2 the bleaching is shown plotted against dose, and is seen to depend on the initial mTHPC fluorescence intensity. Constant laser power is used and the results show that group of cells demonstrating high initial fluorescence (concentration) undergo bleaching at a faster rate then those with lower fluorescence apart from one group of cells (G3). The results observed are in good agreement with the investigations of Bagdonas et al., (2000) [11] where for PpIX a similar dependency upon initial concentration was found and linked to the generation of singlet oxygen, and the subsequent reaction with the PpIX molecule.

In summary, singlet oxygen appears to play a major role in the rapid mTHPC induced fluorescence photobleaching causing the sensitizer degradation rate to vary as a function of the rate of photodynamic $^3O_2$ consumption i.e., photodynamic $^3O_2$ consumption increases with the available drug concentration. Moreover, it results in a fluence rate dependence of photobleaching. Therefore, groups of cells showing high initial fluorescence (concentration) bleach faster than those with lower fluorescence. This is in agreement with a number of in vivo and in vitro studies [12-21].

5. Conclusion:

The results of this study show that sensitivity of mTHPC to photo-oxidation exists alongside the increased PDT efficiency of this compound, compared to porphyrin sensitisers. For example, the localization of mTHPC to cellular lysosomes (rather than mitochondria) has been shown to reduce the light dose necessary for in vitro cell necrosis found by Leung et al, 2002 [22]. Also, inhibiting the intracellular accumulation affected a reduction in PDT efficacy. In our study it was not possible to monitor the oxygen partial pressure, photosensitiser localisation or photodynamic damage. However, it is evident that mTHPC is highly photolabile within a cellular environment, and that oxygen plays a crucial role in the degradation of the fluorescent species. The assumption that the fluorescence bleaching signal can be used as a PDT dosimetry metric relies upon a constant partition between the effect of $^1O_2$ upon sensitiser and cell substrate. New methods of direct detection of the weak near infrared luminescence may ultimately allow parallel monitoring of fluorescence bleaching and $^1O_2$ concentration as described by Niedre et al, 2002 [23].

Acknowledgements:

This project was supported by King Saud University, Deanship of Scientific Research, College of Science Research Center.

References

1. H. H. Wasserman, Introductory remarks. In: Wasserman H. H. ed. Singlet Oxygen. Vol. **40** Academic Press, (1979).
2. H. Barry, M. C. John, Free Radical in Biological and Medicine. Second Edition (1982) Clarwndon Press. Oxford.
3. N. I. Krinsky, Biological roles of singlet oxygen. In: Wasserman H. H; ed. Singlet Oxygen. Vol. **40** Academic Press. pp. 597-641(1979).
4. D. R. Kearns, Chem. Rev. **71**, 395-427 (1971).
5. L. Zhao, Singlet Oxygen. Department of Radiology, The University of Iowa, Iowa City, USA (2001).
6. J. Fuchs and J. Thiele, Free Rad. Bio. and Med. **24,** 835-847 (1998).
7. I. Georgakoudi, M. G. Nichols and T. H. Foster, Photochem. and Photobio. **65,** 135-144 (1997).
8. M. Atif, M. R. Stringer, J.E. Cruse-Sawyer, P.E. Dyer, and S.B. Brown, Photodiagn. Photodyn. Ther. **2**, 235–238 (2005).
9. M. Atif, M. R. Stringer, J. E. Cruse-Sawyer, and S. B. Brown, Photodiagn. Photodyn. Ther. **1**, 173–180 (2004).
10. J. Moan, G. Streckte, S. Bagdonas, O. Bech and K. Berg, International J. of Can. **70,** 90-97 (1997).
11. S. Bagdonas, L. W. Ma, A. Iani, R. Rotomskis, P. Juzenas and J. Moan, Photochem. and Photobio. **72,** 186-192 (2000).
12. D. J. Robinson, H. S. deBruijn, W. J. deWolf, H. J. Sterenborg, and W. M. Star, Photochem. and Photobio. **72,** 794-802 (2000).
13. N. de Veen, K. M. Hebeda, H. S. deBruijn and W. M. Star, Photochem. and Photobio. **70,** 921-929 (1999).
14. I. Van Geel, H. Oppelaar, J. P. Marijnissen and F. A. Stewart, Rad. Res. **145,** 602-9 (1996).
15. J. Schlothauer, S. Hackbarth and B. Röder, Laser Phys. Lett. **6,** 216-221 (2009).
16. D. Chorvat Jr. and A. Chorvatova, Laser Phys. Lett. **6,** 175-193 (2009).

17. R. S. Cavalcante, H. Imasato, V. S. Bagnato and J. R. Perussi, Laser Phys. Lett. **6**, 64-70 (2009).
18. L. G. Sabino, L.M.V. de Negreiros, J.D. Vollet-Filho, J. Ferreira, D.P.C. Tirapelli, P.C. Novais, L.F. Tirapelli, C. Kurachi and V.S. Bagnato, Laser Phys. Lett. **8**, 239-246 (2011).
19. Y. Shen, H. Y. Lin, Z. F. Huang, D. F. Chen, B. H. Li and S. S. Xie, Laser Phy. Lett. **8**, 232-238 (2011).
20. E. S. Tuchina and V. V. Tuchin, Laser Phys. Lett. **7**, 607-612 (2010).
21. L. Buriankova, Z. Nadova, D. Jancura, M. Refregiers, I. Yousef, J. Mikes and P. Miskovsky, Laser Phys. Lett.**7**, 613-620 (2010).
22. W. N. Leung, X. Sun, N. K. Mak, C. M. Yow, Photochem. Photobiol. **75,** 406-11 (2002).
23. M. Niedre, M. S. Patterson, and B. C. Wilson, Photochem. Photobiol. **75,** 382-391 (2002).

Journal of Physics: Conference Series **414** (2013) 012026 doi:10.1088/1742-6596/414/1/012026

Semiconductor A_3B_5 nanostructures for infrared femtosecond lasers

N N Rubtsova[1], S A Kochubei[1], A A Kovalyov[1], V V Preobrazhenskii[1], M A Putyato[1], B R Semyagin[1], T S Shamirzaev[1], G M Borisov[1], V E Kisel'[2], N V Kuleshov[2], S V Kuril'chik[2], O V Buganov[3], S A Tikhomirov[3]

[1]A.V. Rzhanov Institute of Semiconductor Physics of Siberian Branch of Russian Academy of Sciences, 13, academician Lavrentyev prospect, Novosibirsk, 630090, Russia

[2]Scientific Research Institute for Optical Materials and Technologies of Belarus National Technical University, 65, Nezavisimosti avenue, bd. 17, 220013, Minsk, Belarus

[3]B.I. Stepanov Institute of Physics, National Academy of Sciences of Belarus, 68, Nezavisimosti avenue, Minsk, 220072, Belarus

E-mail: rubtsova@isp.nsc.ru

Abstract. Two techniques were suggested and tested for the recovery time shortening of saturable absorbers on a base of A_3B_5 compounds including quantum wells. The first one, proposed by authors, is the sample post-growth treatment by UV laser radiation; it implied generation of point defects, which, in its turn, led to electron-hole recombination acceleration and to recovery time shortening by an order of magnitude and more. Another technique based on special design of barriers gave promising results for the fast saturable absorbers. Semiconductor mirrors designed for Yb^{3+}:KY$(WO_4)_2$ infrared laser mode locking led to 115 fs stable mode-locking regime with average power close to CW operation. Results on fast saturable absorbers for spectral region of 1500 nm are also presented.

1. Introduction

The most suitable way to get the self-mode-locking in compact low-gain lasers is to use the saturable absorber (SA) [1]. The common used design of SA is based on thin layers of semiconductors including the quantum wells (QW) separated by barriers of another semiconductor. Three- or even four- component semiconductor alloys (ternary or quaternary) are the appropriate choice for a wide spectral region. In particular, A_3B_5 ternary and quaternary semiconductor compounds are well suited for short-pulse lasers operating in the spectral region of 1-3 microns [2]. SA including A_3B_5 quantum wells separated by thick barrier show electron -hole recombination rate of about hundreds of picoseconds [3].

Since many applications of ultra-short pulse lasers require high repetition rate of pulses, SA used for laser mode-locking have to meet this requirement. Several approaches to SA recovery time shortening by special technologies of molecular beam epitaxy (MBE) growth are known. Doping of QW layers by Be ions during MBE growth [4] shortened charge carrier recombination time to 0.25 ps. Application of sub-band gap states in semiconductor nitrides [5] resulted in relaxation rate of 12 ps. Intentional lattice mismatching by using metamorphic growth of InGaP on GaAs led to recovery time of 5 ps [6].

Another approach is based on SA post-growth processing. So, the post-growth irradiation of the sample by Ni^+ ions beam [7] resulted in relaxation rate of 2.4 ps.

The goal of this work is to analyze new approaches to SA recovery time shortening which could guarantee minimal losses of optical quality and persistence of excitonic contribution to absorption.

2. UV post-growth treatment

Our first results on UV post-growth treatment were obtained for SA operating in the near-infrared region [8]. In [8] definite dozes of laser UV radiation could accelerate electron-hole recombination by 20 times. Here, we applied UV post-growth irradiation to SA designed for visible light in the vicinity of titanium-sapphire laser operation spectral region.

The sample was grown by MBE technique at the substrate of semi-insulating GaAs with (001) orientation. The rate of growth and layers thickness was controlled by reflective high energy electrons diffraction (RHEED) technique, and the crystal quality of the sample was controlled by X-ray diffraction. The sample consisted of 20 quantum wells of 3 nm thickness GaAs separated by 6 nm $Al_{0.38}Ga_{0.62}As$ barriers.

Pump-probe technique was applied to test the sample reflectivity kinetics. The femtosecond spectrometer operating in the spectral region of 760 - 820 nm provided 150 fs pulses of 1 mJ energy at repetition rate of 10 Hz. The pulsed output radiation at the central wavelength of 780 nm was divided into couple of beams. The most intense beam served as the pump beam. After the passage through the optical delay line, it was directed to the sample at the incidence close to the normal one, and it was focused into the spot of approximately 1 mm diameter. Another part, four times less intense, was used for the super-continuum generation in the cavity filled by water. The super-continuum radiation was divided in two beams — the reference one and the probe one. The reference beam meets the sample before the pump beam arrival. The probe beam arrived after the pump providing just the measured signal. The spectra of both reference and probe pulses reflected from the sample surface were detected by system including polychromator, CCD camera and microprocessor.

The change in the sample reflectivity $\log(R_0/R)$ (here R_0 is the sample reflectivity before pumping, R — after it) was determined as the function of time delay τ_{del} between the probe and the pump, and as the function of the probe radiation wavelength.

The result of UV irradiation was checked via reflectivity kinetics at the wavelength of 750 nm which corresponded to the sample photoluminescence band center.

Excimer XeCl laser operating at the radiation wavelength of 308 nm with pulse duration of 5 ns and repetition rate of 1 Hz was applied for the sample UV treatment. Result of UV radiation action was controlled through photoluminescence (PL) intensity change by registration of PL intensity before and after UV irradiation. In order to find how the irradiation conditions influence the final result, both the energy density of each UV pulse and the number of pulses influenced the sample were changing from one to another area of the sample. Since the energy of UV radiation quantum E=4.03 eV is comparable to the bonding energy of GaAs, we could expect the photochemical modification of the sample surface even at low UV radiation energy density. All the experiments were carried out at room temperature in the atmospheric air ambient.

Excimer laser UV radiation focusing into the spot of approximately 1 micron in diameter led to the laser plume formation over the sample surface. The spectrum of laser plume radiation was controlled by special set up including monochromator and registration system (photomultiplier, peak voltmeter, ADC and personal computer).

Fig. 1 shows sample PL spectra before (bold curve) and after (thin) UV irradiation for different energy densities; the left graphs correspond to 80 mJ/cm^2, and the right one to 350 mJ/cm^2. All the spectra demonstrate PL band centered at 1.66 eV and width at half maximum of about 30 meV. UV irradiation does not change PL spectral shape, but results in PL intensity

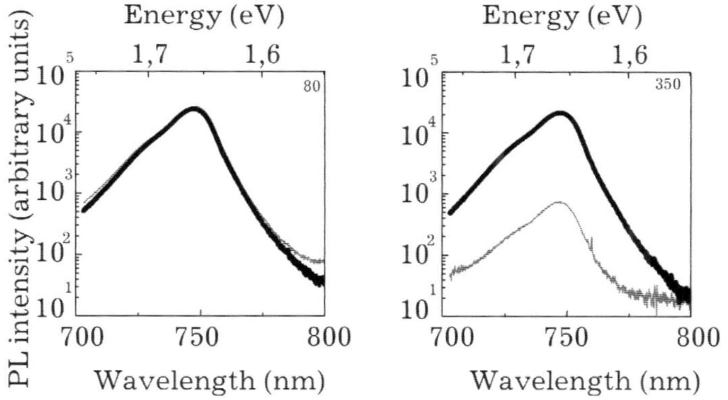

Figure 1. PL spectra of samples before (bold curve) and after (the thin one) UV irradiation for energy density of 80 mJ/cm^2 (left) and 350 mJ/cm^2 (right).

changes. These changes are minimal for UV energy density of 80 mJ/cm^2, as it can be seen in the left graph of Fig. 1; PL intensity increase after irradiation seen at this graph should be attributed to the sample lateral inhomogeneity. Meanwhile, UV irradiation at energy density of 350 mJ/cm^2 results in remarkable (almost two orders of magnitude) decrease of PL intensity.

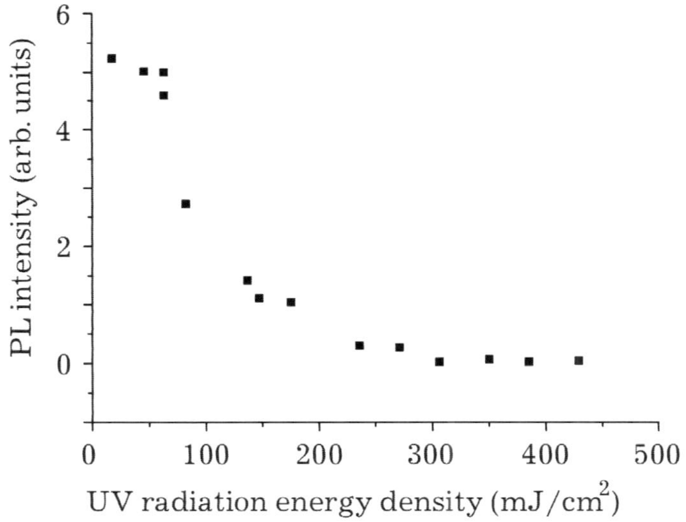

Figure 2. PL intensity of the sample as a function of UV energy density.

PL intensity (at the band center) of the sample treated by 5 UV pulses is shown in Fig. 2 versus UV energy density. PL intensity in the Fig. 2 was corrected to the above mentioned lateral inhomogeneity. As it can be seen from Fig. 2, there are no considerable PL intensity changes for UV energy density less than 100 mJ/cm^2. At UV energy densities higher than 100 mJ/cm^2, the fast decrease of PL intensity was observed. Such behaviour proves the threshold character of the phenomenon. Our results agree with [9] for laser induced point defects generation in bulk GaAs.

PL intensity change due to UV irradiation depends on the number of UV pulses too. This dependence is different for UV energy density under or above the threshold, as it can be seen in Fig. 3. The sample PL intensity practically does not depend on number of UV pulses for low energy density (squares in Fig. 3). For UV energy density over threshold, the PL decreases as number of pulses grows (circles in Fig. 3).

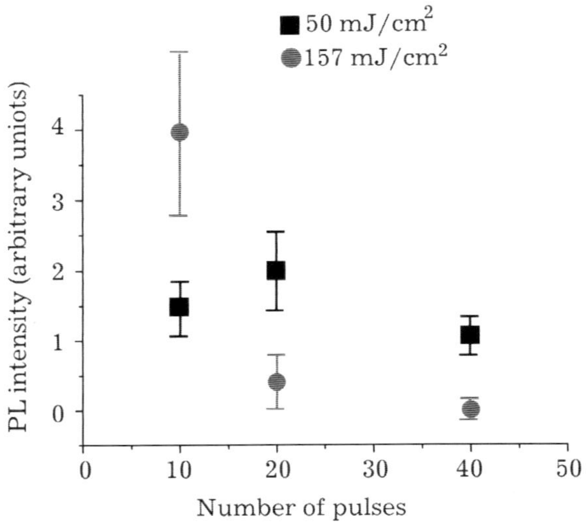

Figure 3. PL intensity dependence on the number of UV pulses may be weak (squares) or quite pronounced (circles) depending on the UV energy being below threshold or over it.

The most important effect of UV irradiation should be seen in the sample reflectivity kinetics. Really, Fig. 4 demonstrates how changes $\log(R_0/R)(\tau_{del}$ kinetics with UV energy density. Results in Fig. 4 were obtained for sample area irradiated by 5 UV pulses with different energy density — from top to bottom correspondingly 17.5 mJ/cm^2, 45.5 mJ/cm^2, and 100 mJ/cm^2. Squares with error bars represent experimental curves, the circles correspond to the model curves. Probe beam wavelength was chosen at PL band maximum (750 nm) for all kinetic curves in Fig. 4.

Evidently, relaxation rates grow as UV energy density increases, and some PL intensity decrease takes place too.

Specificity of a given SA sample is following. The band gap of GaAs at room temperature is 1.424 eV, which corresponds to the absorption edge near the wavelength of 870 nm. Quantum wells transition occurs at the energy 1.65 eV (wavelength 750 nm). The sample was excited by femtosecond pump radiation centered at 780 nm which corresponds to the light quantum energy 1.59 eV. Hence, both sample substrate GaAs and quantum well layers could absorb the pump pulse radiation — not only because of large femtosecond pulse spectral width, but also due to two-photon absorption.

That is why the model of relaxation kinetics accounted for the next processes: 1) excitons formation in quantum wells and increase of non-equilibrium quasi-two-dimensional charge carrier density — which induces fast increase of non-equilibrium reflectivity; 2) decrease of non-equilibrium charge carrier density because of non-radiative recombination due to interaction with defects — which implies non-equilibrium reflectivity relaxation; 3) absorption in the substrate GaAs and increase of non-equilibrium charge carrier density — which increases non-equilibrium reflectivity too; 4) electrons and holes recombination — which should result in reflectivity relaxation to its equilibrium value. The last process is usually the most slow, and it is the process which limits SA recovery time.

The formula accounting for all four above mentioned processes looks as:

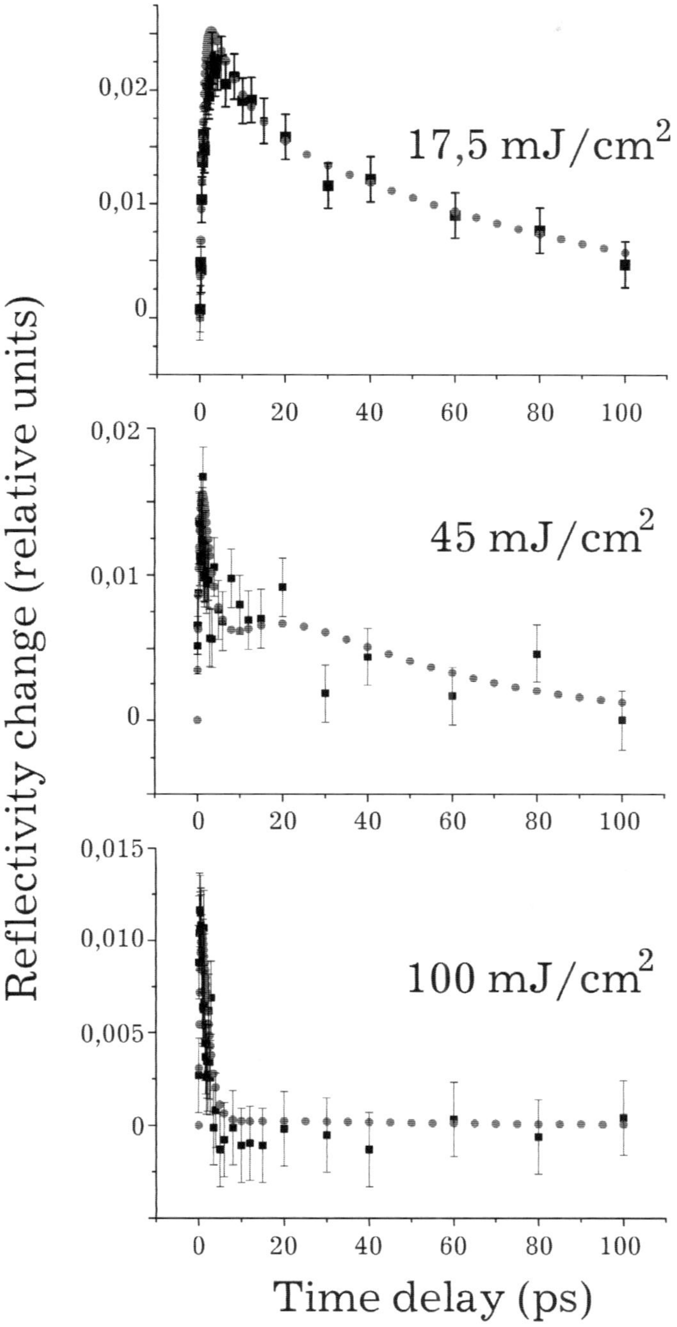

Figure 4. Reflectivity kinetics for the sample area irradiated by UV pulses of different energy density (indicated for each curve).

$$S(t) = a \cdot (1 - \exp(-t/\tau_1)) \cdot \exp(-t/\tau_2) + b \cdot (1 - \exp(-t/\tau_3)) \cdot \exp(-t/\tau_4) \quad (1)$$

Here τ_i are relaxation times of above mentioned processes; the constants a and b reflect relative contribution of quantum wells and of substrate into reflectivity change. The parameters a, b and τ_i were chosen in the course of mean square root fitting of experimental points by the model curves.

Table 1

UV energy density (mJ/cm^2)	17.7	45.5	100
τ_1 (ps)	0.8	0.8	0.8
τ_2 (ps)	10	2.5	1.5
τ_3 (ps)	15	15	15
τ_4 (ps)	80	40	40
a	0.03	0.03	0.03
b	0.02	0.015	0.0015

Table 1 shows relaxation times τ_i and amplitudes a and b for above mentioned processes fitted for three areas of the sample processed by UV radiation of energy density indicated in the first line of the table. The parameters b, τ_2 and τ_4 experience the most remarkable changes. From the viewpoint of general recovery of SA, the relaxation time τ_4 seems to be the most important. The acceleration of electron-hole recombination is also clearly seen from experimental and model curves in Fig. 4.

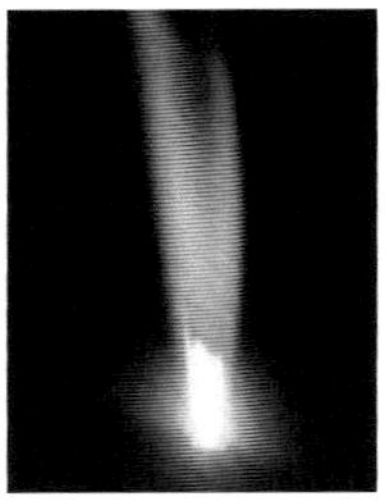

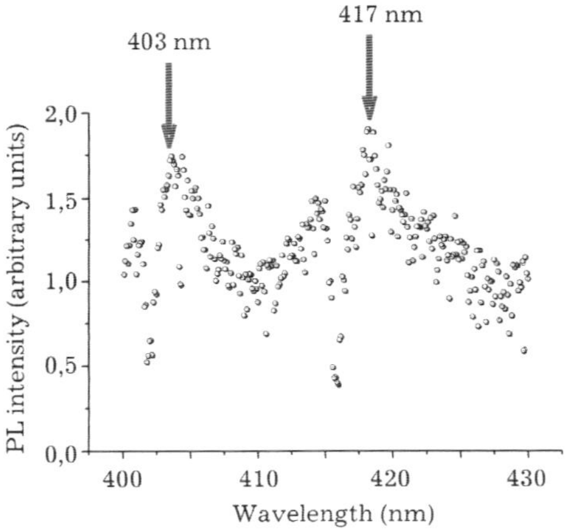

Figure 5. Laser plume over sample surface (left) and its emission spectrum (right).

Fig. 5 shows the laser plume generated over the sample surface as a result of UV focusing into the spot of 1 micron diameter (left) and its emission spectrum (right). Two peaks in laser plume emission spectra centered at wavelength 403 nm and 417 nm should be attributed to radiation of neutral Ga atoms and correspond to transitions ${}^2P^0{}_{1/2}$ - ${}^2S^0{}_{1/2}$ and ${}^2P^0{}_{3/2}$ - ${}^2S^0{}_{1/2}$. Our results agree with [10] where GaAs (001) surface was irradiated by femtosecond (62 fs) focused titanium-sapphire radiation. Like our work, the processing was carried out in the air atmosphere. Experimental conditions with UV radiation focusing into the spot of 1 micron are extremely severe, they are far from conditions necessary for laser generation of point defects. As we hope, future sensitivity improvement of this experiment will be useful in analysis of the cloud over the sample surface at the condition of minimal sample damage.

Journal of Physics: Conference Series **414** (2013) 012026 doi:10.1088/1742-6596/414/1/012026

3. SA with nano-structured barriers for 1040 nm

Our previous experience [3, 11, 12] in design and manufacturing of A_3B_5 semiconductor compounds based saturable absorbers was limited by multi-layer structures including quantum wells with thick barriers. In these cases quantum wells could be considered as isolated parts, and the number of quantum wells allowed to match a necessary level of saturable absorption. Experimental testing of one of our SA was realized in [3] for the home-made Nd^{3+}:$KGd(WO_4)_2$ laser operating at the wavelength of 1067 nm and pumped by diode laser at wavelength of 812 nm. In this case a stable passive mode-locking was achieved at the repetition rate of 70.5 MHz; the duration of individual pulses was estimated as 600 fs. Characterization of SA used in [3] by femtosecond pump-probe gave the "slow" relaxation time of about 188 ps.

The recovery time of SA is very important for obtaining spectral limited ultra-short pulses. The problem of the recovery time shortening is well known for all users and manufacturers of SA. The solutions for shortening of saturable absorber recovery time known in literature are numerous, they were partially cited in the Introduction.

Here we use another approach based on the idea of the coupled quantum wells [13, 14]. The fast tunneling of charge carriers between quantum wells should accelerate the recovery of linear absorption, avoid problems related to critical thickness of semiconductor layers in a presence of mechanical stress, and maintain good crystal quality. As a consequence, a high exciton contribution to the absorption can be expected.

Saturable absorber including multiple quantum wells of InGaAs separated by nano-structured barriers of GaAs was grown by molecular beam epitaxy (MBE) under permanent control of layers crystal quality by using high energy electrons diffraction. Results of the pump-probe testing of this sample are shown in the Fig. 6. Method of pump-probe measurements was described in detail in [3, 8]. The set up is based on femtosecond Yb^{3+}:$KY(WO_4)_2$ laser radiation of 200 fs duration; the beam was divided into two parts forming an exciting beam and a probe beam (at least 50 times lower in intensity and with orthogonal linear polarization). Both pump and probe beams were focused onto the sample surface into the spot of about 10 μm. The lock-in amplifier detected a signal representing the changes in the probe beam reflectivity of a sample induced by the pumping modulated at 880 Hz frequency. Time resolution of the measurements was determined by minimal step of optical delay line variation and was about 4 fs.

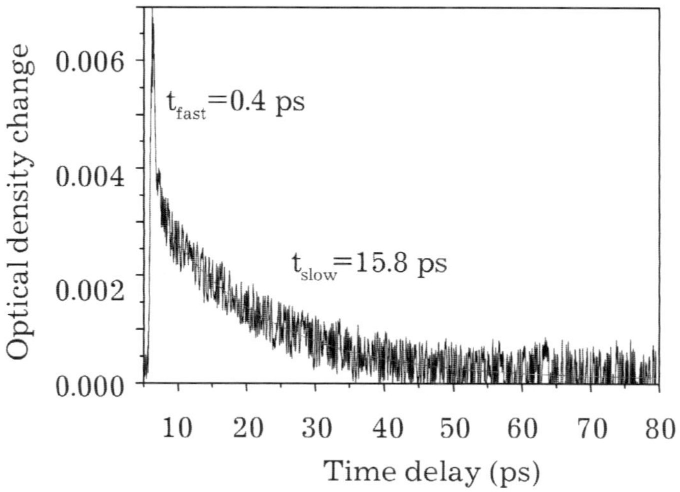

Figure 6. Fast (0.4 ps) and slow (15.8 ps) recovery of SA with nano-structured barriers.

The relative change of optical density of the sample is shown in Fig. 6 versus time delay

Journal of Physics: Conference Series **414** (2013) 012026 doi:10.1088/1742-6596/414/1/012026

between pump and probe pulses. As usually, recovery kinetics contains two parts. The fast kinetics attributed to the ionization of excitons localized in quantum wells shows recovery time of about 0.4 ps. The slow part is related to charge carriers recombination with relaxation time of 15.8 ps. The presence of the fast part in recovery kinetics is a good sign, which confirms that exciton contribution to the absorption still exists. Slow relaxation proved an order of magnitude faster than in our previous SA with isolated quantum wells [3].

SA designed and manufactured by MBE growth on a base of this saturable absorber included conventional wide band total reflector and necessary number of quantum wells separated by nano-structured barriers. Linear spectrum of this SA showed reflectivity not less than 0.96 in the operation area of the Yb^{3+}:$KY(WO_4)_2$ laser of 1040 nm.

All the details for this SA (with nano-structured barriers) testing in Yb^{3+}:$KY(WO_4)_2$ laser can be found in [15]. Here, we describe briefly main results of [15]. The output power of Yb^{3+}:$KY(WO_4)_2$ laser in CW regime reached 1.62 W at the wavelength of 1043 nm for optical pumping power of 5.87 W. So, the optical pumping laser efficiency was 27.6%. The stable mode-locking regime was obtained in the same laser cavity after replacement of total reflector by the SA. The beam spot at the surface of SA was about 165 μm in diameter. The estimate for the saturating energy surface density is 120 $\mu J/cm^2$.

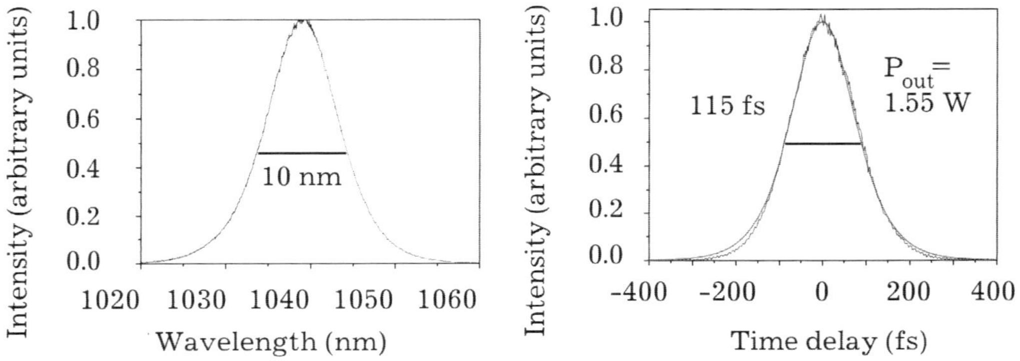

Figure 7. Yb^{3+}:$KY(WO_4)_2$ laser spectrum for the mode-locking regime induced by the fast SA (left graph) and autocorrelation function (noisy curve) with superimposed fitting curve (right graph).

Left part of Fig. 7 shows the Yb^{3+}:$KY(WO_4)_2$ laser spectrum for the mode-locking regime induced by the fast SA. The width at the half maximum is 10 nm, and the spectrum is centered at the wavelength of 1039 nm. Autocorrelation function in the right part of Fig. 7 is represented by the noisy curve. The model curve superimposed onto auto-correlation function is calculated as $sech^2$, and it describes fine the experimental results. Ultra-short pulses duration of 115 fs should be attributed to the application of above described fast SA. Comparison of experimental spectral width and pulse duration gives the time-bandwidth product of 0.319. This means that the pulses are very close to the spectral limited type. The ultra-short pulses repetition rate was 70 MHz, and the peak power reached 171 kW. The maximum average power in the mode-locking regime was 1.57W. This value is very close to above mentioned value of 1.62 W in CW regime. Comparison of these two average power values confirms the low level of non-saturable losses introduced by fast SA.

4. Fast SA for 1500 nm region: preliminary results

Erbium laser operating in the vicinity of 1500 nm is another important laser source for the fast SA applications. We designed SA including thin quantum well layers of $In_{0.5}Ga_{0.5}As$ separated by nano-structured barriers. The transmission spectra of SA is indicated in Fig. 8 for two samples. The lower curve represents SA transmission at GaAs substrate of 400 μm thickness. The upper curve shows transmission for SA with partially removed substrate (mechanically to the thickness of about 10 μm). Both curves demonstrate the wide absorptive deep from 1080 to 1750 nm.

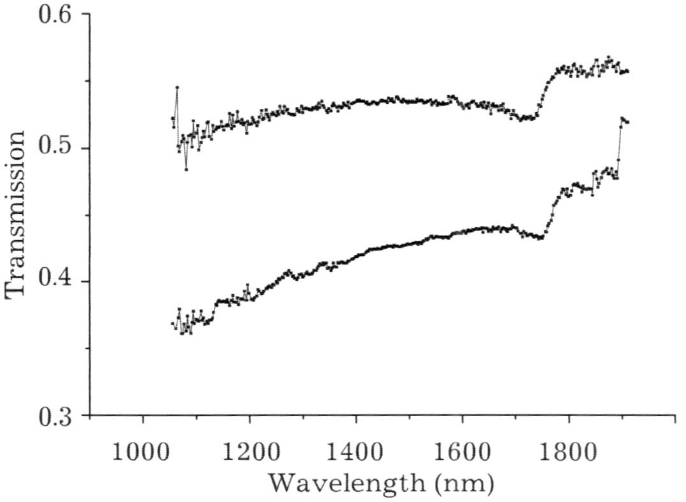

Figure 8. Transmission spectra for SA with nano-structured barriers. SA at GaAs substrate of 400 μm (lower curve) and SA with partially removed substrate to the thickness 10 μm (upper curve).

Reflectivity kinetics was tested at the wavelength of 1514 nm; the pump femtosecond pulse was generated by titanium-sapphire laser at the wavelength of 795 nm. Result is shown in left graph of Fig. 9 and clearly demonstrates very fast relaxation time.

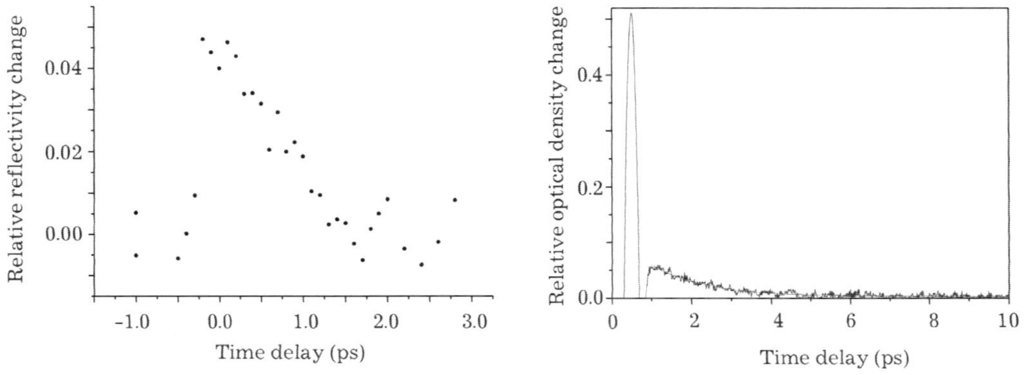

Figure 9. Reflectivity kinetics of fast SA at probe wavelength of 1514 nm (left graph) and at probe wavelength of 1040 nm (right graph).

SA reflectivity kinetics was checked also in another pump-probe set up based on femtosecond Yb^{3+}:KY(WO_4)$_2$ laser operating with 160 fs pulses at wavelength of 1040 nm. The probe beam at the same wavelength was two orders of magnitude lower than the pump one. The excitation intensity in the SA was about 0.5 GW/cm^2. Result is shown in the right graph of Fig. 9. The relaxation time for this kinetic curve equals to 1.3 ps.

So, SA with nanostructured barriers designed for the spectral area of 1500 nm demonstrated short relaxation times in both pump-probe experiments.

5. Discussion and summary

SA incorporating 20 quantum wells of GaAs separated by $Al_{0.38}Ga_{0.62}As$ barriers was irradiated by different number of UV laser pulses with different energy density. It was found experimentally, that PL intensity of SA sample decreases, and the electron-hole recombination rate increases with UV energy density. This effect revealed threshold nature with the threshold value of UV energy density of about 100 mJ/cm^2 — in agreement with [9]. Reflectivity kinetics of the SA sample detected at the probe radiation wavelength corresponding to the PL band maximum showed an increase of electron-hole recombination rate with growth of UV energy density — like in our experience with SA operating in the near-infrared region [8]. At the sharp focusing of UV laser radiation a laser plume formation over the sample surface was observed. The spectrum of the plume contained neutral gallium atoms doublet at wavelengths of 403 nm and 417 nm. This fact partially confirms the mechanism of point defects generation as a result of photo-chemical damage of the surface.

It is worth to stress that no visible damage of SA is observed if UV energy density lies under the threshold. Meanwhile, remarkable SA recovery time shortening was detected. So, UV treatment of semiconductor structures seems to be promising for their optical properties modification.

Another interesting point is the problem of point defects diffusion into the depth of the sample. In this work and in our previous work [8] we operated with SA including absorbing saturable layers placed near the surface. As is known, the penetration depth is extremely low for UV radiation. In the case of our experiments, the depth estimation gives the value of 13 nm (for GaAs at 308 nm UV wavelength) while the thickness of 20 quantum wells layer is about 180 nm. The penetration depth for laser radiation of 750 nm (or 780 nm) exciting photoluminescence (or pumping SA for reflectivity kinetics measurement) of the SA equals to hundreds of nanometers, and this means that PL signal (reflectivity signal) is generated by the whole set of 20 quantum wells. From this viewpoint, it would be interesting to compare results of UV irradiation by nanosecond pulses with the be post-growth treatment of semiconductor structures by femtosecond pulses of UV radiation which should provide non-thermal mechanism of the surface damage like ablation.

It is well known, that SA should satisfy a series of very rigid requirements. First of all, it should provide necessary (for a given laser) level of saturable losses. Besides, SA relaxation time should lie in the range of few tens of picoseconds. Another evident requirement met by good SA is low level of nonsaturable losses. Taking into account all above mentioned, it is not surprising what a definitive role plays a special design and manufacturing of SA. Our result is obtained for $Yb^{3+}:KY(WO_4)_2$ laser stable mode-locking at 70 MHz, at central wavelength of 1039 nm, pulse length of 115 fs, maximum average power 1.57 W by using special fast SA of the own design and manufacturing; the pulses are close to spectral-limited type. Comparative experiments with commercially available SA mirror (BATOP Gmbh, Germany) with close characteristics (absorption of about 3% at 1040 nm, relaxation time of 1 ps, saturating energy density of 50 J/cm^2) was also performed. The shortest pulse duration was increased to 148 fs, and the maximum average output power was decreased down to 1.37 W at the same level of the incident pump power. This also confirms that our SA mirror has lower level of non-saturable losses compared with other similar devices.

In conclusion, two techniques for the recovery time shortening of saturable absorbers on a base of A_3B_5 compounds including quantum wells were tested. The first one, proposed by authors, is based on the post-growth treatment by UV laser radiation; it implied the generation of point defects, which, in its turn, led to the acceleration of electron-hole recombination and recovery time shortening by an order of magnitude and more. Another technique based on special design of barriers gave also promising results for the fast saturable absorbers. The semiconductor mirrors designed for the mode locking of infrared laser Yb^{3+}:$KY(WO_4)_2$ led to 115 fs stable mode-locking with average power close to CW regime. Preliminary results on fast saturable absorbers for spectral region of 1500 nm are also presented.

6. Acknowledgments

This research was supported by the Russian Foundation for Basic Research (grant 12-02-00327), by Presidium of the Russian Academy of Sciences Programme "Quantum mesoscopic and disordered structures" (grant 20.2), by by Presidium of the Russian Academy of Sciences Programme "Fundamental base of nano-structures and nano-materials technologies" (grant 24.20), and by Program of common fundamental research of Belarus National Academy of Science and Siberian Branch of Russian Academy of Science (grant No 14).

References

[1] Pekarek S, Fiebig Ch, Stumpf M Ch, Oehler A E H, Paschke K, Erbert G, Südmeyer T, and Keller U 2010 *Optics Express* **18** 16320
[2] Rubtsova N N, Kochubei S A, Kovalyov A A, Preobrazhenskii V V, Putyato M A, Pchelyakov O P, Semyagin B R, Shamirzaev T S, Kuleshov N V, Kisel V E, and Kurilchik S V 2010 *Proc. SPIE* **7721** 77210G
[3] Rubtsova N N, Kuleshov N V, Kisel' V E, Kovalyov A A, Kurilchik S V, Preobrazhenskii V V, Putyato M A, Pchelyakov O P, Shamirzaev T S 2009 *Laser Physics* **19** 285
[4] Tsuyoshi Okuno, Yasuaki Masumoto 2001 *Appl. Phys. Lett.* **79** 764
[5] Du M Le, Harmand J C, Mauguin O, Largeau L, Travers L, Oudar J L 2006 *Appl. Phys. Lett.* **88** 201110
[6] Soumalainen S, Vainionpaa A, Tengvall O, Hakulinen T, Karirinne S, Guina M, Okhotnikov O G 2005 *Appl. Phys. Lett.* **87** 121106
[7] Mangeney J, Oudar J L, Harmand J C, Meriafec C, Patriarche G, Aubin G, Stelmakh N, Lourtioz J M 2000 *Appl. Phys. Lett.* **76** 1371
[8] Rubtsova N N, Kuleshov N V, Kisel' V E, Kochubei S A, Kovalyov A A, Kurilchik S V, Preobrazhenskii V V, Putyato M A, Pchelyakov O P, Shamirzaev T S 2010 *Laser Physics* **20** 1262
[9] Kashkarov P K, Timoshenko V Yu 1995 *Poverkhnosyt'. Fizika, khimiya, mekhanika* **No 6** 5
[10] Hu Zh, Singha S, Rich D H, and Gordon R G 2012 *Appl. Phys. Lett.* **100** 141102
[11] Kovalyov A A, Pchelyakov O P, Preobrazhenskii V V, Putyato M A, and Rubtsova N N 2007 *Laser Physics* **17** 478
[12] Kovalyov A A, Pchelyakov O P, Preobrazhenskii V V, Putyato M A, Rubtsova N N, Sorokin E, and Sorokina I T 2007 *Int. J. Nanosci.* **6** 315
[13] Tackeuchi A, Muto S, Inata T, and Fujii T 1989 *Jpn. J. Appl. Phys.* **28** L1098
[14] Deveaud B, Chomette A, Clerot F, Auvray P, Regreny A, Ferreira R, and Bastard G 1990 *Superlattices Microstruct.* **8** 73
[15] Kovalyov A A, Preobrazhenskii V V, Putyato M A, Pchelyakov O P, Rubtsova N N, Semyagin B R, Kisel' V E, Kuril'chik S V, Kuleshov N V 2011 *Laser Phys. Lett.* **8** 431

Precise synchronization of qcw pumped active-passive mode locked picosecond lasers

A A Karnaukhov[2], V B Morozov[*1,2], A N Olenin[1] and D V Yakovlev[1]

[1] Faculty of Physics, M.V. Lomonosov Moscow State University, 119992 GSP-2, Moscow, Russia

[2] International Laser Center, M.V. Lomonosov Moscow State University, 119992 GSP-2, Moscow, Russia

[*]morozov@phys.msu.ru

Abstract. We present designing qcw diode pumped Nd:YAG picosecond laser with 50 ps jitter of output pulse relative to active mode locking signal. Adjustment of signal delay in negative feedback circuit provides generation of transform-limited laser pulses. Obtained level of jitter is mainly attributed to phase noise of radio-frequency active mode locking signal generator. Further jitter decrease by means of enhancement of radio-frequency signal filtering and suppressing of phase noise is discussed.

1. Introduction

Precise synchronization of mode-locked pico- and femtosecond lasers integrated in technological and research complexes is a crucial issue for many applications. Jitter of output optical pulse respective to periodic reference signal in continuous wave (cw) lasers can be effectively suppressed up to femtosecond level by means of feedback controlled cavity adjustment [1]. Cross modulation approach gives even more synchronization accuracy [2]. However, high peak power laser systems with precise optical synchronization option based on cw oscillators are overly complex and expensive for most practical applications.

High peak power mode-locked picosecond lasers with pulsed (qcw) pump are applicable for such fields as high precision laser micromachining [3], soft X-ray generation [4], laser photo injection for electron accelerators [5], routine time-resolved spectroscopic measurements [6], laser ranging [7] etc. Pulsed laser systems based on qcw pumped oscillators with generation process development in each laser shot may have definite advantages in comparison with cw ones, providing high pulse energy at the oscillator output, better efficiency of energy consumption, lower thermal loading, relative simplicity and compactness. At the same time, typical duration of generation process development $\sim 10^{-5}$ s is too short both for phase detector operation and for cavity adjustment which are convenient techniques for cw lasers. In paper [8] 130 ps jitter relative to external synchropulse was measured in pulse pumped picosecond laser with active mode locking. This is minimal jitter, to our knowledge, that was achieved in lasers with pulsed pump. Authors of [8] explained such level of jitter by imperfection of radio-frequency (RF) oscillator for active mode locking. Nevertheless, time position of ultrashort pulse which goes several hundreds roundtrips during generation process development in each pump pulse is defined by several factors such as amplification in active element, passive mode-locking mechanism, actions of feedback and active mode locking signal. Minimal jitter is mainly determined by width of "temporal transparency window" which is formed by cumulative action of the above factors in the oscillator.

In the present paper, we discuss experimental results on jitter decreasing in picosecond pulse-diode-pumped Nd:YAG laser within output pulse duration level of several tens of picoseconds.

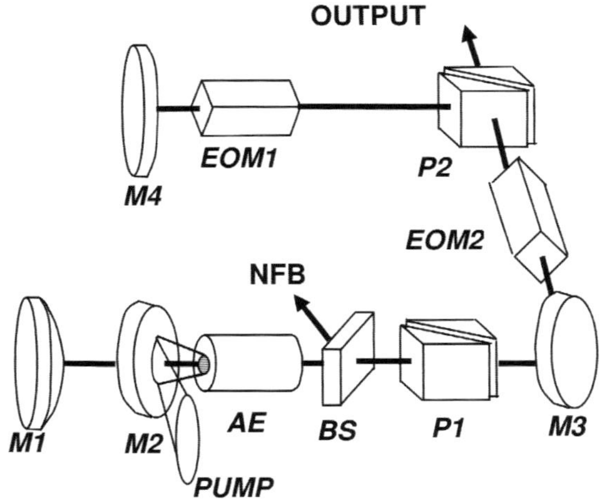

Figure 1. Simplified optical scheme of picosecond laser. M1, M3, M4 – HR@1064 nm mirrors, M2 – HR@808 nm and AR@1064 nm mirror, AE – active element, BS – beam splitter, P1, P2 – polarisers, EOM1, EOM2 – electro-optical modulators, NFB – radiation to fibre-coupled photodiode of negative feedback circuit.

2. Experimental setup

The optical scheme for studying jitter arising during picosecond pulse generation process is shown in figure 1. Resonators formed by HR@1064 nm flat mirrors M1 and M4. Concave mirror M3 compensates beam divergence and provides stability of resonator. Active element AE with length l_{AE} = 11 mm, diameter d_{AE} = 5 mm, Nd concentration of 1.1 % is pumped by a pulsed diode bar with a fiber output (maximum peak power P_{max} = 70 W, wavelength 808 nm). Pump pulses with a width τ_P = 200 μs at repetition frequency f_P equal to 25 Hz from the output of a fiber 0.6 mm in diameter are focused into the AE after reflection from AR@808 nm M2. In the experiments the maximum peak pump power is limited at the level of 55 W.

This highly reflecting resonator is aimed at forming a picosecond pulses with minimal optical losses under signals of negative feedback (NFB) and active mode locking (AML), which are applied during the end part of each pump pulse to a two-crystal thermocompensate electro-optical modulator EOM1 from an RTP crystal [9]. The resonator is dumped and generated picosecond pulse is output by applying a half-wave voltage pulse to electro-optical modulator EOM2. In this case, output pulse energy is only ~1 μJ and pulse duration is ~ 200 ps. Further output pulse energy increase can be obtained using regenerative amplification stage. Minimum value of pulse duration can be achieved using combined active-passive mode-locking scheme [10].

The distance between the mirror Ml and the AE is chosen to be 15 cm to exclude the formation of a standing wave inside pumped region of AE near the resonator end mirror M1 [11] when generating pulses with a width less than 1 ns. Optical length of resonator is 113 cm that corresponds to quarter period of AML signal at frequency f_{AML}~66.4 MHz. In our experiments we use AML frequency equal to half of round trip frequency. This gives us possibility to apply driving signals (bias, harmonical AML signal, NFB signal

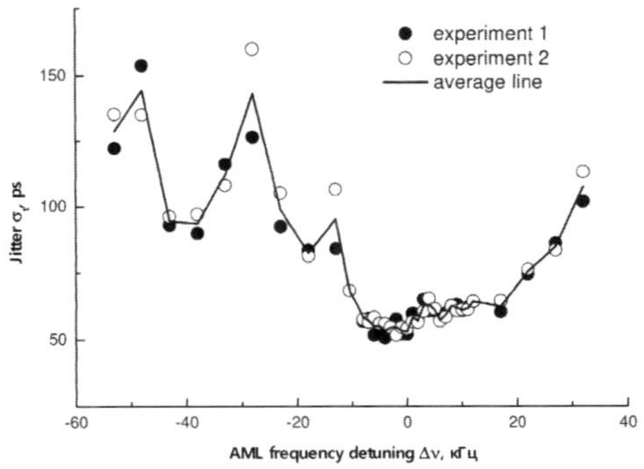

Figure 2 Jitter of output picosecond pulse relative to phase of driving RF oscillator at variation of frequency detuning of AML signal. Zero detuning corresponds to minimal jitter. Amplitude of AML signal is set to maximum

Journal of Physics: Conference Series **414** (2013) 012027 doi:10.1088/1742-6596/414/1/012027

[9]) on EOM1 independently [10]. Radiation of a pulse reflected from the beam splitter (BS) is focused into fiber which is coupled with photodiode. This photodiode signal is used in the NFB circuit on EOM1. Length of the fiber is chosen to provide minimal losses associated with NFB circuit for laser pulse at the next round trip.

Temporal behavior of radiation inside resonator is registered with p-i-n photodiode SFH250V and digital oscilloscope Tektronix DPO4104. Photodiode has response time about 5 ns nevertheless temporal position of its signal can be measured with accuracy about 10 ps by fitting of registered response with proper function having variable time shift parameter. Jitter of laser pulse is measured as standard deviation of time delay of simultaneously registered AML and optical signals. Time shape of output laser pulse is controlled with streak camera.

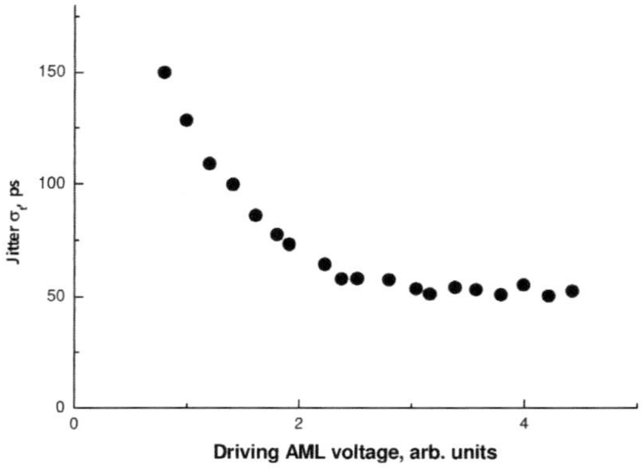

Figure 3 Jitter of output picosecond pulse relative to phase of driving RF oscillator at variation of harmonic AML signal amplitude

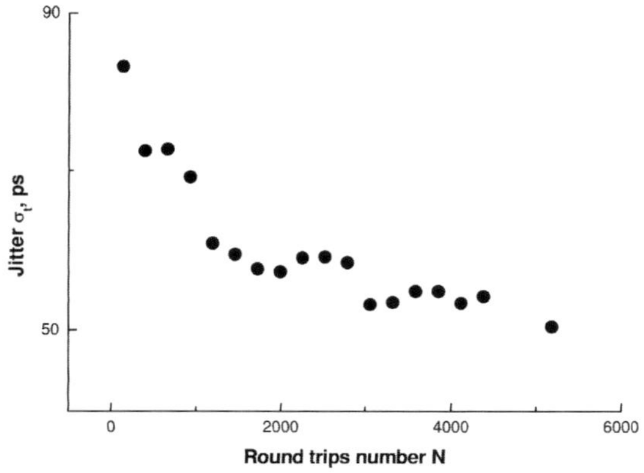

Figure 4 Jitter of output picosecond pulse relative to driving RF oscillator as a function of number of laser pulse round trips in resonator

3. Results and discussion

One of most important parameters that define laser generation behavior is detuning of AML frequency from round trip one. Results on measurements of jitter depending on frequency detuning are shown in figure 2. Unlike to cw pumped lasers, where frequency detuning on the level of sub-1 kHz from round trip frequency about 100 MHz brings to chaotic behavior of generation [12], finite round trips number at pulsed pump approach allows larger range of AML frequency detuning without sufficient jitter increasing.

Efficiency of jitter suppression at variation of amplitude of AML signal and duration of generation process are illustrated by figure 3 and figure 4 correspondingly. Both measurements are fulfilled at AML frequency adjusted to optimal value. Note that amplitude of AML signal directly on EOM1 is very difficult to measure because of very small capacity of RTP crystal and complex structure of driving signal [10]. Although part of AML signal from capacitive divider connected directly to the EOM1 crystal is used for jitter measurements, exact value of division coefficient is known only approximately. Therefore we show in figure 3 jitter dependence on relative units of AML amplitude. Our estimation shows that at maximum AML signal corresponds to almost complete modulator EOM1 closure. It is clear from figures 3 and 4 that jitter goes to value about 50 ps at increasing of amplitude of AML signal and duration of generation process. To find reason of that jitter asymptotic behavior we applied digital synchronous detection procedure to AML signal and obtained phase modulation with regular and noise component. Amplitude of noise component corresponds to jitter with observed value. We attribute origin of this noise to insufficient filtering of digital synthesizer output that is used for AML signal. Enhancing of filtering scheme should provide further decrease of jitter.

Acknowledgments
The work was supported by Russian Foundation for Basic Research (10-02-01277-a) and M.V.Lomonosov Moscow State University Program of Development.

References

[1] Spence D E, Dudley J M, Lamb K, Sleat W E and Sibbett W 1994 *Opt. Lett.* **19** 481
[2] Yoshitomi D, Kobayashi Y, Kakehata M, Takada H and Torizuka K 2006 *Opt. Express* **14** 6359
[3] Gerhard C, Druon F, Blandin P, Hanna M, Balembois F, Georges P and Falcoz F 2008 *Appl. Opt.* **47** 967
[4] Shlyaptsev V N, Luther B M, Wang Y, Larotonda M A, Alessi D, Berrill M, Marconi M C and Rocca J J 2005 *Opt. Lett.* **30** 165
[5] Kobayashi T, Uesaka M, Katsumura Y, Muroya Y, Watanabe T, Ueda T, Yoshii K, Nakajima K, Zhu X and Kando M 2002 *J. Nuc. Sci. and Tec.* **39** 6
[6] Arakcheev V G, Kireev V V, Morozov V B, Olenin A N, Tunkin V G, Valeev A A and Yakovlev D V 2007 *J. Ram. Spectr.* **38** 1038
[7] Busck J and Heiselberg H 2004 *Appl. Opt.* **43** 4705
[8] Zaporozhchenko V A and Tylets N A 1996 *Quantum. Electron.* **26** 512
[9] Zverev G M and Golyaev Yu D 1985 *Lazery na alyumooittrievom granate (Lasers on Aluminium Yttrium Garnet)* (Moscow, Radio i svyaz')
[10] Gorbunkov M V, Konyashkin A V, Kostryukov P V, Morozov V B, Olenin A N, Rusov V A, Telegin L S, Tunkin V G, Shabalin Yu V and Yakovlev D V 2005 *Quantum. Electron.* **35** 2
[11] Cousin A K 1992 *Appl. Opt.* **31** 7259
[12] Eichler H J, Koltchanov I G and Liu B 1995 *Appl. Phys.* B **61** 81

Journal of Physics: Conference Series **414** (2013) 012028 doi:10.1088/1742-6596/414/1/012028

Superfluid to normal phase transition in strongly correlated bosons in two dimensions

J Carrasquilla and M Rigol

Department of Physics, Georgetown University, Washington, DC 20057, USA, and Physics Department, The Pennsylvania State University, 104 Davey Laboratory, University Park, Pennsylvania 16802, USA

E-mail: carrasqu@physics.georgetown.edu

Abstract. Using quantum Monte Carlo simulations, we investigate the finite-temperature phase diagram of hard-core bosons (XY model) in two-dimensional lattices. To determine the phase boundaries, we perform a finite-size-scaling analysis of the superfluid stiffness in two different ways and find that both approaches provide results of comparable accuracy. Furthermore, we discuss how such a phase diagram can be determined by measuring the occupancy of the zero momentum mode in homogeneous and trapped systems. The latter approach can be used in current experiments with ultracold gases.

1. Introduction

Systems of interacting bosons are of interest in a vast range of physical contexts such as in experiments with superfluid helium [1], Josephson-junction arrays [2], elementary excitations in magnetic insulators that can be understood as bosonic quasiparticles [3], and more recently in experiments with ultracold gases loaded in optical lattices [4, 5]. Specifically, the latter ones offer a unique opportunity to study models that are widely considered in statistical and condensed matter physics with unprecedented experimental control over the parameters which, in turn, determine the Hamiltonian describing such systems. The Bose-Hubbard model [6, 7] is perhaps the most successful example of such systems where the match between experiment and theory has been corroborated in widely diverse situations. More concretely, the Mott-insulator transition has been experimentally observed in such a model and in various effective dimensionalities [8, 9, 10, 11]. Despite the fact that it has received considerably less attention, the superfluid to normal transition in the Bose-Hubbard model has been realized in a two-dimensional lattice of Josephson-coupled Bose-Einstein condensates [12, 13], and in experiments with ultracold atoms in two- and three-dimensional optical lattices [14, 15].

One important aspect that determines the nature of the physical phases and their associated order parameters is the dimensionality d. Mermin, Wagner, and Hohenberg rigorously proved that at any nonzero temperature, continuous symmetries cannot be spontaneously broken in systems with sufficiently short-range interactions in dimensions $d \leq 2$ [16, 17]. This implies that, at finite temperature, Bose-Einstein condensation (BEC) cannot occur in one and two dimensions. Two-dimensional Bose systems, however, are marginal in the sense that fluctuations are strong enough to destroy the fully ordered state but are not so strong as to suppress superfluidity. Thus critical behavior develops in the Berezinskii-Kosterlitz-Thouless

(BKT) transition [18, 19], where a superfluid phase with quasi-long-range order competes with thermal fluctuations and induces a continuous phase transition to the normal fluid as the temperature is increased. It was found that the mechanism for the transition in two dimensions is markedly different from the usual finite-temperature phase transitions because it does not involve any spontaneous symmetry breaking. Kosterlitz and Thouless predicted that superfluid transition occurs through the formation of vortex-antivortex pairs in the quasi-long-range ordered superfluid phase that are instead free in the normal phase. Because of the absence of a local order parameter in the low-temperature phase and because of the essential singularity in the divergence of the correlation length, the numerical detection of the superfluid transition in two dimensions is very subtle. However, it was also predicted that in such a transition, the superfluid stiffness jumps from zero to a finite value at the critical temperature. Taking advantage of that observation, several approaches that measure the superfluid stiffness have been introduced. Since most numerical work is based on simulations of finite-sized clusters, several schemes, mainly based on the renormalization group theory of such transition, have been introduced in the literature in order to perform size-scaling analysis of the superfluid stiffness [20, 21, 22].

Our goal here is to focus on the superfluid to normal transition in a system of strongly interacting bosons in two-dimensional lattices. Specifically, we consider the Bose-Hubbard model in the limit of infinite on-site repulsion, i.e., the hard-core boson limit. By the use of exact quantum Monte Carlo simulations, we compute the finite-temperature phase diagram as a function of chemical potential. Accurate results are obtained through finite-size scaling of the the superfluid stiffness. To perform the finite-size scaling, we used two different but related schemes and we show that the results of both methods are consistent. Furthermore, we introduce and test a method to detect such transition that is based on the measurement of the zero-momentum occupation, which can be applied to both homogeneous and inhomogeneous systems. This method can be used in experiments with ultracold atoms even in the presence of unavoidable confining potentials that, in such setups, are needed in order to contain the gas. We note that, in the presence of a confining potential, domains of different phases can coexist in the trap both at zero [23, 24, 25] and finite [26, 27, 28, 29] temperature. A related and more extensive discussion on work discussed here, including an analysis of the three-dimensional case, can be found in Ref. [30].

The paper is organized as follows. In Sec. 2, we introduce the model and its phase diagram in two dimensions. Section 3 is devoted to the discussion of the techniques to obtain the phase boundaries. Finally, in Sec. 4, we draw our conclusions.

2. Model and phase diagram

We consider a system of hard-core bosons on a two-dimensional lattice with L^2 sites. The Hamiltonian can be written as

$$\hat{H} = -t\sum_{\langle i,j\rangle}\left(\hat{a}_i^\dagger\hat{a}_j + \text{H.c.}\right) - \mu\sum_i \hat{n}_i \;, \tag{1}$$

where $\hat{a}_i^\dagger$ ($\hat{a}_i$) is the boson creation (annihilation) operator at a given site i, and $\hat{n}_i = \hat{a}_i^\dagger\hat{a}_i$ is the particle number operator at site i. The hard-core boson creation and annihilation operators satisfy the constraint $\hat{a}_i^{\dagger 2} = \hat{a}_i^2 = 0$, which prohibit multiple occupancy of lattice sites. The first term in equation (1) is the kinetic energy, where t is the hopping amplitude between neighboring sites i and j ($\langle i,j\rangle$). The second term contains the chemical potential μ that controls the total number of particles in the system. In what follows, positions will be given in units of the lattice spacing a and the energy will be given in units of the hopping amplitude t.

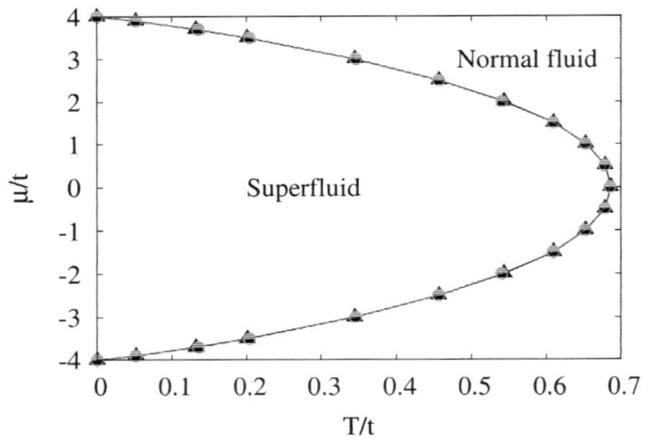

Figure 1. (Color online) Finite temperature phase diagram in two dimensions as obtained by two different size scaling procedures.

We recall that the Hamiltonian in equation (1) can be exactly mapped to the broadly studied quantum XY model [31]

$$\hat{H} = -2t\sum_{\langle i,j\rangle}\left(S_i^x S_j^x + S_i^y S_j^y\right) - \mu\sum_i S_i^z\,, \qquad (2)$$

where S_i^α is the αth component of the spin-1/2 spin operator at site i. In the spin language, the term proportional to t describes a ferromagnetic exchange interaction, while the one proportional to μ describes a magnetic field in the z-direction.

We study the Hamiltonian in equation (1), at finite temperature T, by means of the stochastic series expansion (SSE) quantum Monte Carlo (QMC) method with operator-loop updates [32, 33, 34]. The determination of the phase diagrams is carried out through a finite size scaling of the superfluid stiffness ρ_s using periodic boundary conditions. The numerically exact (QMC) phase diagram in two dimensions (2D) is presented in figure 1. The finite-temperature phase diagram comprises a low-temperature quasi-ordered superfluid lobe surrounded by a high-temperature normal phase with exponentially decaying correlation functions. Specifics on the procedure to obtain the phase boundaries are discussed in the following section.

3. Superfluid to normal phase transition in two dimensions

3.1. Superfluid stiffness and the Kosterlitz-Thouless transition

Our results for the two-dimensional phase diagram in figure 1 are based on the fact that the model in equation (1) undergoes a BKT transition as a function of the temperature. This phase transition has been studied in great detail the context of the two-dimensional quantum XY model in equation (2) in the absence of a magnetic field [35, 36, 37, 21]. Kosterlitz and Thouless predicted that the superfluid stiffness ρ_s jumps from zero to the value $(2/\pi)T_c$ at the critical temperature. Thus, we consider measurements of the superfluid stiffness ρ_s for different system sizes L as a function of temperature. Within the SSE method, the superfluid stiffness is computed by measuring the fluctuations of the winding number W [38]; they are connected through the relation $\rho_s = \langle W^2\rangle/2\beta$, where $\beta = 1/T$ is the inverse temperature.

Figure 2(a) shows results for the superfluid stiffness of 2D hard-core bosons at $\mu = 0$ [or equivalently the spin stiffness of the 2D XY model in equation (2)] as a function of T for several system sizes. The observed slow approach of the superfluid stiffness to the characteristic jump expected for the infinite system is due to strong finite-size effects at the BKT transition. Finite-size scaling relations for the superfluid stiffness can be derived by integrating the Kosterlitz renormalization-group equations [see, for instance, Refs. [21, 39, 20]. This procedure yields

$$\begin{aligned}
\frac{\rho_s(T,L)\,\pi}{2T}-1 &= c\coth 2c\,(\ln L+l_0), \quad T<T_c \\
\frac{\rho_s(T_c,L)\,\pi}{2T}-1 &= \frac{1}{2\,(\ln L+l_0)}, \quad T=T_c \\
\frac{\rho_s(T,L)\,\pi}{2T}-1 &= c\cot 2c\,(\ln L+l_0), \quad T>T_c
\end{aligned} \tag{3}$$

where c measures the distance from the critical point and l_0 depends only weakly on temperature. Close to the critical point $c\sim\sqrt{|T-T_c|}$. In the limit $2c\,(\ln L+l_0)\ll 1$, a scaling form for the superfluid stiffness based on equation (3) can be written as

$$\frac{\rho_s(T,L)\,\pi}{T}-2=\frac{1}{\ln L+l_0}F\left[(\ln L+l_0)^2\,(T-T_c)\right]. \tag{4}$$

From equation (3) in the limit $2c\,(\ln L+l_0)\ll 1$, $F(x)=1-(4/3)\,x$. One can find the scaling function F and critical temperature T_c by computing $x_L=(\ln L+l_0)^2\,(T-T_c)\,/t$ and $y_L=\rho_s(T,L)\,\pi/T-2$ based on our Monte Carlo simulations for different L and T. The adjustment of the constant l_0 and critical temperature T_c, such that the data produce the best possible collapse, yields a numerical estimate of the scaling function F and the critical temperature itself. The result of the determination of the scaling function F is reported in figure 2(b), where a plot of y_L as a function of x_L is presented. Notice that, as expected, the value of F is very close to one for $x_L=0$ and that the function is, to a very good approximation, a straight line with negative slope similar to 4/3. Our result $T_c/t=0.685\pm0.001$ is consistent with the best value reported in Ref. [21], for which $T_c/t=0.6846\pm0.0006$. An analogous procedure to the one just described is carried out for different values of the chemical potential to complete the two-dimensional phase diagram in figure 1 (drawn with green empty circles).

Based on the relations in equation (3) at the critical temperature, one can also use the following scaling Ansatz for the superfluid stiffness as a function of system size and temperature [22]

$$\rho_s(T,L)=\left(1+\frac{1}{2\,[\ln L+l_0]}\right)F'(\xi/L), \tag{5}$$

which is expected to be a good approximation in the vicinity of the critical temperature $T\to T_c^+$. When dealing with equation (5), we assume the essential singularity of the correlation length $\xi\sim e^{b/(T-T_c)^{1/2}}$, where b is a chemical-potential dependent scaling factor. [Note that equation (5) is, strictly speaking, inconsistent with renormalization-group arguments, which lead to equation (4).] F' is a scaling function that, after taking the logarithm of the argument, can be replaced by another function f. Hence, equation (5) can be written as

$$\rho_s(T,L)^*=f\left(x'_L\right), \tag{6}$$

where $x'_L=\ln L-b/\,(T-T_c)^{1/2}$ and $\rho_s(T,L)^*=\rho_s(T,L)\left(1+\frac{1}{2[\ln L+l_0]}\right)^{-1}$. The scaling hypothesis in equation (6) is consistent with the fact that the rescaled superfluid stiffness $\rho_s(T,L)^*$ becomes system-size independent at the critical temperature where the correlation length diverges. The scaling function f can be obtained numerically by finding the constants b and c, as well as the critical temperature T_c, that produce the best possible collapse of the data in the normal phase where the correlation length is finite. The result of the estimation of the scaling function f is portrayed in figure 2(c), where a plot of the rescaled superfluid stiffness as a function of x'_L is presented. As expected, the data is seen to coalesce to a unique curve

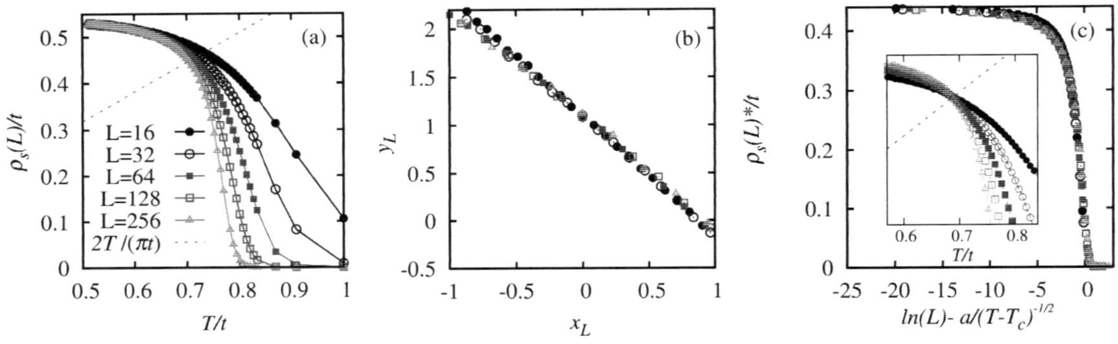

Figure 2. (Color online) (a) Superfluid stiffness for $\mu = 0$ and several values of L. The error bars (not shown) are smaller than the point size used in the plot. (b) Data collapse according to the relation in equation (4). (c) Numerical determination of the scaling function f based on equation (6). The inset in (c) shows the rescaled superfluid stiffness vs T.

representing $f(x'_L)$, which makes apparent the relevance of the scaling relation in equation (6) to the region close to the critical point. As argued before, one expects from equations (3) and (5) that a plot of the rescaled superfluid stiffness $\rho_s(T, L)^*$ as a function of the temperature T should become system-size independent at the critical temperature T_c. This observation is confirmed in the inset of figure 2(c). Remarkably, those curves intersect with the line $(2/\pi)T$ right at the critical temperature, in agreement with the BKT scenario. Our result $T_c/t = 0.687 \pm 0.002$ is again consistent with the values reported in Ref. [21] and with the value obtained through equation (4). This procedure is carried out for different values of the chemical potential to complete the two-dimensional phase diagram in figure 1 (green empty dots). Note that, in figure 1, the two scaling procedures lead to almost identical phase diagrams.

Finally, we should mention that equation (3) predicts the value of the superfluid stiffness in an infinite system at the critical temperature to be $\rho_s(T_c)/T_c = 2/\pi$. However, in Ref. [40], it was shown that the superfluid stiffness at the transition temperature is $\rho_s(T_c)/T_c \simeq 0.63650$, which is very close (and indistinguishable within the present study) to the result based on equation (3) $[2/\pi \simeq 0.63662]$.

3.2. Zero-momentum occupation and trapped systems

We now turn our attention to the behavior of the occupation of the zero momentum state $[n_{k=0} \equiv n_0]$ in the critical region, and address the determination of the transition temperature from it. As mentioned in the introduction, thermal fluctuations in two dimensions prevent the emergence of a Bose-Einstein condensate, which means that and the zero-momentum occupation is not macroscopic. Nevertheless, as the critical temperature is approached from the normal phase, n_0 diverges [see figure 3(a)]. From the Fourier transform of the one-body density matrix in the long-distance limit $\langle \hat{a}_i^\dagger \hat{a}_{i+r} \rangle \propto r^{-1/4} \exp(-r/\xi)$, one can precisely understand the way n_0 diverges as T_c is approached,

$$n_0 \sim \xi^{7/4}. \tag{7}$$

From equation (7), it follows that not only does n_0 diverge at T_c, but also its derivative with respect to T does,

$$\frac{dn_0}{dT} \sim -\frac{\xi^{7/4} \ln^3 \xi}{b^2} \tag{8}$$

In finite clusters, however, when T is close to T_c, the role of the correlation length is taken over by L when $\xi \gtrsim L$. This occurs at a temperature $T^*(L)$ determined by the critical temperature

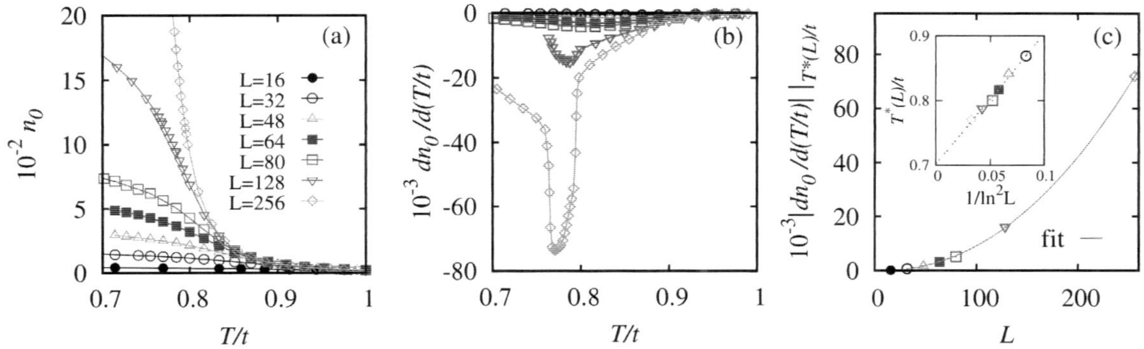

Figure 3. (Color online) (a) Zero-momentum occupation n_0 vs T. (b) Derivative of the zero-momentum occupation n_0 with respect to the temperature for different values of L. (c) Finite-size scaling of the height of the negative peak in dn_0/dT. The continuous line is a fit to the function $g(L) = a_0 + a_1 L^{7/4} \ln^3(a_2 L)$. The inset shows the finite-size scaling of $T^*(L)$.

and the size of the system under consideration, it is given by

$$T^*(L) = T_c + b'/\ln^2 L, \tag{9}$$

where b' is a related to b. At that temperature, the derivative in equation (8) scales with the system size as

$$\left.\frac{dn_0}{dT}\right|_{T^*(L)} \sim -\frac{L^{7/4}\ln^3 L}{b^2}. \tag{10}$$

Below $T^*(L)$, n_0 cannot vary as fast as right above $T^*(L)$ because the exponential increase of the correlation length is truncated by L. Below $T^*(L)$, the variation of n_0 comes mainly from the temperature dependence of the anomalous exponent, which is not as strong as the variation due to the exponential behavior of the correlation length. Accordingly, dn_0/dT should exhibit a sharp minimum at the size-dependent temperature $T^*(L)$. Additionally, in a finite system, n_0 cannot grow indefinitely as the temperature is reduced. In the limit $T \to 0$, n_0 must come closer to its (finite) $T = 0$ value, which implies that the derivative with respect to temperature should eventually vanish for any given finite cluster.

Figure 3(b) shows the derivative of the n_0 for different system sizes as a function of T. The divergence of dn_0/dT is clearly observed. A sharp minimum develops and its location $T^*(L)$ approaches T_c as the system size increases. This is expected from the finite-size relation in equation (9). The scaling of the height of this minimum is studied in figure 3(c), where we plot the absolute value of $dn_0/dT|_{T^*(L)}$ vs L. The data follows the scaling relation in equation (10), as made evident by a fit to the function $g(L) = a_0 + a_1 L^{7/4} \ln^3(a_2 L)$. In the inset in figure 3(c), we show the finite-size scaling of $T^*(L)$. We observe that $T^*(L)$ is consistent with the scaling relation in equation (9), which we use to obtain the critical temperature in the thermodynamic limit. We find $T_c/t = 0.701 \pm 0.007$. This value is compatible with the one found by performing the finite-size scaling of the superfluid stiffness. While this approach is obviously less accurate than the one discussed before for ρ_s, among other things because a numerical derivative is involved, the fact that it works very well is important for current trapped ultracold gases experiments where n_0 can be measured while the superfluid density cannot.

If an additional confining potential is added to the system, the reasoning we have just presented remains valid. The effect of the confining potential can be taken into account by replacing the second term in equation (1) with a spatially varying chemical potential as

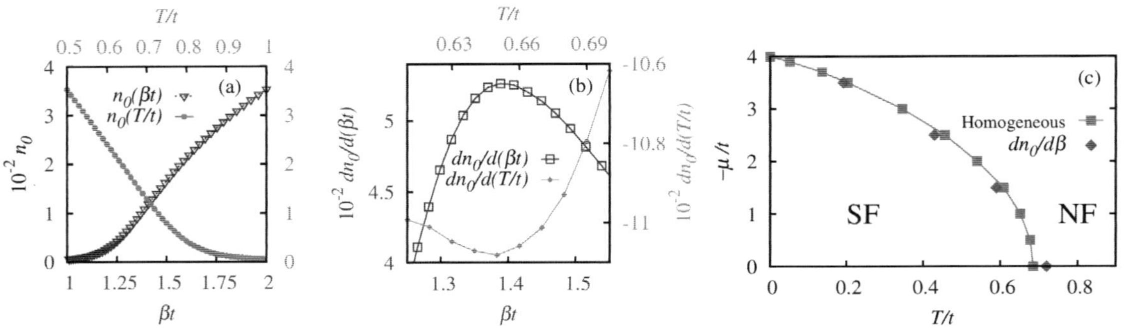

Figure 4. (Color online) (a) n_0 as a function of T and β in a trapped 2D system with $V_0/t = 0.00125$, $\mu = 0$ in the center of the trap, and $L = 128$. (b) Derivatives of n_0 with respect to β and with respect to T. (c) Estimate of the critical points based on the derivative of the zero-momentum occupation with respect to β (red diamonds based on a system with $L = 128$). The phase diagram of the homogeneous system is also shown for comparison.

$\mu \sum_i \hat{n}_i \to \sum_i \mu_i \hat{n}_i$. Here, $\mu_i = \mu - V_0 r_i^2$, where V_0 quantifies the strength of the harmonic potential and μ plays the role of the overall chemical potential. At fixed chemical potential ($\mu \leq 0$), when lowering T, the normal-to-superfluid crossover in the trapped system proceeds through the appearance and enlargement of a superfluid region in the center of the trap. Hence, the zero-momentum state becomes increasingly populated. Exactly as it was argued for finite homogeneous systems, it is expected that as $T \to T_c^+$ for the normal-to-superfluid transition in the center of the system, the rate of growth of n_0 will increase. Below T_c, on the other hand, dn_0/dT will eventually decrease because of the finite extend of the system imposed by the confining potential. If T is lowered well below T_c, almost the entire system will become superfluid and the observables will saturate their (finite) zero-temperature values. This means that the derivative dn_0/dT is expected to exhibit a minimum very close to the critical temperature.

Therefore, a similar approach to detect criticality can be utilized in confined systems. Provided that the trapping potential is weak enough, the accuracy of the estimations based on trapped systems should be comparable with the accuracy obtained in the homogeneous case. Figure 4(a) depicts the evolution of n_0 vs T, as well as vs the inverse temperature β, for a harmonically confined 2D system with $V_0/t = 0.0015$ ($L = 128$) and $\mu = 0$ in the center of the trap. In figure 4(b), we show dn_0/dT which, as expected, exhibits a minimum located at $T/t = 0.66 \pm 0.02$. This temperature is compatible with the value of T_c/t obtained for the homogeneous case where, after a finite-size scaling, we obtained $T_c/t = 0.685 \pm 0.001$. Our estimate derived from the study of a single trapped system is about 4% off the value of the homogeneous system.

The same analysis based on measurements of n_0, but now as a function of the inverse temperature β, can be carried out; in that case, one expects a maximum in the derivative $dn_0/d\beta$ instead of a minimum. Generally, for finite and not very large systems, the position of such maximum β_c will not coincide with $1/T_c$ obtained from the minimum of dn_0/dT [30]. Overall, we find that, for the system sizes available to our QMC simulations, the analysis based on $dn_0/d\beta$ provides more accurate estimates of the critical temperature than the one based on dn_0/dT. This follows from the fact that, the maximum found in $dn_0/d\beta$ is consistently sharper and better defined when compared to the minimum found for dn_0/dT, which instead is shallower and broader, thus leading to results with lower accuracy. Based on measurements of $dn_0/d\beta$ presented in figure 4(b) on the same system with $V_0/t = 0.0015$ ($L = 128$), $\mu = 0$, we find $T_c/t = 0.72 \pm 0.02$. This value is also very close to the critical temperature of the homogeneous system. When the maximum is sharply defined, in the limit of very shallow traps with large numbers of bosons, the two approaches should coincide (i.e., their difference is due to finite size effects) [30]. Therefore, for the determination of the phase diagram based on measurements in harmonically confined systems, we consider only measurements based on $dn_0/d\beta$.

In figure 4(c), we summarize our results for the determination of the critical parameters with the derivative of the zero-momentum occupation with respect to β, and contrast them with the phase diagram of the homogeneous system. In spite of the small differences and the fact that the peaks in $dn_0/d\beta$ are shallower than those of the homogeneous system, the agreement in the calculations of the phase diagram based on both the homogeneous and the trapped systems is remarkably good. We mention that at the tip of the superfluid lobe, where the size effects are expected to be the strongest, we observed that as the size of the system is increased (or the strength of the trap is decreased), keeping constant the chemical potential in the center of the trap, the estimate of the critical temperature decreases approaching the result in homogeneous systems.

4. Conclusions

We have presented a detailed study of the finite temperature phase diagram of strongly correlated bosons in the hard-core limit (or the XY model) in two dimensions. The critical parameters were determined through a finite-size scaling analysis of the superfluid stiffness. Two methods to determine the critical temperature have been used. The first method follows directly from the integration of the Kosterlitz-Thouless renormalization group equations while the second one is based on an Ansatz that is justified only in the vicinity of the critical temperature. The validity of the second method has been, nevertheless, empirically confirmed by the quality of the collapse obtained in our analysis. Both methods provided results with comparable accuracy and the phase diagrams obtained are consistent with each other.

We introduced an approach to estimate the critical temperature from measurements of n_0 in finite systems. It makes use of the behavior of the derivative dn_0/dT and we derived finite-size scaling relations that can be used to extrapolate the results to the thermodynamic limit. This approach can be applied to systems that exhibit a diverging zero-momentum occupation in any dimension, irrespective of the universality class to which the transition belongs to. When an additional confining potential is introduced, we considered measurements of the critical temperature based on measurements of the zero-momentum occupation, out of which we obtained a phase diagram that is in good agreement with the homogeneous counterpart.

Acknowledgments

This work was supported by the US Office of Naval Research under Award No. N000140910966. We thank Nikolai Prokof'ev, Itay Hen, and Rong Yu for useful discussions.

References

[1] Daunt J G and Smith R S 1954 *Rev. Mod. Phys.* **26** 172–236
[2] Bruder C, Fazio R and Schön G 1993 *Phys. Rev. B* **47** 342–347
[3] Giamarchi T, Rüegg C and Tchernyshyov O 2008 *Nature Phys.* **4** 198–204 ISSN 1745-2473
[4] Bloch I, Dalibard J and Zwerger W 2008 *Rev. Mod. Phys.* **80** 885–964
[5] Cazalilla M A, Citro R, Giamarchi T, Orignac E and Rigol M 2011 *Rev. Mod. Phys.* **83** 1405–1466
[6] Fisher M P A, Weichman P B, Grinstein G and Fisher D S 1989 *Phys. Rev. B* **40** 546–570
[7] Jaksch D, Bruder C, Cirac J I, Gardiner C W and Zoller P 1998 *Phys. Rev. Lett.* **81** 3108–3111
[8] Stöferle T, Moritz H, Schori C, Khl M and Esslinger T 2004 *Phys. Rev. Lett.* **92** 130403
[9] Spielman I B, Phillips W D and Porto J V 2007 *Phys. Rev. Lett.* **98** 080404
[10] Jiménez-García K, Compton R L, Lin Y J, Phillips W D, Porto J V and Spielman I B 2010 *Phys. Rev. Lett.* **105** 110401
[11] Greiner M, Mandel O, Esslinger T, Hänsch T W and Bloch I 2002 *Nature* **415** 39–44
[12] Schweikhard V, Tung S and Cornell E A 2007 *Phys. Rev. Lett.* **99** 030401

[13] Trombettoni A, Smerzi A and Sodano P 2005 *New J. Phys.* **7** 57–57
[14] Zhang X, Hung C L, Tung S K and Chin C 2012 *Science* **335** 1070–1072
[15] Trotzky S, Pollet L, Gerbier F, Schnorrberger U, Bloch I, Prokof'ev N V, Svistunov B and Troyer M 2010 *Nature Phys.* **6** 998–1004
[16] Mermin N D and Wagner H 1966 *Phys. Rev. Lett.* **17** 1133–1136
[17] Hohenberg P C 1967 *Phys. Rev.* **158** 383–386
[18] Berezinskii V L 1972 *Sov. Phys. JETP* **34** 610–616
[19] Kosterlitz J M and Thouless D J 1973 *J. Phys. C: Solid State Phys.* **6** 1181–1203
[20] Weber H and Minnhagen P 1988 *Phys. Rev. B* **37** 5986–5989
[21] Harada K and Kawashima N 1997 *Phys. Rev. B* **55** R11949–R11952
[22] Sandvik A W 2010 *AIP Conf. Proc.* **1297** 135–338
[23] Batrouni G G, Rousseau V, Scalettar R T, Rigol M, Muramatsu A, Denteneer P J H and Troyer M 2002 *Phys. Rev. Lett.* **89** 117203
[24] Wessel S, Alet F, Troyer M and Batrouni G G 2004 *Phys. Rev. A* **70** 053615
[25] Rigol M, Batrouni G G, Rousseau V G and Scalettar R T 2009 *Phys. Rev. A* **79** 053605
[26] Diener R B, Zhou Q, Zhai H and Ho T L 2007 *Phys. Rev. Lett.* **98** 180404
[27] Kato Y, Zhou Q, Kawashima N and Trivedi N 2008 *Nature Phys.* **4** 617–621
[28] Pollet L, Prokof'ev N V and Svistunov B V 2010 *Phys. Rev. Lett.* **104** 245705
[29] Mahmud K W, Duchon E N, Kato Y, Kawashima N, Scalettar R T and Trivedi N 2011 *Phys. Rev. B* **84** 054302
[30] Carrasquilla J and Rigol M 2012 arXiv:1205.6484
[31] Matsubara T and Matsuda H 1956 *Progress of Theoretical Phys.* **16** 569–582
[32] Sandvik A W and Kurkijärvi J 1991 *Phys. Rev. B* **43** 5950–5961
[33] Sandvik A W 1992 *J. Phys. A: Math. Gen.* **25** 3667–3682
[34] Dorneich A and Troyer M 2001 *Phys. Rev. E* **64** 066701
[35] Loh E, Scalapino D J and Grant P M 1985 *Phys. Rev. B* **31** 4712–4714
[36] Ding H Q and Makiviĉ M S 1990 *Phys. Rev. B* **42** 6827–6830
[37] Ding H Q 1992 *Phys. Rev. B* **45** 230–242
[38] Pollock E L and Ceperley D M 1987 *Phys. Rev. B* **36** 8343–8352
[39] Olsson P 1995 *Phys. Rev. B* **52** 4526–4535
[40] Hasenbusch M 2005 *J. Phys. A: Math. Gen.* **38** 5869–5883

Journal of Physics: Conference Series **414** (2013) 012029 doi:10.1088/1742-6596/414/1/012029

Dynamics of the Mott Insulator to Superfluid quantum phase transition in the truncated Wigner approximation

Marek Tylutki

E-mail: marek.tylutki@uj.edu.pl
Institute of Physics and Center for Complex Systems Research, Jagiellonian University, Reymonta 4, 30-059 Kraków, Poland

Jacek Dziarmaga

Institute of Physics and Center for Complex Systems Research, Jagiellonian University, Reymonta 4, 30-059 Kraków, Poland

Wojciech H. Zurek

Theory Division, Los Alamos National Laboratory, Los Alamos, NM 87545, USA

Abstract. The quantum phase transition from the Mott insulator state to the superfluid in the Bose-Hubbard model is investigated. We research one, two and three dimensional lattices in the truncated Wigner approximation. We compute both kinetic and potential energy and they turn out to have a power law behaviour as a function of the transition rate, with the power equal to 1/3. The same applies to the total energy in a system with a harmonic trap, which is usually present in the experimental set-up. These observations are in agreement with the experiment of [8], where such scalings were also observed and the power of the decay was numerically close to 1/3. The results confirm the Kibble-Zurek (adiabatic-impulse-adiabatic approximation) scenario for this transition.

1. Introduction

Excitations resulting from crossing the gapless quantum critical points are a serious problem for quantum simulations with ultracold atomic gases or ion traps. There, one would like to prepare a simple ground state of a simple initial Hamiltonian and then drive the system adiabatically to an interesting final ground state. This general observation has been recently supported by a more quantitative theory [1, 2], that is by now confirmed by several numerical studies [3]. The theory is a quantum generalization of the classical Kibble-Zurek mechanism (KZM) [4, 5]. The theory predicts that density of excitations or excitation energy decay with a power of quench timescale τ_Q. Experiments dedicated to the quantum theory were made in Refs. [6, 7, 8]. In Ref. [8] ultracold spinless bosonic atoms in a three-dimensional (3D) optical lattice were driven from the Mott insulator phase to the superfluid phase across a quantum phase transition. The transition was non-adiabatic and the excitation energy was reported to scale algebraically with the transition time, and the power was numerically equal to one third. Similarily, adiabaticity of loading atoms into an optical lattice [9] or releasing them from the lattice confinement [10] has been questioned recently. The 1/3-scaling reported in the experiment [8] coincides with the earlier prediction in Ref. [11] for 1D. In [21] it was shown, that the same scalings also hold for 3D.

2. The Bose-Hubbard model

The Bose-Hubbard (BH) model is a lattice model that describes bosonic atoms in a field of an optical lattice [8, 10, 12, 13]. Its Hamiltonian assumes the following form

$$H_{\mathrm{BH}} = -J \sum_{\langle \mathbf{i},\mathbf{j} \rangle} \left(a^\dagger_{\mathbf{i}} a_{\mathbf{j}} + a^\dagger_{\mathbf{j}} a_{\mathbf{i}} \right) + \frac{1}{2n} \sum_{\mathbf{i}} a^\dagger_{\mathbf{i}} a^\dagger_{\mathbf{i}} a_{\mathbf{i}} a_{\mathbf{i}} + \sum_{\mathbf{i}} V_{\mathbf{i}} a^\dagger_{\mathbf{i}} a_{\mathbf{i}} \ , \tag{1}$$

where $a_{\mathbf{i}}$ is the annihilation operator at site $\mathbf{i}$ and $a^\dagger_{\mathbf{i}}$ is the corresponding creation operator. Here we do not specify the dimensionality of a lattice, which is three dimensional in real systems, but can be effectively made two- or one dimensional when some dimensions of the system are very small compared to the other.

The BH Hamiltonian has three terms: (1) the kinetic term, which corresponds to the hopping of atoms between the neighbouring sites of the lattice, (2) the on-site interaction term, which is responsible for on-site repulsion between the atoms and possibly (3) the interaction with external potential, which could be a harmonic trap considered later in the article. Here, the coefficient $\frac{1}{2n}$ in front of the on-site interaction term is a result of our choice of units.

When the kinetic term dominates over the on-site repulsion, the system is in the superfluid regime which is characterized by strong spatial correlations. On the other hand, when the on-site interaction term is much larger than the kinetic part, the system is in the Mott Insulator state, where there is a definite and constant number of atoms n at each site,

$$|\mathrm{MI}\rangle = |n, n, n, \cdots\rangle \ . \tag{2}$$

Clearly, between these two phases there is a boundary, crossing which causes the system to undergo a phase transition. The phase diagram of (1) is depicted in Fig. 1, where Mott Insulator phase forms the famous lobes, each characterized by a distinct value of atom density n. Those lobes are surrounded by a sea of the superfluid phase.

The relative strength of the two terms is controlled by the coefficient J. To cross the phase boundary we vary this quantity in time. In this article, we assume that initially the system is in the Mott state and we drive it into the superfluid regime by increasing $J(t)$. This driven transition will be called here a *quench*, and we choose it to be linear in time:

$$J = \frac{t}{\tau_Q} \ , \tag{3}$$

where τ_Q is the inverse of the quench rate.

3. The truncated Wigner method

To investigate the model (1), we adopt an approach known as the Truncated Wigner Method, where the full quantum dynamics of the original Hamiltonian is approximated by an evolution of a statistical ensemble of complex lattice fields ϕ_i. This approximation is valid in the regime of large atom densities, $n \gg 1$, which we now assume. In the Truncated Wigner Method, we make an identification $a_i \approx \sqrt{n}\,\phi_i$, $a^\dagger_i \approx \sqrt{n}\,\phi^\star_i$. With this identification, the equation of motion for a_i becomes the nonlinear equation for ϕ_i

$$i\,\partial_t \phi_{\mathbf{i}} = -J\nabla^2 \phi_{\mathbf{i}} + \left(|\phi_{\mathbf{i}}|^2 - 1 \right) \phi_{\mathbf{i}} \ , \tag{4}$$

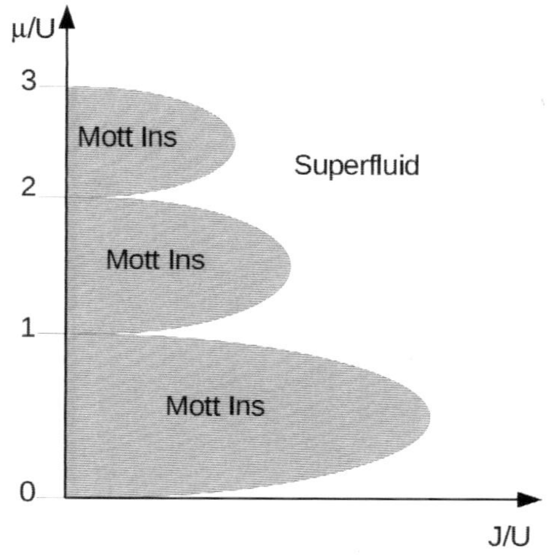

Figure 1. The schematic phase diagram of the system defined by (1). The lobes of the Mott insulator for $J \ll 1$ characterize with a constant number of atoms per site. The surrounding superfluid phase has a vanishing energy gap and non-zero correlations in phase between sites.

which is known as the Gross-Pitaevskii equation (GPE). The quantum evolution of the original system (1) is represented by the whole set of trajectories of ϕ_i and the quantum expectation values become statistical averages of relevant observables over the ensemble. The initial condition for the evolution becomes the distribution for ϕ_i. The initial state (2) translates to

$$\phi_i(0) = e^{i\theta_i} , \tag{5}$$

where θ's are uniformly distributed over $(-\pi, \pi]$. This reflects the uncertainty between the atom number and phase: n is well defined and θ is completely random.

The kinetic and potential energies take the form:

$$E_{\mathrm{kin}} = J \sum_{\mathbf{i}} \overline{\nabla\phi_{\mathbf{i}}^* \nabla\phi_{\mathbf{i}}} , \tag{6}$$

$$E_{\mathrm{pot}} = \frac{1}{2} \sum_{\mathbf{i}} \overline{(|\phi_{\mathbf{i}}|^2 - 1)^2} . \tag{7}$$

The behaviour of energy with the rate of the transition will be at the center of our interest.

4. The impulse-adiabatic scenario

The next ingredient that we add to our analysis is the impulse-adiabatic scenario [5]. It simplifies the picture of the evolution across the critical point. In the vicinity of that point we can treat the evolution as impulse, which means that the transition rate is much faster than the pace of the change of the system. Far from the critical point, on the contrary, the relaxation rate of the system exceeds the driving pace and the evolution is adiabatic. The idea of the impulse-adiabatic scenario is to say, that there is a certain value of the parameter, in our case $J = \hat{J}$ where the two rates become comparable, and which separates the impulse stage of the evolution from the adiabatic stage. In other words, below $\hat{J}$ the system undergoes the impulse evolution, and above - adiabatic. This idea is summarized in the Fig. 2.

In our units, the critical value of J is $J_{\mathrm{cr}} \approx n^{-2}$ and tends to zero for large densities, $J \approx 0$.

Journal of Physics: Conference Series **414** (2013) 012029 doi:10.1088/1742-6596/414/1/012029

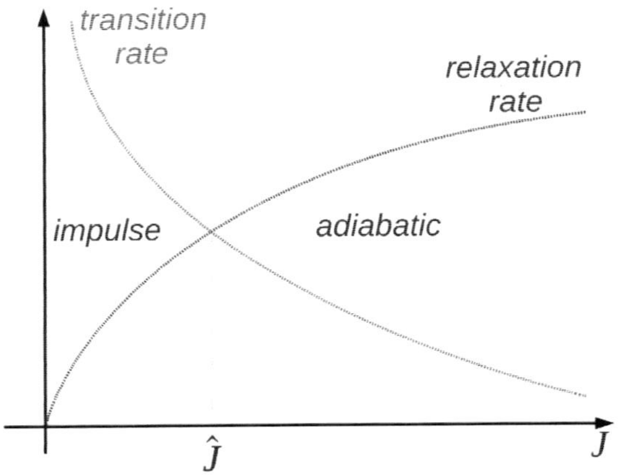

Figure 2. The schematic depiction of the impulse-adiabatic approximation to the dynamics of the considered model. As long as the relative transition rate, $\dot{J}/J$, is larger than the relaxation rate, the dynamics is said to be impulse, i.e. the state of the system is frozen in spite of the changing Hamiltonian. When the relaxation rate begins to dominate, the state has more and more time to follow the Hamiltonian, and we assume that the evolution is adiabatic.

5. The Josephson regime

First, we restrict ourselves to the regime, where $J \ll 1$. In this regime the density fluctuations are small, $|\phi_{\mathbf{i}}|^2 \approx 1$; the field amplitude does not deviate much from the Mott state's amplitude. Hence, it is convenient to parametrize

$$\phi_{\mathbf{i}} = (1 + f_{\mathbf{i}})\ e^{i\theta_{\mathbf{i}}}\ , \tag{8}$$

with real amplitude correction $f_{\mathbf{i}}$ and phase $\theta_{\mathbf{i}}$. This gives us equations of motion for $f_{\mathbf{i}}$ and $\theta_{\mathbf{i}}$, where the initial conditions are $f_{\mathbf{i}} = 0$ and random $\theta_{\mathbf{i}}$. After elimination of $f_{\mathbf{i}} \ll 1$ in Eq. (4) we get the Josephson equations

$$\frac{d^2}{dt^2}\theta_{\mathbf{i}} = 2J \sum_{\mathbf{j},\,\mathrm{n.n.}\,\mathbf{i}} \sin\left(\theta_{\mathbf{j}} - \theta_{\mathbf{i}}\right) \tag{9}$$

with random initial $\theta_{\mathbf{i}}(0)$ and $\frac{d\theta_{\mathbf{i}}}{dt}(0) = 0$. The summation in Eq. (9) extends over the nearest neighbours of $\mathbf{i}$.

The energies defined by Eqs. (6, 7) in the approximation (8) yield

$$E_{\mathrm{kin}} \approx J \sum_{\mathbf{i}} \overline{\nabla\theta_{\mathbf{i}} \cdot \nabla\theta_{\mathbf{i}}}\ ,$$
$$E_{\mathrm{pot}} \approx 2 \sum_{\mathbf{i}} \overline{f_{\mathbf{i}}^2}\ . \tag{10}$$

6. Thermalization

Since the parameter J could be eliminated from Eq. (9) by introducing a rescaled time variable $u = J^{1/2}t$, the equations have a characteristic time-scale

$$\tau \simeq J^{-1/2}\ . \tag{11}$$

This timescale is a relaxation time towards thermal equilibrium in the Josephson regime. In one dimension the system has finite correlation length and quickly reaches its equilibrium. The thermalization for 1D is depicted in Fig. 3. For two and three dimensions the situation is

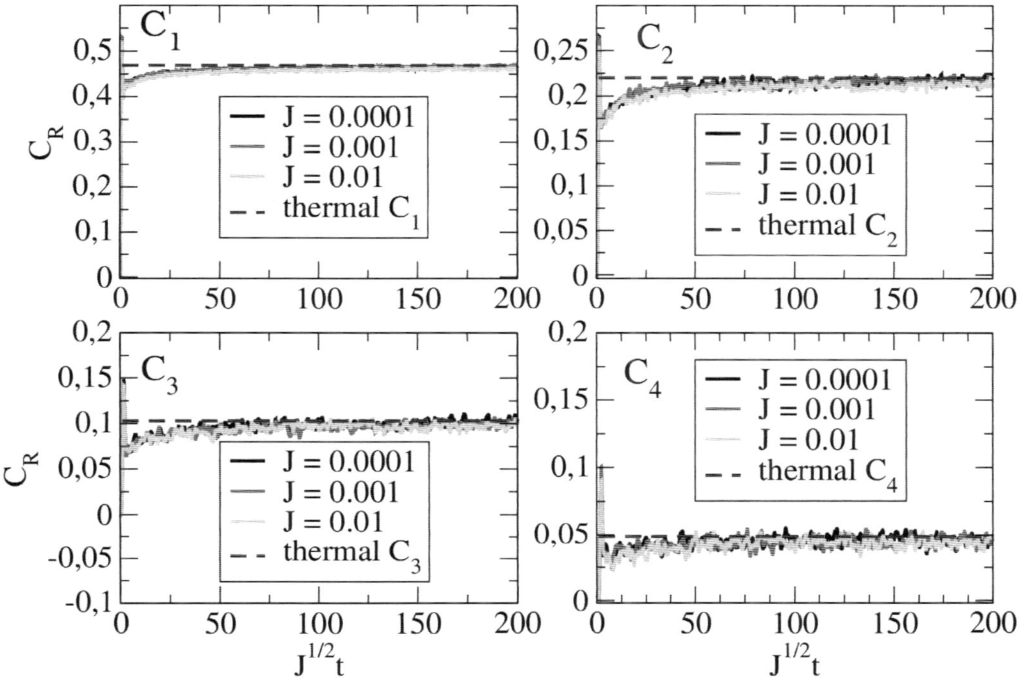

Figure 3. Thermalization in 1D after a sudden quench, where J abruptly jumps to its final value at $t = 0$. The short-range correlations quickly reach their equilibrium, whose numerical value was predicted in detail in [21]. In each panel, the plots for various J's against time $J^{1/2}t$ collapse to a single curve, proving, that the thermalization time scales like $\tau \sim J^{-1/2}$

different. In 2D the correlations of the system display quasi-long-range order and in 3D they are of a long range order. Therefore, the two- or three dimensional system cannot equilibrate in finite time. However, short range correlation functions can reach equilibrium in finite time, which is shown in Fig 4. The correlations will be discussed in more detail in section 10.

Since the kinetic and potential energies are quadratic in their degrees of freedom, Eq. (10), we can apply the equipartition principle and write

$$E_{\text{kin}} = \frac{1}{2}TV \,,$$
$$E_{\text{pot}} = \frac{1}{2}TV \,, \tag{12}$$

where we introduced temperature T which is a function of J in the adiabatic quench ($\tau_Q \gg 1$) and $V = L^D$ is the volume of the system, i.e. the number of sites. Eqs. (12) are valid for the Josephson regime, $J \ll 1$. The total energy is, obviously, $E = E_{\text{kin}} + E_{\text{pot}} = TV$

For adiabatic processes, where the system remains in equilibrium for every instant of time, both $T(t)$ and $J(t)$ are functions of time t. Hence we can write

$$\frac{d}{dt}E = \frac{dT}{dt}V \,, \tag{13}$$

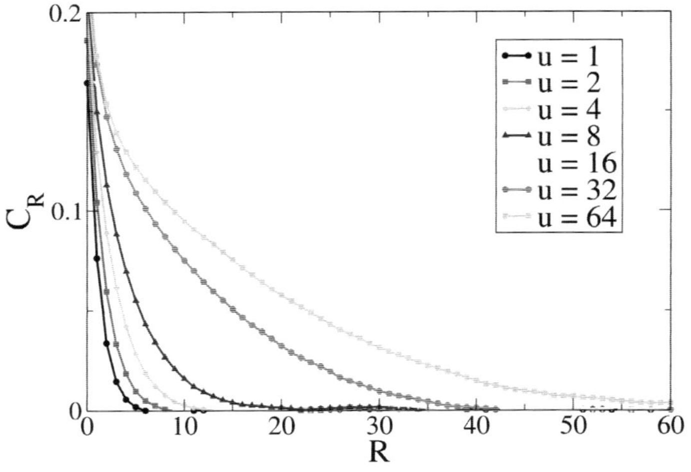

Figure 4. Correlations in 3D. The correlation function $C_R = \overline{\phi_s^* \phi_{s+R}}$ for several values of the rescaled time $u = J^{1/2}t$. The initial state was a result of an evolution from $J = 0$ to $J = 0.01$ with $\tau_Q = 512$. Than it was allowed to thermalize. The expected long range order is not reached in finite time. However, the short range correlations do thermalize quickly, and so do the local observables, such as the energy. Thus, we can apply the equipartition principle.

where we assume the energy to be a function of a time dependent temperature and

$$\frac{d}{dt}E = \frac{dJ}{dt} E_{\text{kin}} J^{-1} = \frac{dJ}{dt}\frac{TV}{2J} \ . \tag{14}$$

Equating (13) with (14) and solving for $T(J)$ we get the equation characterizing the thermodynamic process induced by driving J:

$$T = A\sqrt{J} \ , \tag{15}$$

which we will call the adiabate equation. Again, it is valid for an adiabatic processes in the Josephson regime.

7. Excitation energy in the Josephson regime

Having the adiabate equation (15), we may now attempt to predict the scalings of energies (10) using the impulse-adiabatic approximation introduced in section 4. The transition rate is

$$\frac{\dot{J}}{J} = \frac{1}{t} \tag{16}$$

and the relaxation rate of the system

$$\tau^{-1} = \sqrt{J} = \left(\frac{t}{\tau_Q}\right)^{1/2} \ . \tag{17}$$

The cross-over takes place at the instant of time $\hat{t}$ (corresponding to the parameter value $\hat{J}$) where these two rates become equal. Thus,

$$\hat{t}^{-3/2} \simeq \tau_Q^{-1/2} \tag{18}$$

or

$$\hat{J} \simeq \tau_Q^{-2/3} \; . \tag{19}$$

The idea behind the impulse-adiabatic scenario is that the initial state remains frozen throughout the impulse stage of the evolution. Thus, the adiabatic process begins at $\hat{J}$ with random phases $\theta_{\mathbf{i}}$,

$$E_{\rm kin}|_{\hat{J}} \approx \hat{J}\overline{\sum_{\mathbf{i}} \nabla\theta_{\mathbf{i}} \cdot \nabla\theta_{\mathbf{i}}|_{\theta-{\rm random}}} = \frac{2\pi^2}{3}\hat{J}VD \; . \tag{20}$$

The above equation gives the initial temperature for the adiabatic evolution

$$T|_{\hat{t}} \simeq \frac{4\pi^2}{3}\hat{J}D \; . \tag{21}$$

Comparing this formula with the adiabate equation allows us to determine the coefficient $A = \frac{4\pi^2}{3}\sqrt{\hat{J}}D$. The adiabate equation becomes

$$T \simeq \frac{4\pi^2}{3}D\sqrt{\hat{J}\,J} \simeq \sqrt{J}\tau_Q^{-1/3} \; , \tag{22}$$

where in the last step we used the scaling for $\hat{J}$ (Eq. (19)). As a consequence, we receive the algebraic scaling for kinetic and potential energy

$$E_{\rm kin} \simeq J^{1/2}\tau_Q^{-1/3}V \; , \quad E_{\rm pot} \simeq J^{1/2}\tau_Q^{-1/3}V \tag{23}$$

in the Josephson regime.

The predicted scaling agrees perfectly with the numerical simulation of the full Gross-Pitaevskii equation, (4), where we also get algebraic scalings with the exponent numerically close to 1/3. The calculated $E_{\rm kin}$ and $E_{\rm pot}$ are presented in Fig. 5 and the fitted exponents are collected in Table 1.

J	$\langle E^{\rm 1D}_{\rm kin}\rangle$	$\langle E^{\rm 2D}_{\rm kin}\rangle$	$\langle E^{\rm 3D}_{\rm kin}\rangle$	$\langle E^{\rm 1D}_{\rm pot}\rangle$	$\langle E^{\rm 2D}_{\rm pot}\rangle$	$\langle E^{\rm 3D}_{\rm pot}\rangle$
0.1	0.33	0.33	0.33	0.33	0.30	0.30
1.0	0.33	0.33	0.33	0.33	0.31	0.31
10.0	0.33	0.33	0.33	0.32	0.31	0.31

Table 1. The best fits to α in $\langle E_{\rm kin/pot}\rangle \sim \tau_Q^{-\alpha}$ are consistent with $\alpha = 1/3$. Lattice sizes as in Fig. 5.

8. Excitation energy in the Rabi regime

The data presented in Fig. 5 suggest, that the $\tau_Q^{-1/3}$ scaling extends also to the Rabi regime, $J \gg 1$. This can also be explained in terms of the impulse-adiabatic scenario. Since, for $J \gg 1$, the kinetic energy dominates, $\langle E\rangle \approx \langle E_{\rm kin}\rangle$, and (14) becomes

$$\frac{d}{dt}\langle E_{\rm kin}\rangle = \frac{dJ}{dt}\frac{\langle E_{\rm kin}\rangle}{J} \; . \tag{24}$$

This has a simple solution $\langle E_{\rm kin}\rangle = BJ$, where B is a proportionality constant. The crossover from the Josephson to the Rabi regime is at $J \approx 1$ and thus we can determine the constant B:

$$\langle E_{\rm kin}\rangle\Big|_{J\approx 1} = J^{1/2}\tau_Q^{-1/3}V\Big|_{J\approx 1} = \tau_Q^{-1/3}V = B \; . \tag{25}$$

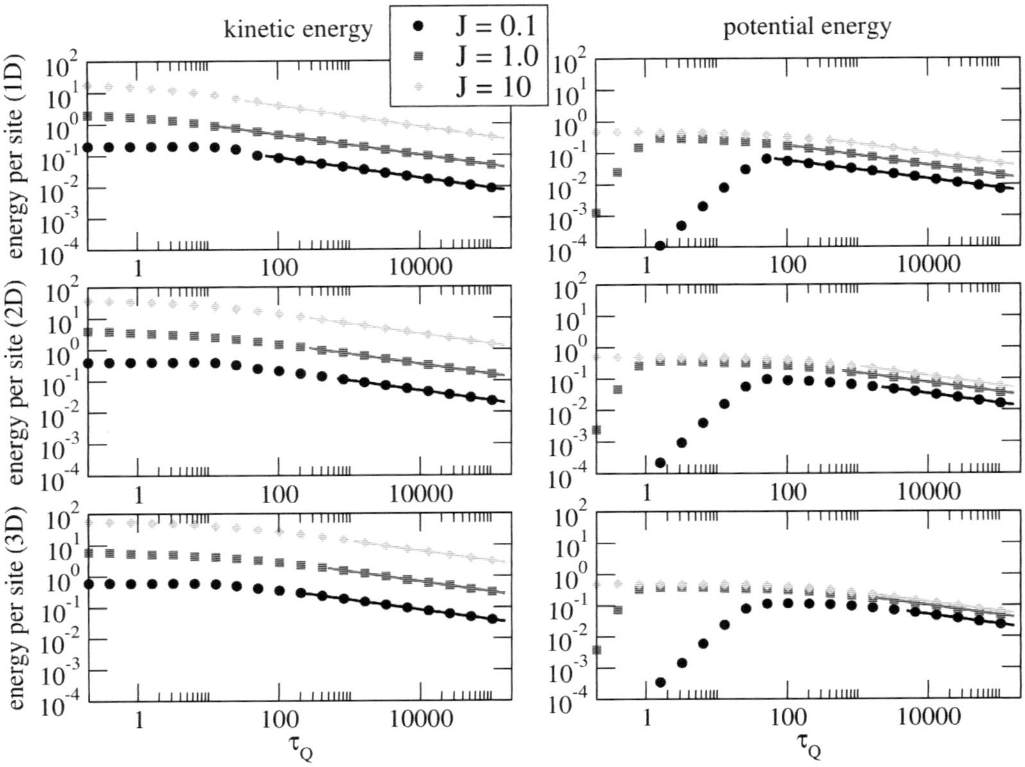

Figure 5. The numerical evolution of the Eq. (4) leads to the kinetic and potential energies decay algebraically with τ_Q. The exponent is numerically equal to $-\frac{1}{3}$, see Tab. 1, in agreement with the prediction (23). The lattice sizes were $L = 4096$, 256, 128 for 1, 2 and 3 dimensions respectively.

Hence, in the Rabi regime, $\langle E \rangle \approx \langle E_{\text{kin}} \rangle \approx J\tau_Q^{-1/3}V$. This is true, as long as the crossover takes place in the Josephson regime, which according to $\hat{J} \sim \tau_Q^{-2/3}$, happens for quenches slow enough.

For fast quenches, the impulse-adiabatic crossover takes place in the Rabi regime. The thermalization time is set by the strength of the nonlinearity in (4) ($\tau \sim 1$) and, consequently,

$$\left. \frac{1}{J}\frac{dJ}{dt} \right|_{\hat{J}} = \frac{1}{\hat{t}} \approx \tau^{-1} \sim 1 \ . \tag{26}$$

Therefore, $\hat{t} \approx 1$. Since at J phases remain random, $\langle E_{\text{kin}} \rangle \big|_{\hat{J}} \simeq \hat{J}V$. Finally,

$$\langle E_{\text{kin}} \rangle = \frac{J}{\hat{J}} \left. \langle E_{\text{kin}} \rangle \right|_{\hat{J}} = \frac{2\pi^2}{3} DJV \ . \tag{27}$$

For fast quenches, energy does not depend on τ_Q, consistently with the data from Fig. 5.

9. Excitation energy for the system in a harmonic trap

In order to make our analysis even more realistic and corresponding to the experimental conditions, we place our system in an external harmonic trap, i.e. quadratic potential. This amounts to setting the $V_{\mathbf{i}}$ in the Bose-Hubbard Hamiltonian (1) to

$$V_{\mathbf{i}} = \frac{1}{2}\omega^2 \mathbf{i}^2 , \tag{28}$$

where ω is the frequency of the trap, which sets its width, and we assume $\mathbf{i} = 0$ at the center of the lattice. Now, the Gross-Pitaevskii equation becomes

$$i\,\partial_t \phi_{\mathbf{i}} = -J\nabla^2 \phi_{\mathbf{i}} + \left(|\phi_{\mathbf{i}}|^2 - 1\right)\phi_{\mathbf{i}} + \frac{1}{2}\omega^2 \mathbf{i}^2 \phi_{\mathbf{i}} , \tag{29}$$

where the initial conditions are again random phases $\theta_{\mathbf{i}}$ and the density is distributed according to

$$|\phi_{\mathbf{i}}(0)|^2 = \frac{\omega^2}{2}(R_{\mathrm{TF}}^2 - \mathbf{i}^2) , \tag{30}$$

for $\mathbf{i}^2 < R_{\mathrm{TF}}^2$ and $|\phi_{\mathbf{i}}(0)|^2 = 0$ for $\mathbf{i}^2 \geq R_{\mathrm{TF}}^2$. This is an equilibrium solution to (29) for $J = 0$. In the above distribution R_{TF} is a radius (Thomas-Fermi radius) of a sphere inside which the distribution is non-zero. We choose $\omega^2 = \frac{2}{R_{\mathrm{TF}}^2}$, so that in the center of the trap initial field density is equal to 1, $|\phi(0)|^2 = 1$. This makes comparison with the uniform case easier.

In order to examine scaling of excitation energy in case of the presence of the harmonic trap, we have to subtract the ground state energy from the total energy calculated in the numerical simulation. The ground state energy can be computed as a solution to

$$\frac{\delta E[\phi]}{\delta \phi_{\mathbf{i}}} = 0 , \tag{31}$$

which minimizes the energy functional

$$E[\phi_{\mathbf{i}}] = \sum_{\mathbf{i}} \left\{ J\nabla\phi_{\mathbf{i}}^* \nabla\phi_{\mathbf{i}} + \frac{1}{2}\phi_{\mathbf{i}}^*\phi_{\mathbf{i}}^*\phi_{\mathbf{i}}\phi_{\mathbf{i}} + \frac{\omega^2}{2}\mathbf{i}^2\phi_{\mathbf{i}}^*\phi_{\mathbf{i}} \right\} , \tag{32}$$

subject to the constraint

$$\sum_{\mathbf{i}} |\phi_{\mathbf{i}}|^2 = \text{const.} , \tag{33}$$

i.e. the norm should be kept constant throughout the evolution. The minimalization of (32) can be done by evolving the following equation,

$$\frac{\partial \phi_{\mathbf{i}}}{\partial t} = -\frac{\delta E[\phi]}{\delta \phi_{\mathbf{i}}} , \tag{34}$$

until $\phi_{\mathbf{i}}$ reaches the steady state, which corresponds to the solution of (31) and therefore is the ground state.

The evolving cloud in the trap is initially centered in the form of a Thomas-Fermi distribution, but subsequently expands and asymptotically tends to the Gaussian distribution for large J. This is depicted in the Fig 7.

The numerical simulation of (29) indicates, that as in the uniform case, the excitation energy decays like $\tau_Q^{-1/3}$ with the quench rate. This data is shown in the Fig. 8. The scaling obtained for a uniform system survives also in case of a trap, because the kinetic energy accumulated in random phase is much larger than the energy related to the localization of the cloud in the center of the trap.

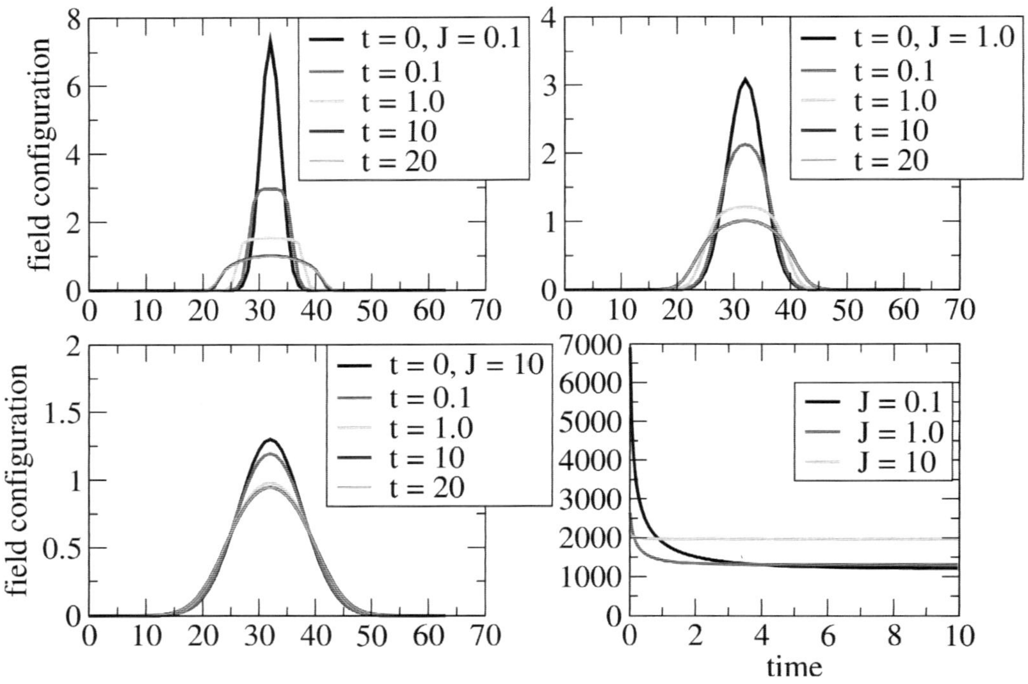

Figure 6. Relaxation to a ground state of the energy functional $E[\phi_{\mathbf{i}}] = \sum_{\mathbf{i}} \left\{ J \nabla \phi_{\mathbf{i}}^* \nabla \phi_{\mathbf{i}} + \frac{1}{2} \phi_{\mathbf{i}}^* \phi_{\mathbf{i}}^* \phi_{\mathbf{i}} \phi_{\mathbf{i}} + \frac{\omega^2}{2} \mathbf{i}^2 \phi_{\mathbf{i}}^* \phi_{\mathbf{i}} \right\}$ according to the evolution of the equation $\frac{\partial \phi_{\mathbf{i}}}{\partial t} = -\frac{\delta E[\phi]}{\delta \phi_{\mathbf{i}}}$. Three panels represent subsequent field configuration as it converges to the energy minimum and the bottom right panel shows the corresponding convergence of the ground state energy. The initial configuration was an educated guess: we assumed it is proportional to the ground state of the equation in the $J \gg 1$, ie. the harmonic oscillator equation. The larger J the final ground state has more resemblance to the Gauss function as the harmonic oscillator approximation becomes more accurate.

10. Correlations and Vortices

In one dimension, the correlation length is finite, i.e. the correlation function is exponential, see derivation in [21],

$$C_R = \frac{1}{n} \langle a_{\mathbf{i}}^{\dagger} a_{\mathbf{i}} \rangle = e^{-\frac{R}{\xi}} , \tag{35}$$

with ξ depending on J and τ_Q through

$$\xi = \frac{4J}{T} \simeq J^{1/2} \tau_Q^{1/3} . \tag{36}$$

With growing τ_Q the correlation length decays like $\tau_Q^{-1/3}$, in agreement with the numerical data, presented in Fig. 9. This is not the case, however, in 2 and 3D, where there is a (quasi-)long-range order. Nevertheless, due to the finite rate, with which the system is driven, the correlations also spread with a finite rate and their range is limited, see Fig. 10.

The finite rate of the correlation build-up results in the domain formation and appearance of topological vortices. Such vortices, in case of a 2D system, are presented in Fig. 11. For fast quenches (left panel), the there is little time for the correlations to spread, so the resulting pattern

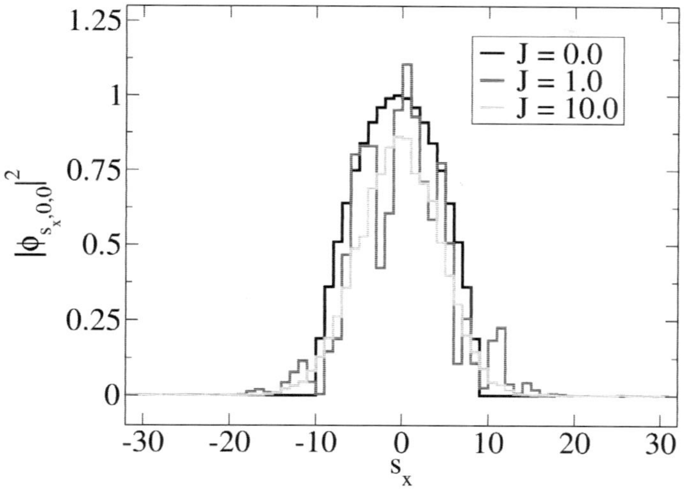

Figure 7. The evolution of the cloud of atoms in a harmonic trap. The figure shows a cross-section through the center of a 3D lattice. The initial Thomas-Fermi profile at $J = 0$ expands and for large J (here $J = 10$) tends to a Gaussian. The simulation for $\tau_Q = 102.4$.

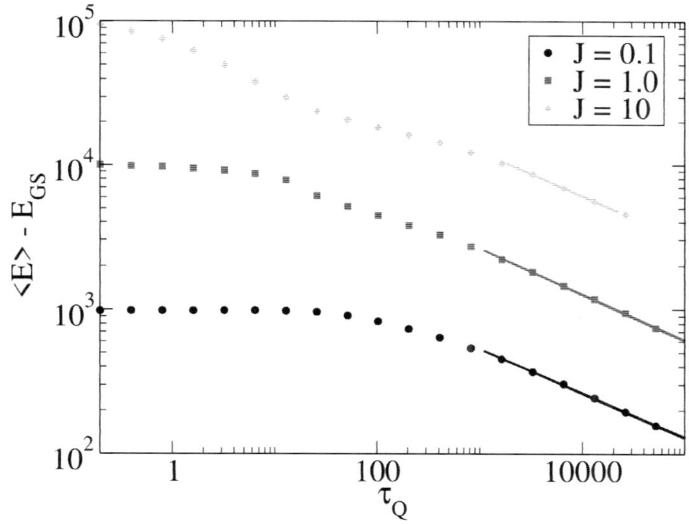

Figure 8. The excitation energy $\langle E \rangle - E_{\mathrm{GS}}$ for the atoms in a trap. The energy scales algebraically with τ_Q and the powers are 0.32 for $J = 1.0$ and 0.31 for $J = 0.1$ and $J = 10.0$. The initial radius of the Thomas-Fermi profile was $R = 10$.

contains small domains and many vortices. When the system is allowed to cool down during a slow quench (right panel), the domains are larger and there are fewer, but more pronounced vortices.

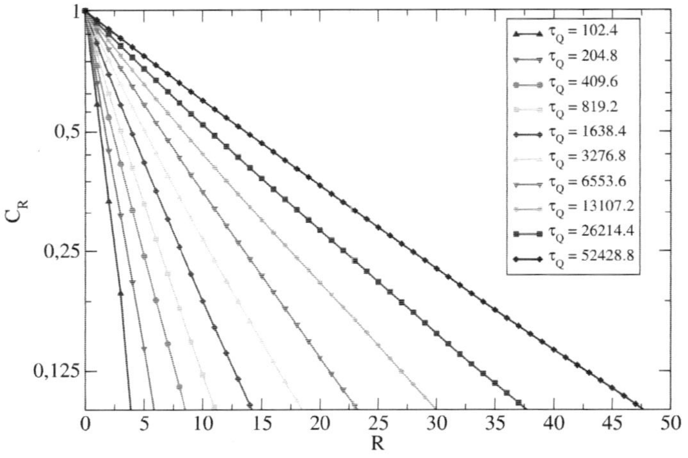

Figure 9. The correlation functions in 1D at $J = 0.1$ are exponential, with the correlation length numerically scaling as $\xi \sim \tau_Q^{-1/3}$, consistently with (36).

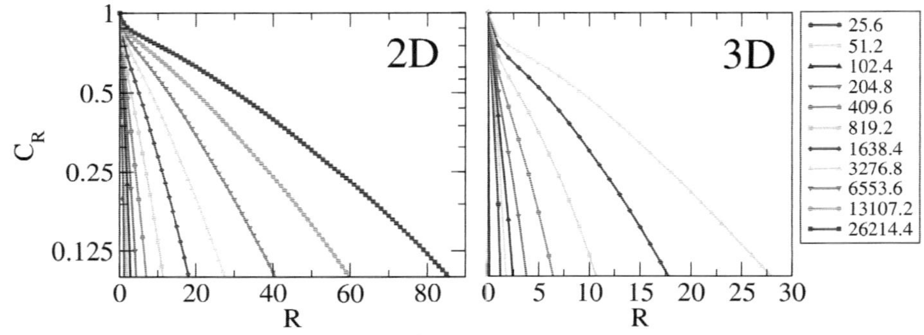

Figure 10. Correlations at $J = 0.1$ in 2 and 3D respectively. The correlations have a finite range due to a limited pace of their growth. The local degrees of freedom have equilibrated, but at large scales the (quasi-)long-range order cannot be reached in finite time.

11. Conclusion

The physics beyond the linear quench from the Mott insulator phase to the superfluid turns out to be well approximated by the impulse-adiabatic scenario, where the phases remain frozen at random values, as for the initial state, but at some point the evolution becomes adiabatic and the system is allowed to thermalize. In 2 and 3 dimensions, however, the finite rate of the transition does prevent the (quasi-)long-range order to fully develop, and the resulting correlations have finite range. This is spectacularly manifested by the formation of vortices in phase. The simple picture of the impulse-adiabatic approximation allows for derivation of the algebraic decay of energies with the quench time τ_Q, i.e. $E_{\mathrm{kin/pot}} \sim \tau_Q^{-1/3}$. This result is reproduced also when the system is subject to the external harmonic potential. This is in agreement with the experimental data obtained in [8].

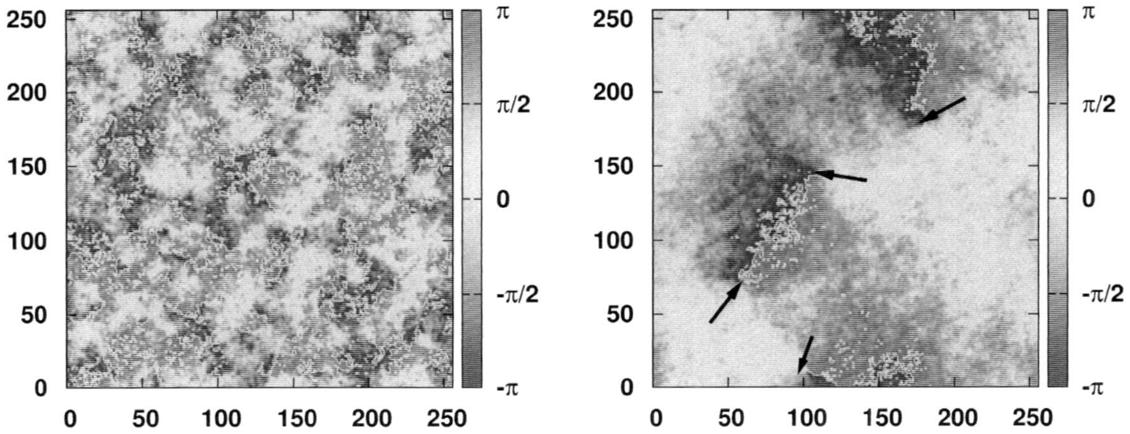

Figure 11. The phase θ_s for a 2D 256×256 system. The top panel shows the phase at $J = 0.1$ after a relatively fast quench with $\tau_Q = 1638.4$. In this case the domains that formed are small, as is the resulting correlation length $\xi \sim \tau_Q^{1/3}$. The bottom panel shows the situation after a quench with $\tau_Q = 52428.8$. The correlation length is much larger, the domains more prominent and the resulting vortices clearly visible (pointed with arrows).

Acknowledgments

This work was supported in part by the NCN grant 2011/01/B/ST3/00512 (JD and MT) and the PL-Grid Infrastructure (MT).

References

[1] B. Damski, Phys. Rev. Lett. **95**, 035701 (2005); W.H. Zurek, U. Dorner, and P. Zoller, Phys. Rev. Lett. **95**, 105701 (2005); A. Polkovnikov, Phys. Rev. B **72**, R161201 (2005; J. Dziarmaga, Phys. Rev. Lett. **95**, 245701 (2005);

[2] J. Dziarmaga, Adv. in Phys. **59**, 1063 (2010); A. Polkovnikov, K. Sengupta, A. Silva, and M. Vengalattore, Rev. Mod. Phys. **83**, 863883 (2011).

[3] L. Cincio *et al.*, Phys. Rev. A **75**, 052321 (2007); T. Caneva, R. Fazio, G. E. Santoro, Phys. Rev. B **76**, 144427 (2007); B. Damski, H. T. Quan, W. H. Zurek, Phys. Rev. A **83**, 062104 (2011); J. Sabbatini, W. H. Zurek, M. J. Davis, Phys. Rev. Lett. 107, 230402 (2011);

[4] T.W.B. Kibble, J. Phys. A **9**, 1387 (1976); Phys. Rep. **67**, 183 (1980); Physics Today, 60, 47 (2007).

[5] W.H. Zurek, Nature **317**, 505 (1985); Acta Phys. Polon. B **24**, 1301 (1993); Phys. Rep. **276**, 177 (1996).

[6] L. E. Sadler, J. M. Higbie, S. R. Leslie, M. Vengalattore, and D. M. Stamper-Kurn, Nature (London) 443, 312 (2006).

[7] D. R. Scherer, C. N. Weiler, T. W. Neely, and B. P. Anderson, Phys. Rev. Lett. **98**, 110402 (2007); R. Carretero-Gonzalez, B. P. Anderson, P. G. Kevrekidis, D. J. Frantzeskakis, and C. N. Weiler, Phys. Rev. A **77**, 033625 (2008).

[8] D. Chen, M. White, C. Borries, and B. DeMarco, Phys. Rev. Lett. **106**, 235304 (2011)

[9] J. Zakrzewski, Phys. Rev. A **71**, 043601 (2005); J. Zakrzewski and D. Delande, Phys. Rev. A **80**, 013602 (2009).

[10] S. S. Natu, D. C. McKay, B. DeMarco, and E. J. Mueller, arXiv:1201.3154.

[11] J. Dziarmaga, A. Smerzi, W. H. Zurek, and A. Bishop, Phys. Rev. Lett. **88**, 167001 (2002); F. Cucchietti, B. Damski, J. Dziarmaga, and W. H. Zurek, Phys. Rev. A **75**, 023603 (2007); J. Dziarmaga, J. Meisner, and W. H. Zurek, Phys. Rev. Lett. **101**, 115701 (2008); J. Dziarmaga, M. Tylutki, and W. H. Zurek, Phys. Rev. B **84**, 094528 (2011) .

[12] C. Orzel, A. K. Tuchman, M. L. Fenselau, M. Yasuda, and M. A. Kasevich, Science **291**, 2386 (2001); A. K. Tuchman, C. Orzel, A. Polkovnikov, and M. A. Kasevich, Phys. Rev. A **74**, 051601 (2006); W. Li, A. K. Tuchman, H.-C. Chien, and M. A. Kasevich , Phys. Rev. Lett. **98**, 040402 (2007).

[13] M. Greiner *et al.*, Nature **415**, 39 (2002); Nature **419**, 51 (2002).

[14] L. Amico and V. Penna, Phys. Rev. Lett. **80**, 2189 (1998); P. Buonsante and V. Penna, J. Phys. A **41**, 175301 (2008).

[15] K. Goral, M. Gajda, and K. Rzazewski, Opt. Express **8**, 92 (2001); M. J. Davis and S. A. Morgan, Phys. Rev. A **68**, 053615 (2003); A. Sinatra, C. Lobo, and Y. Castin, Phys. Rev. Lett. **87**, 210404 (2001); J. Phys. B **35**, 3599 (2002); P. B. Blakie, A. S. Bradley, M. J. Davis, R. J. Ballagh, and C. W. Gardiner, Adv. Phys. **57**, 363 (2008); A. P. Itin and P. Törmä, Phys. Rev. A 79, 055602 (2009); A. Polkovnikov, Phys. Rev. A **68**, 053604 (2003); Ann. Phys. **325**, 1790 (2010).

[16] P. Navez and R. Schützhold, Phys. Rev. A **82**, 063603 (2010); C. Trefzger and K. Sengupta, Phys. Rev. Lett. **106**, 095702 (2011); A. Dutta, C. Trefzger, and K. Sengupta, arXiv:1111.5085; J.-S. Bernier, D. Poletti, P. Barmettler, G. Roux, and C. Kollath, arXiv:1111.4214; M. P. Kennett and D. Dalidovich, Phys. Rev. A **84**, 033620 (2011); M. Lacki, D. Delande, and J. Zakrzewski, arXiv:1106.4906; B. Sciolla and G. Biroli, Phys. Rev. Lett. **105**, 220401 (2010); C.-L. Hung, X. Zhang, N. Gemelke, and C. Chin, Phys. Rev. Lett. **104**, 160403 (2010); T. Venumadhav, M. Haque, and R. Moessner, Phys. Rev. B **81**, 054305 (2010); A. Collin, J.-P. Martikainen, and J. Larson, Phys. Rev. A **81**, 013603 (2010); H. M. Cataldo and D. M. Jezek, Phys. Rev. A **84**, 013602 (2011); A. Lazarides and M. Haque, arXiv:1112.3657.

[17] A. Trombettoni, A. Smerzi, and P. Sodano, New J. Phys. 7, 57 (2005).

[18] M. Cerruti-Sola, C. Clementi, and M. Pettini, Phys. Rev. E **61**, 5171 (2000).

[19] F. E. Zimmer, M. Haque, arXiv:1012.4492.

[20] M. Cramer, C. M. Dawson, J. Eisert, and T. J. Osborne, Phys. Rev. Lett. **100**, 030602 (2008); S. Trotzky, Yu-Ao Chen, A. Flesch, I. P. McCulloch, U. Schollwck, J. Eisert, I. Bloch, Nature Physics **8**, 325 (2012)

[21] J. Dziarmaga, M. Tylutki, and W. H. Zurek, Phys. Rev. B **86**, 144521 (2012).

Mixing dipolar condensates: a new opportunity for enhancing superfluid pairing in a spin-polarized Fermi gas

Ben Kain[1] **and Hong Y. Ling**[2]

[1] Department of Physics, College of the Holy Cross, Worcester, MA 01610, USA
[2] Department of Physics and Astronomy, Rowan University, Glassboro, NJ 08028, USA

E-mail: bkain@holycross.edu, ling@rowan.edu

Abstract. The recent experimental progress in realizing Bose-Fermi mixtures and dipolar quantum gases has created the exciting possibility of creating dipolar Bose-Fermi mixtures with intriguing and unique properties. In a dipolar condensate, the dipole-dipole interaction represents a control knob inaccessible to nondipolar Bosons. Thus, mixing dipolar bosons with fermions may open up new possibilities for creating superfluids with unconventional pairings. We consider a mixture of a spin-polarized Fermi gas and a dipolar Bose-Einstein condensate in a three-dimensional setting in which s-wave scattering between fermions and the quasiparticles of the dipolar condensate can result in an effective attractive Fermi-Fermi interaction anisotropic in nature and tunable by the dipolar interaction. We develop a procedure which allows us to determine the parameters such as densities and scattering lengths that are required to achieve optimal critical temperature before the system starts to phase separate. We perform a systematic investigation, finding that a superfluid with a critical temperature, that is orders of magnitude higher than achievable in nondipolar mixtures, can be realized in dipolar mixtures at experimentally accessible densities and scattering lengths before the system phase separates.

In this proceeding paper, we report our recent work concerning the p-wave (or triplet, to be precise) Bardeen-Cooper-Schrieffer (BCS) superfluid pairing in a spin-polarized Fermi gas mixed with a dipolar condensate. To begin with, we note that experimentally accessible samples of quantum gases or liquids used to be limited only to a few isotopes such as ^{3}He and ^{4}He because most substances in nature solidify before the temperature can reach the regime where the macroscopic quantum nature of gases or liquids can be manifested. The rapid technological advancement in cooling and trapping of neutral atoms has completely turned the situation around. The experimentally realized Bose-Fermi mixtures now include ^{7}Li - ^{6}Li [1, 2], ^{23}Na - ^{6}Li [3], ^{87}Rb - ^{40}K [4, 5, 6, 7], ^{6}Li -^{87}Rb [8], and ^{23}Na - ^{40}K [9]. The possible quantum gas mixtures under this new technology seem endless in view of the rich existence of atomic elements and their isotopes in nature.

More recently even mixtures containing *two*-component Fermi gases were realized in cold-atom experiments [10, 11]. Another development that motivates our interest in cold-atom mixtures involving dipolar condensates has been the recent upsurge of efforts aimed at expanding an ever-growing list of the experimentally realized dipolar quantum gases. Included are the heteronuclear $^{40}K^{87}Rb$ molecules with electric dipoles (led by Deborah Jin and Jun Ye at JILA [12, 13, 14]) and spinor ^{87}Rb condensates with magnetic dipoles (led by Dan Stamper-Kurn at Berkeley [15]). This list has now expanded from alkali elements to the transition element, ^{52}Cr, with magnetic dipoles [16]. The latest members that joined this list are ^{164}Dy [17] and ^{168}Er [18], two lanthanide elements with such large magnetic moment and large atomic mass that promise to open a new frontier in the study of strongly dipolar quantum gases. Thus, it is conceivable that the dipolar Bose-Fermi mixtures can be realized in cold-atom experiments not too far down the road.

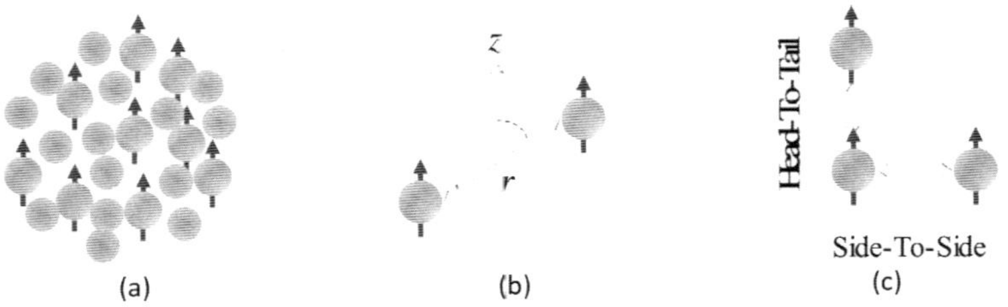

Figure 1. (Color online) (a) A mixture between a spin-polarized Fermi gas of nondipolar fermions (blue spheres) and a condensate of dipolar bosons (orange spheres with arrows). (b) The dipole-dipole interaction between two dipolar bosons separated by a distance r. (c) The side-to-side (head-to-tail) arrangement where the dipole-dipole interaction is the most repulsive (attractive).

A distinctive advantage of ultracold atomic systems over solid-state systems is that with the untracold atomic systems, key system parameters, including the dimensionality and the interaction between particles of same and different species, can be tuned precisely. In a dipolar condensate, the dipole-dipole interaction represents a control knob inaccessible to nondipolar Bosons. Thus, mixing dipolar bosons with fermions opens up new possibilities. Of particular relevance to this proceeding paper is the question whether mixing dipolar bosons can help to enhance the opportunity for two fermions of same spin to form highly correlated p-wave Cooper pairs. For a typical pair of ground state atoms in a partial wave with a finite angular momentum $l\hbar$, the two-body scattering amplitude f_l scales, according to the celebrated Wigner threshold law [19], as $\left(k_F^0\right)^{2l}$, where $\hbar k_F^0$ is the Fermi momentum, so that the cross section for the p-wave scattering is vanishingly small in the ultracold limit where $k_F^0 r_0 \ll 1$ with r_0 the range of interparticle potential. Thus, for ultracold quantum gases consisting of typical ground state atoms, the critical temperature for p-wave BCS superfluid is at such a low level that it becomes essentially inaccessible to current technology. Numerous schemes have been proposed, using, for example, dipolar Fermi gases [20, 21, 22, 23, 24, 25] and p-wave Feshbach resonance [26, 27, 28], in order to overcome the limitation set by the Wigner threshold law (the latter appears not feasible since Feshbach molecules are short lived [29, 30, 31, 32]).

In our recent studies [33, 34], we have paid particular attention to the schemes that seek to achieve the same goal but using cold-atom quantum gas mixtures (see, references [35, 36, 37, 38] for several examples). In particular, we have considered, in Ref. [34], a uniform mixture of a spin-polarized Fermi gas and a dipolar condensate, shown in Fig. 1(a), in a three-dimensional (3D) setting, with the goal of extending the work by Dutta and Lewenstein [36] from 2D to 3D dipolar Bose-Fermi mixture. In our model, a boson of mass m_B interacts with a fermion of mass m_F via a contact interaction of strength $U_{BF} = 4\pi\hbar^2 a_{BF}/m_{BF}$, and two bosons interact via both a short-range s-wave interaction of strength $U_{BB} = 4\pi\hbar^2 a_{BB}/m_B$ and a long-range dipole-dipole interaction [Fig. 1(b)] of the form $U_{DD}(\mathbf{k}) = 8\pi d^2 P_2(\cos\theta_{\mathbf{k}})$ in momentum space, where a_{BF} and a_{BB} are the corresponding s-wave scattering lengths, d the dipole-dipole interaction strength, $P_2(x)$ the second-order Legendre polynomial, $\theta_{\mathbf{k}}$ the polar angle of wavevector $\mathbf{k}$, and $m_{BF} = 2m_B m_F/(m_B + m_F)$.

At very low temperatures, spontaneous symmetry breaking of the global gauge [U(1)] symmetry results in an ordered phase, known as the Bose-Einstein condensate (BEC), where the order parameter $<\hat{\psi}_B>$ has a finite value ($<\hat{\psi}_B>\neq 0$), where $\hat{\psi}_B$ is the field operator for annihilating a dipolar boson and $n_B = |<\hat{\psi}_B>|^2$ is the number density for the dipolar condensate. Nambu and Goldstone theorem then asserts that the condensate must contain a gapless excitation. Using the Bogoliubov perturbative ansatz and diagonalization procedure, one can show that this gapless branch of the excitation corresponds simply to the phonon spectrum given by

$$E_{\mathbf{k},B} = v_B\hbar k\sqrt{1 + (\xi_B k)^2 + 2\varepsilon_{dd}P_2(\cos\theta_{\mathbf{k}})} \tag{1}$$

where $\varepsilon_{dd} = 4\pi d^2/(3U_{BB})$ [39] measures the dipolar interaction relative to the s-wave collision, $v_B = \sqrt{n_B U_{BB}/m_B}$ is the phonon speed, and $\xi_B = \hbar/\sqrt{4m_B n_B U_{BB}}$ is the healing length. In a 3D dipolar condensate, the side-to-side arrangement where the dipole-dipole interaction is the most repulsive coexists with the head-to-tail configuration where the dipole-dipole interaction is the most attractive [see Fig. 1(c)]. This renders the phonon spectrum anisotropy and for the same reason, $\varepsilon_{dd} = 4\pi d^2/(3U_{BB}) < 1$ has to be less than 1, meaning that the short-range repulsive interaction has to be sufficiently large, in order for the condensate to be stable against the collapse due to the attractive part of the dipole-dipole interaction.

In a spin-polarized model, although the s-wave contact interaction is inaccessible to two identical atoms, two identical fermions can interact indirectly by exchanging virtual phonons with the help of the intermediate states containing one phonon without violating the Pauli exclusion principle. One atom can emit a virtual phonon which is later captured by another atom or vise versa. This phonon-induced Fermi-Fermi interaction in momentum space is given by

$$U(\mathbf{k}) = -\frac{U_{BF}^2}{U_{BB}}\frac{1}{1 + (\xi_B k)^2 + 2\varepsilon_{dd}P_2(\cos\theta_{\mathbf{k}})}, \tag{2}$$

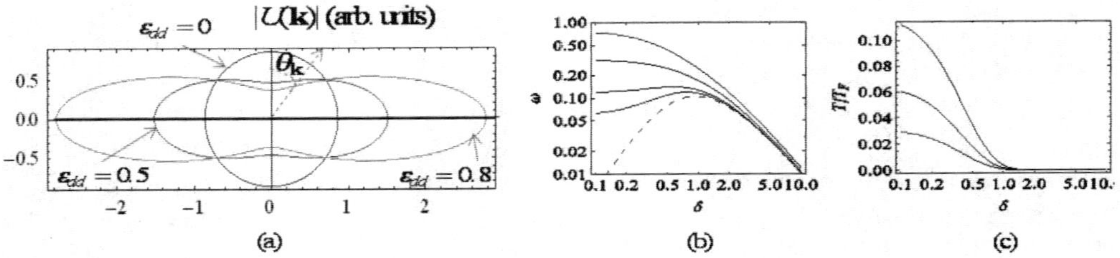

Figure 2. (Color online) (a) A polar plot of the scaled $|U(\mathbf{k})|$ [Eq. (2)] when $k/k_F^0 = 1$ and $\delta(\equiv \xi_B k_F^0) = 0.4$ shows that $\theta_{\mathbf{k}} = 90^\circ$ is the most attractive direction. (b) The largest positive eigenvalue ω (scaled to $-U_{BF}^2/U_{BB}$) of the triplet interaction potential matrix, $U_t(\mathbf{k}, \mathbf{k}')$, on the Fermi surface as a function of δ [34]. The dashed curve corresponds to $\varepsilon_{dd} = 0$. The solid curves, from bottom to top, corresponds to ε_{dd} $= 0.1$, 0.2, 0.5, and 0.8. The superfluid critical temperature, T, is proportional to $\exp(-1/\omega\lambda)$ according to Eq. (7). (c) T (scaled to the Fermi temperature ϵ_F^0/k_B) is plotted as a function of δ [34] when, from bottom to top, $\varepsilon_{dd} = 0.6, 0.7$ and 0.8. Throughout, $\lambda = 0.87$. [T in the absence of the dipolar interaction is too low to be visible in the scale of Fig. 2(c)]

in the static limit [40, 35, 41]. A polar plot of the induced interaction at a fixed momentum, Fig. 2(a), indicates that in the absence of dipolar interaction, this induced interaction is isotropic, and it becomes increasingly anisotropic as the dipolar interaction, ε_{dd}, increases with $\theta_{\mathbf{k}} = 90^\circ$ being the most attractive direction.

Integrating out the phonon degrees of freedom and taking the Hartree-Fock mean-field approach, we arrive at the mean-field Hamiltonian

$$\hat{H}_F^M = \sum_{\mathbf{k}} \xi_{\mathbf{k}} \hat{a}_{\mathbf{k}}^\dagger \hat{a}_{\mathbf{k}} + \frac{1}{2} \sum_{\mathbf{k}} \left[\Delta(\mathbf{k}) \hat{a}_{\mathbf{k}}^\dagger \hat{a}_{-\mathbf{k}}^\dagger + h.c \right], \tag{3}$$

where $\xi_{\mathbf{k}} = \epsilon_{\mathbf{k}} - \mu_F + \Sigma(\mathbf{k})$, $\hat{a}_{\mathbf{k}}$ is the fermionic field operator, μ_F is the chemical potential. By diagonalizing the Hamiltonian in Eq. (3) using the Bogoliubov transformation, we obtain the equation for the Fock potential

$$\Sigma(\mathbf{k}) = -\frac{1}{V} \sum_{\mathbf{k}'} U(\mathbf{k} - \mathbf{k}') \frac{1}{2} \left[1 - \tanh \frac{\beta E_{\mathbf{k}'}}{2} \right], \tag{4}$$

and the equation for the gap parameter,

$$\Delta(\mathbf{k}) = -\frac{1}{V} \sum_{\mathbf{k}'} U_t(\mathbf{k}, \mathbf{k}') K(\mathbf{k}') \Delta(\mathbf{k}'), \tag{5}$$

which are to be solved simultaneously with the particle number equation

$$n_F = \frac{1}{V} \sum_{\mathbf{k}'} \frac{1}{2} \left(1 - \tanh \frac{\beta E_{\mathbf{k}'}}{2} \right), \tag{6}$$

to determine $\Sigma(\mathbf{k}), \mu_F$ and $\Delta(\mathbf{k})$ self-consistently for a given set of Fermi density n_F and temperature T. In Eq. (5), the integration kernel $K(\mathbf{k}) = \tanh(\beta E_{\mathbf{k}}/2)/(2E_{\mathbf{k}}) - 1/(2\epsilon_{\mathbf{k}})$ is defined in terms of the temperature T via $\beta = 1/k_B T$ and the quasi particle spectrum $E_{\mathbf{k}} = \sqrt{\xi_{\mathbf{k}}^2 + \Delta(\mathbf{k})^2}$, and $U_t(\mathbf{k}, \mathbf{k}') = [U(\mathbf{k}+\mathbf{k}') - U(\mathbf{k}-\mathbf{k}')]/2$ is the triplet interaction potential symmetrized so that the gap parameter becomes the antisymmetric function of the relative position or momentum.

The external field introduced to establish the dipole direction breaks the spherical symmetry so that the matrix, $U_t(\mathbf{k}, \mathbf{k}')$, is not fully diagonalized in the space spanned by the spherical harmonic functions. However, the system under the external field still possesses the cylindrical symmetry, which means that $U_t(\mathbf{k}, \mathbf{k}')$ remains diagonalized with respect to the magnetic quantum number but not the angular momentum quantum number. As a result, the superfluid resulting from anisotropic attractive interactions like the one shown in Fig. 2(a) is a triplet with a gap parameter characterized with a coherent superposition of all the odd partial waves [22, 34]. In the weak-coupling limit, one can single out the states near the Fermi surface as the dominating contribution to the gap equation. The critical temperature (correct up to a preexponential factor) then becomes

$$T = \frac{8\epsilon_F^0 e^{\gamma-2}}{\pi k_B} \exp\left(-\frac{1}{\omega\lambda}\right), \tag{7}$$

where γ is the Euler's constant, k_B the Boltzmann gas constant, ϵ_F^0 the Fermi energy of a non-interacting Fermi gas with number density n_F, $\lambda = N(\epsilon_F^0) U_{BF}^2/U_{BB}$ a unitless parameter with $N(\epsilon_F^0)$ being the density of states, and finally ω the largest positive eigenvalue of the triplet potential matrix, $U_t(\mathbf{k}, \mathbf{k}')$, on the Fermi surface. Figure 2(b) shows how the largest positive eigenvalue ω changes with δ for different dipolar interaction ε_{dd}. The dotted curve represents the value of ω in the absence of the dipolar interaction. As δ decreases (or equivalently the Fermi momentum reduces), ω begins to decrease to zero after reaching its peak at $\delta \approx 1$. In contrast, when the dipolar interaction is sufficiently strong, ω not only asymptotes to a finite value but this value increases appreciably with ε_{dd} in the limit $\delta \to 0$, in clear defiance of Wigner's threshold law. As a result, we see from Fig. 2(c) that the critical temperature increases dramatically with the increase in the dipolar interaction.

There is an issue of whether bosons and fermions can stay miscible or stay phase separated. An illuminating way to examine this issue is to construct the phase diagram in the chemical potential space from the grand canonical thermodynamical potential density:

$$\Omega = \frac{1}{2} U_{BB} n_B^2 + U_{BF} n_B n_F + \frac{3}{5} A n_F^{5/3} - \mu_B n_B - \mu_F n_F. \tag{8}$$

where μ_B (μ_F) is the chemical potential of bosons (fermions), $n_B(n_F)$ the boson (fermion) number density, and $A = \hbar^2 (6\pi^2)^{2/3}/2m_F$. Phase separation manifests itself in the chemical potential space as a first-order phase transition line where different phases share the same energy. Figure 3 is the phase diagram in the chemical potential. Here, the space is divided into a pure Bose phase, a pure Fermi phase, and a Bose-Fermi mixture. Our proposal works only in the region indicated as BS where the bosons and fermions can coexist in the same spatial volume. Notice that there exists a first-order phase transition indicated by the dashed line between the mixed phase and the pure

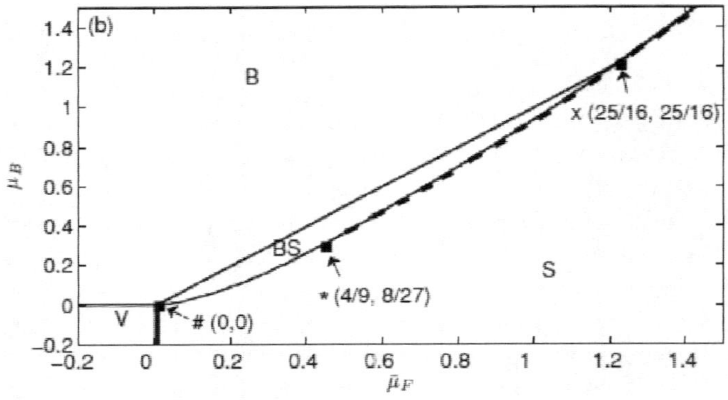

Figure 3. The phase diagram in the chemical potential [42] where the scaled chemical potentials are defined as $\bar{\mu}_B = U_{BF}^2 \mu_B / U_{BB}^3 A^3$ and $\bar{\mu}_F = U_{BF}^4 \mu_F / U_{BB}^2 A^3$ with $A = \hbar^2 \left(6\pi^2\right)^{2/3} / 2m_F$. B, S, and BS stand for pure Bose, pure Fermi, and Bose-Fermi mixed phases, respectively.

Fermi phase. This suggests that at sufficiently high densities the system will phase separate into a pure Fermi phase and a mixed Bose-Fermi system.

Phase separation limits the densities and therefore limits the achievable critical temperature. In our study, we have developed a procedure which allows us to determine the parameters such as densities and scattering lengths that are required to achieve the optimal critical temperature before the system starts to phase separate. A systematic investigation that takes into consideration the phase separation [42] finds that a superfluid with a critical temperature, that is orders of magnitude higher than achievable in nondipolar mixtures, can be realized in dipolar mixtures at experimentally accessible densities and scattering lengths before the system phase separates [34]. In conclusion, the recent experimental progress in realizing Bose-Fermi mixtures and dipolar quantum gases has created the exciting possibility of creating dipolar Bose-Fermi mixtures with intriguing and unique properties. In a dipolar condensate, the dipole-dipole interaction represents a control knob inaccessible to nondipolar Bosons. Thus, mixing dipolar bosons with fermions may open up new possibilities for creating superfluids with unconventional pairings. In this proceeding paper, we have reported our recent work and have shown that mixing dipolar bosons can induce an effective Fermi-Fermi attractive interaction which, owning to its long-range and anisotropic nature, can significantly increase the prospect of realizing a superfluid with a gap parameter which is a coherent superposition of all odd partial waves.

Acknowledgments

H. Y. L acknowleges support from the US National Science Foundation and the US Army Research Office

References

[1] Truscott A G, Strecker K E, McAlexander W I, Partridge G B and Hulet R G 2001 *Science* **291** 2570–2572
[2] Schreck F, Khaykovich L, Corwin K L, Ferrari G, Bourdel T, Cubizolles J and Salomon C 2001 *Phys. Rev. Lett.* **87**(8) 080403
[3] Hadzibabic Z, Stan C A, Dieckmann K, Gupta S, Zwierlein M W, Görlitz A and Ketterle W 2002 *Phys. Rev. Lett.* **88**(16) 160401
[4] Ferrari G, Inguscio M, Jastrzebski W, Modugno G, Roati G and Simoni A 2002 *Phys. Rev. Lett.* **89**(5) 053202
[5] Roati G, Riboli F, Modugno G and Inguscio M 2002 *Phys. Rev. Lett.* **89**(15) 150403
[6] Inouye S, Goldwin J, Olsen M L, Ticknor C, Bohn J L and Jin D S 2004 *Phys. Rev. Lett.* **93**(18) 183201
[7] Ferlaino F, D'Errico C, Roati G, Zaccanti M, Inguscio M, Modugno G and Simoni A 2006 *Phys. Rev. A* **73**(4) 040702
[8] Deh B, Marzok C, Zimmermann C and Courteille P W 2008 *Phys. Rev. A* **77**(1) 010701
[9] Park J W, Wu C H, Santiago I, Tiecke T G, Will S, Ahmadi P and Zwierlein M W 2012 *Phys. Rev. A* **85**(5) 051602
[10] Shin Y I, Schirotzek A, Schunck C H and Ketterle W 2008 *Phys. Rev. Lett.* **101**(7) 070404
[11] Wu C H, Santiago I, Park J W, Ahmadi P and Zwierlein M W 2011 *Phys. Rev. A* **84**(1) 011601
[12] Ospelkaus S, Peér A, Ni K K, Zirbel J J, Neyenhuis B, Kotochigova S, Julienne P S, Ye J and Jin D S 2008 *Nature Phys.* **4**(8) 622–626
[13] Ni K K, Ospelkaus S, de Miranda M H G, Pe'er A, Neyenhuis B, Zirbel J J, Kotochigova S, Julienne P S, Jin D S and Ye J 2008 *Science* **322** 231–235
[14] Ospelkaus S, Ni K K, de Miranda M H G, Neyenhuis B, Wang D, Kotochigova S, Julienne P S, Jin D S and Ye J 2009 *Faraday Discuss.* **142** 351–359
[15] Vengalattore M, Leslie S R, Guzman J and Stamper-Kurn D M 2008 *Phys. Rev. Lett.* **100**(17) 170403
[16] Stuhler J, Griesmaier A, Koch T, Fattori M, Pfau T, Giovanazzi S, Pedri P and Santos L 2005 *Phys. Rev. Lett.* **95**(15) 150406
[17] Lu M, Burdick N Q, Youn S H and Lev B L 2011 *Phys. Rev. Lett.* **107**(19) 190401
[18] Aikawa K, Frisch A, Mark M, Baier S, Rietzler A, Grimm R and Ferlaino F 2012 *Phys. Rev. Lett.* **108**(21) 210401
[19] Wigner E P 1948 *Phys. Rev.* **73**(9) 1002–1009
[20] Marinescu M and You L 1998 *Phys. Rev. Lett.* **81**(21) 4596–4599
[21] You L and Marinescu M 1999 *Phys. Rev. A* **60**(3) 2324–2329
[22] Baranov M A, Mar'enko M S, Rychkov V S and Shlyapnikov G V 2002 *Phys. Rev. A* **66**(1) 013606
[23] Bruun G M and Taylor E 2008 *Phys. Rev. Lett.* **101**(24) 245301
[24] Wu C and Hirsch J E 2010 *Phys. Rev. B* **81**(2) 020508
[25] Shi T, Zhang J N, Sun C P and Yi S 2010 *Phys. Rev. A* **82**(3) 033623
[26] Bohn J L 2000 *Phys. Rev. A* **61**(5) 053409
[27] Ohashi Y 2005 *Phys. Rev. Lett.* **94**(5) 050403
[28] Ho T L and Diener R B 2005 *Phys. Rev. Lett.* **94**(9) 090402
[29] Regal C A, Ticknor C, Bohn J L and Jin D S 2003 *Phys. Rev. Lett.* **90**(5) 053201
[30] Zhang J, van Kempen E G M, Bourdel T, Khaykovich L, Cubizolles J, Chevy F, Teichmann M, Tarruell L, Kokkelmans S J J M F and Salomon C 2004 *Phys. Rev. A* **70**(3) 030702
[31] Ticknor C, Regal C A, Jin D S and Bohn J L 2004 *Phys. Rev. A* **69**(4) 042712
[32] Günter K, Stöferle T, Moritz H, Köhl M and Esslinger T 2005 *Phys. Rev. Lett.* **95**(23) 230401
[33] Kain B and Ling H Y 2011 *Phys. Rev. A* **83**(6) 061603
[34] Kain B and Ling H Y 2012 *Phys. Rev. A* **85**(1) 013631
[35] Efremov D V and Viverit L 2002 *Phys. Rev. B* **65**(13) 134519
[36] Dutta O and Lewenstein M 2010 *Phys. Rev. A* **81**(6) 063608
[37] Nishida Y 2009 *Annals of Physics* **324** 897 – 919
[38] Nishida Y and Tan S 2010 *Phys. Rev. A* **82**(6) 062713
[39] O'Dell D H J, Giovanazzi S and Eberlein C 2004 *Phys. Rev. Lett.* **92**(25) 250401

[40] Bijlsma M J, Heringa B A and Stoof H T C 2000 *Phys. Rev. A* **61**(5) 053601
[41] *The retardation effect which occurs in the non-static regime is currently under investigation*
[42] Fodor M and Ling H Y 2010 *Phys. Rev. A* **82**(4) 043610

Quasicondensation reexamined

Przemysław Bienias[1], **Krzysztof Pawłowski**[12], **Mariusz Gajda**[34], **Kazimierz Rzążewski**[124]

[1] Center for Theoretical Physics, Polish Academy of Sciences, Aleja Lotników 32/46, 02-668 Warsaw, Poland

[2] 5.Physikalisches Institut, Universität Stuttgart, Pfaffenwaldring 57, 70550 Stuttgart, Germany

[3] Institute of Physics, Polish Academy of Sciences, Aleja Lotników 32/46, 02-668 Warsaw, Poland

[4] Faculty of Mathematics and Sciences, Cardinal Stefan Wyszyński University, ulica Dewajtis 5, 01-815, Warsaw, Poland

E-mail: pbienias@cft.edu.pl

Abstract. We study in detail the effect of quasicondensation. We show that this effect is strictly related to dimensionality of the system. It is present in one dimensional systems independently of interactions – exists in repulsive, attractive or in non-interacting Bose gas in some range of temperatures below characteristic temperature of the quantum degeneracy. Based on this observation we analyze the quasicondensation in terms of a ratio of the two largest eigenvalues of the single particle density matrix for the ideal gas. We show that in the thermodynamic limit in higher dimensions the second largest eigenvalue vanishes (as compared to the first one) with total number of particles as $\simeq N^{-\gamma}$ whereas goes to zero only logarithmically in one dimension. We also study the effect of quasicondensation for various geometries of the system: from quasi-1D elongated one, through spherically symmetric 3D case to quasi-2D pancake-like geometry.

1. Introduction

Properties of the interacting, harmonically trapped, ultracold gas are much more interesting in one dimension than in two and three dimensions. In 2D and 3D a phase transition occurs whereas in 1D does not (surprisingly in quasi-1D systems two-step condensation is possible [24]). Moreover, in symmetric 3D systems it was shown experimentally [22, 11] that nearly all the way up to critical temperature the phase of the cloud is spatially uniform, equivalently: a coherence length of the system is equal to its size. On the contrary, Petrov et al. have shown that the observation of the quasicondensate (condensate with fluctuating phase) is possible in 1D [19], 2D [18] and very elongated 3D [20] repulsive Bose gas. Those predictions were confirmed experimentally [9, 21, 12]. Recently density and phase properties of elongated systems were investigated in a number of experiments: [10, 23, 3, 15]. In spite of many theoretical attempts [2, 1, 7, 8] the theory of quasicondesates is still not as mature as the theory of condensates. In this paper we shed a new light on the quasicondensation phenomenon. In [6] we stressed that shortening of the coherence length is not only the property of a repulsive gas but also of an attractive one. There exists a direct connection between the quasicondensation and the spectrum of a one-body density matrix [1]. More precisely, the quasicondensation occurs when occupation of more than one eigenmode is comparable with the total population [17]. In some sense it is similar to the "fragmented" spinor condensate [14]. In Section 2 we show that properties of one dimensional Bose gas are similar regardless the sign of the interaction. In Section 3 we explore coherence properties of the ideal gas in 1D, 2D, 3D and with arbitrary ratio of the trap frequencies.

2. Interacting gas

Firstly, we study a one dimensional, weakly interacting Bose gas confined in a harmonic trap. Thus, our Hamiltonian of the one dimensional Bose gas has a form:

$$\begin{aligned} H &= \int \hat{\Psi}^\dagger(x) \left(\frac{p^2}{2m} + \frac{1}{2} m\, \omega_0^2 x^2 \right) \hat{\Psi}(x)\, \mathrm{d}x + \\ &+ \frac{g}{2} \int \hat{\Psi}^\dagger(x) \hat{\Psi}^\dagger(x) \hat{\Psi}(x) \hat{\Psi}(x)\, \mathrm{d}x. \end{aligned} \tag{1}$$

The Hamiltonian is a sum of the single particle oscillator energy with mass m and angular frequency ω and a conventional contact interaction with the coupling constant g. Throughout this paper we use the oscillator units of position, energy and temperature, $\sqrt{\frac{\hbar}{m\omega_0}}$, $\hbar\omega_0$ and $\frac{\hbar\omega_0}{k_B}$ respectively. Hence a dimensionless coupling g is in units of $\sqrt{\frac{\hbar^3\omega_0}{m}}$. Almost all results are calculated for 1000 atoms.

At the beginning we study a one dimensional, repulsive, weakly interacting Bose gas confined in a harmonic trap. Our results are for the canonical statistical ensemble, thus the temperature is a control parameter. In previous works we showed that in a wide range of temperatures such a system can be efficiently described using the so called classical field approximation [6, 5]. To analyze the coherence properties of the system we use the first-order correlation function defined as:

$$\begin{aligned} g_1(-x,x) &= \frac{\langle \Psi^*(-x)\Psi(x)\rangle}{\langle |\Psi(x)|^2\rangle} \\ &= \frac{\langle \sqrt{n(-x)}e^{-i\phi(-x)}\sqrt{n(x)}e^{i\phi(x)}\rangle}{\langle n(x)\rangle}. \end{aligned} \tag{2}$$

Where $\langle \ldots \rangle$ is an ensemble average, $\phi(x)$ and $\sqrt{n(x)}$ are respectively: the phase and the absolute value of the wave function $\Psi(x)$ describing the whole system. As a coherence length l_ϕ of the system we take the full-width at half maximum of the $g_1(-x,x)$. Analogously the width of the atomic cloud is a full-width at half maximum of $|\Psi(x)|^2$. In recent papers we showed that in both repulsive [6] and attractive [5] gas two regimes exist. One between the zero temperature and T_{ph} in which the size of the condensate is smaller than the coherence length. The second one, above T_{ph}, in which the opposite condition occurs - a quasicondensate regime. Such a situation is shown in Figure 1 for weakly interacting repulsive gas. To check how important are fluctuations of the phase in the behavior of l_ϕ we compared g_1 with:

$$\tilde{g}_1(-x,x) = \langle e^{-i\phi(-x)} e^{i\phi(x)} \rangle \tag{3}$$

Function $\tilde{g}_1$ unlike g_1 accounts for phase fluctuations only. In Figure 1 we see that in our regime of parameters $g_1(-x,x)$ is nearly the same as $\tilde{g}_1(-x,x)$, so density fluctuations do not contribute to the first-order correlation function regardless the sign of the interaction. Moreover, density

[1] In the description of partially coherent light spatial coherence modes are used also as eigenfunctions of the first order correlation function of the light field. See for instance B. Saleh *Photoelectron Statistics: With Applications to Spectroscopy and Optical Communication* (Springer-Verlag, New York, 1978)

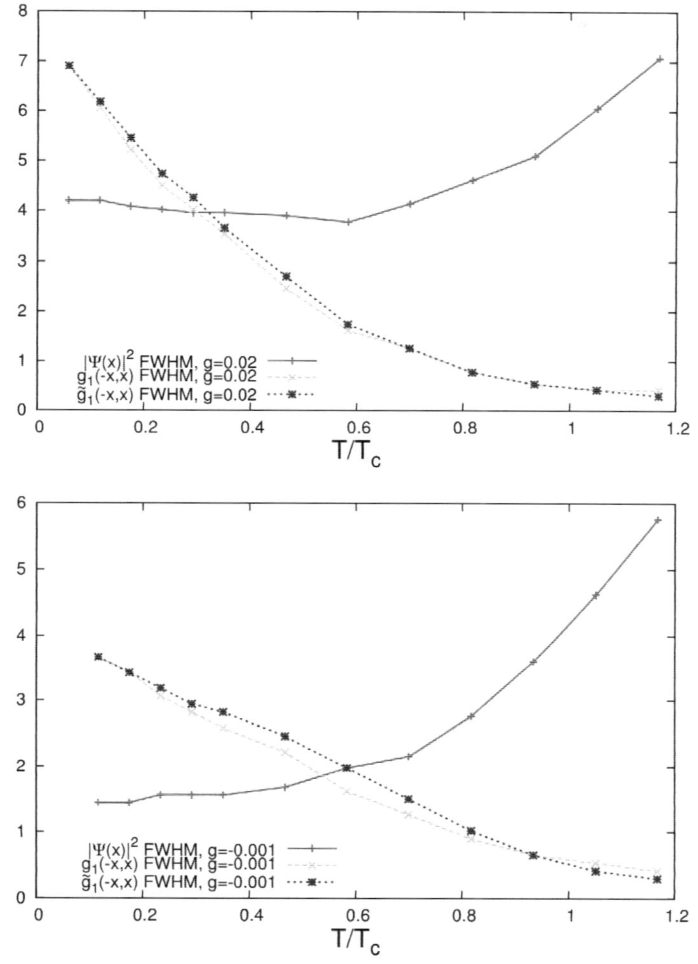

Figure 1. (Color online) Correlation length, width of $\tilde{g}_1$ and size of the whole cloud. All lengths are defined as full width at half maximum. Results for (a) repulsive and (b) attractive gas. 1000 atoms considered.

fluctuations are very similar for both attractive and repulsive cases. In Figure 2 we show $\delta^2 n/n$ (square of the density fluctuations divided by the density of atoms) at the center of the trap for changing g. We see that for the repulsive gas the fluctuations are smaller than for the attractive one. The reason is that for the repulsive case the mean-field interparticle interaction energy is positive and proportional to the density squared. Because of this, the system prefers even distribution of the density rather than concentration of atoms. In the attractive gas we have the opposite situation. The interaction energy is lower for local bunching of atoms. Nevertheless, we don't see any qualitative difference between positive and negative g, as the dependence is nearly linear. What is more: the value of the fluctuations divided by the square of the density is the same for every g what is presented in Figure 2. Finally, we see that coherence properties as well as density fluctuations are very similar for attractive and repulsive gas. Therefore next, we take a closer look at the properties of the ideal gas.

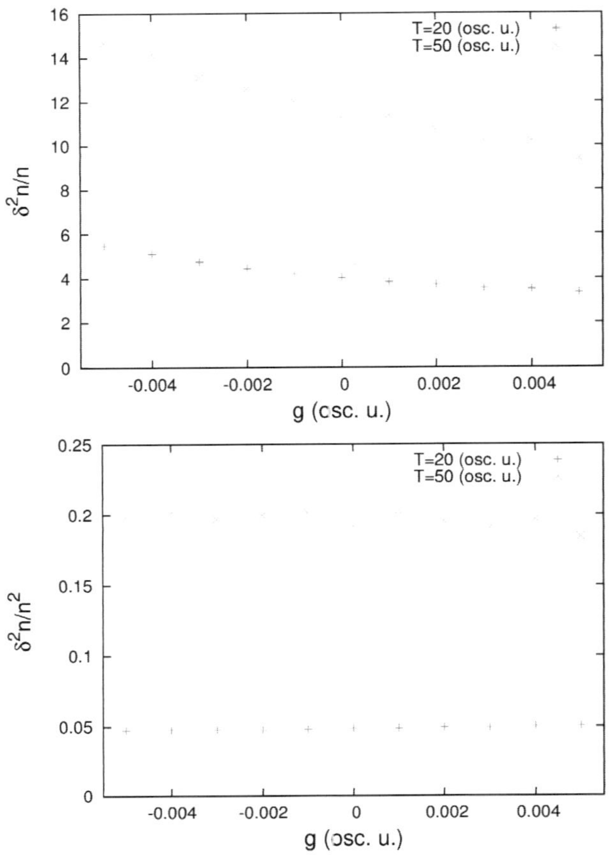

Figure 2. (Color online) Behavior of the density fluctuations at the center of the trap versus interaction for 1000 atoms. In (a) fluctuations divided by the density and in (b) divided by the square of density.

3. Ideal gas

Detailed exact calculations are possible for the ideal gas. In the grand canonical ensemble results can be found in [16, 4, 25].

We know that $\phi_i(x)$ in (8) are the eigenstates of harmonic oscillator. In the canonical ensemble we used the recursion derived in [26] to calculate the partition function of N atoms:

$$\begin{aligned} Z_0 &= 1, \\ Z_1(\beta) &= \textstyle\sum_\nu exp(-\beta\epsilon_\nu), \\ Z_N(\beta) &= \frac{1}{N}\textstyle\sum_{n=1}^{N} Z_1(n\beta)Z_{N-n}(\beta), \end{aligned} \tag{4}$$

where ϵ_ν is a single particle energy of the state ν, N is a number of atoms. Knowing Z_N we can calculate probability of finding n atoms in a state of energy ϵ_ν. Probability of finding at least n atoms there is:

$$P_\nu^{\geq}(n|N) = e^{-n\beta\epsilon_\nu}\frac{Z_{N-n}}{Z_N}. \tag{5}$$

Therefore:

$$P_\nu(n|N) = e^{-n\beta\epsilon_\nu}\frac{Z_{N-n}}{Z_N} - e^{-(n+1)\beta\epsilon_\nu}\frac{Z_{N-n-1}}{Z_N}. \tag{6}$$

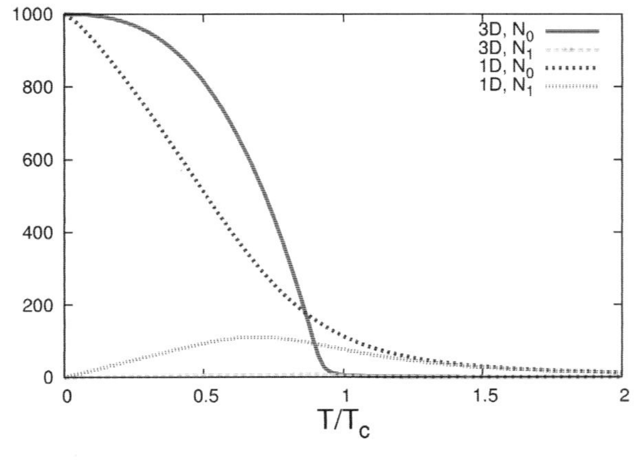

Figure 3. (Color online) Comparison of 1D and 3D ideal Bose gas in a harmonic trap. Populations of the ground state and the first excited state in both geometries for 1000 atoms are shown.

Then we can find the average number of atoms in the ν mode:

$$N_\nu = \sum_n P_\nu(n|N)n. \tag{7}$$

For deeper understanding of coherence properties lets go back to the definition (2) of $g_1(-x,x)$. We know that the one-body density matrix has the form:

$$\rho(x,y) = \frac{1}{N}\langle\Psi^*(x)\Psi(y)\rangle = \sum_{i=0} \frac{N_i}{N}\phi_i^*(x)\phi_i(y), \tag{8}$$

where N_i/N and $\phi_i(x)$ are eigenvalues and eigenvectors of $\rho(x,y)$. Hence,

$$g_1(-x,x) = \frac{\sum_{i=0}\frac{N_i}{N}\phi_i^*(-x)\phi_i(x)}{\sum_{i=0}\frac{N_i}{N}\phi_i^*(x)\phi_i(x)}. \tag{9}$$

We see a strong connection between the coherence length of the system and the spectrum of one-body density matrix. Existence of a few comparable eigenvalues of the one-body density matrix signifies fragmented condensate [27]. It follows from the Penrose-Onsager criterion that the quasicondensate transition is caused by "sticking" of one-body density matrix eigenvalues - the situation when N_0 is comparable to N_1.

There are two important remarks which enable us to gain better intuition before further analysis. First, when all atoms are in the condensate, this means that $N_0 = N$, then $g_1(-x,x)$ is everywhere equal to one so we have a fully coherent system. On the other hand if all eigenvalues are equal then $g_1(-x,x) \simeq \delta(x-x')$ and coherence length is equal to zero. To show a notable difference in the form of density matrix spectrum in 1D and 3D for the ideal gas we present in Figure 3 a population of two lowest modes (N_0, N_1 respectively) as a function of temperature. We see a qualitative difference between 1D and 3D. In 3D N_0 is much bigger than N_1 all the way up to T_c. Additionally the "sticking" of N_0 and N_1 takes place for almost vanishing N_0, $N_0/N \ll 1$, i.e. almost at T_c. On the other hand, in 1D we see the "sticking" for lower relative temperatures and, what is important, for much higher values of N_0/N.

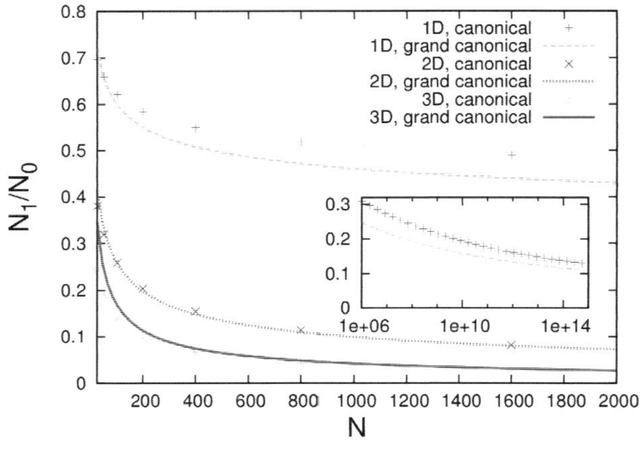

Figure 4. (Color online) Ratio of populations of the ground state and the first excited state for 1D, 2D and 3D in canonical ensemble (points) for $N_0/N = 0.2$. Comparison with grand canonical results is presented (lines). Note a good agreement in 2D and 3D. Grand canonical results for 1D are calculated numerically using (12). In the inset is shown, that the discrepancy between both ensembles in 1D exists even for huge number of atoms.

Using the intuition gained, we analyze more precisely one, two and three dimensions. The most intuitive way of comparing temperature effects for different regular geometries is to present temperature in T_c units:

$$\begin{aligned} 1D: \quad T_c &= \frac{\hbar\omega}{k_B}\frac{N}{\log(2N)}, \\ 2D: \quad T_c &= \frac{\hbar\omega}{k_B}\left(\frac{N}{\zeta(2)}\right)^{1/2}, \\ 3D: \quad T_c &= \frac{\hbar\omega}{k_B}\left(\frac{N}{\zeta(3)}\right)^{1/3}. \end{aligned} \tag{10}$$

For 2D and 3D T_c is equal to the critical temperature in thermodynamical limit. In 1D the phase transition doesn't occur. This is why only characteristic temperature is used [13]. However to make a comparison for different aspect ratios with finite number of particles it is more convenient to keep a relative population of the condensate N_0/N constant rather than the corresponding relative temperature.

In Figure 4 we show N_1/N_0 for different numbers of atoms in 1, 2 and 3 dimensions. All results are for $N_0/N = 0.2$. In 1D 'sticking' is definitely much stronger than in 2D and 3D. Results in canonical ensemble are presented up to 1600 atoms - the number used in the experiment [3]. Because of numerical limitations the ratio of N_1/N_0 for large N can be obtained in the grand canonical ensemble only. We see a good agreement between canonical and grand canonical ensembles in crossover range of N for 2D and 3D.

The agreement is worse in 1D. This is because the ensembles are equivalent in the thermodynamic limit only. Results of both ensembles (inset in Figure 4) approach each other on the logarithmic scale. Even for extremely large number of atoms the ratio N_1/N_0 in 1D is above 0.1, thus one would expect quasicondensation even for very large one dimensional system. Evidently, the finite size corrections in 1D systems are quite important. The asymptotic behavior is not reached even for $N \simeq 10^{15}$ particles.

Moreover, the grand canonical approach allows also for finding the asymptotic value of the ratio of two dominant eigenvalues of the one-body density matrix. The quantum partition function for a system of noninteracting bosons in the grand canonical ensemble may be written as follows:

$$\mathcal{Z}(z,T) = \sum_{n_0=0}^{\infty} \sum_{n_1=0}^{\infty} \dots \prod_{\lambda} z^{n_\lambda} \exp(-E_\lambda n_\lambda / k_B T), \tag{11}$$

where E_λ is the energy of the single particle state, and $z = \exp(\mu/k_B T)$ is a fugacity. For atoms trapped in a harmonic potential with frequency ω we have in 1D: $E_\lambda = \hbar\omega\lambda$, 2D: $E_\lambda = \hbar\omega(\lambda_x + \lambda_y)$ and in 3D: $E_\lambda = \hbar\omega(\lambda_x + \lambda_y + \lambda_z)$. Knowing that $\langle N \rangle = z\frac{\partial}{\partial z} \ln \mathcal{Z}(z,T)$ we get an expression for the average number of atoms in the system:

$$N = \frac{z}{1-z} + \sum_{\lambda \neq 0} \frac{z e^{-\beta E_\lambda}}{1 - z e^{-\beta E_\lambda}}, \tag{12}$$

where we have extracted the term corresponding to the lowest energy $E_\lambda = 0$. In 2D and 3D we can safely set $z = 1$ in the sum over all excited states (12). Next, we can expand the formula (12) considering only leading terms in a small parameter $\beta\hbar\omega$ [28], it leads to the expression for the atom number:

$$N = \frac{z}{1-z} + \zeta(D) \left(\frac{k_B T}{\hbar\omega} \right)^D, \tag{13}$$

where $D = 2$ (3) in 2D (3D). Remembering that we are interested in results for the constant value $N_0 = CN = \frac{z}{1-z}$ we get $z = \frac{CN}{1+CN}$ and $\frac{k_B T}{\hbar\omega} = \left(\frac{N(1-C)}{\zeta(D)} \right)^{1/D}$. In the 1D in the thermodynamic limit we have

$$N = \frac{z}{1-z} - \ln(1-z) \frac{k_B T}{\hbar\omega}. \tag{14}$$

Finally, we get that $\frac{k_B T}{\hbar\omega}$ is equal to $\frac{N(1-C)}{\ln(CN+1)}$. The asymptotic behavior of N_1/N_0 can be obtained from the relation:

$$\lim_{N\to\infty} \frac{N_1}{N_0} = \frac{z e^{-\hbar\omega\beta}}{1 - z e^{-\hbar\omega\beta}} \frac{1-z}{z}. \tag{15}$$

Finally, in 1D we get the following scaling of the 'sticking' ratio:

$$\lim_{N\to\infty} \frac{N_1}{N_0} \sim \frac{1}{\ln(N)}, \tag{16}$$

while in higher dimensions we have:

$$\lim_{N\to\infty} \frac{N_1}{N_0} \sim \frac{1}{N^{(D-1)/D}}; \tag{17}$$

that is

$$\frac{1}{N^{1/2}} \quad \text{for} \quad D = 2$$

and

$$\frac{1}{N^{2/3}} \quad \text{for} \quad D = 3\,.$$

The sticking ratio has totally different behavior in the asymptotic limit depending on the dimensionality of the system. In higher dimensions it goes to zero with dimension-depend power while in 1D the decay of the sticking ratio is logarithmically slow. This is the reason while in 1D system there is always a finite range of temperatures below the quantum degeneracy temperature

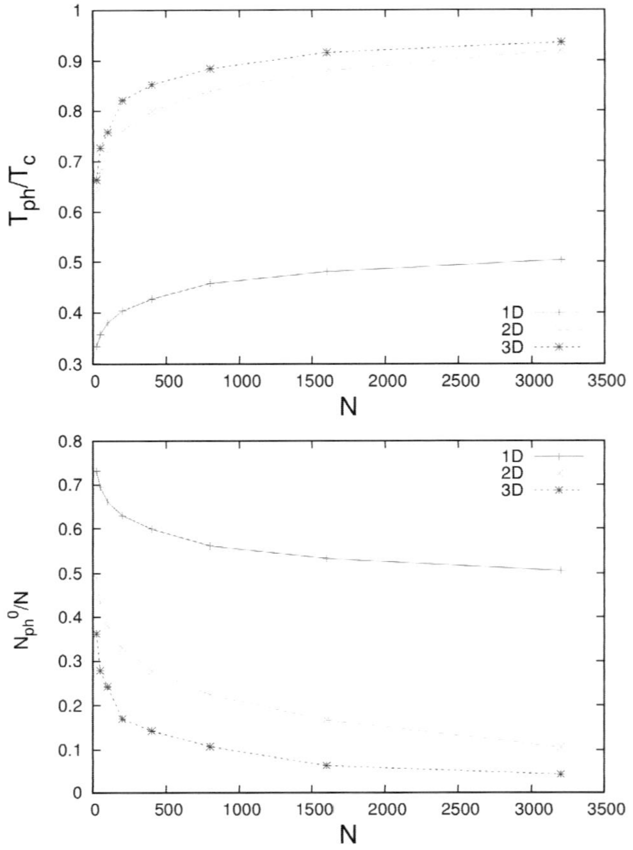

Figure 5. (Color online) Parameters of the system at the point of the transition from true condensate to quasicondensate versus number of atoms. In (a) temperature of the transition is presented, in (b) fraction of atoms in the ground state at the transition temperature.

where occupation of the higher orbitals of the one-body density matrix is relatively high. The spectrum of one-body density matrix enables us to look at the dependence of T_{ph} (the temperature when the width of the system is equal to the coherence length) as a function of N. In Figure 5(a) for better comparison we divided T_{ph} by T_c (10). Once more a significant difference between 1D and higher dimensions is seen. For 2D and 3D the temperature of quasicondensation is really near T_c whereas for 1D the temperature T_{ph} is much less than T_c. Because of the ambiguity of the T_c definition in 1D we looked at N_{ph}^0 (a number of atoms in the ground state at the temperature T_{ph}). In Figure 5 (b) we show N_{ph}^0/N as a function of N. The ratio N_{ph}^0/N is a fast decreasing function tending to zero in 2D and 3D, while for 1D it decreases much slower exceeding the value 0.5 in the whole presented region. Having analyzed 1D, 2D and 3D symmetric cases we study the ratio of N_1/N_0 and N_2/N_0 versus trap aspect ratio $\omega_Z/\omega_\perp$. In Figure 6 we present results for fixed: $N = 1000$ and $N_0/N = 0.4$. This way we can analyze transition between different geometries of the system. When $\omega_Z \ll \omega_\perp$ the system is nearly 1D, when $\omega_Z \approx \omega_\perp$ is nearly 3D, and when $\omega_z \gg \omega_\perp$ is nearly 2D. Once more we see a clear difference between dimensions. However, we do not observe any sharp transition between them. The left vertical line corresponds to $k_B T_{ph} = \hbar\omega_\perp$ and the right to $k_B T_{ph} = \hbar\omega_z$, so to the situations when system

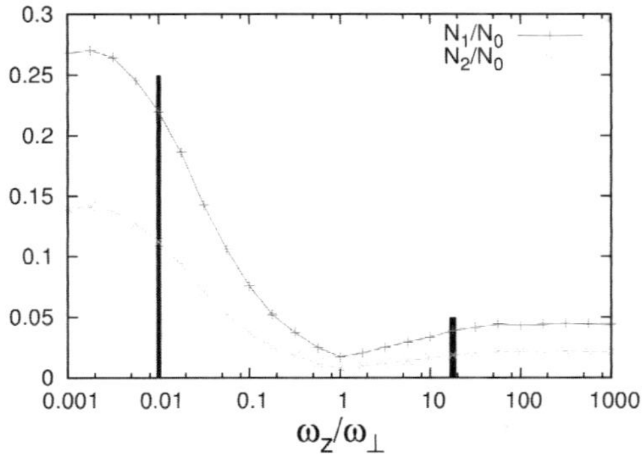

Figure 6. (Color online) Relative number of atoms in the ground state and in the first excited state versus the aspect ratio $\omega_z/\omega_\perp$ for $N = 1000$ and $N_0/N = 0.4$. The left vertical line denotes point when $k_B T_{ph} = \hbar\omega_\perp$ and the right one when $k_B T_{ph} = \hbar\omega_z$.

can be considered respectively as one and two dimensional. We see that these points agree with the beginning of the flattening of N_1/N_0 and N_2/N_0. Moreover 2D case is much more similar to 3D than to 1D. In experiments we never have exactly symmetric systems, however relatively broad range of aspect ratios around $\omega_z = \omega_\perp$ corresponds to systems which can be considered as 3D because of the flattening of N_1/N_0 in this region. We checked that this region gets broader with increasing N.

Summarizing, we have shown that quasicondensation phenomenon is strongly related to the dimensionality of the system rather then to interactions. Using the sticking ratio N_1/N_0 as the quasicondensation criterion we found the asymptotic behavior of this quantity in the ideal gas case. The 1D asymptotic of the sticking ratio differs drastically from asymptotic in higher dimensions. Vanishing of the second largest eigenvalue of the one-body density matrix is logarithmically slow in 1D. Some decreasing of the correlation length in 2D and 3D can be also observed but this effect results from the finite number of atoms in the system and occurs just below the critical temperature.

Acknowledgments

This work was supported by Polish Government Funds for the years 2010-2012. Two of us (K.P.and K.Rz.) acknowledge financial support of the project "Decoherence in long range interacting quantum systems and devices" sponsored by the Baden-Würtenberg Stiftung".

References

[1] U. Al Khawaja, N. P. Proukakis, J. O. Andersen, M. W. J. Romans, and H. T. C. Stoof. Dimensional and temperature crossover in trapped bose gases. *Phys. Rev. A*, 68:043603, Oct 2003.

[2] J. O. Andersen, U. Al Khawaja, and H. T. C. Stoof. Phase fluctuations in atomic bose gases. *Phys. Rev. Lett.*, 88:070407, Feb 2002.

[3] J. Armijo, T. Jacqmin, K. Kheruntsyan, and I. Bouchoule. Mapping out the quasicondensate transition through the dimensional crossover from one to three dimensions. *Phys. Rev. A*, 83:021605, Feb 2011.

[4] S. M. Barnett, S. Franke-Arnold, A. S. Arnold, and C. Baxter. Coherence length for a trapped bose gas. *J. Phys. B: At. Mol. Opt. Phys.*, 33(19):4177–4191, 2000.

[5] P. Bienias, K. Pawłowski, M. Gajda, and K. Rzążewski. Statistical properties of one-dimensional attractive bose gas. *Europhys. Lett.*, 96:10011, 2011.

[6] P. Bienias, K. Pawłowski, M. Gajda, and K. Rzążewski. Statistical properties of one-dimensional bose gas. *Phys. Rev. A*, 83(3):033610, Mar 2011.

[7] I. Bouchoule, K. V. Kheruntsyan, and G. V. Shlyapnikov. Interaction-induced crossover versus finite-size condensation in a weakly interacting trapped one-dimensional bose gas. *Phys. Rev. A*, 75:031606, Mar 2007.

[8] S. P. Cockburn, D. Gallucci, and N. P. Proukakis. Quantitative study of quasi-one-dimensional bose gas experiments via the stochastic gross-pitaevskii equation. *Phys. Rev. A*, 84:023613, Aug 2011.

[9] S. Dettmer, D. Hellweg, P. Ryytty, J. J. Arlt, W. Ertmer, K. Sengstock, D. S. Petrov, G. V. Shlyapnikov, H. Kreutzmann, L. Santos, and M. Lewenstein. Observation of phase fluctuations in elongated bose-einstein condensates. *Phys. Rev. Lett.*, 87(16):160406, Oct 2001.

[10] J. Esteve, J.-B. Trebbia, T. Schumm, A. Aspect, C. I. Westbrook, and I. Bouchoule. Observations of density fluctuations in an elongated bose gas: Ideal gas and quasicondensate regimes. *Phys. Rev. Lett.*, 96(13):130403, Apr 2006.

[11] E. W. Hagley, L. Deng, M. Kozuma, M. Trippenbach, Y. B. Band, M. Edwards, M. Doery, P. S. Julienne, K. Helmerson, S. L. Rolston, and W. D. Phillips. Measurement of the coherence of a bose-einstein condensate. *Phys. Rev. Lett.*, 83:3112–3115, Oct 1999.

[12] D. Hellweg, L. Cacciapuoti, M. Kottke, T. Schulte, K. Sengstock, W. Ertmer, and J. J. Arlt. Measurement of the spatial correlation function of phase fluctuating bose-einstein condensates. *Phys. Rev. Lett.*, 91:010406, Jul 2003.

[13] W. Ketterle and N. J. van Druten. Bose-Einstein condensation of a finite number of particles trapped in one or three dimensions. *Phys. Rev. A*, 54(1):656–660, Jul 1996.

[14] A. J. Leggett. Bose-einstein condensation in the alkali gases: Some fundamental concepts. *Rev. Mod. Phys.*, 73:307–356, Apr 2001.

[15] S. Manz, R. Bücker, T. Betz, C. Koller, S. Hofferberth, I. E. Mazets, A. Imambekov, E. Demler, A. Perrin, J. Schmiedmayer, and T. Schumm. Two-point density correlations of quasicondensates in free expansion. *Phys. Rev. A*, 81:031610, Mar 2010.

[16] M. Naraschewski and R. J. Glauber. Spatial coherence and density correlations of trapped bose gases. *Phys. Rev. A*, 59:4595–4607, Jun 1999.

[17] O. Penrose and L. Onsager. Bose-einstein condensation and liquid helium. *Phys. Rev.*, 104:576–584, Nov 1956.

[18] D. S. Petrov, M. Holzmann, and G. V. Shlyapnikov. Bose-einstein condensation in quasi-2d trapped gases. *Phys. Rev. Lett.*, 84:2551–2555, Mar 2000.

[19] D. S. Petrov, G. V. Shlyapnikov, and J. T. M. Walraven. Regimes of quantum degeneracy in trapped 1d gases. *Phys. Rev. Lett.*, 85(18):3745–3749, Oct 2000.

[20] D. S. Petrov, G. V. Shlyapnikov, and J. T. M. Walraven. Phase-fluctuating 3d bose-einstein condensates in elongated traps. *Phys. Rev. Lett.*, 87:050404, Jul 2001.

[21] S. Richard, F. Gerbier, J. H. Thywissen, M. Hugbart, P. Bouyer, and A. Aspect. Momentum spectroscopy of 1d phase fluctuations in bose-einstein condensates. *Phys. Rev. Lett.*, 91:010405, Jul 2003.

[22] J. Stenger, S. Inouye, A. P. Chikkatur, D. M. Stamper-Kurn, D. E. Pritchard, and W. Ketterle. Bragg spectroscopy of a bose-einstein condensate. *Phys. Rev. Lett.*, 82:4569–4573, Jun 1999.

[23] A. H. van Amerongen, J. J. P. van Es, P. Wicke, K. V. Kheruntsyan, and N. J. van Druten. Yang-yang thermodynamics on an atom chip. *Phys. Rev. Lett.*, 100:090402, Mar 2008.

[24] N. J. van Druten and W. Ketterle. Two-step condensation of the ideal bose gas in highly anisotropic traps. *Phys. Rev. Lett.*, 79:549–552, Jul 1997.

[25] B. P. van Zyl. Off-diagonal long-range order in a harmonically confined two-dimensional bose gas. *Journal of Low Temperature Physics*, 132:207–222, 2003. 10.1023/A:1024409617342.

[26] M. Wilkens and C. Weiss. Particle number fluctuations in an ideal bose gas. *Journal of Modern Optics*, 44:10:1801–1814, 1997.

[27] N. K. Wilkin, J. M. F. Gunn, and R. A. Smith. Do attractive bosons condense? *Phys. Rev. Lett.*, 80:2265–2268, Mar 1998.

[28] C. W. Y. Castin, edited by R. Kaiser and F. David. *'Coherent atomic matter waves'*. EDP Sciences and Springer-Verlag, 2001.

Effective time-reversal via periodic shaking

Christoph Weiss[1,2]

[1] Joint Quantum Centre (JQC) Durham–Newcastle, Department of Physics, Durham University, Durham DH1 3LE, United Kingdom
[2] Institut für Physik, Carl von Ossietzky Universität, D-26111 Oldenburg, Germany

E-mail: `Christoph.Weiss@durham.ac.uk`

Abstract. For a periodically shaken optical lattice, effective time-reversal is investigated numerically. For interacting ultra-cold atoms, the scheme of [J. Phys. B 45, 021002 (2012)] involves a quasi-instantaneous change of both the shaking-amplitude and the sign of the interaction. As the wave function returns to its initial state with high probability, time-reversal is ideal to distinguish pure quantum dynamics from the dynamics described by statistical mixtures.

1. Introduction

In the classical world our intuition is based on, there are many processes which cannot easily be reversed. For many cases a film run backwards will be easily identifiable as such: a broken glass does not reassemble itself and jump on the table; a mixture of milk and tea does not unmix in such a way that the milk flows back into the milk bottle. However, the same is not always true for quantum mechanics. This paper investigates effective time-reversal [1] via periodic shaking (Fig. 1) for ultra-cold bosonic atoms in an optical lattice [2, 3, 4].

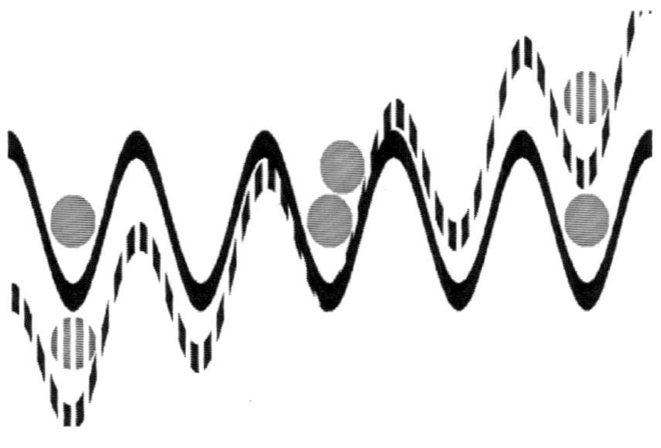

Figure 1. (Colour online) Ultra-cold atoms in an optical lattice combined with periodic shaking. The full quantum dynamics are modelled by the time-dependent Bose-Hubbard Hamiltonian (1); many aspects can be understood by the effective Hamiltonian (2).

Tunnelling control via periodically shaking an optical lattice has already been applied in several experimental setups. The topics investigated experimentally range from control of the superfluid-to-insulator transition [5] (cf. [6, 7]) over frustrated classical magnetism [8] to photon-assisted tunnelling [9, 10, 11] (cf. [12, 13, 14, 15, 16]). Research on periodically shaken systems also includes destruction of tunnelling [17, 18, 19, 20], generation of quantum superpositions [21, 22], robust dynamical recurrences [23] and reflection-less defects [24].

Experiments with periodically shaken optical lattices [5, 25, 10, 11] can often be modelled via a time-periodic Bose-Hubbard Hamiltonian like

$$\hat{H}(t) = - J \sum_{\ell=-\infty}^{\infty} \left(\hat{c}_\ell^\dagger \hat{c}_{\ell+1} + \hat{c}_{\ell+1}^\dagger \hat{c}_\ell\right) + \frac{U}{2} \sum_{\ell=-\infty}^{\infty} \hat{c}_\ell^\dagger \hat{c}_\ell^\dagger \hat{c}_\ell \hat{c}_\ell + \sum_{\ell=-\infty}^{\infty} 2\ell\hbar\mu \cos(\omega t + \alpha)\hat{c}_\ell^\dagger \hat{c}_\ell \,. \quad (1)$$

The operators $\hat{c}_\ell/\hat{c}_\ell^\dagger$ annihilate/create a boson in well ℓ; J describes the hopping of particles in the lattice and $2\hbar\mu$ is the driving amplitude. The on-site pair interaction is denoted by U. In the high-frequency limit ($\hbar\omega \gg U$, $\hbar\omega \gg J$), many aspects of experiments can be understood by an effective-Hamiltonian approach for which Eq. (1) is replaced by [5]

$$\hat{H}_{\text{eff}} = - J_{\text{eff}} \sum_{\ell=-\infty}^{\infty} \left(\hat{c}_\ell^\dagger \hat{c}_{\ell+1} + \hat{c}_{\ell+1}^\dagger \hat{c}_\ell\right) + \frac{U}{2} \sum_{\ell=-\infty}^{\infty} \hat{c}_\ell^\dagger \hat{c}_\ell^\dagger \hat{c}_\ell \hat{c}_\ell \,, \quad (2)$$

where the hopping J is multiplied by the $\mathcal{J}_0$ Bessel-function [26]

$$J_{\text{eff}} \equiv J\mathcal{J}_0\left(\tfrac{2\mu}{\omega}\right). \quad (3)$$

For larger interactions, the model (2) has to be modified (cf. [7, 16]). Note that the effective Hamiltonian and in particular the ground state are independent of the phase α (see Sec. 4.2). In order to realise complex hopping matrix elements which leads to phase-dependent ground states (cf. [27]), other approaches are necessary [28, 29].

In experiments, systems modelled by time-periodic Hamiltonians can be switched on quasi-instantaneously. If – as in Eq. (1) – the Hamiltonian contains a term $\propto \cos(\omega t+\alpha)$, it does make a difference if the cosine is at a minimum or maximum. Counter-intuitively, switching it on at a maximum is the best thing to do if one does not want to change the wave function (cf. [30, 31]). For periodically shaken systems of ultra-cold atoms, sudden changes of shaking have been used to cover topics as diverse as super Bloch oscillations [25, 32, 33], directed transport [34], effective magnetic fields [35] and time-reversal for a Bose-Einstein condensate in a double well [1].

The topic of this paper is to apply the time-reversal scheme of Ref. [1] to periodically shaken lattices. Section 2 motivates time-reversal as a method to distinguish quantum dynamics from similar looking classical diffusion. The model used to describe the shaken lattice is introduced in Sec. 3 (and in the Appendix). Section 4 introduces the time-reversal scheme. The results are presented in Sec. 5.

2. Quantum dynamics vs. classical diffusion

Figure 2 shows both the solution of the one-dimensional diffusion equation and of the one-dimensional Schrödinger equation for a free particle. For Fig. 2 both the classical diffusion equation in one dimension [36],

$$\partial_t \varrho(x,t) = D\partial_x^2 \varrho(x,t) \,, \quad (4)$$

where $D > 0$ is the diffusion constant, and the Schrödinger equation for a free particle of mass m in one dimension,

$$i\hbar\partial_t \Psi(x,t) = -\frac{\hbar^2}{2m}\partial_x^2 \Psi(x,t) \,, \quad (5)$$

have been solved for the initial conditions $\varrho(x,t=0) = \exp(-x^2)/\sqrt{\pi}$ and $\Psi(x,t=0) = \exp(-0.5\,x^2)/\sqrt[4]{\pi}$ using dimensionless variables ($D = 1$, $m = 0.5$, $\hbar = 1$). Both the solution of the diffusion equation [36] and the solution of the Schrödinger equation [37] are known analytically. The rate at which the width of these distributions spreads is a valuable way to distinguish quantum dynamics from classical dynamics [38].

Journal of Physics: Conference Series **414** (2013) 012032 doi:10.1088/1742-6596/414/1/012032

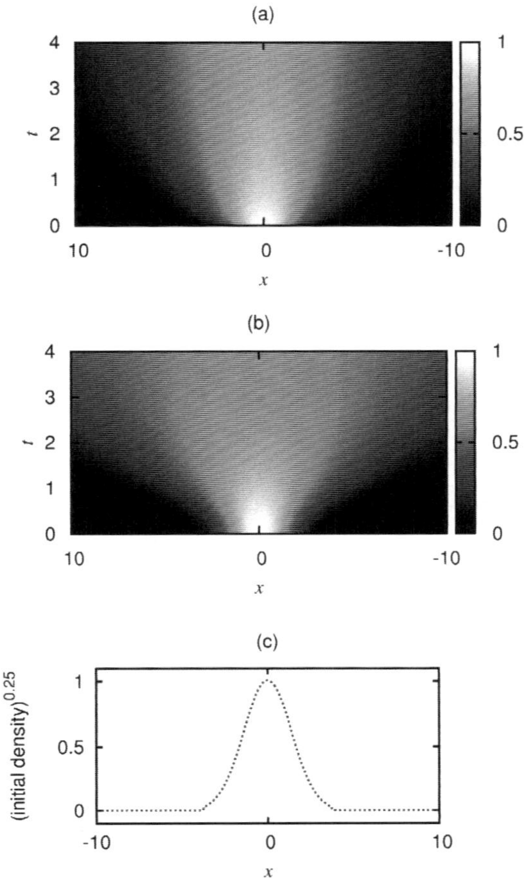

Figure 2. Quantum dynamics versus classical dynamics in one dimension. **(a)** Two-dimensional projection of the solution $\varrho(x,t)$ of the classical diffusion equation (4). To increase the visibility, $\varrho^{0.25}$ was plotted (in arbitrary units). **(b)** Two-dimensional projection of the probability density $|\Psi(x,t)|^2$ of a free particle [cf. Eq. (5)]. Again, $(|\Psi|^2)^{0.25}$ is plotted. **(c)** The initial conditions for both cases are chosen to be the same.

On the quantum level it is, in principle, possible to construct an initial wave function for which the width initially decreases by, e.g., numerically preparing a wave function for which the probability function reverses the motion shown in Fig. 2 (b). There is even an example where such a case was realised experimentally: by cleverly changing the wave function [39], the quantum motion has been turned backwards in time for atom-field interaction in a cavity quantum electrodynamics experiment [40]. Time-reversal schemes on the level of the Gross-Pitaevskii equation can be found in Refs. [41, 42].

3. Model

This section derives the effective Hamiltonian (2) for a single particle in a double well; the extension to an optical lattice is given in the Appendix. Hamiltonian (1) reduces to

$$\hat{H}(t) \;=\; -J\left(\hat{c}_1^\dagger\hat{c}_2+\hat{c}_2^\dagger\hat{c}_1\right)+\hbar\mu\cos(\omega t+\alpha)\left(\hat{c}_2^\dagger\hat{c}_2-\hat{c}_1^\dagger\hat{c}_1\right)\;; \tag{6}$$

one thus has to solve the corresponding Schrödinger equation:

$$i\partial_t\psi_1(t) \;=\; -\mu\cos(\omega t+\alpha)\psi_1(t)-\frac{J}{\hbar}\psi_2(t)\;;\quad i\partial_t\psi_2(t)+\mu\cos(\omega t+\alpha)\psi_2(t)-\frac{J}{\hbar}\psi_1(t) \tag{7}$$

The ansatz

$$\psi_1(t) \;=\; \varphi_1(t)\exp\left[i\frac{\mu}{\omega}\sin(\omega t+\alpha)\right]\;;\quad \psi_2(t)\varphi_2(t)\exp\left[-i\frac{\mu}{\omega}\sin(\omega t+\alpha)\right] \tag{8}$$

leads to

$$i\partial_t\varphi_1(t) = -\frac{J}{\hbar}\varphi_2(t)\exp\left[i\frac{2\mu}{\omega}\sin(\omega t+\alpha)\right] \; ; \quad i\partial_t\varphi_2(t) = -\frac{J}{\hbar}\varphi_1(t)\exp\left[-i\frac{2\mu}{\omega}\sin(\omega t+\alpha)\right] .$$

In order to solve this in the high-frequency limit $\hbar\omega \gg J$ (which may already be reached for $\hbar\omega \approx 10J$), the expansion in Bessel functions [26] is particularly useful

$$e^{iz\sin(\omega t+\alpha)} = \sum_{k=-\infty}^{\infty} \mathcal{J}_k(z) e^{ik(\omega t+\alpha)} . \tag{9}$$

For high frequencies, only the time-independent part of this sum is relevant in the above equations, leading to:

$$i\partial_t\varphi_1(t) \simeq -\frac{J_{\rm eff}}{\hbar}\varphi_2(t) \; ; \quad i\partial_t\varphi_2(t) \simeq -\frac{J_{\rm eff}}{\hbar}\varphi_1(t), \tag{10}$$

where $J_{\rm eff}$ is given by Eq. (3).

The effective tunnelling can be both zero and negative. Note that sometimes the physics is described by a zero [5] whereas in other cases the dynamics is governed by the other Bessel functions [16]. Equations (10) and (10) yield the effective, time-independent Hamiltonian:

$$\hat{H}_{\rm eff} = -J_{\rm eff}\left(\hat{c}_1^\dagger\hat{c}_2 + \hat{c}_2^\dagger\hat{c}_1\right) . \tag{11}$$

Repeating the above derivation for the case of an optical lattice (see Appendix) gives the effective Hamiltonian (2) for non-interacting particles. It remains valid even for interacting particles, as long as the interaction is not too large.

Strictly speaking, in the periodically shaken case there is no lowest energy: energies are only defined up to integer multiples of $\hbar\omega$. Thus, the quasi-energies for high frequencies read:

$$\epsilon = \pm J_{\rm eff} + j\hbar\omega , \tag{12}$$

with integer j. For negative effective tunnelling, the ground state wave function in the effective model is antisymmetric, which would be impossible without shaking. Despite the infinite number of quasi-energies, there are only two Floquet-states for the case discussed here:

$$\psi_\pm(t) \simeq \begin{pmatrix} \varphi_1(t) \\ \varphi_2(t) \end{pmatrix}_\pm = \frac{e^{\pm iJ_{\rm eff}t/\hbar}}{\sqrt{2}} \begin{pmatrix} \exp[i\frac{\mu}{\omega}\sin(\omega t+\alpha)] \\ \pm\exp[-i\frac{\mu}{\omega}\sin(\omega t+\alpha)] \end{pmatrix} .$$

In order to calculate the time-averaged energy in the time-dependent Hamiltonian,

$$\overline{\langle\hat{H}(t)\rangle}_\pm \equiv \frac{1}{T}\int_0^T \langle\psi_\pm(t)|\hat{H}|\psi_\pm(t)\rangle , \tag{13}$$

one needs

$$\begin{aligned}\langle\psi_\pm(t)|\hat{H}|\psi_\pm(t)\rangle &= -\hbar\mu\cos(\omega t+\alpha)\psi_1^*(t)\psi_1(t) - J\psi_1^*(t)\psi_2(t) \\ &\quad +\hbar\mu\cos(\omega t+\alpha)\psi_2^*(t)\psi_2(t) - J\psi_2^*(t)\psi_1(t) .\end{aligned}$$

The second step involves the time-average over one oscillation period, which removes the parts proportional to $\psi_j^*(t)\psi_j(t)$, $j = 1,2$ and leaves:

$$\overline{\langle\hat{H}(t)\rangle}_\pm = -\frac{1}{T}\int_0^T J\psi_2^*(t)\psi_1(t) - \frac{1}{T}\int_0^T J\psi_1^*(t)\psi_2(t) = \mp J_{\rm eff}, \tag{14}$$

which uses Eq. (9). Note that the only approximation in the above equation involves the high-frequency approximation for the Floquet-states $\psi_\pm(t)$.

4. Sudden changes of the Hamiltonian

4.1. Forcing time-reversal via changes to the shaking amplitude

If the Hamiltonian is changed quasi-instantaneously, this may involve changing, for example, either the amplitude or the phase or both at $t = t_0$:

$$\mu \to \widetilde{\mu}_1 \; ; \quad \alpha \to \widetilde{\alpha}. \tag{15}$$

To understand the effect of sudden changes of the Hamiltonian, the example of a single particle in a double well is again a good starting point. In this case, the eigenfunctions will be given by

$$\psi_\pm(t) = \begin{cases} \frac{e^{\pm iJ_{\rm eff}t/\hbar}}{\sqrt{2}} \begin{pmatrix} \exp\left[i\frac{\mu}{\omega}\sin(\omega t+\alpha)\right] \\ \pm\exp\left[-i\frac{\mu}{\omega}\sin(\omega t+\alpha)\right] \end{pmatrix} & : \; t \le t_0 \\ \frac{e^{\pm iJ_{\rm eff}t/\hbar}}{\sqrt{2}} \begin{pmatrix} \exp\left[i\frac{\widetilde{\mu}_1}{\omega}\sin(\omega t+\widetilde{\alpha})\right] \\ \pm\exp\left[-i\frac{\widetilde{\mu}_1}{\omega}\sin(\omega t+\widetilde{\alpha})\right] \end{pmatrix} & : \; t > t_0 \end{cases} \tag{16}$$

Choosing the example $\frac{\mu}{\omega} = \frac{\pi}{2}$ shows that a particle which is initially in the ψ_+-state might be in the ψ_--state at $t > t_0$ if the jump takes place for $\sin(\omega t_0 + \alpha) = 1$ and if $\widetilde{\alpha}$ is chosen such that $\sin(\omega t_0 + \widetilde{\alpha}) = 0$. Thus, although the wave function is continuous at $t = t_0$, the average energy can change even if the effective Hamiltonian is the same before and after the jump (cf. [31, 43]).

In the following, a quasi-instantaneous change of the Hamiltonian is used to switch the sign of the time-independent Hamiltonian (2). The sign can be switched by quasi-instantaneously changing both the tunnelling term by switching the shaking amplitude, e.g. [1],

$$\mathcal{J}_0(1.692) \simeq 0.403 \quad \text{and} \quad \mathcal{J}_0(3.832) \simeq -0.403 \; , \tag{17}$$

and the sign of the interaction via a Feshbach-resonance [44].

$$\hat{H}_{\rm ideal} \equiv \begin{cases} +\hat{H}_{\rm eff} & : \; 0 \le \tau < \tau_0 \\ -\hat{H}_{\rm eff} & : \; \tau \ge \tau_0 \end{cases} , \quad \tau \equiv \frac{tJ}{\hbar} \; . \tag{18}$$

The corresponding unitary time-evolution is given by $U(0,\tau) = \exp[i(\tau - 2\tau_0)\hat{H}_{\rm eff}/J]$ for $\tau > \tau_0$ with perfect return to the initial state at $\tau = 2\tau_0$. The turning point τ_0 however has to be chosen with care: only by taking τ_0 close to the maximum of the shaking, can unwanted excitations be excluded (cf., e.g., Ref. [31]). In the following, the time-reversal is demonstrated by numerically solving the full, time-dependent Hamiltonian (1) using the Shampine-Gordon routine [45].

4.2. Complex $J_{\rm eff}$ induced via phase-jumps

As the ansatz in Eq. (8) is not unique, one can also derive effective equations for which $J_{\rm eff}$ becomes complex [34, 35, 16]. Choosing, e.g., complex phase-factors such that the exponential in the ansatz is 1, i.e.,

$$\psi_1(t) = \varphi_1(t)e^{i\frac{\mu}{\omega}\sin(\omega t+\alpha) - i\frac{\mu}{\omega}\sin(\alpha)} \; ; \quad \psi_2(t) = \varphi_2(t)e^{-i\frac{\mu}{\omega}\sin(\omega t+\alpha) + i\frac{\mu}{\omega}\sin(\alpha)} \; ,$$

would replace the effective Hamiltonian (11) by

$$\hat{G}_{\rm eff} = -J_{\rm eff}\left(e^{i\frac{\mu}{\omega}\sin(\alpha)}\hat{c}_1^\dagger\hat{c}_2 + e^{-i\frac{\mu}{\omega}\sin(\alpha)}\hat{c}_2^\dagger\hat{c}_1\right) \tag{19}$$

where $J_{\rm eff}$ still is the real number defined in Eq. (3).

The corresponding 2×2-matrix has the same eigenvalues but different eigenvectors than Eq. (11). However, the full time-dependent solution is again identical to Eq. (13). Thus, as before, the ground state does not depend on the phase of the driving. Choosing

$$e^{i\frac{\mu}{\omega}\sin(\alpha)} = -1 \tag{20}$$

rather than 1 would, however, give the impression that the sign of the effective Hamiltonian has changed and ground and excited states had swapped their places (this was the aim of Sec. 4.1). Using the full Hamiltonian shows that is not the case. In the following, the jump is modelled at $t = 0$ under the assumption that before the jump one has $\alpha = 0$ and $J_{\text{eff}} > 0$. Introducing the phase-jump (20) quasi-instantaneously changes the Floquet-states

$$|\psi_+(t)\rangle_{\alpha=0} \longrightarrow |\psi_-(t)\rangle_{\alpha\neq 0}\ ; \quad |\psi_-(t)\rangle_{\alpha=0} \longrightarrow |\psi_+(t)\rangle_{\alpha\neq 0}\ , \tag{21}$$

where $\alpha \neq 0$ is defined by Eq. (20), but leaves the time-averaged energies (13) unchanged:

$$\overline{\langle \hat{H}(t)\rangle}_+ = \text{const.}\ ; \quad \overline{\langle \hat{H}(t)\rangle}_- = \text{const.} \tag{22}$$

Contrary to this, the sudden change of the amplitude discussed in Sec. 4.1 leaves the *wave functions unchanged* but *quasi-instantaneously changes the sign of the average energy.* This leads, in particular, to having a ground state which is antisymmetric. Another example of a related system with a counter-intuitive ground state is discussed in [46]. As the ground-state properties do not change by the jump (20), this paper avoids calling Eq. (19) an effective Hamiltonian.

As long as one is primarily interested in the dynamics (rather than, e.g., the ground state) it would nevertheless be perfectly valid to use such equations; the decision if phase-jumps or jumps in the shaking amplitude are preferable is a question of available technical laboratory resources and thus can not be discussed by the present paper. While this paper focuses on sudden changes for which $\sin(\alpha) = 0$, discussing $\sin(\alpha) \neq 0$ can be easier for Eq. (19) rather than Eq. (11) as the latter will involve complicated changes in the effective wave function at the time of the jump.

Truly complex J can also be realised in periodically shaken systems [28, 29].

5. Results

Figure 3 shows the spreading of the probability density for two interacting particles initially localised near the centre of a periodically shaken optical lattice. The initial state was chosen to be the ground state of a three-lattice-site version of the effective Hamiltonian (2) where the potential energy of the two outer wells was increased by $10J$ compared to the middle well, thus mimicking the strong harmonic confinement of a Bose-Einstein condensate released into such a shaken lattice in Ref. [47]. To visualise the probability density (which is a single number for each lattice site), a Gaussian density profile of the Wannier-function was modelled.

The dynamics displayed in Fig. 3 (a) are similar to the text-book case of a free particle in one-dimension [37]. At first glance, the main difference to the classical diffusion equation seems to be the rate at which the probability density expands (cf. [38]). The time-reversal visible in Fig. 3 (b) demonstrates deeper differences between classical and quantum physics. To quantify the quality of time-reversal, the "return probability", i.e., the probability to return to the original state,

$$p \equiv \left|\langle\Psi(\tau=0)|\Psi(\tau=2\tau_0)\rangle\right|^2\ , \tag{23}$$

is used (Fig. 4). This figure shows that, as for the case of the Bose-Einstein condensate in a periodically shaken double well [1], the time-reversal is not too sensitive to the precise choice of experimentally relevant parameters like interaction [Fig. 4 (a)], point of switching [Fig. 4 (b)] and shaking frequency [Fig. 4 (c)].

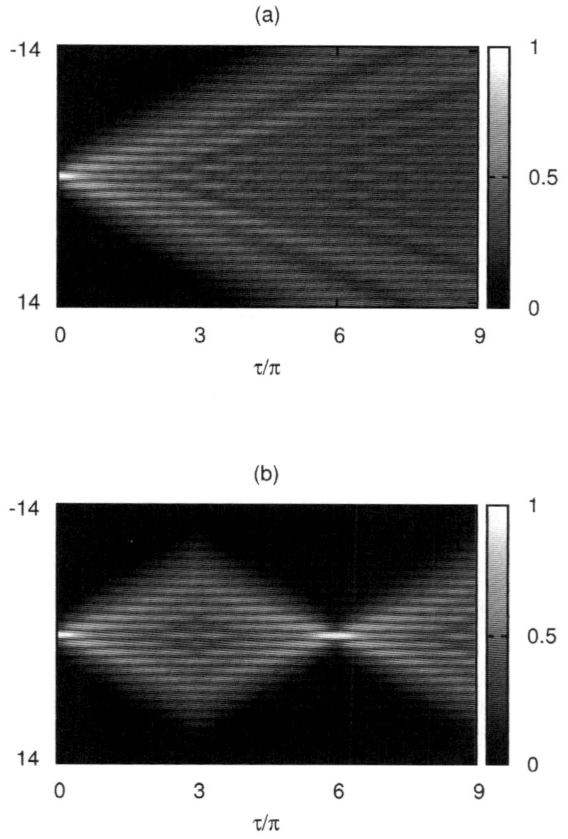

Figure 3. (Colour online) Time-reversal of the quantum dynamics for two interacting particles in a periodically shaken optical lattice. **(a)** Two-dimensional projection of the probability density (to the power 1/4, in arbitrary units) if both particles initially are close to the middle of the lattice ($U = 1.6J$, $\hbar\omega = 20J$, $2\mu/\omega = 1.692\,J$). **(b)** After applying coherently controlled time-reversal at $\tau = \tau_0$, the "echo" is visible near $\tau \approx 6\pi$ ($\tau_0 = 3\pi$; all parameters as in a except for $\tau \geq \tau_0$: $2\mu/\omega = 3.832\,J$ and $NU = -1.6J$).

The high probability to return to the initial state observed in the numerics offers an ideal method to distinguish between statistical mixtures and pure quantum states (cf. [1]). For larger particle numbers, experiments could observe the "echo" by studying the width of the wave function with the existing experimental setups of Refs [25, 47].

6. Conclusion

To conclude, effective time-reversal was numerically induced via periodic shaking to an optical lattice for two indistinguishable interacting bosons. The time-reversal was ideally described by an effective Hamiltonian and subsequently tested numerically by using the full, time-dependent model. Time-reversal was realised by switching the sign of the effective Hamiltonian by quasi-instantaneously changing the strength of the shaking. At the same time, the scheme suggests a change of the sign of the interaction via a Feshbach resonance. The Floquet-states remain unchanged by this; however their time-averaged mean energy switches its sign. This reflects the change of the sign of the effective Hamiltonian and supports the claim that the effective equations used to describe the dynamics can be labelled as an effective Hamiltonian.

The high probability to return to the initial state shows that the time-reversal can be used to distinguish quantum mechanics from statistical mixtures induced by decoherence.

The numerical simulations indicate that the effective time-reversal should be realisable with present day technology. To further optimise suitable time-reversal in an experiment, the feed-back loops of optimal control theory could be used [48, 49].

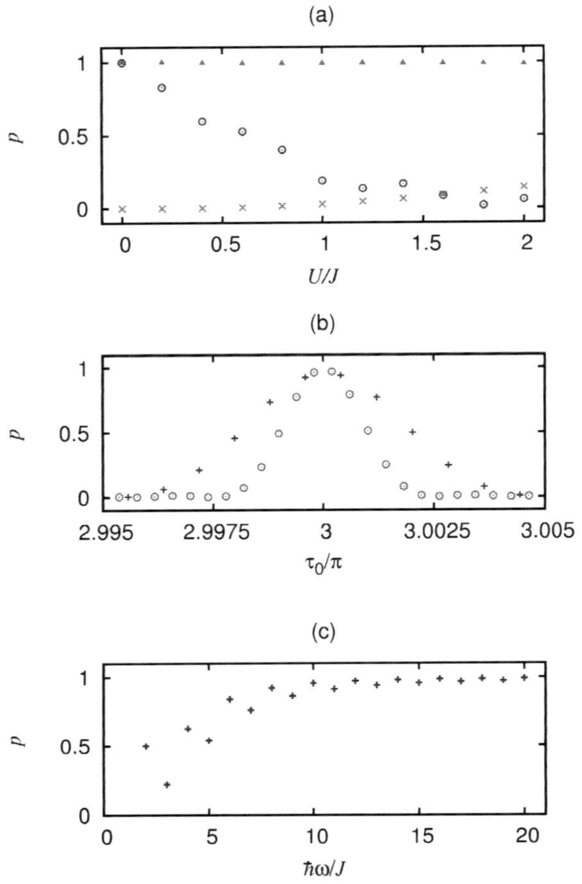

Figure 4. (Colour online) Time-reversal in a periodically shaken optical lattice **(a)** The return probability (23) at $\tau = 2\tau_0$ as a function of interaction. It is only large if both interaction and amplitude are changed (red/triangles). Switching only the shaking amplitude (blue/circles) as expected only works for non-interacting particles, switching only the interaction (magenta/crosses) may work for high interactions. **(b)** The return probability as a function of switching time has a broader distribution for low frequencies (blue/crosses, $\hbar\omega = 10J$) than for high frequencies (red/circles, $\hbar\omega = 20J$). All other parameters are as in panel (b). **(c)** The return probability as a function of shaking frequency decreases for lower angular frequencies (all other parameters as in b). While in the high frequency regime [cf. Eqs. (13) and (A.8)] instantaneously changing the Hamiltonian should be equally reliable for any zero of $\sin(\omega t + \alpha)$; for lower frequencies multiples of 2π lead to better return probabilities than π, 3π, ... (cf. [31]).

Acknowledgments
I thank S. A. Gardiner, E. Haller, M. Holthaus, S. Trotzky and C. Vaillant for discussions.

Appendix A. Effective Hamiltonian for a single particle in a tight-binding lattice

The Schrödinger equation for a single particle in a shaken lattice reads:

$$i\partial_\ell \psi_\ell(t) \quad = \quad 2\ell\mu\cos(\omega t + \alpha)\psi_\ell(t) - \frac{J}{\hbar}\left[\psi_{\ell+1}(t) + \psi_{\ell-1}(t)\right] \ , \tag{A.1}$$

with integer ℓ. Choosing the ansatz

$$\psi_\ell(t) \quad = \quad \varphi_\ell(t)[A(t)]^\ell \ , \tag{A.2}$$

where $A(t) = \exp\left[-i\frac{2\mu}{\omega}\sin(\omega t + \alpha)\right]$, leads to

$$i\partial_\ell \varphi_\ell(t) = -\frac{J}{\hbar}\left[A(t)\varphi_{\ell+1}(t) + [A(t)]^{-1}\varphi_{\ell-1}(t)\right] . \tag{A.3}$$

Within the high-frequency regime, one gets:

$$i\partial_\ell \varphi_\ell(t) \simeq -\frac{J_{\text{eff}}}{\hbar}\left[\varphi_{\ell+1}(t) + \varphi_{\ell-1}(t)\right] , \tag{A.4}$$

where $J_{\rm eff}$ is again given by Eq. (3). Thus, the effective Hamiltonian now reads:

$$\hat{H}_{\rm eff} = -J_{\rm eff} \sum_{\ell=-\infty}^{\infty} \left(\hat{c}_\ell^\dagger \hat{c}_{\ell+1} + \hat{c}_{\ell+1}^\dagger \hat{c}_\ell \right) \tag{A.5}$$

The eigenfunctions of this effective Hamiltonian are known:

$$|\varphi_k(t)\rangle = \exp\left(-i\frac{\varepsilon(k)}{\hbar}t\right) \sum_{\ell=-\infty}^{\infty} \exp(ikd\ell)|\ell\rangle\ , \tag{A.6}$$

where d is the lattice constant, k the quasi-momentum and the energy is

$$\varepsilon(k) = -2J_{\rm eff}\cos(kd)\ . \tag{A.7}$$

The notation $\varepsilon(k)$ was chosen to indicate that it will become the quasi-energy of the time-dependent problem. Within the high-frequency approximation, the Floquet-states are now:

$$|\psi_k(t)\rangle = \exp\left(-i\frac{\varepsilon(k)}{\hbar}t\right) \sum_{\ell=-\infty}^{\infty} [A(t)]^\ell \exp(ikd\ell)|\ell\rangle. \tag{A.8}$$

In order to calculate the mean energy, time-averaged over one oscillation period T, where $\omega T = 2\pi$ (i.e., the energy associated with this Floquet-state) we can replace the sum $\sum_{\ell=-\infty}^{\infty}$ by $\frac{1}{\sqrt{N}}\sum_{\ell=1}^{N}$ and take the limit $N \to \infty$ at the end. Thus, with this normalised version of $|\psi_k(t)\rangle$ defined on N lattice sites we can write:

$$\begin{aligned} \overline{E_k} &= \frac{1}{T}\int_0^T {\rm d}t\, \langle\psi_k(t)|\hat{H}(t)|\psi_k(t)\rangle \qquad (A.9) \\ &= \lim_{N\to\infty}\frac{1}{N}\sum_{\ell=1}^{N}\frac{1}{T}\int_0^T {\rm d}t\, \left\{2\ell\hbar\mu\cos(\omega t+\alpha) - J\left(A(t)\exp(ikd) + [A(t)\exp(ikd)]^{-1}\right)\right\}. \end{aligned}$$

Without any further approximation this leads to:

$$\overline{E_k} = -2J\mathcal{J}_0\left(\tfrac{2\mu}{\omega}\right)\cos(kd) = -2J_{\rm eff}\cos(kd) = \varepsilon(k)\ . \tag{A.10}$$

References

[1] Weiss C 2012 *J. Phys. B* **45** 021002
[2] Lewenstein M, Sanpera A, Ahufinger V, Damski B, Sen A and Sen U 2007 *Advances in Physics* **56** 243
[3] Bloch I, Dalibard J and Zwerger W 2008 *Rev. Mod. Phys.* **80** 885
[4] Yukalov V I 2009 *Laser Phys.* **19** 1
[5] Zenesini A, Lignier H, Ciampini D, Morsch O and Arimondo E 2009 *Phys. Rev. Lett.* **102** 100403
[6] Eckardt A, Weiss C and Holthaus M 2005 *Phys. Rev. Lett.* **95** 260404
[7] Creffield C E and Monteiro T S 2006 *Phys. Rev. Lett.* **96** 210403
[8] Struck J, Ölschläger C, Le Targat R, Soltan-Panahi P, Eckardt A, Lewenstein M, Windpassinger P and Sengstock K 2011 *Science* **333** 996–999
[9] Sias C, Lignier H, Singh Y P, Zenesini A, Ciampini D, Morsch O and Arimondo E 2008 *Phys. Rev. Lett.* **100** 040404
[10] Chen Y A, Nascimbène S, Aidelsburger M, Atala M, Trotzky S and Bloch I 2011 *Phys. Rev. Lett.* **107** 210405
[11] Ma R, Tai M E, Preiss P M, Bakr W S, Simon J and Greiner M 2011 *Phys. Rev. Lett.* **107** 095301
[12] Creffield C E, Sols F, Ciampini D, Morsch O and Arimondo E 2010 *Phys. Rev. A* **82** 035601
[13] Weiss C and Breuer H P 2009 *Phys. Rev. A* **79** 023608
[14] Teichmann N, Esmann M and Weiss C 2009 *Phys. Rev. A* **79** 063620
[15] Xie Q, Rong S, Zhong H, Lu G and Hai W 2010 *Phys. Rev. A* **82** 023616

[16] Esmann M, Pritchard J D and Weiss C 2012 *Laser Phys. Lett.* **9** 160
[17] Grossmann F, Dittrich T, Jung P and Hänggi P 1991 *Phys. Rev. Lett.* **67** 516
[18] Della Valle G, Ornigotti M, Cianci E, Foglietti V, Laporta P and Longhi S 2007 *Phys. Rev. Lett.* **98** 263601
[19] Kierig E, Schnorrberger U, Schietinger A, Tomkovic J and Oberthaler M K 2008 *Phys. Rev. Lett.* **100** 190405
[20] Gong J, Morales-Molina L and Hänggi P 2009 *Phys. Rev. Lett.* **103** 133002
[21] Creffield C E 2007 *Phys. Rev. Lett.* **99** 110501
[22] Stiebler K, Gertjerenken B, Teichmann N and Weiss C 2011 *J. Phys. B* **44** 055301
[23] Ayub M and Saif F 2012 *Phys. Rev. A* **85** 023634
[24] Longhi S and Della Valle G 2011 *Phys. Rev. B* **84** 193105
[25] Haller E, Hart R, Mark M J, Danzl J G, Reichsöllner L and Nägerl H C 2010 *Phys. Rev. Lett.* **104** 200403
[26] Abramowitz M and Stegun I A 1984 *Pocketbook of Mathematical Functions* (Thun: Verlag Harri Deutsch)
[27] Martikainen J P 2012 *Phys. Rev. A* **86** 033630
[28] Struck J, Ölschläger C, Weinberg M, Hauke P, Simonet J, Eckardt A, Lewenstein M, Sengstock K and Windpassinger P 2012 *Phys. Rev. Lett.* **108** 225304
[29] Sacha K, Targońska K and Zakrzewski J 2012 *Phys. Rev. A* **85** 053613
[30] Ridinger A and Davidson N 2007 *Phys. Rev. A* **76** 013421
[31] Ridinger A and Weiss C 2009 *Phys. Rev. A* **79** 013414
[32] Kudo K and Monteiro T S 2011 *Phys. Rev. A* **83** 053627
[33] Arlinghaus S and Holthaus M 2011 *Phys. Rev. B* **84** 054301
[34] Creffield C E and Sols F 2011 *Phys. Rev. A* **84** 023630
[35] Kolovsky A R 2011 *EPL (Europhys. Lett.)* **93** 20003
[36] Mathews J and Walker R L 1970 *Mathematical Methods of Physics* 2nd ed (Redwood City: Addison-Wesley Publishing Company)
[37] Flügge S 1990 *Rechenmethoden der Quantentheorie* (Berlin: Springer)
[38] Steinigeweg R, Breuer H P and Gemmer J 2007 *Phys. Rev. Lett.* **99** 150601
[39] Morigi G, Solano E, Englert B G and Walther H 2002 *Phys. Rev. A* **65** 040102(R)
[40] Meunier T, Gleyzes S, Maioli P, Auffeves A, Nogues G, Brune M, Raimond J M and Haroche S 2005 *Phys. Rev. Lett.* **94** 010401
[41] Martin J, Georgeot B and Shepelyansky D L 2008 *Phys. Rev. Lett.* **101** 074102
[42] Tsukada N, Yoshida H and Suzuki T 2008 *Phys. Rev. A* **78** 015601
[43] Cleary P W, Hijmans T W and Walraven J T M 2010 *Phys. Rev. A* **82** 063635
[44] Bauer D M, Lettner M, Vo C, Rempe G and Durr S 2009 *Nature Phys.* **5** 339
[45] Shampine L F and Gordon M K 1975 *Computer Solution of Ordinary Differential Equations* (San Francisco: Freeman)
[46] Eckardt A, Hauke P, Soltan-Panahi P, Becker C, Sengstock K and Lewenstein M 2010 *EPL (Europhys. Lett.)* **89** 10010
[47] Lignier H, Sias C, Ciampini D, Singh Y, Zenesini A, Morsch O and Arimondo E 2007 *Phys. Rev. Lett.* **99** 220403
[48] Judson R S and Rabitz H 1992 *Phys. Rev. Lett.* **68** 1500
[49] Assion A, Baumert T, Bergt M, Brixner T, Kiefer B, Seyfried V, Strehle M and Gerber G 1998 *Science* **282** 919

21st International Laser Physics Workshop IOP Publishing
Journal of Physics: Conference Series **414** (2013) 012033 doi:10.1088/1742-6596/414/1/012033

Expansion after a geometric quench of an atomic polarized attractive Fermi gas in one dimension

C. J. Bolech,[1] F. Heidrich-Meisner,[2,3] S. Langer,[2,4] I. P. McCulloch,[5] G. Orso,[6] and M. Rigol[7]

[1]Department of Physics, University of Cincinnati, Cincinnati, Ohio 45221-0011, USA
[2]Department of Physics and Arnold Sommerfeld Center for Theoretical Physics, Ludwig-Maximilians-Universität München, D-80333 München, Germany
[3] Friedrich-Alexander-Universität Erlangen-Nürnberg, Institute for Theoretical Physics II, Staudtstraße 7, 91058 Erlangen, Germany
[4]Department of Physics and Astronomy, University of Pittsburgh, Pittsburgh, Pennsylvania 15213, USA
[5]Centre for Engineered Quantum Systems, School of Mathematics and Physics, The University of Queensland, St Lucia, Brisbane 4072, Australia
[6]Laboratoire Matériaux et Phénomènes Quantiques, Université Paris Diderot-Paris 7 and CNRS, UMR 7162, 75205 Paris Cedex 13, France
[7]Department of Physics, The Pennsylvania State University, University Park, Pennsylvania 16802, USA

Abstract. The expansion of an ultracold lattice gas of fermions in a one-dimensional optical lattice after a geometric quench in which the atoms are released from a box (or parabolic) trap is studied numerically using the time-dependent density matrix renormalization group method. It is found that the momentum distributions of the two atomic species present in the simulations quickly converge toward stationary values dictated by the integrals of motion of the lattice gas at the expansion onset. An account of the central results and some of the technical details is given along with explicit examples.

1. Introduction

Developments in atomic laser cooling and trapping allow researchers to capture clouds of atomic gases and reach the limit of quantum degeneracy. The combined use of these techniques and Feshbach resonances, can be tailored to produce systems in which both the confining potentials and the interactions are controllable and yield interesting quantum behaviors like, for instance, exotic superfluid phases. Recent experimental efforts on systems of trapped cold atomic Fermi gases were aimed at studying the effects of polarization on superfluid pairing in low-dimensional systems [1]. The findings for atomic gases confined by a strong 2D optical lattice to move preferentially in one direction were consistent with the predictions for 1D systems of dilute polarized attractive gases [2, 3, 4, 5, 6, 7, 8, 9, 10, 11, 12, 13, 14]. The measurements are in quantitative agreement with theoretical calculations (using a wide array of numerical and analytical techniques) in which a partially polarized region is found to be the 1D analogue of the so called Fulde-Ferrell [15] or Larkin-Ovchinnikov [16] states (1D-FFLO), —which are superfluid states produced by the condensation of Cooper pairs with non-zero center-of-mass momentum. This unusual paired states are not found in standard metals, but are thought to be present in certain organic conductors [17] or heavy fermions [18], and even in nuclear and quark matter [19]. To go beyond the recent round of experiments, it is arguable that modifying the trapping or the optical-lattice strength could yield practical ways to better characterize and understand this system (for a recent review, discussing these and other ideas, see [20]). In this proceeding article, we will focus on the expansion dynamics after a sudden quench and make predictions for possible such experiments.

For 1D gases, interaction effects in the presence of a lattice cannot be neglected and they modify the expansion dynamics [21, 22, 23]. We use time-dependent density-matrix renormalization group (t-DMRG) [24, 25] to study the expansion process and we interpret our findings by referring to analytical (Bethe-Ansatz) results. We show that a process similar to the so called *quantum distillation* effect [26] takes place and the momentum distribution functions (MDFs) of majority and minority atoms become stationary after a relatively short expansion time, while the 1D-FFLO typifying characteristics are lost during the expansion.

2. Modeling

We shall model the system using the standard Hubbard model (see [27]), which with the usual second-quantized notation is given by:

$$H = -J \sum_{\sigma=\uparrow,\downarrow} \sum_{\ell=1}^{L-1} (c^{\dagger}_{\ell+1,\sigma} c_{\ell,\sigma} + \text{H.c.}) + U \sum_{\ell=1}^{L} n_{\ell\uparrow} n_{\ell\downarrow} \, . \tag{1}$$

As the initial pre-quench state, we take the ground state of a trapped system (here we focus on the case of a centered box-trap of size L_0 for which results can be more easily interpreted in terms of Bethe-Ansatz rapidities). We consider lattices with L sites and $N = N_\uparrow + N_\downarrow$ particles distributed with a global polarization given by $p = (N_\uparrow - N_\downarrow)/N$; positions are given in units of the lattice spacing and momenta in inverse units of the lattice spacing (we use the convention that $\hbar = 1$). The type of quench we shall consider consists of the sudden turning off of the confining potential, this will be described in the next sections.

3. Computational Results

Before proceeding further in the description of the process and into considerations of analytical nature, we first show some representative results for the expansion of a polarized attractive gas obtained by direct calculation using t-DMRG.

3.1. Density Distributions

We prepare the system in the ground state of an infinitely-tall-box trap of size L_0 and, at time $t = 0$, we perform a geometric quench (*i.e.* a sudden change of boundary conditions which does not modify the strength of the interactions or other parameters that define the system) consisting of turning off the box-trap potential and letting the atoms expand in a larger Hubbard chain with L sites. $L \gg L_0$ is chosen large enough that during the maximum expansion times considered none of the atoms reach the outer edges of the larger chain. In the initial ground state, all the minority atoms are paired up and it is thus natural to study the expansion of the pairs as well as that of the excess majority atoms that remain unpaired (the latter identified with the spin density). These two time-dependent density distributions are displayed in Fig. 1 for typical values in the range of parameters we considered.

It is interesting to notice in Fig. 1(b) that the pair density is seen to first focus in during the initial stages of the expansion (for $t \leq 5/J$) before it starts growing in size. This is the period during which the pairs are drawn to the center of the trap as the unpaired fermions that were initially distributed throughout move to the outer edges of the atom cloud and expand faster.

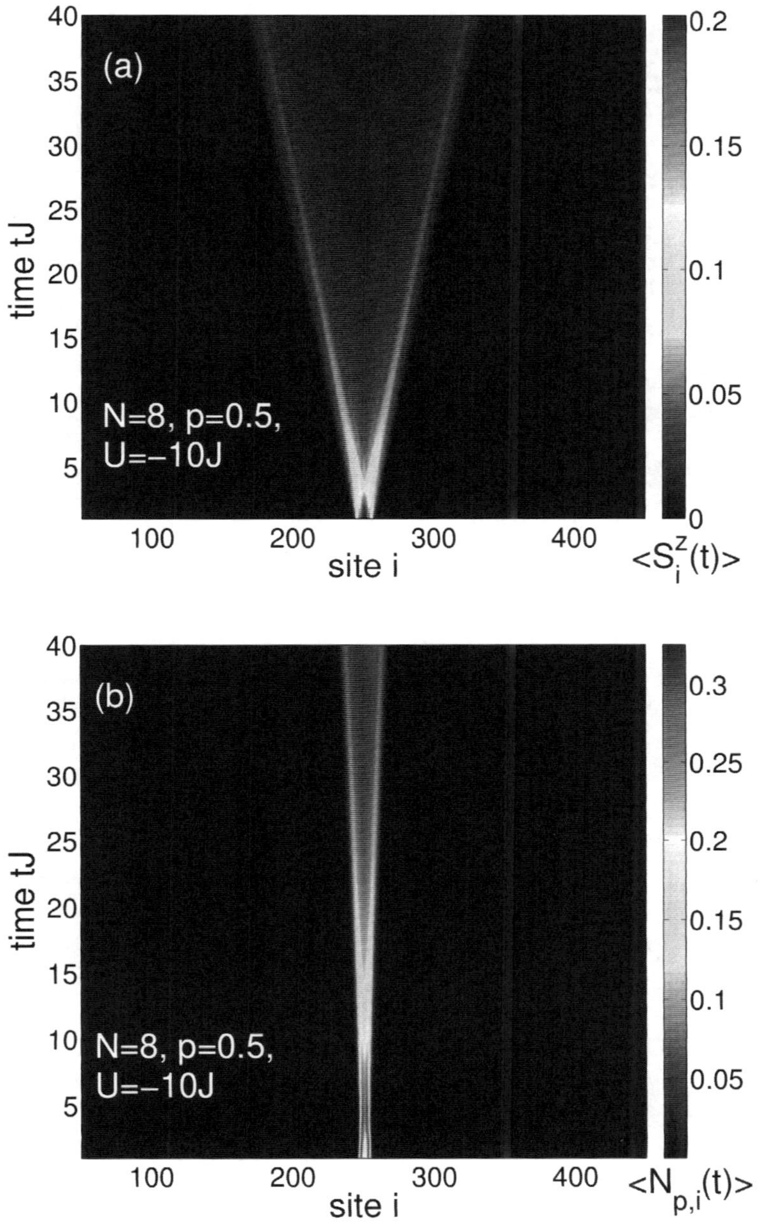

Figure 1. Typical contour plots of the excess-fermion (or spin) density $\langle S_i^z(t)\rangle$ and the pair density $\langle N_{p,i}(t)\rangle$, panels (a) and (b) respectively, during the expansion starting from the ground state of a polarized attractive Fermi gas ($|U| = 10J$, $N = 8$, $p = 1/2$ and $L_0 = 10$). In here, $N_{p,i}$ is defined as the site double occupancy ($n_{\uparrow,i}n_{\downarrow,i}$), which is only an approximate proxy for the local pair density but has the advantages of being experimentally accessible and constituting an appropriate measure to illustrate the quantum distillation process. The slanted enveloping contours indicate the different maximum speeds at which the excess fermions and pairs, respectively, propagate during the expansion.

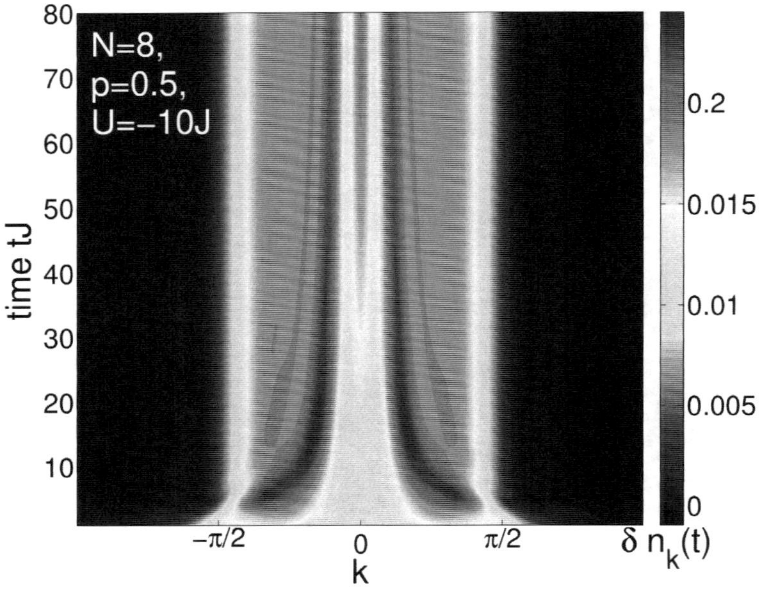

Figure 2. Contour plots for the momentum distribution function of the excess (unpaired) fermions defined as the spin-sector of the time-dependent Fourier transforms of the single-particle density matrices, $\delta n_k(t) = n_{k,\uparrow}(t) - n_{k,\downarrow}(t)$ (cf. Ref. [28]). The parameters are the same as in Fig. 1 above (*i.e.*, $|U| = 10J$, $N = 8$, $p = 1/2$ and $L_0 = 10$). Contour lines becoming vertical indicate the convergence of that part of the distribution to its asymptotic large-time limit.

This mechanism is possible thanks to a quantum-distillation-like process that takes place driven by the dynamics of the interacting system (for more details and further results see [28]). It is during this initial stage that the 1D-FFLO characteristic spatial coexistence of pairs and unpaired atoms is lost and ulterior direct probing of the modulated superfluid order is no longer possible.

3.2. Momentum Distributions

After reaching a *steady* expansion regime, the real-space densities can be scaled into fixed-in-time distributions that correspond to asymptotic momenta. Numerically, we can see this more clearly by going into reciprocal space and studying the expansion of the MDFs. This procedure clearly shows that the distributions quickly converge towards an asymptotic limit as illustrated by Fig. 2. After an initial complicated transient regime, we find that the outer shape of the MDF of the excess fermions becomes stationary for $t \geq 10/J$ and, slowly, that convergence moves towards the center of the Brillouin zone. This is because the faster particles quickly order themselves according to their relative speeds and effectively stop interacting, while for pairs and unpaired atoms with smaller momentum this process takes longer.

4. Analytical Results

We turn now to analytical considerations and exploit our choice of model (a Hubbard chain) by making use of its integrability. We consider the case of N fermions in a lattice with open boundary conditions (OBCs) and label their *rapidities* as $\{\kappa_j\}_{j=1}^{j=N}$. These *charge* rapidities (augmented by a second set of *spin* rapidities, $\{\lambda_m\}_{m=1}^{m=M}$ with $M = N_\downarrow$; which turn out to play only an auxiliary role in the ground state of the attractive case) constitute the integrals of motion that fully characterize any eigenstate of the system. To determine them, it is possible to write down a set of $N+M$ coupled algebraic equations known as the Bethe-Ansatz Equations (BAEs). These equations for the case of the Hubbard model are "trigonometric" and can be conveniently written in terms of the phase parameters $z_j = e^{i\kappa_j}$. For the case of OBCs, the equations are (cf. Ref. [29, 30, 31]):

$$z_j^{2L+2} = \prod_n \frac{\left(z_j^2-1\right)+z_j\left(i\lambda_n-c\right)}{\left(z_j^2-1\right)+z_j\left(i\lambda_n+c\right)} \frac{\left(z_j^2-1\right)-z_j\left(i\lambda_n+c\right)}{\left(z_j^2-1\right)-z_j\left(i\lambda_n-c\right)} \tag{2}$$

$$\prod_{m\neq n} \frac{\lambda_n+\lambda_m+2ic}{\lambda_n+\lambda_m-2ic}\frac{\lambda_n-\lambda_m+2ic}{\lambda_n-\lambda_m-2ic} = \prod_j \frac{\left(i\lambda_n-c\right)z_j+\left(z_j^2-1\right)}{\left(i\lambda_n+c\right)z_j+\left(z_j^2-1\right)}\frac{\left(i\lambda_n-c\right)z_j-\left(z_j^2-1\right)}{\left(i\lambda_n+c\right)z_j-\left(z_j^2-1\right)} \tag{3}$$

which can be seen to have a $z \to 1/z$ and $\lambda \to -\lambda$ symmetry (and the solutions need to be restricted accordingly to avoid double counting). Starting from this form of the BAEs, one can readily multiply out by the denominators and get high-order polynomial equations, which are more suitable for practical numerical solution using computer-algebra methods, provided that N and L are not too large. A direct solution confirms (as expected by continuity with the $L \to \infty$ thermodynamic limit) that the charge rapidities for the ground state appear as two sets: (i) $N-2M$ real rapidities corresponding to the unpaired fermions, and (ii) M pairs of complex rapidities corresponding to the paired fermions. An example of these rapidities for the same values of the parameters as those used in the figures above is given in Table 1.

Table 1. Results for the analytical parametrization of the initial pre-quench ground state of the system. The system parameters are the same as in Figs. 1 and 2 above (*i.e.*, $|U| = 10J$, $N = 8$, $p = 1/2$ and $L_0 = 10$). The Bethe-Ansatz and t-DMRG eigenstate solutions are matched by confirming that their eigenenergies coincide within a small numerical error.

Rapidity Type	Order	Numerical Value
Real	κ_1	0.325953
	κ_2	0.646688
	κ_3	0.958605
	κ_4	1.260180
Complex	$\kappa'_{5,6}$	$0.347184 \pm i1.70459$
	$\kappa'_{7,8}$	$0.351801 \pm i1.70616$

The ground state wave function for an L_0-size chain when the box trap is present, has to be projected into the Bethe-Ansatz basis for an L-size chain after the geometric quench takes place. In the limit of large L, this produces a continuous distribution that is in one-to-one correspondence with the asymptotic momentum distribution of the system. As time progresses during the expansion that follows the quench, the MDFs of the paired and unpaired atoms gradually approach the limiting value given by these two rapidity distributions (for the case of the complex valued rapidities, one focuses on the real parts). These rapidity distributions encode the expectation values of N integrals of motion of the system and are thus constrained by the quantum dynamics to remain invariant after the quench (since the Hamiltonian is constant and commutes with them). Given the structure of the Bethe-Ansatz wave function, they thus constitute the necessary asymptotic limit for the MDFs in the *steady-expansion* regime during

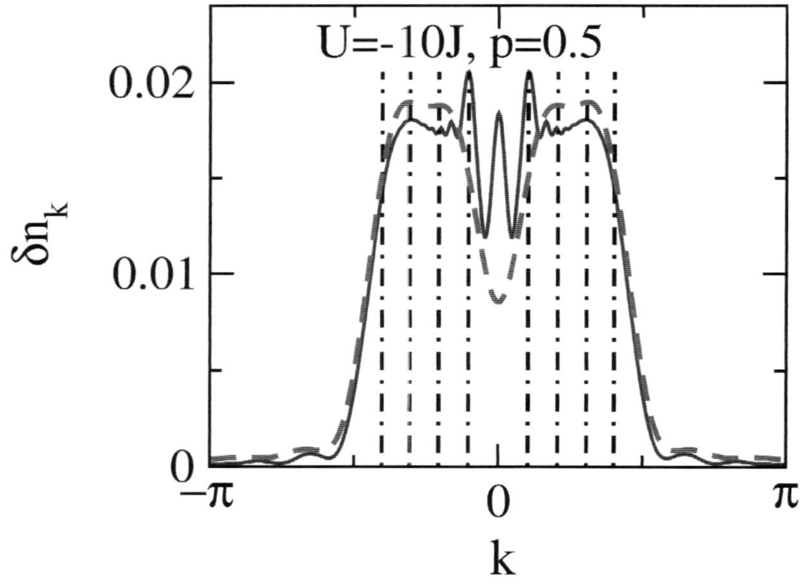

Figure 3. Comparison of the MDF of the excess-fermion gas δn_k (blue solid line) and the predicted asymptotic limit obtained from modeling the geometric quench (red dashed line). The MDF is obtained after a long-time expansion ($t \approx 80/J$) starting from a box-trap potential of size $L_0 = 10$ that was turned off at $t = 0$ and the system was allowed to expand and modeled using t-DMRG. The rapidity distribution after the geometric quench was calculated starting from the Bethe-Ansatz data of Table 1 indicated in the figure by the vertical lines (black dot-dashed lines) and using a convolution formula with no fitting parameters (see [28] for more details). The system parameters are the same as in the previous figures (*i.e.*, $|U| = 10J$, $N = 8$ and $p = 1/2$).

which boundary conditions play no role (formally corresponding to $L \to \infty$). A direct detailed comparison of these computational and analytic results is given below, while a demonstration of the rapid convergence of the former to the latter was already presented elsewhere [28].

4.1. Comparison with the Computational Results

To illustrate the convergence of the MDFs to the distributions of Bethe-Ansatz rapidities obtained after the quench, we show in Fig. 3 results for the excess-fermion distribution obtained as a single time slice of Fig. 2 for a sufficiently large time and the projected rapidity distribution using the $\{\kappa_j\}_{j=1}^{j=4}$ values from Table 1. The convergence can be convincingly observed in the tails of the distributions, but is not yet present at the center of the Brillouin zone. We point out, however, that the agreement improves systematically with increasing time, as expected from the analytic considerations given above.

An analytic calculation of the effect of the geometric quench in smearing the distribution of rapidities in the initial ground state inside the box trap proceeds via a single-particle approximation. We observe that the overlap between the post-quench state and the pre-quench eigenstate has a maximum amplitude for components of the former with the same set of rapidities as the latter (cf. [32]). We proceed to identify the distribution of real-valued charge rapidities obtained from the BAEs with that of unpaired fermions (δn_k) and model the quench by convolving the pre-quench rapidities with the (periodized) kernel $L_0 \mathrm{sinc}^2(kL_0/2)$. This procedure is motivated by the exact result for the release of a single particle from a box and is expected to work better the larger the polarization of the system is. An exact calculation is also possible, in principle, by using the exact Bethe-Ansatz states, but a comparison with the asymptotic MDFs going beyond the present level of detail would require the simulation of much larger times (going beyond the current state of the art in t-DMRG). The convergence of momentum distributions into rapidity distributions during an expansion has also been seen for infinitely repulsive bosons [21, 22, 23], for which it is possible to calculate much longer expansion times (due to the availability of a mapping to free fermions).

5. Conclusions

To summarize, we have studied the expansion dynamics of attractive spin-imbalanced ultracold atomic clouds of fermions in 1D geometries. We have shown that while the signatures of the 1D-FFLO state are quickly lost during the expansion, the integrals of motion are respected and emerge as asymptotic momentum distributions. Since one finds that the MDFs of paired and unpaired atoms rapidly take a stationary form (specially so for the case of the excess unpaired atoms), they should be accessible on typical experimental time scales. Since these integrals of motion fully characterize the state of the 1D Hubbard model, they constitute an indirect probe of the character of the ground state of the system before the quench.

Here we concentrated on geometric quenches which are easier to handle in the Bethe Ansatz because the information about the boundary conditions enters only through the rapidities. Other types of quenches are also calculable, but are more involved. One would like to consider, for instance, interaction quenches. In particular, it would be interesting to study the effects of turning off the interactions (or even switching their sign). This could be combined with a simultaneous release of the system to study its expansion in the absence of interactions. The challenge would be to identify a protocol that would best lend itself for the probing of the 1D-FFLO state (see Refs. [33, 34] for examples of such earlier attempts). The resulting synergy between experiments and theory is likely to bring in new insights into the properties of polarized superfluid matter, that could in turn be applicable in many different contexts.

Acknowledgments

We would like to thank the organizers for giving us the opportunity to present this work during the 21st annual International Laser Physics Workshop (LPHYS'12) which took place in Calgary, Canada, in July 2012. We (CJB, FHM, and MR) thank the KITP at Santa Bárbara, for its hospitality (supported by NSF under grant No. PHY05-51164) during the initial stages of this work. We acknowledge support from the DARPA OLE program through ARO W911NF-07-1-0464 (CJB), the Deutsche Forschungsgemeinschaft through FOR 801 (FHM and SL) and FOR 912 (SL), the Office of Naval Research (MR), and the Australian Research Council Centre of Excellence for Engineered Quantum Systems and the Discovery Projects funding scheme under project number DP1092513 (IPM).

References

[1] Liao Y-a, Rittner A S C, Paprotta T, Li W, Partridge G B, Hulet R G, Baur S K and Mueller E J 2010 *Nature* **467** 567–569
[2] Yang K 2001 *Phys. Rev. B* **63** 140511
[3] Orso G 2007 *Phys. Rev. Lett.* **98** 070402
[4] Hu H, Liu X J and Drummond P D 2007 *Phys. Rev. Lett.* **98** 070403
[5] Feiguin A and Heidrich-Meisner F 2007 *Phys. Rev. B* **76** 220508(R)
[6] Guan X W, Batchelor M T, Lee C and Bortz M 2007 *Phys. Rev. B* **76** 085120
[7] Batrouni G G, Huntley M H, Rousseau V G and Scalettar R T 2008 *Phys. Rev. Lett.* **100** 116405
[8] Casula M, Ceperley D M and Mueller E J 2008 *Phys. Rev. A* **78** 033607
[9] Lüscher A, Noack R M and Läuchli A 2008 *Phys. Rev. A* **78** 013637
[10] Rizzi M, Polini M, Cazalilla M, Bakhtiari M, Tosi M and Fazio R 2008 *Phys. Rev. B* **77** 245105

[11] Tezuka M and Ueda M 2008 *Phys. Rev. Lett.* **100** 110403
[12] Kakashvili P and Bolech C J 2009 *Phys. Rev. A* **79** 041603
[13] Heidrich-Meisner F, Orso G and Feiguin A 2010 *Phys. Rev. A* **81** 053602
[14] Sun K and Bolech C J 2012 *Phys. Rev. A* **85** 051607(R)
[15] Fulde P and Ferrell R A 1964 *Phys. Rev.* **135** A550–A563
[16] Larkin A I and Ovchinnikov Y N 1965 *Sov. Phys. JETP* **20** 762–769
[17] Singleton J *et al* 2000 *J. Phys.: Condens. Matter* **12** L641 ; Yonezawa S *et al* 2008 Phys. Rev. Lett. **100**, 117002. See also : Bolech C J and Giamarchi T 2004 Phys. Rev. Lett. **92** 127001 ; Bolech C J and Giamarchi T 2005 Phys. Rev. B **71** 024517
[18] Radovan H A *et al* 2003 *Nature* **425** 51 ; Miclea C F *et al* 2006 Phys. Rev. Lett. **96** 117001
[19] Sedrakian A, Clark J W and Alford M G (eds) 2006 *Pairing in Fermionic Systems: Basic Concepts and Modern Applications* (Singapore: World Scientific)
[20] Feiguin A, Heidrich-Meisner F, Orso G and Zwerger W 2011 *Lect. Not. Phys.* **836** 503
[21] Rigol M and Muramatsu A 2005 *Phys. Rev. Lett.* **94** 240403
[22] Minguzzi A and Gangardt D M 2005 *Phys. Rev. Lett.* **94** 240404
[23] Rigol M and Muramatsu A 2005 *Mod. Phys. Lett.* **19** 861
[24] Daley A, Kollath C, Schollwöck U and Vidal G 2004 *J. Stat. Mech.: Theory Exp.* P04005
[25] White S R and Feiguin A E 2004 *Phys. Rev. Lett.* **93** 076401
[26] Heidrich-Meisner F, Manmana S R, Rigol M, Muramatsu A, Feiguin A E and Dagotto E 2009 *Phys. Rev. A* **80** 041603
[27] Essler F H L, Frahm H, Göhmann F, Klümper A and Korepin V E 2004 *The One-Dimensional Hubbard Model* (Cambridge University Press)
[28] Bolech C J, Heidrich-Meisner F, Langer S, McCulloch I P, Orso G and Rigol M 2012 *Phys. Rev. Lett.* **109** 110602
[29] Sklyanin E K 1988 *J. Phys. A* **21** 2375
[30] Zhou H Q 1996 *Phys. Rev. B* **54** 41–43
[31] Guan X 2000 *J. Phys. A: Math. Gen.* **33** 5391
[32] Mossel J, Palacios G and Caux J S 2010 *J. Stat. Mech* L09001
[33] Kajala J, Massel F and Törmä P 2011 *Phys. Rev. A* **84** 041601(R)
[34] Lu H, Baksmaty L O, Bolech C J and Pu H 2012 *Phys. Rev. Lett.* **108** 225302

Vortex Phases of Rotating Superfluids

Michele Correggi[1], Florian Pinsker[2], Nicolas Rougerie[3] and Jakob Yngvason[4]

[1]Dipartimento di Matematica, Università degli Studi Roma Tre, Largo San Leonardo Murialdo 1, 00146, Roma, Italy,
[2]DAMTP, University of Cambridge, Wilbertforce Road, Cambridge CB3 0WA, UK,
[3]Université Grenoble 1 & CNRS, LPMMC (UMR 5493), B.P. 166, 38 042 Grenoble, France,
[4]Faculty of Physics, University of Vienna, Boltzmanngasse 5, and Erwin Schrödinger Institute for Mathematical Physics, Boltzmanngasse 9, A-1090 Vienna, Austria

E-mail: [1]michele.correggi@gmail.com, [2]florian.pinsker@gmail.com, [3]nicolas.rougerie@grenoble.cnrs.fr, [4]jakob.yngvason@univie.ac.at

Abstract. We report on the first mathematically rigorous proofs of a transition to a giant vortex state of a superfluid in rotating anharmonic traps. The analysis is carried out within two-dimensional Gross-Pitaevskii theory at large coupling constant and large rotational velocity and is based on precise asymptotic estimates on the ground state energy. An interesting aspect is a significant difference between 'soft' anharmonic traps (like a quartic plus quadratic trapping potential) and traps with a fixed boundary. In the former case vortices persist in the bulk until the width of the annulus becomes comparable to the size of the vortex cores. In the second case the transition already takes place in a parameter regime where the size of vortices is very small relative to the width of the annulus. Moreover, the density profiles in the annulus are different in the two cases. In both cases rotational symmetry of the density in a true ground state is broken, even though a symmetric variational ansatz gives an excellent approximation to the energy.

1. Introduction

A superfluid confined in a rotating anharmonic trap, where the rotation speed can in principle be arbitrarily large, undergoes several phase transitions as the speed increases. At first the fluid is vortex free [22, 3] but then quantized vortices emerge, eventually forming a vortex lattice [6, 16, 2, 23, 1, 7, 17] that may persist even when the speed is so large that the centrifugal force creates a 'hole' with strongly depleted density in the middle of the trap [19, 8, 9, 10]. Above a certain rotation speed a transition to a *giant vortex state* takes place. In this state the vortices disappear from the annulus where the bulk of the superfluid is concentrated while a macroscopic phase circulation remains. This phenomenon has been studied theoretically by variational and numerical methods in the past [17, 18, 19, 20, 24, 26, 15] but mathematically rigorous proofs of the giant vortex transition have been obtained only very recently [31, 11, 12, 13, 14, 32]. An experimental realization of this transition appears to be still out of reach although anharmonic traps have been available already for some time [4, 34, 21, 37]. In the following we report on the main findings of this analysis with emphasis on [13, 14].

2. Setting the Stage

2.1. *The basic many-body Hamiltonian*

The quantum mechanical Hamiltonian for N spinless bosons with an external potential, V, and a pair interaction potential, v, in a rotating frame with angular velocity $\mathbf{\Omega}_{\rm rot}$ is

$$H=\sum_{j=1}^{N}\left(-\tfrac{1}{2}\nabla_j^2+V(\mathbf{x}_j)-\mathbf{L}_j\cdot\mathbf{\Omega}_{\rm rot}\right)+\sum_{1\leq i<j\leq N}v(|\mathbf{x}_i-\mathbf{x}_j|).$$

Here $\mathbf{x}_j\in\mathbb{R}^3$ and $\mathbf{L}_j=-{\rm i}\mathbf{x}_j\times\nabla_j$ is the angular momentum of the jth particle. The Hamiltonian can alternatively be written in the 'magnetic' form

$$H=\sum_{j=1}^{N}\left\{\tfrac{1}{2}({\rm i}\nabla_j+\mathbf{A}(\mathbf{x}_j))^2+V(\mathbf{x}_j)-\tfrac{1}{2}\Omega_{\rm rot}^2r_j^2\right\}+\sum_{1\leq i<j\leq N}v(|\mathbf{x}_i-\mathbf{x}_j|)\qquad(1)$$

with the vector potential

$$\mathbf{A}(\mathbf{x})=\mathbf{\Omega}_{\rm rot}\times\mathbf{x}=\Omega_{\rm rot}r\,\mathbf{e}_\vartheta$$

where r denotes the distance from the rotation axis and $\mathbf{e}_\vartheta$ the unit vector in the angular direction. This way of writing the Hamiltonian corresponds to the splitting of the rotational effects into Coriolis and centrifugal forces.

2.2. *Harmonic vs. anharmonic traps*

If V is a harmonic oscillator potential in the direction $\perp$ to $\mathbf{\Omega}_{\rm rot}$, i.e.,

$$V(\mathbf{x})=\tfrac{1}{2}\Omega_{\rm trap}r^2+V^{\|}(z)$$

then stability requires $\Omega_{\rm rot}<\Omega_{\rm trap}$. Rapid rotation means here that

$$\Omega_{\rm rot}\to\Omega_{\rm trap}$$

from below. On the other hand, if V is anharmonic and increases faster than quadratically in the direction $\perp$ to $\Omega_{\rm rot}$, e.g. $V(\mathbf{x})\sim r^s+V^{\|}(z)$ with $s>2$, then $\Omega_{\rm rot}$ can in principle be as large as one pleases and rapid rotation means simply $\Omega_{\rm rot}\to\infty$.

These two cases are quite different both physically and mathematically. The former leads to an effective many-body Hamiltonian in the lowest Landau level of the magnetic kinetic energy term in (1) and bosonic analogues of the Fractional Quantum Hall Effect (see [39, 38, 7, 29]). In the case of rapid rotation in an anharmonic trap, as considered here, it is usually sufficient to employ Gross-Pitaevskii (GP) theory for an effective description. We remark, however, that a small anharmonic term appropriately tuned can also lead to interesting modifications of the Quantum Hall states of harmonic traps [33].

2.3. *The Gross-Pitaevskii limit theorem*

The following basic fact about the many-body Hamiltonian (1) for $N\to\infty$ with Na and $\Omega_{\rm rot}$ *fixed*, where a is the scattering length of the (repulsive, short range) interaction potential v was proved in [27]:

There is (possibly fractionated) Bose-Einstein condensation in the ground state as $N\to\infty$, and the wave function of the condensate ("superfluid order parameter") is a minimizer (in general not unique) of the GP energy functional

$$\mathcal{E}^{\rm GP}_{\rm 3D}[\Psi]=\int_{\mathbb{R}^3}\left\{\tfrac{1}{2}\left|({\rm i}\nabla+\mathbf{A})\,\Psi\right|^2+(V-\tfrac{1}{2}\Omega_{\rm rot}^2r^2)|\Psi|^2+2\pi Na\,|\Psi|^4\right\}\qquad(2)$$

Journal of Physics: Conference Series **414** (2013) 012034 doi:10.1088/1742-6596/414/1/012034

with $\int |\Psi|^2 = 1$. The Gross-Pitaevskii partial differential equation

$$\left\{ (\mathrm{i}\nabla + \mathbf{A})^2 + (V - \tfrac{1}{2}\Omega_{\mathrm{rot}}^2 r^2) + 4\pi Na|\Psi|^2 \right\}\Psi = \mu\Psi$$

with the chemical potential μ is the variational equation corresponding to this minimization problem.

The rigorous proof of this theorem is far from simple, as can be seen from the fact that a Hartree variational ansatz for the Hamiltonian (1) (that would anyhow only lead to an upper bound) is meaningless if the interaction potential has a hard core. Even for 'soft' potentials a naive computation would not lead to (2) with the scattering length as parameter, but rather the integral $\int v$ (that gives only the lowest Born approximation to the scattering length). For the mathematical background of this and related results [28] may be consulted. A limit theorem that holds uniformly the parameters Ω_{rot} and Na as $N \to \infty$, but is restricted to the leading order, was proved in [5].

2.4. 2D Gross-Pitaevskii theory in anharmonic traps

The GP minimization problem has two parameters, Ω_{rot} and Na. We shall be concerned with phenomena that occur in anharmonic traps in the asymptotic regime where both Ω_{rot} and Na are large. For convenience introduce

$$\varepsilon \equiv (2\pi Na)^{-1/2}$$

which is small if Na is large. (In appropriate units ε is the 'healing length'.)

For traps that are sufficiently elongated along the rotational axis (z-direction) the properties of the condensate are to a good approximation independent of z and we may consider a 2D[1] energy functional

$$\mathcal{E}_{\mathrm{2D}}^{\mathrm{GP}}[\Psi] = \int_{\mathbb{R}^2} \left\{ \tfrac{1}{2} |(\mathrm{i}\nabla + \mathbf{A})\,\Psi|^2 + (V - \tfrac{1}{2}\Omega_{\mathrm{rot}}^2 r^2)|\Psi|^2 + \frac{|\Psi|^4}{\varepsilon^2} \right\}$$

with a trap potential of the form (for simplicity)

$$V(r) = kr^s \tag{3}$$

with $s > 2$, $k > 0$. Here Ω_{rot} can be arbitrary large.

The limiting case $s \to \infty$ corresponds to a 'flat' trap with fixed boundary at $r = 1$. The effective potential is then simply $-\frac{1}{2}\Omega_{\mathrm{rot}}^2 r^2$ and the integration is limited to the unit disc in $\mathbb{R}^2$. A word of caution: The limit $s \to \infty$ can not be interchanged with the limits $\varepsilon \to 0$, $\Omega_{\mathrm{rot}} \to \infty$ as discussed in Section 3.6 below.

The analysis of the GP minimizer is guided by the following heuristics:

- A vortex, i.e., a zero of the wave function $\Psi(\mathbf{x}) = |\Psi(\mathbf{x})| \exp(\mathrm{i}\theta(\mathbf{x}))$ with an accompanying nonzero winding number of the phase factor, reduces the kinetic energy because the associated current $\sim \nabla\theta(\mathbf{x})$ compensates partly the velocity field generated by $\mathbf{A}(\mathbf{x}) = \mathbf{\Omega}_{\mathrm{rot}} \times \mathbf{x}$.
- A vortex causes also a change in the density, however, (mass is moved from the vortex core to the bulk) and this increases the interaction energy that depends on the density at the potential location of the vortex. The energy balance decides whether or not a vortex is favorable, and if that is the case, the size of the vortex core.
- A vortex is the more costly the higher the density. At sufficiently high rotational velocities the compression due to centrifugal forces creates a 'hole' and the density in the bulk increases until, at some point, vortices become too costly.

[1] A 2D description is, of course, also appropriate in thin traps where the motion along the z-axis is 'frozen' [36].

2.5. Scaling of the energy functional

The effective potential $(kr^s - \frac{1}{2}\Omega_{\mathrm{rot}}^2 r^2)$ has a unique minimum at $r = (\Omega_{\mathrm{rot}}^2/(sk))^{1/(s-2)}$. Taking this as a length unit we obtain the scaled energy functional

$$\mathcal{E}^{\mathrm{GP}}[\psi] = \int_{\mathbb{R}^2} \left\{ \tfrac{1}{2}|(\mathrm{i}\nabla + \Omega x \mathbf{e}_\vartheta)\psi|^2 + \Omega^2 W(x)|\psi|^2 + \varepsilon^{-2}|\psi|^4 \right\}$$

where $x = |\mathbf{x}|$, $\Omega \sim \Omega_{\mathrm{rot}}^{(s+2)/(s-2)}$, and

$$W(x) = \left(\tfrac{1}{s}x^s - \tfrac{1}{2}x^2\right).$$

The scaled potential has a minimum at $x = 1$, independent of Ω.

3. Analysis of the GP Minimizers

3.1. Critical velocities

The basic facts for traps of the form (3) with $2 < s < \infty$ can be summarized as follows.

As Ω increases there are **three critical velocities:**

- $\Omega'_{c1} \sim |\log \varepsilon|$ marking the appearance of the first vortex.[2]
- $\Omega_{c2} \sim \varepsilon^{-1}$ marking the appearance of a 'hole' due to the centrifugal forces.
- $\Omega_{c3} \sim \varepsilon^{-4}$ marking the transition to a 'giant vortex'.

For the first transition we refer to [2, 22, 1, 30]. For $\Omega_{c1} \ll \Omega \ll \Omega_{c3}$ the ground state energy is well approximated by assuming a triangular vortex lattice in the bulk.[3] In the limit $\varepsilon \to 0$ the vorticity becomes uniformly distributed with density Ω [14]. For $\Omega > \Omega_{c3}$ the bulk is free of vortices but a macroscopic circulation around the origin remains [13, 14].

3.2. The vortex lattice regime

The ground state energy for $\Omega_{c1} \ll \Omega \ll \Omega_{c3}$ can be computed exactly to subleading order [14]:

Theorem 1 (Energy between Ω_{c2} and Ω_{c3}) *If $\varepsilon^{-1} \lesssim \Omega \ll \varepsilon^{-4}$ as $\varepsilon \to 0$, then*

$$E^{\mathrm{GP}} = E^{\mathrm{TF}} + \tfrac{1}{6}\Omega|\log(\varepsilon^4\Omega)|(1 + o(1)).$$

Here E^{TF} denotes the energy without the kinetic term. Below Ω_{2c} a similar formula holds (with a different scaling, $\varepsilon^{-2/(s+2)}$ as length unit):

Theorem 2 (Energy between Ω_{c1} and Ω_{c2}) *If $|\log \varepsilon| \ll \Omega' \lesssim \varepsilon^{-1}$ as $\varepsilon \to 0$, then*

$$E^{\mathrm{GP}'} = E^{\mathrm{TF}'} + \tfrac{1}{2}\Omega'|\log(\varepsilon^2\Omega')|(1 + o(1)).$$

3.3. Vortices reduce kinetic energy

The potential term $\sim \Omega^2$ and the interaction term $\sim \varepsilon^{-2}$ become comparable when

$$\Omega \sim \varepsilon^{-1}.$$

This is the order of the second critical speed Ω_{c2} above which the centrifugal force creates a 'hole'. In the sequel we shall focus on rotation speeds around and above Ω_{c2} which means that $\Omega \gtrsim \varepsilon^{-1}$. In this regime the kinetic energy term $\frac{1}{2}|(\mathrm{i}\nabla + \mathbf{A})\Psi|^2$ is formally also of order $1/\varepsilon^2$ if $\Omega \sim 1/\varepsilon$. Its contribution to the energy is, however, of lower order, namely $\sim \Omega|\log \varepsilon|$, because a lattice of vortices emerges as $\varepsilon \to 0$ and reduces the kinetic energy as remarked in 2.4.

[2] Here $\Omega' \sim \varepsilon^{-4/(s+2)}\Omega_{\mathrm{rot}}$. This scaling is more convenient than $\Omega \sim \Omega_{\mathrm{rot}}^{(s+2)/(s-2)}$ for $\Omega \ll 1/\varepsilon$.

[3] The reason why a triangular arrangement with hexagonal unit cells is optimal amongst regular lattices can be made plausible by appealing to an electrostatic analogy and Newton's theorem [10]: Hexagonal cells are as close to being circular as possible and thus have smaller multipole moments and lower interaction energy than other cells.

3.4. The giant vortex regime

Consider a variational ansatz for the wave function of the form[4]

$$\psi(\mathbf{x}) = g(\mathbf{x})\exp(\mathrm{i}\Omega\vartheta)$$

with a real valued function g, normalized such that $\int g^2 = 1$.This gives

$$\mathcal{E}^{\mathrm{GP}}[\psi] = \int_{\mathbb{R}^2} \left\{\tfrac{1}{2}|\nabla g|^2 + \tfrac{1}{2}\Omega^2(x - x^{-1})^2 g^2 + \Omega^2\left(\tfrac{1}{s}x^s - \tfrac{1}{2}x^2\right)g^2 + \varepsilon^{-2}g^4\right\} \equiv \mathcal{E}^{\mathrm{gv}}[g].$$

The unique positive minimizer g_{gv} of $\mathcal{E}^{\mathrm{gv}}$ is rotationally symmetric and we denote the corresponding energy by E^{gv}.

The following results are proved in [13, 14].

Theorem 3 (Energy in the giant vortex regime) *There is a constant* $0 < \bar{\Omega}_0 < \infty$ *such that for* $\Omega = \Omega_0\,\varepsilon^{-4}$ *with* $\Omega_0 > \bar{\Omega}_0$ *the ground state energy is*

$$E^{\mathrm{GP}} = E^{\mathrm{gv}} + O(|\log\varepsilon|^{9/2}).$$

Theorem 4 (Absence of vortices in the bulk) *There is a constant* $c > 0$ *such that for* $\Omega = \Omega_0\,\varepsilon^{-4}$ *with* $\Omega_0 > \bar{\Omega}_0$ *and* ε *sufficiently small the minimizer* ψ^{GP} *is free of zeros in the annulus*

$$\mathcal{A} = \{\mathbf{x} : |1 - x| \le c\Omega^{-1/2}|\log\varepsilon|^{1/2}\}.$$

3.5. On the proof of the GV transition

The main issue is a precise lower bound to the energy. We restrict $\mathcal{E}^{\mathrm{gv}}$ to the annulus $\mathcal{A}$, obtaining a positive minimizer g_0. Define $u(\mathbf{x})$ on the annulus by writing

$$\psi^{\mathrm{GP}}(\mathbf{x}) = g_0(x)u(\mathbf{x})\exp(\mathrm{i}\Omega\vartheta).$$

The function u contains all possible zeros of ψ^{GP} in the annulus.

The variational equation for g leads to the lower bound

$$E^{\mathrm{GP}} \ge E^{\mathrm{gv}}_{\mathcal{A}} + \mathcal{E}_{\mathcal{A}}[u]$$

with a functional of Ginzburg-Landau type with g_0^2 as weight:

$$\mathcal{E}_{\mathcal{A}}[u] = \int_{\mathcal{A}} g_0^2\left\{\tfrac{1}{2}|\nabla u|^2 - \mathbf{B}\cdot\mathbf{J}(u) + \varepsilon^{-2}g_0^2(1 - |u|^2)^2\right\}$$

where $\mathbf{B} = \Omega\,(x - x^{-1})\,\mathbf{e}_\vartheta$ and $\mathbf{J}(u) = \frac{\mathrm{i}}{2}(u\nabla u^* - u^*\nabla u)$. The main task is to estimate the negative term $-\int g_0^2\,\mathbf{B}\cdot\mathbf{J}(u)$.

For this purpose one writes $g^2\mathbf{B} = \nabla^\perp F$ with $\nabla^\perp = (-\partial_{x_2}, \partial_{x_1})$ and a potential function F. Integration by parts and estimates of F (this is the key point!) give

$$\int_{\mathcal{A}} g^2\left\{\tfrac{1}{2}|\nabla u|^2 - \mathbf{B}\cdot\mathbf{J}(u)\right\} \ge -C\Omega_0^2|\log\varepsilon|^{3/2}$$

leading to the lower energy bound.

A consequence of this bound, combined with the variational upper bound $E^{\mathrm{gv}}_{\mathcal{A}} \le 0$ is an upper bound on the interaction term for large Ω_0:

$$\int_{\mathcal{A}} \varepsilon^{-2}g_0^4(1 - |u|^2)^2 \le C\Omega_0^2|\log\varepsilon|^{3/2}.$$

Together with the upper bound to the kinetic energy and standard inequalities this implies that u must be close to 1, in particular free of zeros.

[4] For simplicity of notation we assume that Ω is an integer which is justified since $\Omega \to \infty$.

3.6. Comparison with the 'flat' case

The flat case, $s = \infty$, that is treated in detail in [11, 12], differs from the case $s < \infty$ in several respects:

- The GV transition takes place at $\Omega \sim \varepsilon^{-2}|\log\varepsilon|^{-1}$ rather than $\Omega \sim \varepsilon^{-4}$.
- The density profile in the GV regime is of 'Thomas-Fermi' type in the 'flat' case, but for $s < \infty$ it is gaussian around $x = 1$.
- The 'last' vortices before the GV transition have size $\sim \varepsilon^{3/2}$ that is much smaller than the thickness of the annulus $\sim \varepsilon|\log\varepsilon|$. For $s < \infty$ the size of vortices, $\sim \varepsilon^2$ and the size of the annulus, $\sim \varepsilon^2|\log\varepsilon|^{1/2}$, are almost comparable.

The techniques of proof in the two cases are also by necessity different: While *vortex ball constructions* and subsequent *jacobian estimates* (see [35]) for the potential function are applicable for the 'small' vortices in a 'flat' trap they are useless for $s < \infty$ and new ideas are required.

3.7. Circulation and symmetry breaking

At low rotation speeds below the onset of the second vortex the GP minimizer has rotationally symmetric density, but a vortex lattice clearly breaks the symmetry. On the other hand, the giant vortex variational ansatz, that gives an excellent approximation to the energy and circulation for $\Omega_0 > \bar{\Omega}_0$, is an eigenfunction of angular momentum. A true minimizer does not have this property, however:

Theorem 5 (Circulation and rotational symmetry breaking) *In the giant vortex regime* $\Omega_0 > \bar{\Omega}_0$ *the circulation of any GP minimizer is* $2\pi\,\Omega + O(\Omega_0\,|\log\varepsilon|^{9/4})$, *but no minimizer is an eigenfunction of angular momentum.*

Theis result holds both for $s < \infty$ and $s = \infty$ [11]–[14].

4. Summary

The study of the GP equation for dilute Bose gases in rotating, anharmonic reveals a surprising rich landscape, both from the mathematical and physical point of view. Detailed analysis can be carried out in an asymptotic regime where both the coupling constant and the rotational speed are large. Among the results found are:

- Energy asymptotics corresponding to a distribution of vorticity in a lattice of vortices for $\Omega_{c1} \ll \Omega \ll \Omega_{c3}$.
- Emergence of a 'hole' with strongly depleted density above a critical rotation speed $\Omega_{c2} \sim \varepsilon^{-1}$.
- Transition to a 'giant vortex' state above $\Omega_{c3} \sim \varepsilon^{-4}$ where the vortex lattice disappears from the bulk and all vorticity resides in the 'hole', creating a macroscopic circulation in the bulk.
- Breaking of rotational symmetry, also in the giant vortex regime.

Acknowledgments

M.C. gratefully acknowledges financial support from the European Research Council under the European Unions Seventh Framework Programme ERC Starting Grant CoMBoS (grant agreement n. 239694). N.R. was partially supported by the same programme (grant agreement MNIQS-258023).

References

[1] Aftalion A 2006 *Vortices in Bose-Einstein Condensates* (Basel: Birkhäuser)
[2] Aftalion A and Du Q 2001 *Phys. Rev.* A **64** 063603a
[3] Aftalion A, Jerrard R L and Royo-Letelier J 2011 *J. Func. Anal.* **260** 2387–2460
[4] Bretin V, Stock S, Seurin Y and Dalibard J 2004 *Phys. Rev. Lett.* **92** 050403
[5] Bru J B, Correggi M, Pickl P and Yngvason J 2008 *Commun. Math. Phys.* **280**, 517–544
[6] Castin Y and Dum R 1999 *Eur. Phys. J.* D **7** 399
[7] Cooper N R 2008 *Adv. Phys.* **57** 539–616
[8] Correggi M, Rindler-Daller J and Yngvason J 2007 *J. Math. Phys.* **48** 042104
[9] Correggi M, Rindler-Daller J and Yngvason J 2007 *J. Math. Phys.* **48** 102103.
[10] Correggi M and Yngvason J 2008 *J. Phys. A: Math. Theor.* **41** 445002
[11] Correggi M, Rougerie N and Yngvason J 2011 *Commun. Math. Phys.* **303** 451–508
[12] Correggi M, Pinsker F, Rougerie N and Yngvason J 2011 *J. Stat. Phys.* **143** 261–305
[13] Correggi M, Pinsker F, Rougerie N and Yngvason J 2011 *Phys. Rev.* A **84** 053614
[14] Correggi M, Pinsker F, Rougerie N and Yngvason J 2012 *J. Math. Phys.* **53** 095203
[15] Danaila I and P Kazemi P 2010 *SIAM J. Sci. Comp.* **32** 2447
[16] Fetter A L 2001 *Phys. Rev. A* **64** 063608
[17] Fetter A L 2009 *Rev. Mod. Phys.* **81** 647–691
[18] Fetter A L, Jackson B and Stringari S 2005 *Phys. Rev.* A **71** 013605
[19] Fischer U R and Baym G 2003 *Phys. Rev. Lett.* **90** 140402
[20] Fu H and Zaremba E 2006 *Phys. Rev.* A **73** 013614
[21] Henderson K, Ryu C, Mac Cormick C and Boshier M G 2009 *New J. Phys.* **11** 043030
[22] Ignat R and Millot V 2006 *J. Funct. Anal.* **233** 260–306
[23] Ignat R and Millot V 2006 *Rev. Math. Phys.* **18** 119–162
[24] Kasamatsu K, Tsubota M and Ueda M 2002 *Phys. Rev.* A **66** 053606
[25] Kavoulakis G M and Baym G 2003 *New J. Phys.* **5** 51.1-11
[26] Kim J K and Fetter A L 2005 *Phys. Rev.* A **72** 023619
[27] Lieb E H and Seiringer R 2006 *Comm. Math. Phys.* **264** 505–537
[28] Lieb E H, Seiringer R, Solovej J P and Yngvason J 2005 *The Mathematics of the Bose Gas and its Condensation* (Basel: Birkhäuser)
[29] Lewin M and Seiringer R 2009 *J. Stat. Phys.* **137** 1040–1062
[30] Rindler-Daller T 2008 *Physica* A **387** 1851–1874
[31] Rougerie N 2011 *J. math. pures et appl.* **95** 296
[32] Rougerie N 2012 *Arch. Rat. Mech. Anal.* **203** 69
[33] Rougerie N, Serfaty S and Yngvason J 2012 Quantum Hall states of bosons in rotating anharmonic traps *Preprint* arXiv:1212.1085
[34] Ryu C, Andersen M F , Clade P, Natarajan V, Helmerson K and Phillips W D 2007 *Phys. Rev. Lett.* **99** 260401
[35] Sandier E and Serfaty S 2007 *Vortices in the Magnetic Ginzburg-Landau Model* (Basel: Birkhäuser)
[36] Schnee S and Yngvason J 2007 *Commun. Math. Phys.* **269** 659–691
[37] Sherlock B E, Gildemeister M, Owen E, Nugent E and Foot C J 2011 *Phys. Rev. A* **83** 043408
[38] Viefers S 2008 *J.Phys.* C **12** 123202
[39] Wilkin N K and Gunn J M F 2000 *Phys. Rev. Lett.* **84** 6-9

Composite vortices in two-component Bose-Einstein condensates

J Tempere, S Ceuppens and E Vermeyen

TQC, Universiteit Antwerpen, Universiteitsplein 1, B-2610 Antwerpen, Belgium

E-mail: jacques.tempere@ua.ac.be

Abstract. Motivated by the recent observation of composite vortices in two-gap superconductors, we investigate the occurrence and stability of composite vortices in a two-component Bose condensate. As a function of the inter-component interaction strength, the vortices in the two components of the Bose mixture can overlap and form a composite vortex object. However, when two such composite vortices approach each other, their mutual interaction may cause the composite object to dissociate in its component vortices. We derive the phase diagram for the stability of composite vortices as a function of the interaction strengths of the Bose condensates, based on a variational calculation. The effective forces between the component vortices of the composite object are also derived within this approach, and allow for an interpretation of the phase diagram.

1. Introduction

The discovery of superconductors with multiple superconducting gaps provides new prospects for the study of vortices. Magnesium diboride is an example of a two-gap superconductor, which has been named "type 1.5 superconductor" [1] since one gap corresponds to much more strongly bound pairs than the other gap. In magnesium diboride, vortices have been observed to form stripe patterns, which do not come about due to pinning centers, defects or edge effects [2]. This seems to indicate that the interactions between the vortices may have a dipolar component, due to the *composite* nature of vortices in these materials. The precise interpretation of the observed patterns in these materials with two superconducting components is still under discussion.

Motivated by this debate on composite vortices, we turn our attention to the simpler system of a two-component atomic superfluid. Recent advances in ultracold atom physics have allowed for an unprecedented level of control over the experimental parameters of ultracold atomic gases. The number of particles, temperature, confinement geometry, dimensionality and interaction strength can be tuned accurately over a broad range of values. Due to this experimental flexibility, ultracold atoms are also known as a quantum simulator for many-body systems and theories [3].

In this contribution, we consider a mixture of two Bose gases, with tunable interactions, and with vortices in both components. These two components reflect the presence of the two types of Cooper pairs present in two-gap superconductors. We consider the regime where a vortex in this system is actually not a single vortex, but consists of two vortices: one in each component. This we will call a composite vortex. Interactions between composite vortices are more complicated than interactions between single vortices and have not yet been fully understood. In this contribution, we first investigate the influence of the inter-component interaction on the core size of the component vortices. Next the interaction between two composite vortices is studied as a function of the distance between them. Finally, we will study the stability of two interacting composite vortices against dissociation (separation of both vortices in a composite vortex).

2. Gross-Pitaevskii description of the two-component condensate

A mixture of two Bose condensed atomic gases can be described by two order parameters Ψ_1 and Ψ_2. These reduce to the individual order parameters of each component of the mixture, in the absence of interactions between atoms of different components. The number of atoms of each component can be experimentally tuned, and to describe this we introduce chemical potentials μ_1 and μ_2 for components 1 and 2, respectively. The Gross-Pitaevskii energy functional for a mixture of two bosonic gases in the regime of constant chemical potentials is given by

$$\begin{aligned} E[\Psi_1,\Psi_2] &= \mathcal{E}[\Psi_1,\Psi_2] - \mu_1 N[\Psi_1] - \mu_2 N[\Psi_2] \\ &= \int d\mathbf{r} \left\{ \frac{\hbar^2}{2m} \left|\nabla_{\mathbf{r}}\Psi_1(\mathbf{r})\right|^2 + (V_1(\mathbf{r}) - \mu_1)\left|\Psi_1(\mathbf{r})\right|^2 \right\} \\ &+ \int d\mathbf{r} \left\{ \frac{\hbar^2}{2m} \left|\nabla_{\mathbf{r}}\Psi_2(\mathbf{r})\right|^2 + (V_2(\mathbf{r}) - \mu_2)\left|\Psi_2(\mathbf{r})\right|^2 \right\} \\ &+ \int d\mathbf{r} \left\{ \frac{1}{2} g_{11} \left|\Psi_1(\mathbf{r})\right|^4 + g_{12}\left|\Psi_1(\mathbf{r})\right|^2 \left|\Psi_2(\mathbf{r})\right|^2 + \frac{1}{2} g_{22}\left|\Psi_2(\mathbf{r})\right|^4 \right\}. \end{aligned} \tag{1}$$

Here, m is the mass of the atoms, taken to be the same for both components. Typically, different hyperfine spin states can be trapped simultaneously in a optical trap, and brought into the Bose-Einstein condensed regime. In this way, a mixture of Bose condensates has been realized experimentally[4, 5]. The trapping potentials for components 1 and 2 are indicated by $V_1(\mathbf{r})$ and $V_2(\mathbf{r})$, respectively. The interatomic interactions, in the ultracold regime where the de Broglie wavelength is much larger than the range of the interatomic potential, is well characterized by a contact pseudopotential with given strength g_{ij} for the interactions between an atom of component i with an atom of component j. The strength of the contact potentials is related to the s-wave scattering lengths a_{ij} through $g_{ij} = 4\pi\hbar^2 a_{ij}/m$.

It helps to rewrite the Gross-Pitaevskii functional in a dimensionless form. For this purpose, we write

$$\Psi_1(\mathbf{r}) = \rho_1 f_1(\mathbf{r}) e^{i\theta_1(\mathbf{r})}, \tag{2}$$

$$\Psi_2(\mathbf{r}) = \rho_2 f_2(\mathbf{r}) e^{i\theta_2(\mathbf{r})}, \tag{3}$$

with real functions $f_j(\mathbf{r})$ for the amplitudes and $\theta_j(\mathbf{r})$ for the phase fields. The constants ρ_1^2 and ρ_2^2 represent the homogeneous bulk particle density of the components, far away from any perturbation like a vortex core. Hence, we require $f_j(r \to \infty) = 1$. These bulk densities are naturally related to the chemical potentials. Extremizing $E_{bulk} = E[\rho_1, \rho_2]$ one straightforwardly finds

$$\rho_1^2 = \frac{\mu_1 g_{22} - \mu_2 g_{12}}{g_{11} g_{22} - g_{12}^2}, \tag{4}$$

$$\rho_2^2 = \frac{\mu_2 g_{11} - \mu_1 g_{12}}{g_{11} g_{22} - g_{12}^2}. \tag{5}$$

The mixture is stable (with respect to unmixing and component separation) when the extremum in E_{bulk} is a minimum, requiring

$$g_{11} g_{22} > g_{12}^2, \tag{6}$$

and $g_{11} > 0$ or $g_{22} > 0$. In order to obtain a dimensionless form for the Gross-Pitaevskii potential, we must choose length and energy scales. We will relate those to the first component, and set

$$\tilde{\xi}_1 = \sqrt{\frac{\hbar^2}{2mg_{11}\rho_1^2}} \tag{7}$$

as length scale. If inter-component interactions are absent, this represents the coherence length of the first component condensate. The energy scale that we use is $\epsilon_0 = \hbar^2/(2m\tilde{\xi}_1^2)$. With these conventions we obtain for the energy $\Delta E = (E[\Psi_1, \Psi_2] - E_{bulk})/\epsilon_0$ to be extremized

$$\begin{aligned} \Delta E[f_1, \theta_1, f_2, \theta_2] = \int d\mathbf{r} \ \Big\{ & (\nabla_{\mathbf{r}} f_1)^2 + \delta (\nabla_{\mathbf{r}} f_2)^2 + f_1^2 (\nabla_{\mathbf{r}} \theta_1)^2 + \delta f_2^2 (\nabla_{\mathbf{r}} \theta_2)^2 \\ & + \frac{1}{2}(1-f_1^2)^2 + \gamma\delta \left(1-f_1^2\right)\left(1-f_2^2\right) + \frac{1}{2}\beta\delta^2(1-f_2^2)^2 \Big\}, \end{aligned} \tag{8}$$

with

$$\beta = \frac{g_{22}}{g_{11}}, \ \gamma = \frac{g_{12}}{g_{11}}, \ \delta = \frac{\rho_2^2}{\rho_1^2}. \tag{9}$$

For experiments with mixtures of ^{87}Rb condensates in the $F = 2, m_F = +1$ and $F = 1, m_F = -1$ states[5], the experimental values of the constants are $\beta = 1.062(12)$ for the ratio of intraspecies scattering lengths[6] and $\gamma = 1.030(10)$ for the dimensionless interspecies interaction strength[7]. Feshbach resonances are used experimentally to tune the interaction strengths[8]. In the current context these are magnetically tunable resonances between scattering states of the atoms in the open scattering channel, and a bound (molecular) state in a closed scattering channel. Typically, it is difficult to independently fix β and γ to an arbitrary value, as the changes in both β and γ will be coupled when sweeping the magnetic field through the Feshbach resonance. Nevertheless, at least one of these parameters can be made large (increasing the s-wave scattering length up to 10^4 Bohr radii) through the use of the resonance.

3. Composite vortex

The first question to be answered is how much the vortex cores, seen as holes in the density of the atomic clouds, are renormalized by the coupling. For this, we take as variational trial functions

$$f_j(\mathbf{r}) = \tanh\left[r/\left(\sqrt{2}\xi_j\right)\right], \tag{10}$$

$$\theta_j(\mathbf{r}) = \phi, \tag{11}$$

where cylindrical coordinates $\mathbf{r} = \{r, \phi, z\}$ are used. Firstly, note that both single-component vortices are placed on the z-axis, i.e. on top of each other in the otherwise homogeneous situation. If there are no inter-component interactions, there is (in the homogeneous case) no need to place the phase singularity in the first component on top of the phase singularity in the second component. Only if the inter-component interaction is attractive, $\gamma < 0$, does this lower the energy. Indeed, each component will have a hole in the atom density at the vortex core, and since atoms of different components attract, the lowest energy configuration will occur when the holes in the density overlap. Secondly, note that we have introduced ξ_j as a typical size of the vortex core in component j: these are now adjustable variational parameters and they can differ

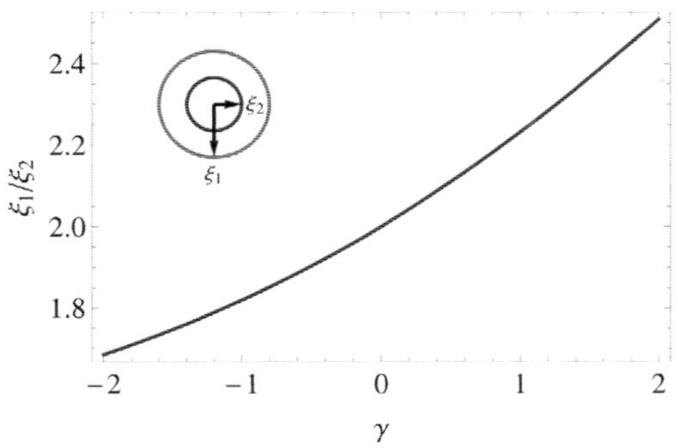

Figure 1. The ratio between the core sizes in component 1 and 2, as obtained from the variational calculation, is shown as a function of the inter-component interaction strength. The composite vortex considered in this calculation is such that the component vortices have coinciding centers.

from the unrenormalized healing lengths $\tilde{\xi}_j$ introduced before. The variational energy becomes

$$\Delta E(\xi_1, \xi_2) = \ln(1/\xi_1) + \delta \ln(1/\xi_2) + \frac{\ln(16) - 1}{3} \left(\xi_1^2 + \beta\delta\xi_2^2\right) + \delta\gamma \int_0^\infty r \operatorname{sech}^2\left(\frac{r}{\sqrt{2}\xi_1}\right) \operatorname{sech}^2\left(\frac{r}{\sqrt{2}\xi_2}\right) dr. \tag{12}$$

In order to get a grasp of the typical behavior to be expected for vortices in two-component condensates, we study as a concrete example the case with $\delta = 1$ (the same bulk particle density for both components), and $\beta = 4$. In the absence of inter-component interactions ($\gamma = 0$), the renormalised healing lengths are equal to the unrenormalised healing lengths, $\xi_j = \tilde{\xi}_j$. In that case, the unrenormalised healing lengths are by definition (7) related as $\tilde{\xi}_2 = \tilde{\xi}_1/\sqrt{\delta\beta}$ and the vortex core of component 2 is half the size of the vortex core in component 1. To get to a nontrivial example, we need the inter-component coupling. We allow the coupling to vary between $\gamma = -2, ..., 2$, satisfying the stability criterion (6) which can be rewritten as $\gamma^2 < \beta$. The result for these parameter values is shown in Figure 1. We find that the effect of the interaction-induced renormalization on the ratio ξ_2/ξ_1 is relatively small, of the order of 20% in the full range of possible γ values. In comparison, $\xi_2/\tilde{\xi}_2$ ranges from 0.68 to 1.23, and $\xi_1/\tilde{\xi}_1$ ranges from 0.87 to 1.12 (where in both cases the size of the core is largest for $\gamma = -2$). So, the individual core sizes get influenced more strongly than the ratio between core sizes, but all in all, the estimates of the core sizes based on the non-interacting component analysis is roughly right.

4. Interactions between composite vortices

Now we consider two composite vortices, approaching each other. We fix the distance a between the vortex cores in the first component. This means that for the first component, we have

$$f_1(\mathbf{r}) = \tanh\left[\frac{r}{\sqrt{2}\xi_1}\right] \tanh\left[\frac{\sqrt{r^2 + a^2 - 2ra\cos\phi}}{\sqrt{2}\xi_1}\right], \tag{13}$$

$$\theta_1(\mathbf{r}) = \phi + \arctan\left(\frac{r\sin\phi}{r\cos\phi - a}\right). \tag{14}$$

The amplitude of the order parameter of the first component has two holes drilled in it, one at the origin and one at a distance a away on the x-axis. The phase field is the sum of the angle around the origin, and the angle around the center of the second vortex. In the second component, the positions of the vortex cores are adapted variationally, and we have

$$f_2(\mathbf{r}) = \tanh\left[\frac{|\mathbf{r}-\mathbf{r}_1|}{\sqrt{2}\xi_2}\right]\tanh\left[\frac{|\mathbf{r}-\mathbf{r}_2|}{\sqrt{2}\xi_2}\right], \tag{15}$$

where $\mathbf{r}_1$ and $\mathbf{r}_2$ are the positions of the vortices in component 2. The phase in component 2 will likewise be a sum of the angle around the first vortex plus the angle around the second vortex. When these expressions are introduced in the energy functional, a rather cumbersome expression appears. Insight can be gained by realizing that the variational energy functional is dominated by three contributions: firstly there is a repulsion between the vortices in the first component; secondly there is the analogous repulsion between vortices in the second component; and thirdly there is an attraction between vortices from component 1 and component 2. These contributions, for our case with $\beta = 4$, $\gamma = -2$ are shown in Figure 2.

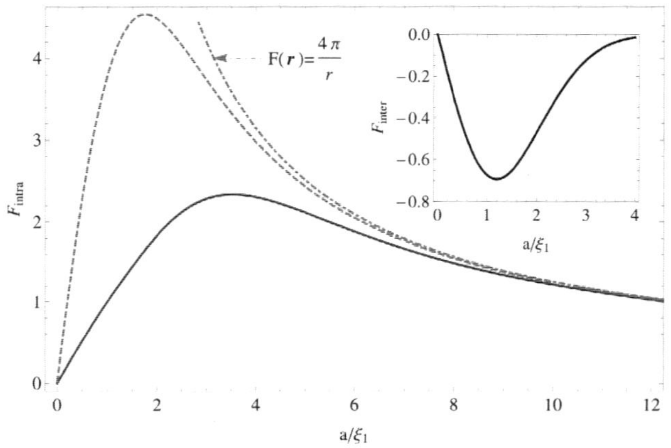

Figure 2. The individual contributions to the force between the vortices at $\beta = 4$, $\gamma = -2$. Full blue: the repulsion between component 1 vortices. Dashed red: the repulsion between component 2 vortices. Inset: inter-component attraction.

This figure depicts the (negative) gradients in the energy, i.e. the forces, as a function of distance. When the vortices within a same component are far apart, they repel each other with an inverse distance force $F(\mathbf{r}) = 4\pi/r$ that corresponds to the Lorentz force felt by a vortex due to the superfluid velocity field of the other vortex. At shorter distances, the cores start to overlap and the force is reduced. The repulsive force becomes maximal at a distance near 2ξ where ξ is representative for the vortex core radius. The renormalization of the vortex core size (the healing length) at short intervortex distance is taken into account, but does not change these results significantly. With only the intra-component vortex-vortex interaction, there is a repulsive force at all length scales. The inter-component vortex-vortex interaction, on the other hand, is attractive, and proportional to γ, the interatomic interaction strength between atoms of the two components. This force is due not to the kinetic energy of superfluid velocity fields, but it is due to the overlap of the component order parameters. Hence, it is short range as the vortices of different components "feel" each other only when the cores are overlapping. The attraction is maximal when the vortex cores (of sizes represented by ξ_1, ξ_2) are just touching, at an intervortex distance of the order of $\xi_1 + \xi_2$. Once the "smaller" vortex core is captured inside the "larger" vortex core, the force diminishes again: the "small vortex core" is free to rattle around in the well provided by the "large" vortex core. For the parameters used in Figure 2 (with a modest γ), the attraction provided by the inter-component force is relatively weak in comparison to the repulsion; however it scales with γ.

What does this mean for the interaction between composite vortices? Two composite vortices at a large distance from each other will repel each other with a force inversely proportional to their distance; and the force for the constituent component 1 and component 2 vortices are of the same size. So, the composite vortices move apart, but retain their character as composite objects (i.e. the component 2 vortex stays bound to the component 1 vortex). We can now envisage the following experiment: pin the component 1 vortices, and slowly move them closer together while allowing the component 2 vortices to adapt their position so as to minimize the energy. The inter-component attraction will favour keeping component 2 and component 1 vortices together in a composite object. The intra-component repulsion will try to push the component 2 vortices away from each other, regardless of the position of component 1 vortices. At some distance, the intra-component repulsion will overcome the inter-component attraction and the component 2 vortices will unbind from their component 1 vortex. The composite vortices, when moved closer than a critical distance, can decompose. We calculated the phase diagram for this process using the variational calculation outlined in Sec. II and the results are summarized in Figure 3.

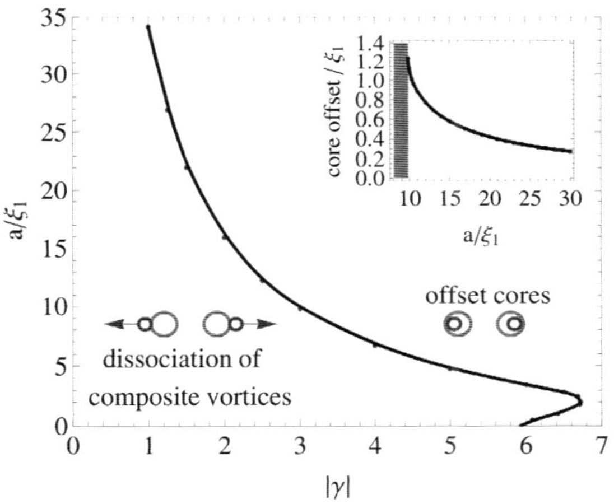

Figure 3. Phase diagram at $\beta = 4$ of the composite vortex interaction as function of the distance a between the cores in component 1 and ratio $|\gamma|$ of the attractive inter-component interaction strength to the interaction strength in component 1. Inset: Core offset of the component 2 vortices relative to the center of the component 1 vortices as function of distance a.

Figure 3 shows which phase is realized, for a set of distances a separating the component 1 vortices and for a set of inter-component interaction strengths $|\gamma|$, assumed to correspond to an attractive interaction [1]. Left of the phase separation line, the composite vortices decompose as the component 2 vortices unbind from their component 1 partner and move apart. Right

[1] We show results also for the regime $|\gamma| > \sqrt{\beta}$, which according to our criterion (6) leads to unmixing. However, this criterion was derived in essence in the grand canonical ensemble (working with fixed chemical potentials rather than with fixed numbers of atoms during the unmixing transition), whereas for a closed system, strong attractive interspecies interactions need not lead to unmixing.

of the curve, the composite vortices stay composite in the sense that the component 2 vortex remains bound to the component 1 vortex, even though it will be offset from the center of the component 1 vortex. This offset is due to the fact that the "small" component 2 vortex can move around in the "large" component 1 core without this costing too much energy. The offset gradually increases as the composite vortices are forced closer together, as shown in the inset of Figure 3. When one wants to be able to bring the composite vortices closer together (a smaller), a stronger attraction γ is needed to glue the component 2 vortex to its component 1 partner. But this seems not to be the case at the smallest distances: note the bending over of the phase separation line in Figure 3. What happens there is that, as the vortex cores of the component 2 vortices start to overlap, the repulsion between the component 2 vortices weakens, whereas the attraction keeping them bound to the component 1 cores only weakens at an even shorter scale, as can be seen in Figure 2. The result is twofold: for certain inter-component interaction strengths there is reentrant behavior, and there exists a critical inter-component interaction strength that results in component vortices remaining stable with respect to dissociation at all distance scales. For the situation in Figure 3 this value is approximately $\gamma = -6.72$.

5. Conclusion

In this article, the interaction between two composite vortices in a two-component Bose-Einstein condensate was discussed. First it was shown that the coupling between the two components implies a change in both the size of the vortex cores and their ratio, with the latter experiencing the smallest amount of modification. This renormalization was taken into account in the subsequent calculations, but was found to have no essential qualitative influence. Next, the force between two composite vortices was analysed by considering the three contributions separately. We found two strong, long range repulsive contributions (corresponding to the intra-component vortex interaction) and one weak, short range attractive contribution for the inter-component interaction, due to the overlap of the component order parameters. This results finally in a phase diagram with two distinct areas: a first where the composite vortices dissociate and a second where they remain stable despite their offset cores. It was found that for a sufficiently strong inter-component interaction, the system will be stable with respect to dissociation at all distance scales, meaning that the composite vortices will remain intact.

Acknowledgments

E.V. acknowledges financial support in the form of a Ph. D. fellowship of the Research Foundation - Flanders (FWO). This work is supported financially by the Research Foundation - Flanders (FWO) under Projects No. G.0180.09N, G.0370.09N, G.0115.12N and No. G.0119.12N

References

[1] Moshchalkov V V, Menghini M, Nishio T, Chen Q H, Silhanek A V, Dao V H, Chibotaru L F, Zhigadlo N D and Karpinski J 2009 *Phys. Rev. Lett.* **102** 117001
[2] Gutierrez J, Raes B, Silhanek A V, Li L J, Zhigadlo N D, Karpinski J, Tempere J and Moshchalkov V V 2012 *Phys. Rev.* B **85** 094511
[3] Bloch I, Dalibard J and Zwerger W 2008 *Rev. Mod. Phys.* **80** 885–964
[4] Stamper-Kurn D M, Andrews M R, Chikkatur A P, Inouye S, Miesner H-J, Stenger J and Ketterle W 1998 *Phys. Rev. Lett.* **80** 2027–30
[5] Matthews M R, Anderson B P, Haljan P C, Hall D S, Wieman C E and Cornell E A 1999 *Phys. Rev. Lett.* **83** 2498–501
[6] Matthews M R, Hall D S, Jin D S, Ensher J R, Wieman C E and Cornell E A 1998 *Phys. Rev. Lett.* **81** 243–7
[7] Hall D S, Matthews M R, Ensher J R, Wieman C E and Cornell E A 1998 *Phys. Rev. Lett.* **81** 1539–42
[8] Inouye S, Andrews M R, Stenger J, Miesner H-J, Stamper-Kurn D M and W. Ketterle W 1998 *Nature* **392** 151–4

Roton Instability and Phonon Collapse of Two-dimensional Tilted Dipoles

Aleksey K Fedorov[1,2,3], **I. L Kurbakov**[4], **Yu. E Lozovik**[4]

[1] Center for Photonics and Infrared Engineering, Bauman Moscow State Technical University, 2nd Baumanskaya str., 5, 105005, Moscow, RUSSIA
[2] Russian Quantum Center, Novaya str., 100, 143025, Skolkovo, Moscow region, RUSSIA
[3] Visiting Research Associate at Institute for Quantum Information Science, University of Calgary, 2500 University Dr. NW, T2N 1N4, Calgary, Alberta, CANADA
[4] Institute for Spectroscopy RAS, Fizicheskaya str., 5, 142190, Troitsk, Moscow region, RUSSIA

E-mail: lozovik@isan.troitsk.ru, akfedorov@student.bmstu.ru

Abstract. Bose-Einstein condensed gas of two-dimensional tilted dipoles in a thin layer is under consideration. The problem of stability is resolved. Phase diagram for system of two-dimensional dipole atoms in a thin layer is obtained. A formation of density waves and their orientation control possibility via polarizing field are considered. These effects can be experimentally observed for magnetically dipolar atoms at the Feshbach resonance or polar molecules in one-dimensional optical lattices.

1. Introduction

Anisotropy and attraction of dipole-dipole interaction potential are reasons for set of wonderful phenomena. There are density waves (DWs) formation [1], [2], [3], [4], phonon collapse [5], supersolid [7], [8], [9] and anisotropy superfluidity [10] among them. In this regard, great attention is recently paid to the dipole-dipole bosonic systems [11], [12].

In this note we consider two-dimensional (2D) *tilted* dipole atoms in a thin layer at zero temperature. There are two experimental realizations for the model. The first one is atoms with magnetic dipole moment [13], [14] at the Feshbach resonance [15] in a 1D optical lattice [16]. The second one is polar molecules with electrically induced moment [17] in a 1D optical lattice [18]. Also, problem statement (from experimental point of view) is illustrated in the Fig. 1.

Stability criteria defines the phase diagram. The stability criteria is non-negativity of the Bogoliubov spectrum square

$$\varepsilon_{\mathbf{p}}^2 = \frac{p^2}{2m}\left(\frac{p^2}{2m} + 2n_0 U_{2d}(\mathbf{p})\right) \geq 0, \tag{1}$$

where n_0 is 2D condensate density, m is dipole mass, $U_{2d}(\mathbf{p})$ is Fourier transform of 2D interaction potential of tilted dipoles.

An exposition of a part of results of this note is presented in Ref. [19].

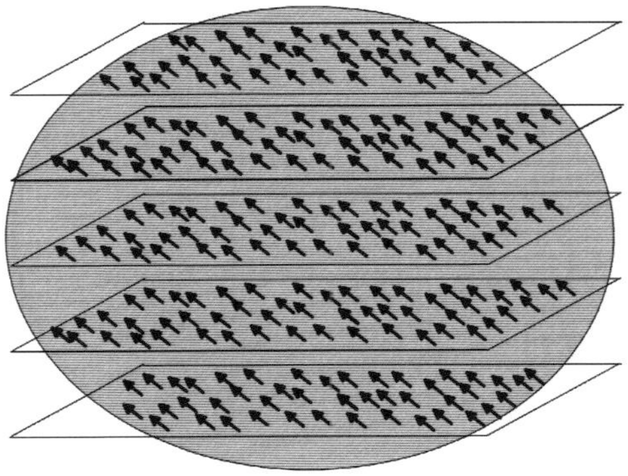

Figure 1. Bose-Einstein condensed gases of 2D tilted dipoles in thin layers, which are formed for a 3D condensate in a 1D optical lattice.

2. Interaction potential

Bogoliubov spectrum calculation is necessary to investigate a stability problem for the system. Therefore, it is necessary to calculate effective part of 2D interaction potential $U_{2d}(\mathbf{r})$ of dipoles in a thin layer. To this aim, let us start from 3D Hamiltonian of the system

$$\hat{H} - \mu_{3d}\hat{N} = \int d\vec{\mathrm{r}}\hat{\Psi}(\vec{\mathrm{r}})\hat{\Psi}^{+}(\vec{\mathrm{r}})\left(-\frac{\hbar^2}{2m}\Delta_3 + V(\mathbf{r}) + V_{\mathrm{ti}}(z) - \mu_{3d}\right) +$$

$$+\frac{1}{2}\int d\vec{\mathrm{r}}d\vec{\mathrm{r}}'(V_{dd}(\vec{\mathrm{r}}-\vec{\mathrm{r}}') + g_{3d}\delta(\vec{\mathrm{r}}-\vec{\mathrm{r}}'))\hat{\Psi}^{+}(\vec{\mathrm{r}})\hat{\Psi}^{+}(\vec{\mathrm{r}}')\hat{\Psi}(\vec{\mathrm{r}}')\hat{\Psi}(\vec{\mathrm{r}}), \quad (2)$$

where Δ_3 is 3D Laplace operator, $\hat{\Psi}(\vec{\mathrm{r}})$ is 3D field operator, $g_{3d}\delta(\vec{\mathrm{r}})$ is contact van der Waals pseudopotential, μ_{3d} is chemical potential, $V_{\mathrm{ti}}(z)$ is confinement potential in the tight direction, $V(\mathbf{r})$ is confinement potential in thin layer plane,

$$V_{dd}(\vec{\mathrm{r}},\theta) = \frac{d^2}{\vec{\mathrm{r}}^5}(\mathrm{r}^2 - 3(x\sin\theta + z\cos\theta)^2)$$

is dipole-dipole potential, d is dipole moment, and dipoles are tilted at angle θ to z axis. Arrows is used for 3D vectors and bold type is used for 2D vectors

$$\vec{\mathrm{r}} = \{\mathbf{r}, z\}, \quad \mathbf{r} = \{x, y\}, \quad r = |\mathbf{r}|, \quad \mathrm{r} = |\vec{\mathrm{r}}| = \sqrt{r^2 + z^2}.$$

In the sufficiently thin layer the motion in the tight direction is frozen at the lowest level, the excited levels being practically not populated. Therefore, in the expansion of 3D field operator on basis $\{\varphi_k^{\mathrm{ti}}(z)\}$ in the tight direction

$$\hat{\Psi}(\vec{\mathrm{r}}) = \sum_{k=0}^{\infty}\varphi_k^{\mathrm{ti}}(z)\hat{\Psi}_k(\mathbf{r}), \quad \hat{\Psi}_k(\mathbf{r}) = \int dz\varphi_k^{\mathrm{ti}*}(z)\hat{\Psi}(\vec{\mathrm{r}})$$

we can neglect all $\hat{\Psi}_k(\mathbf{r})$ for $k > 0$. Here the eigenfunctions $\varphi_k^{\mathrm{ti}}(z)$ and eigenenergies E_k^{ti} are determined from the following equation

$$\left(-\frac{\hbar^2}{2m}\frac{d^2}{dz^2} + V_{\mathrm{ti}}(z)\right)\varphi_k^{\mathrm{ti}}(z) = E_k^{\mathrm{ti}}\varphi_k^{\mathrm{ti}}(z).$$

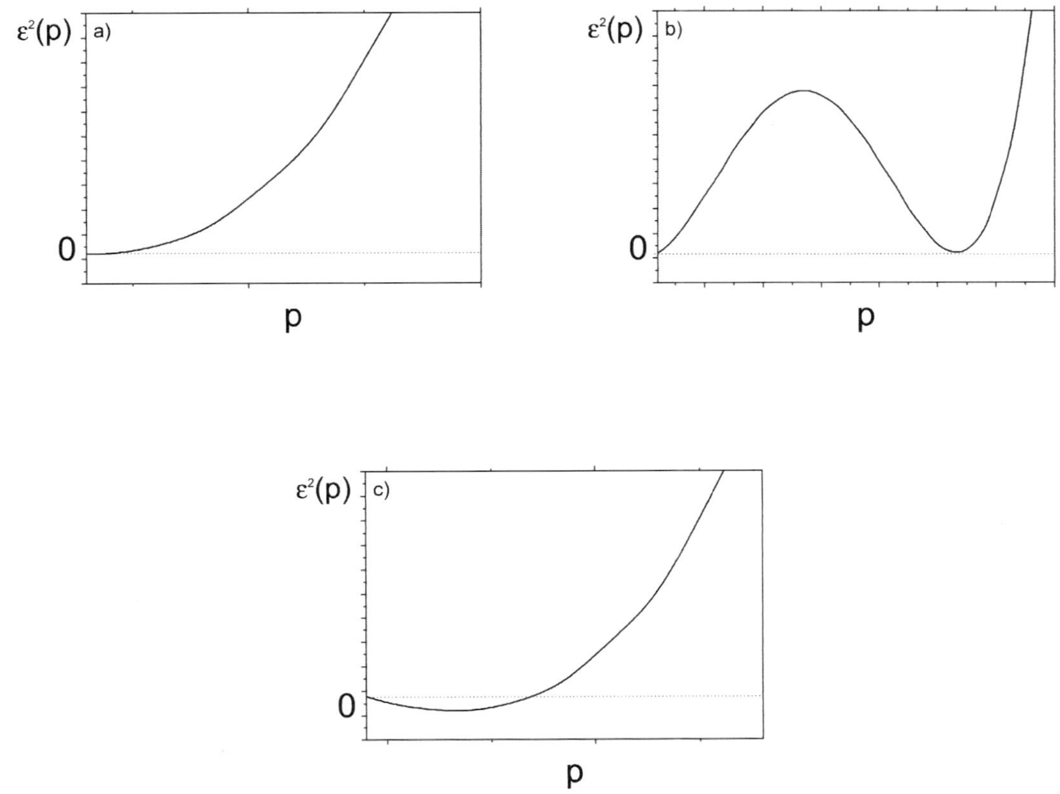

Figure 2. Bogoliubov spectrums for three phases. a) Stable homogeneous gas. b) Phase with roton instability with respect to density wave formation. c) Phase with phonon instability with respect to long-wave-lenght collapse.

Thus, we can consider $k = 0$ element

$$\hat{\Psi}(\vec{r}) \approx \varphi_0^{\mathrm{ti}}(z)\hat{\Psi}(\mathbf{r}), \quad \hat{\Psi}(\mathbf{r}) = \int dz \varphi_0^{\mathrm{ti}*}(z)\hat{\Psi}(\vec{r}), \tag{3}$$

where $\hat{\Psi}(\mathbf{r})$ is effective 2D field operator, which is satisfied to bosonic commutation relations.

After substituting (3) in (2), effective 2D Hamiltonian for thin-layer motion is obtained in the following form

$$\hat{H} - \mu\hat{N} = \int \hat{\Psi}^{+}(\mathbf{r}) \left(-\frac{\hbar^2}{2m}\Delta_2 - \mu + V(\mathbf{r}) \right) \hat{\Psi}(\mathbf{r}) d\mathbf{r} +$$

$$+ \int U_{2d}(\mathbf{r} - \mathbf{r}')\hat{\Psi}^{+}(\mathbf{r})\hat{\Psi}^{+}(\mathbf{r}')\hat{\Psi}(\mathbf{r}')\hat{\Psi}(\mathbf{r}) d\mathbf{r}\, d\mathbf{r}' \tag{4}$$

with effective 2D interaction

$$U_{2d}(\mathbf{r} - \mathbf{r}') = \int dz dz' |\varphi_0^{\mathrm{ti}}(z)\varphi_0^{\mathrm{ti}}(z')|^2 (g_{3d}\delta(\vec{r} - \vec{r}) + V_{dd}(\vec{r} - \vec{r})) \tag{5}$$

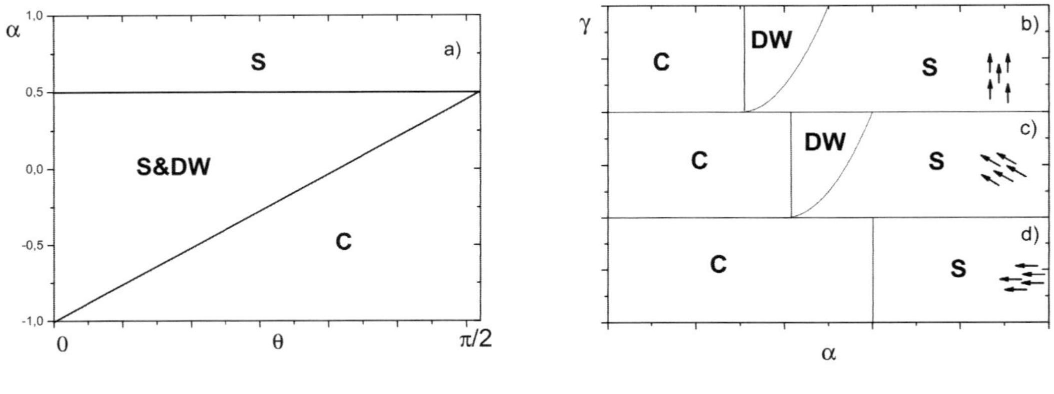

Figure 3. Phase diagrams. a) A phase diagram for two-dimension homogeneous layer is shown; b) homogeneous superfluid phase (**S** on the figure); c) phonon instability phase with long-wave collapse (**C** on the figure); and d) roton instability phase with DWs generation (**DW** on the figure).

and chemical potential $\mu = \mu_{3d} - E_0^{\rm ti}$ (where Δ_2 is 2D Laplace operator).

Note, that one of the instability possible reasons is negativity of effective 2D potential Fourier transform. Let us calculate the potential $U_{2d}(\mathbf{r})$ and define a condition of it's negativity. We suppose the presence of harmonic trap along tight direction, i.e., $V_{\rm ti}(z) = (m/2)\omega^2 z^2$, therefore

$$\varphi_{\rm ti}(z) = \exp(-z^2/2z_0^2)/\sqrt{\sqrt{\pi}z_0} \tag{6}$$

and $E_0^{\rm ti} = \hbar\omega/2$. Here $z_0 = \sqrt{\hbar/m\omega}$ is tight-confinement oscillator length and ω is the oscillator frequency.

After substituting (6) in (5), the Fourier transform of the effective 2D potential in first Born approximation is obtained in the following form

$$U_{2d}(\mathbf{p}) = g_s - \frac{g_d}{2} + U_h(\mathbf{p})\sin^2\theta + U_v(\mathbf{p})\cos^2\theta, \tag{7}$$

where

$$U_h(\mathbf{p}) = \frac{2d^2}{\hbar}\int_{-\infty}^{+\infty}\exp\left(-\frac{p_z^2 z_0^2}{2\hbar^2}\right)\frac{p_x^2 dp_z}{p_x^2+p_y^2+p_z^2},$$

$$U_v(\mathbf{p}) = \frac{2d^2}{\hbar}\int_{-\infty}^{+\infty}\exp\left(-\frac{p_z^2 z_0^2}{2\hbar^2}\right)\frac{p_z^2 dp_z}{p_x^2+p_y^2+p_z^2},$$

$$g_s = \frac{2\sqrt{2\pi}\hbar^2 a_s}{m z_0} = \frac{g_{3d}}{\sqrt{2\pi}z_0},\ g_d = \frac{2\sqrt{2\pi}\hbar^2 a_d}{m z_0} = \frac{2\sqrt{2\pi}d^2}{3z_0}.$$

a_s is 3D s-wave scattering length and a_d is the length scale characterizing the dipole-dipole interaction.

After substituting (7) in (1), Bogoliubov spectrums for the systems are obtained. There are Bogoliubov spectrums in the Fig. 2.

Some assumptions was used for obtaining relation (7). First, it is the condition of weak interaction $a_s, a_d \ll z_0$ [20]. Second, the condition of the tight-confinement quantization, i.e.,

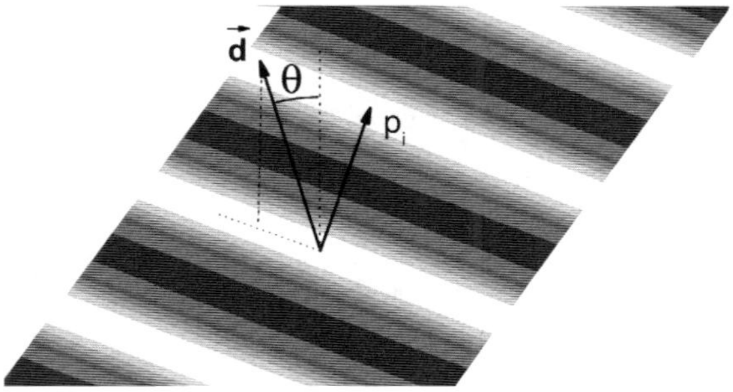

Figure 4. Bose-condensed gas of 2D dipole atoms in a thin layer.

z_0 to be small. And third, the condition of independence of layers formed by the optical lattice, i.e., small interlayer tunneling and screening interlayer interaction of dipoles. A more precise calculation for independent 2D layers must consider the summation of the ladder diagrams [21], [22] and dipole-dipole scattering problem [23].

3. Phase diagram

In accordance with (1) there are following phase for system:

1. Homogeneous superfluid phase (Fig. 2a), where $\varepsilon_{\mathbf{p}}^2 > 0$ is positive for all momentums $\mathbf{p}$.
2. Phase with phonon instability and long-wave-length collapse (Fig. 2b), where $U_{2d}(0) < 0$ and $\varepsilon_{\mathbf{p}}^2 < 0$ is negative for momentums $p < p_{\text{ph}}$.
3. Phase with roton instability and density wave (DW) formation (Fig. 2c), where $U_{2d}(0) > 0$, but there is a certain momentum range $p_1 < p < p_2$ in which $U_{2d}(\mathbf{p}) < 0$, so $\varepsilon_{\mathbf{p}}^2$ touch to zero with *roton momentum* $\mathbf{p} = \mathbf{p}_{\text{r}}$.

A phase diagram with variables $\theta - \alpha - \gamma$ (where $\gamma = 2\sqrt{2\pi} z_0 a_d n_0$ is density parameter and $\alpha = a_s/a_d$) for 2D layer is shown on Fig. 3a. There are a) homogeneous superfluid phase, b) phase with phonon instability and long-wave-length collapse, and c) phase with roton instability and DWs formation. This phases with parameters $\gamma - \alpha$ is shown on fig. 3b-3d for fixed angle θ. There are no DWs if dipoles are parallel to layer plane ($\theta = \pi/2$).

DWs orientation control possibility via (magnetic or electric) field is considered in Fig. 4. Lines of equal phase DWs directed along the projection of the dipole moment on the plane layer. Therefore, it's principally possible to control the orientation of the wave.

4. Experiment

Experimental threshold roton instability in our assessment in implementing following dipole systems:

1) Dysprosium atoms ^{164}Dy [14]: $m = 164$ u, $z_0 = 150$ nm, $\hbar\omega = 130$ nK, $\omega/2\pi = 2.7$ kHz, $\theta = 72°$, $a_d = 14$ nm, $a_s = 5.5$ nm, $a = 0.5$ nm, $n_0 = 2.8 \times 10^9$ sm^{-2}, $\alpha = 0.393$, $\gamma = 0.292$, $\mu = 1.4$ nK, $T_0/4 = 130$ nK.

2) Polar molecules ^{41}K^{87}Rb [28]: $m = 128$ u, $z_0 = 150$ nm, $\hbar\omega = 168$ nK $\omega/2\pi = 3.5$ kHz, $\theta = 56.8°$, $a_d = 30$ nm, $a_s = 5.5$ nm, $a = 3$ nm, $n_0 = 4.65 \times 10^9$ sm^{-2}, $\alpha = 0.183$, $\gamma = 0.262$, $\mu = 4.4$ nK, $T_0/4 = 69$ nK.

Here $T_0 = 2\pi\hbar^2/m$ is degeneracy temperature. The quantity $T_0/4$ is of order of Berezinskii-Kosterlitz-Thouless superfluid transition temperature [24], [25], [26] $T_{BKT} = \xi_s^c T_0/4$ [27] (where $\xi_s^c \equiv n_s(T_{BKT})/n \sim 1$ is superfluid fraction at the transition).

5. Conclusion

We have considered the stability problem for Bose-Einsten condensed gases of 2D tilted dipolar atoms or polar molecules in a thin layers formed for a 3D condensate in a 1D optical lattice. Based on our stability criteria we have obtained the phase diagram in which a stable homogeneous superfluid as well as phonon-collapsed and density-wave phases take place. Rotation of polarizing (magnetic or electric) field around the axis of the optical lattice yields controlling orientation of formed density wave. The effects can be observed for Dy atoms or polar molecules at nK temperatures.

Acknowledgments

This work is supported by Russian Foundation for Basic Research (11-02-00858).

References

[1] Santos L, Shlyapnikov G V, Lewenstein M 2003 Roton-maxon Spectrum and Stability of Trapped Dipolar Bose-Einstein Condensates, Phys. Rev. Lett. **90** 250403.
[2] Fischer U R 2006 Stability of Quasi-two-dimensional Bose-Einstein Condensates with Dominant Dipole-dipole Interactions, Phys. Rev. A **73** 031602.
[3] Wang D.W., Demler E 2008 Collective Excitations and Instabilities in Multi-layer Stacks of Dipolar Condensates, arXiv:0812.1838.
[4] Klawunn M, Santos L 2009 Hybrid Multisite Excitations in Dipolar Condensates in Optical Lattices, Phys. Rev. A **80** 013611
[5] Santos L, et al. 2000 Bose-Einstein Condensation in Trapped Dipolar Gases, Phys. Rev. Lett. **85** 1791.
[6] Koch L, et al. 2008 Stabilization of a Pure Dipolar Quantum Gas Againist Collapse, Nat. Phys. **4** 218.
[7] Kurbakov I L, Lozovik Yu E, Astrakharchik G E, Boronat J 2010 Quasiequilibrium Supersolid Phase of a Two-dimensional Dipolar Crystal, Phys. Rev. B **82** 014508.
[8] Lozovik Yu E, Volkov S Yu, Willander M 2004 Structural Properties of the Condensate in Two-dimensional Mesoscopic Systems of Strongly Correlated Excitons, JEPT Lett. **79** 473.
[9] Pino J M, et. al. 2004 Photon Counting for Bragg Spectroscopy of Quantum Fases, JEPT Lett. **83** 033615.
[10] Ticknor C, Wilson R M, Bohn J L 2011 Anisotropic Superfluidity in a Dipolar Bose Gas, Phys. Rev. Lett. **106** 065301.
[11] Baranov M A 2008 Theoretical Progress in Many-body Physics with Ultracold Dipolar Gases, Phys. Rep. **464** 71.
[12] Lozovik Yu E, Mandelshtam V A 1989 2D Dipole System in a Strong Field, Phys. Lett. A **138** 204.
[13] Griesmaier A, Werner J, Hensler S, et al. 2005 Observation of Feshbach Resonances in an Ultracold Gas of ^{52}Cr, Phys. Rev. Lett. **94** 160401.
[14] Lu M, Burdick N Q, Youn S H, Lev B L 2011 Strongly Dipolar Bose-Einstein Condensate of Dysprosium, Phys. Rev. Lett. **107** 190401.
[15] Lahaye T, Koch T, Frohlich B, et al. 2007 Strong Dipolar Effects in a Quantum Ferrofluid, Nature **448** 672.
[16] Müller S, Billy J, Henn E A L, et al. 2011 Stability of a Dipolar Bose-Einstein Condensate in a One-dimensional Lattice, Phys. Rev. A **84** 053601.
[17] Carr L D, DeMille D, Krems R V, Jun Ye 2009 Cold and Ultracold Molecules: Science, Technology, and Applications, New J. Phys. **11** 055049.
[18] de Miranda M H G, Chotia A, Neyenhuis B 2011 Controlling the Quantum Stereodynamics of Ultracold Bimolecular Reactions, Nat. Phys. **7** 502.
[19] Fedorov A K, Kurbakov I L Lozovik Yu E 2012 TBP.
[20] Petrov D S, Holzmann M, Shlyapnikov G V 2000 Bose-Einstein Condensation in Quasi-2D Trapped Gases, Phys. Rev. Lett. **84** 2551.
[21] Belyaev S T 1958 Application of the Methods of Quantum Field Theory to a System of Bosons, JETPh **34** 417 [in Russian].
[22] Lozovik Yu E. Yudson V I 1978 Coherent Transfer of Photoassociated Molecules into the Rovibrational Ground State, Phys. A **93** 493.

[23] Bolda E L, Tiesinga E, Julienne P S 2003 Pseudopotential Model of Ultracold Atomic Collisions in Quasi-one- and Two-dimensional Traps, Phys. Rev. A **68** 032702.

[24] Berezinskii V L 1972 Destruction of Long-range Order in One-dimensional and Two-dimensional Systems with a Continuous Symmetry Group I. Classical Systems, Sov. Phys. JETP **32** 493.

[25] Berezinskii V L 1971 Destruction of Long-range Order in One-dimensional and Two-dimensional Systems with a Continuous Symmetry Group. II. Quantum systems, Sov. Phys. JETP **34** 610.

[26] Kosterlitz J. M, Thouless D J 1973 Ordering, Metastability and Phase Transitions in Two-dimensional Systems, J. Phys. C **6** 1181.

[27] Nelson D R. Kosterlitz D R 1977 Coherent Transfer of Photoassociated Molecules into the Rovibrational Ground State, Phys. Rev. Lett. **39** 1201.

[28] Aikawa K, Akamatsu D, Hayashi D, *et al.* 2010 Coherent Transfer of Photoassociated Molecules into the Rovibrational Ground State, Phys. Rev. Lett. **105** 203001.

21st International Laser Physics Workshop IOP Publishing
Journal of Physics: Conference Series **414** (2013) 012037 doi:10.1088/1742-6596/414/1/012037

Ghost Imaging of Space Objects

Dmitry V. Strekalov, Baris I. Erkmen and Nan Yu

Jet Propulsion Laboratory, California Institute of Technology, 4800 Oak Grove Drive, Pasadena, California 91109-8099 USA

E-mail: dmitry.v.strekalov@jpl.nasa.gov

Abstract. The term "ghost imaging" was coined in 1995 when an optical correlation measurement in combination with an entangled photon-pair source was used to image a mask placed in one optical channel by raster-scanning a detector in the other, empty, optical channel. Later, it was shown that the entangled photon source could be replaced with thermal sources of light, which are abundantly available as natural illumination sources. It was also shown that the bucket detector could be replaced with a remote point-like detector, opening the possibility to remote-sensing imaging applications. In this paper, we discuss the application of ghost-imaging-like techniques to astronomy, with the objective of detecting intensity-correlation signatures resulting from space objects of interest, such as exo-planets, gas clouds, and gravitational lenses. An important aspect of being able to utilize ghost imaging in astronomy, is the recognition that in interstellar imaging geometries the object of interest can act as an effective beam splitter, yielding detectable variations in the intensity-correlation signature.

1. Introduction

In observation astronomy, a direct intensity measurement technique is often complimented by other types of measurements, a well-known example of which is intensity interferometry. The first and perhaps the most famous application of this technique in astronomy is the measurement of the angular size of a star, which was first demonstrated by R. Hanbury Brown and R.Q. Twiss in 1957 [1]. Fifty years after the pioneering work, a space-deployable version of this technique has been proposed [2].

Intensity correlation measurements also find common use in a very different field, namely in quantum optics, where the field fluxes incident on the photodetectors are typically at single-photon levels, and the correlation processing simplifies to a photon-coincidence measurement [3]. The focus of our paper is an intensity-correlation imaging technique referred to as *ghost imaging*, which was invented in the quantum optics community, although since then it has found broader range of applicability. Ghost imaging refers to a two-arm imaging system in which an object of interest is located in one arm, as shown in Fig. 1. A single-pixel bucket detector is placed in the signal arm, which collects all light that has been transmitted through (or reflected off of) the object, offering no spatial resolution. A high-resolution detector array (or a scanning pinhole detector) is placed in the reference arm (the arm without the object), but the incident field has had no interaction with the object of interest.

Historically, both entangled photon pairs (biphotons) obtained from spontaneous parametric down conversion (see Fig. 2) [4], and classical sources with pseudo thermal [5, 6] or thermal [7] radiation have been used to perform ghost imaging. The crux of virtually all ghost imaging schemes is that a source with low spatial coherence is configured such that correlated

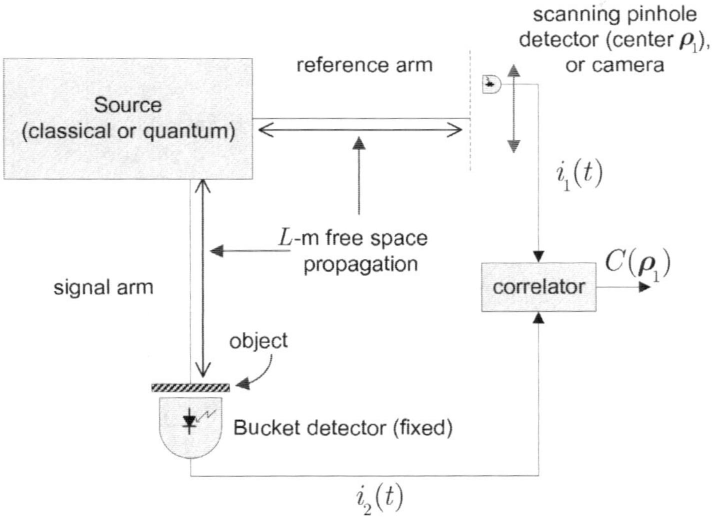

Figure 1. A generic two-arm ghost imaging setup.

intensity fluctuations (speckle patterns) are incident on both the object and the reference-arm photodetector array. Then, by cross-correlating the photodetector outputs (i.e., the photon counts in the single-photon regime or the photocurrents in the high-flux-imaging regime) one is able to reconstruct the transverse spatial transmissivity profile of the object of interest.

Besides its significance in quantum optics, ghost imaging has a few apparent practical advantages. Since no spatial resolution in the object channel is required, a very primitive single-pixel optical sensor could be placed in this channel, with the more advanced optics responsible for the image quality placed in the reference channel. This is convenient for imaging objects that are hard to access. The ghost imaging has also been demonstrated with distinct center wavelengths in the two arms, rendering it beneficial for tuning the object-arm wavelength

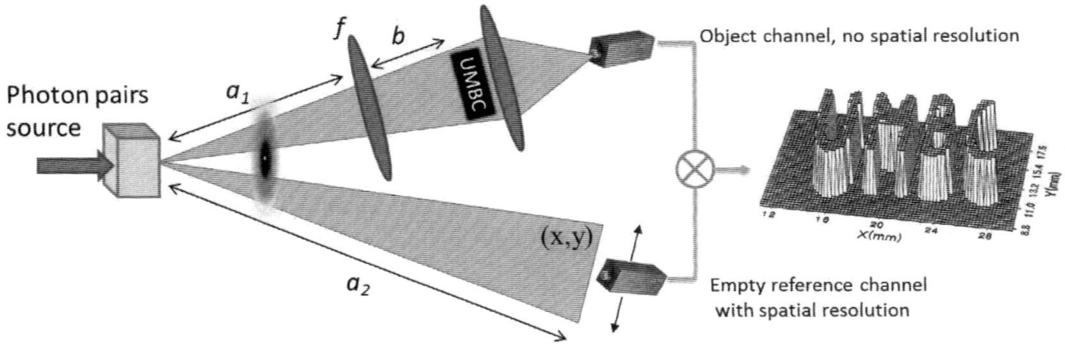

Figure 2. Simplified illustration of the original ghost imaging experiment [4]. A mask with letters "UMBC" is placed in the object channel, where all light is collected by a "bucket detector". Nonetheless, an image is reconstructed by correlating this detector's photo counts with those from another detector, raster-scanning the *empty* reference channel. A sharp image is observed when a modified thin lens equation is fulfilled: $1/(a_1 + a_2) + 1/b = 1/f$.

independently from that employed in the reference channel [8]. Furthermore, under certain background and photodetection-noise constraints, the coincidence measurement technique can offer higher robustness than a direct intensity measurement.

Despite the apparent potential advantages, the early laboratory demonstrations of ghost imaging have been far from the practical requirements arising in astronomy. However, several important advances in the past decade allow us to consider this application. First, the ability to perform with ghost imaging with thermal light [5, 6, 7, 9] was a critical breakthrough, as most natural illumination sources in our universe (e.g., stars) generate thermal light. Second, it was shown that the bucket detector in the reference arm—which is nominally expected to collect most of the light scattered from the object—could be replaced with one that collects a fraction of the field scattered by the object [10, 11]. This permits an object to be placed at a large distance from the observer, which is critical for an astronomical application, wherein the interstellar distances result in large speckles sizes on the detection planes, and subsequently the detectors in both arms resemble point-like detectors. However, using a thermal light source brings about a new complication: in general, a beam splitter has to be placed between the source and the object to create correlated speckle patterns in the signal and reference arms. While this is not a challenge in the laboratory environment, it becomes problematic when the source and object are located far in space.

In this work we investigate the possible application of ghost imaging in space astronomy, by addressing the beam splitter issue. Our approach is illustrated in Fig. 3. Its main idea is that an object which partially transmits (or diffracts) and partially reflects (or scatters) light, can itself play the role of the beam splitter. Moreover, even a perfectly opaque scattering object may under certain conditions create coherence between the transmitted and scattered light, which may be utilized for the intensity correlation imaging. Since the object is present in both channels, this approach leads us away from "canonical" ghost imaging towards intensity interferometry of Hanbury Brown - Twiss type. The analogy and distinctions between these two types of imaging have been discussed in literature [12]. In our case, an important distinction is that we will assume the source properties are known, instead focusing on the parametric variation of the intensity-correlation signature as a function of the object's geometry and location relative to the source. We will attempt to restore these parameters from the correlation measurements and show that these results could provide important information in addition to conventional direct observations.

In addition to the configuration shown in Fig. 3(b), we will consider the case when both detectors receive transmitted and scattered light. This case is more realistic when both observers are ground-based and the distant source and object cannot be optically resolved. We will develop

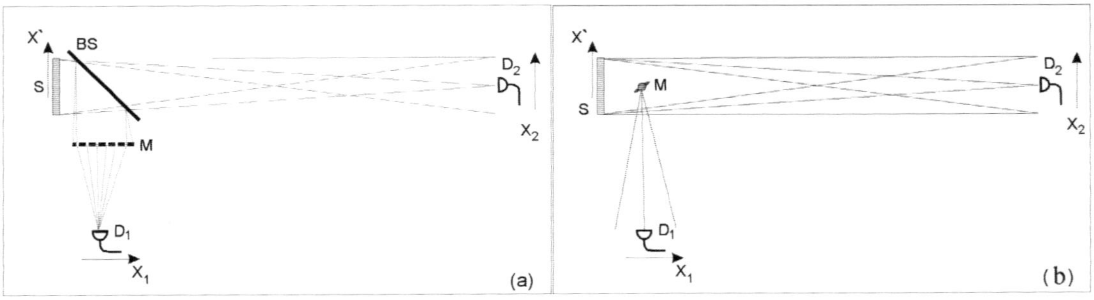

Figure 3. Conceptual schematic of conventional thermal light ghost imaging setup (a) and of our approach (b)). S is a thermal light source, M object, $D_{1,2}$ detectors and BS (not present in case (b)), a beam splitter.

and test a simple analytical model that will allow us to study the intensity-correlation signatures of the simplest test objects. Based on this model we will make predictions concerning the observability of various space objects, and concerning their parameters that can be inferred from such observations. Potential space objects of interest will include Earth-like planets (including those near bright stars), gravitational lenses formed by black holes or other massive objects, dust or gas clouds.

The rest of this paper is organized as follows. In Section 2 we will discuss the simplified geometry of the physical system under study and introduce its physical model. In Section 3 we will narrow this model to a particular but very important case of equal distances from the object to the detectors. This will allow us to carry out this analysis in the analytical form and gain the insight in its physical aspects by considering two simple model objects. We will also test some of our theory predictions by comparing them to the actual astronomy observation data, available from the Kepler mission. In Section 4 we briefly discuss the consequences of departing from the equal arms limit. In Section 5 we will discuss the signal to noise ratio in the correlation measurement and compare it to a direct intensity measurement. The results of our analysis will be summarized in Section 6.

2. The model and approach

We will consider a transverse (2D) extended source and a similar object placed in the source and object planes in which we introduce the local transverse coordinates $\vec{\rho}$ and $\vec{\rho_o}$, respectively. Let L_s be the distance between the source and object planes, and let $L_{1,2}$ denote the distances between the object plane and the planes of point-like detectors 1 and 2. The local transverse positions of these detectors are $\vec{\rho_1}$ and $\vec{\rho_2}$, respectively, as shown in Fig. 4.

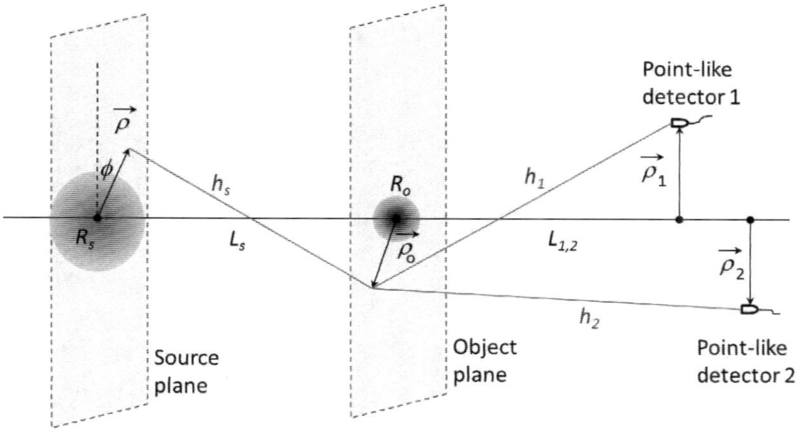

Figure 4. Relative position of the source, object and detectors in the flat paraxial model.

Let us further assume that the source field is bound by a Gaussian envelope with the width R_s and can be written as $E(\vec{\rho},t)e^{-\frac{\rho^2}{2R_s^2}}$. This model approximates the source with a diameter (intensity-distribution width) equal to $\sqrt{2}R_s$. In the paraxial approximation, the field at detector 1 or 2 is related to the field of the source as the following:

$$E_{1,2}(\vec{\rho_1},t) = \int\int d^2\rho d^2\rho_o e^{-\frac{\rho^2}{2R_s^2}} E(\vec{\rho},t - \frac{L_1 + L_s}{c}) h_{L_s}(\vec{\rho} - \vec{\rho_o}) T(\vec{\rho_o}) h_{L_{1,2}}(\vec{\rho_o} - \vec{\rho_{1,2}}). \quad (1)$$

Although it looks complicated, expression (1) is actually straightforward. It is expressed in terms of the field propagation functions

$$h_Z(\vec{x}) = \frac{e^{ikZ}}{i\lambda Z} e^{ik\frac{|\vec{x}|^2}{2Z}} \tag{2}$$

relating the electric field at two spatial points separated by a distance Z in the longitudinal direction and by a transverse displacement $\vec{x}$ (assuming that $|\vec{x}| \ll Z$) in the transverse directions. A field produced by an extended source at a remote point is then given by a convolution of (2) with the source-field distribution. This way we find the field in the object plane. We multiply this field by the object's transmission function $T(\vec{\rho_o})$ which may be real for a purely absorbing object and imaginary for a phase object, e.g. a lens. Then we repeat the propagation and integration steps to obtain the fields at the detectors (1).

The observable in case of the correlation-based ghost imaging is proportional to the product of local intensities, i.e. to

$$\langle I_1(\vec{\rho}_1, t_1) I_2(\vec{\rho}_2, t_2)\rangle = \langle E_1^\dagger(\vec{\rho}_1, t_1) E_2^\dagger(\vec{\rho}_2, t_2) E_1(\vec{\rho}_1, t_1) E_2(\vec{\rho}_2, t_2)\rangle, \tag{3}$$

where the fields can be treated as quantum-mechanical operators, or as classical values. For thermal light, the phase-sensitive term in (3) vanishes [13], and we arrive at

$$\langle I_1(\vec{\rho}_1, t_1) I_2(\vec{\rho}_2, t_2)\rangle = |\langle E_1^\dagger(\vec{\rho}_1, t_1) E_2(\vec{\rho}_2, t_2)\rangle|^2 + \langle E_1^\dagger(\vec{\rho}_1, t_1) E_1(\vec{\rho}_1, t_1)\rangle \langle E_2^\dagger(\vec{\rho}_2, t_2) E_2(\vec{\rho}_2, t_2)\rangle, \tag{4}$$

where the first term describes the possible ghost image and the second term gives the uncorrelated "background" intensity product, which also describes the object's shadow. To separate these effects it is convenient to introduce the normalized Glauber correlation function [14]

$$g^{(2)}(\vec{\rho}_1, t_1; \vec{\rho}_2, t_2) = 1 + \frac{|\langle E_1^\dagger(\vec{\rho}_1, t_1) E_2(\vec{\rho}_2, t_2)\rangle|^2}{\langle E_1^\dagger(\vec{\rho}_1, t_1) E_1(\vec{\rho}_1, t_1)\rangle \langle E_2^\dagger(\vec{\rho}_2, t_2) E_2(\vec{\rho}_2, t_2)\rangle} = 1 + \frac{|G_{12}|^2}{G_{11} G_{22}}. \tag{5}$$

Glauber correlation function will be our main observable. However let us mention that other types of measurements are possible. In particular, one can measure higher-order correlation functions $g^{(m,n)}$, or the variance of intensity difference (rather than a product) [15, 16, 17]. The analysis based on the field propagation equation (1) can be easily extended to these measurements. Such measurements will have different dependencies on the optical mode structure and on the detector's quantum efficiencies, and may offer interesting resolution-SNR trade-off opportunities. These measurement strategies depart from conventional intensity interferometry, which exclusively focuses on inferences from the $g^{(2)}$-function measurement.

To evaluate Eq. (5) above, we substitute $E_1(\vec{\rho}_1, t_1)$ and $E_2(\vec{\rho}_2, t_2)$, given by (1), into (5) and we take into account the correlation property of the source field $E(\vec{\rho}, t)$

$$\langle E^\dagger(\vec{\rho}, t) E(\vec{\rho'}, t')\rangle \propto \delta(\vec{\rho} - \vec{\rho'}) \Gamma(t - t'), \tag{6}$$

where Γ is a δ-like function whose width corresponds to the optical coherence time. The latter may be determined by the spectral filters' bandwidth. Unless we are interested in color imaging or spectroscopy, using narrow-band filters is undesirable because they reduce the signal. On the other hand, short coherence time requires fast optical detectors and electronics in order to ensure single longitudinal mode detection. Therefore to carry out a fair comparison between the direct intensity measurement and the correlation measurement, we need to take into account the photon flux reduction due to that minimal spectral filtering, required in the latter case. Let us assume a 1 ps timing accuracy (which exceeds the state of the art but may be attainable in the future) and the central wavelength of 1 micron. At this wavelength, the 1 ps coherence time corresponds to a 3.3 nm wide spectral band. Comparing the optical power detected within this band to the total power within the typical band of a silicon photo detector (see Fig. 5), we find that for a correlation measurement we have 0.5% of the power available for the broadband intensity measurement. This reduces the signal to noise ratio (SNR) in a shot-noise limited narrowband measurement. The relative SNR will be more favorable for the correlation measurement when compared to a color-resolved intensity measurement. In particular, no flux loss will be suffered in a very narrowband measurement, e.g. a measurement with a specific spectral line. A more detailed discussion of the direct-intensity vs. correlation measurement SNRs will be given below in Section 5.

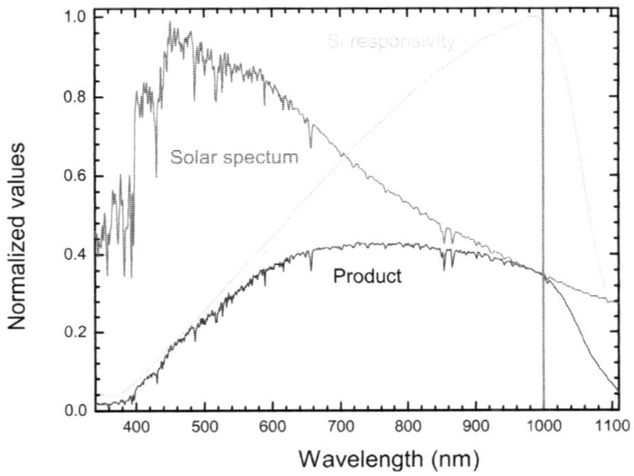

Figure 5. The solar radiation normalized spectral density (from [18]), spectral sensitivity of a typical silicon photo diode, and their product.

Let us point out that in addition to the high speed and low jitter requirements put on the photodetectors and correlation electronics, a broadband correlation measurement places stringent requirements on the clock synchronization between the two detectors, as well as on the knowledge of their relative position $L_1 - L_2$.

To continue our analysis we will assume that perfect synchronization between the detectors has been achieved and $\Gamma(t-t') = 1$ in (6). We then suppress the temporal part of the problem. For the numerator in (5) we derive

$$G_{12}(\vec{\rho_1},\vec{\rho_2}) = \int\int\int d^2\rho d^2\rho_o' d^2\rho_o \, e^{-\frac{\rho^2}{R_s^2}} T_2(\vec{\rho_o})T_1^*(\vec{\rho_o'}) \, h_{L_s}^*(\vec{\rho}-\vec{\rho_o'})h_{L_s}(\vec{\rho}-\vec{\rho_o})h_{L_1}^*(\vec{\rho_o'}-\vec{\rho_1})h_{L_2}(\vec{\rho_o}-\vec{\rho_2}). \tag{7}$$

In (7) we have introduced $T_{1,2}(\vec{\rho_o})$ to allow the transmission functions to be different for detectors 1 and 2, which will be important in the following discussion.

For the following analysis it will be convenient to introduce a correlation function

$$G_{12}^{(R_s)}(Z_a,\vec{\rho_a};Z_b,\vec{\rho_b}) = \int d^2\rho e^{-\frac{\rho^2}{R_s^2}} h_{Z_a}^*(\vec{\rho}-\vec{\rho_a})h_{Z_b}(\vec{\rho}-\vec{\rho_b}) \tag{8}$$

for the fields emitted by an extended Gaussian source of thermal light that has a width R_s and is located at $Z=0$, propagating to locations $(Z_a,\vec{\rho_a})$ and $(Z_b,\vec{\rho_b})$. Equivalently, from the *advanced wave* perspective [19], it describes time-reversed propagation of a photon from $(-Z_a,\vec{\rho_a})$ to the source, and then forward in time to $(Z_b, \vec{\rho_b})$. If the source is infinitely large, $R_s \to \infty$, the aperture-limited propagation function (8) becomes equal to a point-source propagation function from one detector to the other:

$$G_{12}^{(\infty)}(Z_a, \vec{\rho_a}; Z_b, \vec{\rho_b}) = h_{Z_b - Z_a}(\vec{\rho_b} - \vec{\rho_a}). \tag{9}$$

The two-point propagation function arises in (5):

$$G_{12}(\vec{\rho_1}, \vec{\rho_2}) = \int\int d^2\rho_o d^2\rho_o' h_{L_2}(\vec{\rho_o} - \vec{\rho_2}) h_{L_1}^*(\vec{\rho_o'} - \vec{\rho_1}) T_2(\vec{\rho_o}) T_1^*(\vec{\rho_o'}) G_{12}^{(R_s)}(L_s, \vec{\rho_o'}; L_s, \vec{\rho_o}), \tag{10}$$

and likewise for $G_{11}(\vec{\rho_1})$ and $G_{22}(\vec{\rho_2})$. To evaluate $G_{12}^{(R_s)}(Z_a, \vec{\rho_a}; Z_b, \vec{\rho_b})$ in a general form, we introduce polar coordinates such that

$$\int d^2\rho = \int_0^\infty \rho d\rho \int_0^{2\pi} d\varphi \quad \text{and} \quad |\vec{\rho_a} - \vec{\rho_b}|^2 = \rho_a^2 + \rho_b^2 - 2\rho_a\rho_b\cos(\varphi_a - \varphi_b). \tag{11}$$

The angular integration in Eq. (8) yields

$$G_{12}^{(R_s)}(Z_a, \vec{\rho_a}; Z_b, \vec{\rho_b}) = 2\pi \frac{e^{ik(Z_b - Z_a)}}{\lambda^2 Z_a Z_b} e^{i\frac{k}{2}(\rho_b^2/Z_b - \rho_a^2/Z_a)} \int_0^\infty \rho d\rho e^{-\rho^2\left[\frac{1}{R_s^2} + ik\frac{Z_b - Z_a}{2Z_a Z_b}\right]} J_0\left(k\rho\left|\frac{\vec{\rho_b}}{Z_b} - \frac{\vec{\rho_a}}{Z_a}\right|\right). \tag{12}$$

Then integrating over the radius we obtain

$$G_{12}^{(R_s)}(Z_a, \vec{\rho_a}; Z_b, \vec{\rho_b}) = 2\pi \frac{R_s^2}{\lambda^2} \frac{e^{ik(Z_b - Z_a)}}{2Z_a Z_b + ikR_s^2(Z_b - Z_a)} e^{i\frac{k}{2}(\rho_b^2/Z_b - \rho_a^2/Z_a)} e^{-\frac{k^2}{2}\left|\frac{\vec{\rho_b}}{Z_b} - \frac{\vec{\rho_a}}{Z_a}\right|^2 \frac{R_s^2 Z_a Z_b}{2Z_a Z_b + ikR_s^2(Z_b - Z_a)}}. \tag{13}$$

In the case of interest (10) we have $Z_a = Z_b = L_s$, which leads to

$$G_{12}^{(R_s)}(L_s, \vec{\rho_o'}; L_s, \vec{\rho_o}) = q^2\pi^{-1} e^{i\frac{q}{R_s}(\rho_o^2 - \rho_o'^2)} e^{-q^2|\vec{\rho_o} - \vec{\rho_o'}|^2}, \tag{14}$$

where $q^{-1} = 2L_s/(kR_s)$ is the speckle size.

3. Balanced arms configuration

In this section we consider the special balanced case when $L_1 = L_2 = L$. While this case limits possible observation scenario, it allows us to carry out exact analytical calculations in many cases of interest, and to assess the practical utility of our approach. To carry out these calculations it will be convenient to introduce the new coordinates $\vec{x} = (\vec{\rho_o} + \vec{\rho_o'})/\sqrt{2}$ and $\vec{y} = (\vec{\rho_o} - \vec{\rho_o'})/\sqrt{2}$. Then substituting (14) into (10) we obtain

$$G_{12}(\vec{\rho_1}, \vec{\rho_2}) = A \int\int S(\vec{x}, \vec{y}) e^{-2q^2y^2} e^{i\vec{\Delta}\cdot\vec{x}} e^{i\vec{\Sigma}\cdot\vec{y}} e^{i\gamma\vec{x}\cdot\vec{y}} d^2x d^2y, \tag{15}$$

where

$$\begin{aligned}
A &= \pi\left(\frac{R_s}{\lambda^2 L L_s}\right)^2 e^{\frac{ik}{2L}(\rho_2^2 - \rho_1^2)}, \\
S(\vec{x}, \vec{y}) &= T_2\left(\frac{\vec{x} + \vec{y}}{\sqrt{2}}\right) T_1^*\left(\frac{\vec{x} - \vec{y}}{\sqrt{2}}\right), \\
\vec{\Delta} &= \frac{k}{\sqrt{2}L}(\vec{\rho_1} - \vec{\rho_2}), \\
\vec{\Sigma} &= \frac{k}{\sqrt{2}L}(\vec{\rho_1} + \vec{\rho_2}), \\
\gamma &= k(1/L + 1/L_s).
\end{aligned} \tag{16}$$

The Gaussian term in (15) arises from Fourier transform of the source field distribution. This suggests that (15) could be generalized for any such distribution. However at this stage we will limit our consideration to a Gaussian source.

As a sanity check we notice that if we "turn off" the object by setting $S(\vec{x}, \vec{y}) = 1$, the integral over d^2x in (15) yields $(2\pi)^2\delta(\vec{\Delta} + \gamma\vec{y})$. Then the d^2y integral yields, quite expectedly, the correlation function of a Gaussian source (14) with increased free-space propagation length $L_s \to L + L_s$:

$$G_{12}(\vec{\rho_1}, \vec{\rho_2}) \to G_{12}^{(R_s)}(L + L_s, \vec{\rho_1}; L + L_s, \vec{\rho_2}). \quad (17)$$

Let us now consider a few example objects and discuss their possible relevance for the astronomy applications.

3.1. *A Gaussian absorber*

This case can represent, e.g., a spherical dust or gas cloud of roughly uniform density. It also can be used as a crude model for a planet occluding a star. The transmission function of such an object can be modeled as

$$T(\vec{\rho}_o) = 1 - T_0 e^{-\frac{\rho_o^2}{2R_o^2}}, \quad (18)$$

which gives rise to four terms in the two-channel transmission function:

$$S(\vec{x}, \vec{y}) = S_0 + S_{1a} + S_{1b} + S_2 = 1 - T_0 e^{-\frac{(\vec{x}+\vec{y})^2}{4R_o^2}} - T_0 e^{-\frac{(\vec{x}-\vec{y})^2}{4R_o^2}} + T_0^2 e^{-\frac{x^2+y^2}{2R_o^2}}. \quad (19)$$

In (18) and (19) T_0 is the amplitude transmission of the central part of the object. Consequently, the correlation function also will consist of four terms: $G_{12} = G_{12}^{(0)} + G_{12}^{(1a)} + G_{12}^{(1b)} + G_{12}^{(2)}$, where the zero-order term corresponds to free-space propagation (17): $G_{12}^{(0)} = G_{12}^{(R_s)}(L + L_s, \vec{\rho_1}; L + L_s, \vec{\rho_2})$. A straightforward but cumbersome calculations gives both the first and the second order terms in the form

$$G_{12}^{(1,2)}(\vec{\rho_1}, \vec{\rho_2}) = \frac{(-T_0)^n}{\pi(\tilde{q}^2 + R_o^2\tilde{\gamma}^2)} \left(\frac{k^2 R_s R_o}{2LL_s}\right)^2 e^{-i\frac{k^2 R_o^2 \tilde{\gamma}}{2L^2}\frac{\rho_2^2-\rho_1^2}{\tilde{q}^2+R_o^2\tilde{\gamma}^2}} e^{-\frac{k^2(\vec{\rho}_2+\vec{\rho}_1)^2}{8L^2(\tilde{q}^2+R_o^2\tilde{\gamma}^2)}} e^{-\frac{k^2 R_o^2 \tilde{q}^2}{2L^2}\frac{(\vec{\rho}_2-\vec{\rho}_1)^2}{\tilde{q}^2+R_o^2\tilde{\gamma}^2}}. \quad (20)$$

To obtain $G_{12}^{(1a),(1b)}$, we substitute in (20)

$$n = 1, \quad \tilde{q}^2 = 2q^2 + \frac{1}{4R_o^2}, \quad \tilde{\gamma} = \gamma \pm \frac{i}{2R_o^2}.$$

For $G_{12}^{(2)}$, we substitute

$$n = 2, \quad \tilde{q}^2 = 2q^2 + \frac{1}{2R_o^2}, \quad \tilde{\gamma} = \gamma.$$

Note that in the multi-mode case when the speckle size on the object greatly exceeds the object size, we have $\tilde{q}^2 \approx 2q^2$. However we will proceed without this approximation.

First, let us investigate the result (20) for a set of parameters that can be easily implemented on an optical bench. We simulate an opaque ($T_0 = 1$) object placed between the source ($R_s = 1$ cm) and the detector plane, so that $L_s = L = 50$ cm. The object size R_o is varied from zero to 1, 2 and 3 mm. In Fig. 6(a) we show the correlation function $g^{(2)}(|\vec{\rho}_1 - \vec{\rho}_2|)$ measured at the line of sight (i.e., for $\vec{\rho}_1 + \vec{\rho}_2 = 0$). In Fig. 6(b) we show the intensity profile, i.e. the object's shadow. We see that as the object becomes larger its shadow becomes deeper, and the speckle size becomes smaller. For larger objects the speckle shape also becomes distorted. However since we are interested in detection and characterization of small objects that distort the speckle

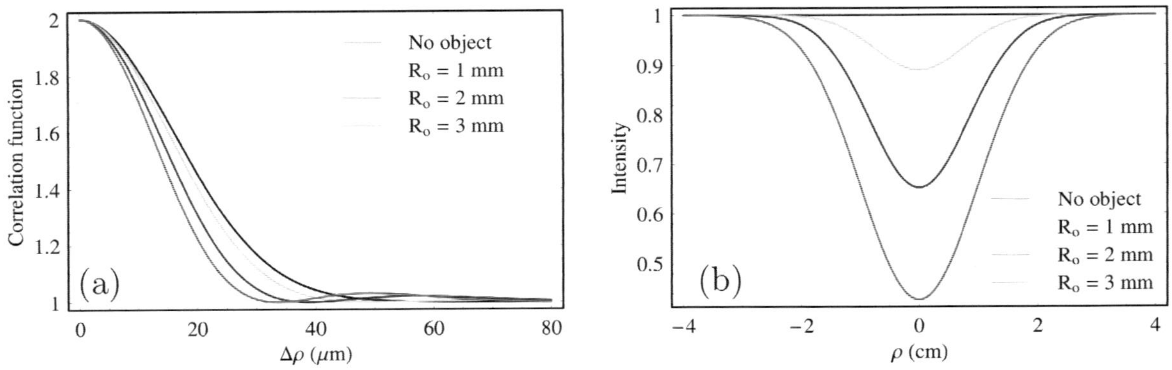

Figure 6. (a) The correlation function $g^{(2)}(\vec{\rho}_1, \vec{\rho}_2)$ vs. the distance $\Delta\rho = |\vec{\rho}_1 - \vec{\rho}_2|$; and (b) the intensity profile, for a lab parameter set ($T_0 = 1$, $L_s = L = 50$ cm and $R_s = 1$ cm). The object size R_o is varied from zero to 1, 2 and 3 mm.

shape only by a little amount, the *speckle width* can be conveniently introduced and used as the main observable in the following discussion. In practice, one can measure the speckle width by a linear array of photo detectors. Depending on the system's parameters and the desired measurement resolution, as few as two detectors could be sufficient.

The speckle width undergoes a non-intuitive evolution as an object crosses the line of sight, imitating a planet passing across the star. In Fig. 7(a) we see that the speckle width measured along the direction of the transient first gets broader, and then narrower, reaching the minimum when the object is exactly on the line of sight. In Fig. 7(b) we show the corresponding variation of the intensity, or the photon flux. Remarkably, the fractional variation of both values due to the transient object is approximately 9%.

Let us now apply our model to an actual astronomical observation carried out at Kepler space telescope [20, 21]. Substituting the Kepler-20f parameters [21] into our model we find relative intensity variation on the order of 2×10^{-4}, which is consistent with the actual observation [21], see Fig. 8. Based on the earlier discussion, we set R_o equal to $\sqrt{2}$ times the planet radius.

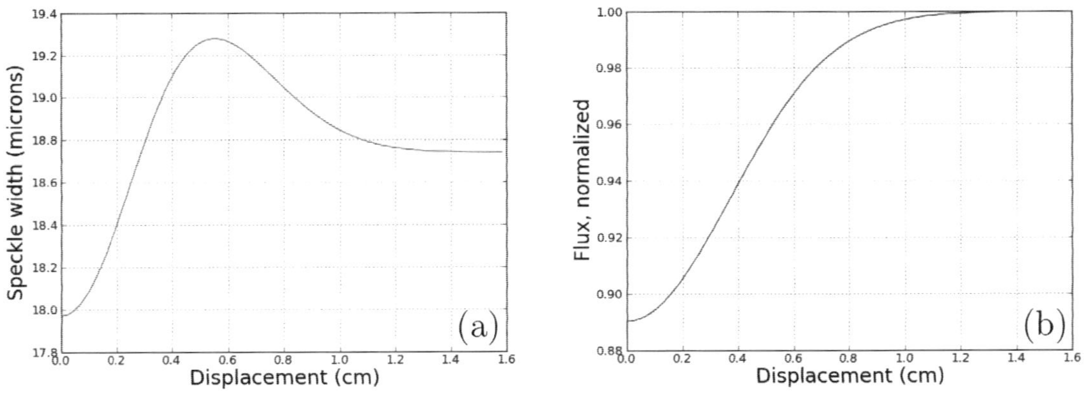

Figure 7. Dependence of the correlation function $g^{(2)}(\Delta\rho)$ width (a) and of the normalized photon flux (b) on displacement of the object from the line of sight ρ_o, for a lab parameter set ($T_0 = 1$, $L_s = L = 50$ cm, $R_s = 1$ cm, $R_o = 1$ mm).

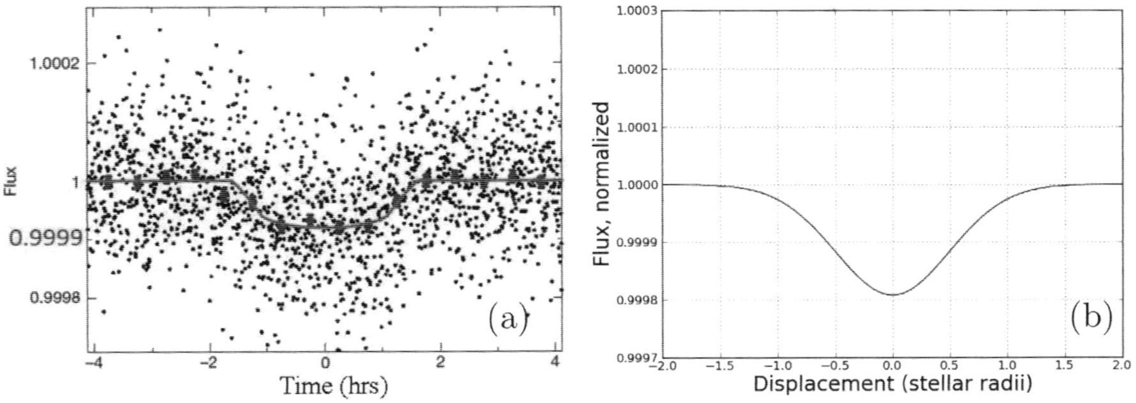

Figure 8. The intensity variation for Kepler-20e observed in [21] (a) and computed based on our model (b).

The disagreement in the dip shape shown in Fig. 8(a) and Fig. 8(b) arises from using the Gaussian absorber model while the actual planet is, of course, better approximated by an opaque disk. However the numerical agreement with the experiment shows that our simplistic, yet fully analytical, Gaussian model is sufficiently powerful. Following this model we predict that the speckle size will increase from 3,603.6 m to 3,604.2 m when the planet is centered on in the line of sight, which corresponds to a relative variation of 1.7×10^{-4}. Again, the magnitude of this variation is very close to the magnitude of the photon flux variation. Of course, a detailed SNR analysis is required in order to conclude which type of planetary detection will be more efficient. Let us however point out that the measurement of the speckle width does not preclude conventional intensity measurement, such as has been carried out in the Kepler experiment, and should be considered as an extension of such a measurement rather than its substitute.

3.2. Phase objects.
Another model object allowing for the fully analytical treatment is a thin lens. The motivation for studying of this example is the possibility to observe a phase object where direct intensity measurements may not be as efficient. Examples of such objects in space may be gravitational lenses or dilute gas clouds.

For an infinite thin lens with focal distance f,

$$S(\vec{x},\vec{y}) = e^{-\frac{ik}{2f}(\rho_o^2-\rho_o'^2)} = e^{-i\frac{k}{f}\vec{x}\vec{y}}. \tag{21}$$

The effect of such a lens can be absorbed into $\gamma \to \gamma_f = k(1/L + 1/L_s - 1/f)$. Then we quickly arrive at

$$\begin{aligned} G_{11} = G_{22} &= \frac{q^2}{\pi D^2}, \\ G_{12}(\vec{\rho_1},\vec{\rho_2}) &= \frac{q^2}{\pi D^2} e^{\frac{ik}{2L}\frac{D-1}{D}(\rho_2^2-\rho_1^2)} e^{-(q/D)^2|\vec{\rho_2}-\vec{\rho_1}|^2}, \\ g_{12}(\vec{\rho_1},\vec{\rho_2}) &= 1 + e^{-2(q/D)^2|\vec{\rho_2}-\vec{\rho_1}|^2}, \end{aligned} \tag{22}$$

where $D \equiv L(1/L + 1/L_s - 1/f)$ is a dimensionless "out-of-focus" factor. If the lens images the source plane onto the detector plane then $D = 0$ and the speckles become infinitely small while the intensity goes to infinity. Both observations are a consequence of our assumption in (6) of a delta-correlated source field. If the source is in focus then the light propagates as a collimated beam and consequently we have $D = 1$. In this case, the speckle size as well as the intensity in the detector plane will be the same as in the lens plane.

Let us consider another example of a lens-like phase *and* amplitude object. This object has a finite extent and allows for the analytical treatment. Its transmission function is

$$T(\vec{\rho}_o) = 1 - e^{-\frac{\rho_o^2}{2R_o^2}}(1 - e^{-\frac{ik\rho_o^2}{2f}}). \tag{23}$$

This object can produce shadows very similar to those from a Gaussian absorber, see Fig. 9 (a) and (b). Therefore it would be very difficult to distinguish these two objects based on the intensity measurement. However the behavior of the speckle widths in these two cases is clearly distinct, as we see from Fig. 9 (c) and (d) (notice the difference in the signal magnitude, as well as in its qualitative behavior). This example demonstrates the advantages offered by a correlation measurement for object *characterization*, beyond its mere detection.

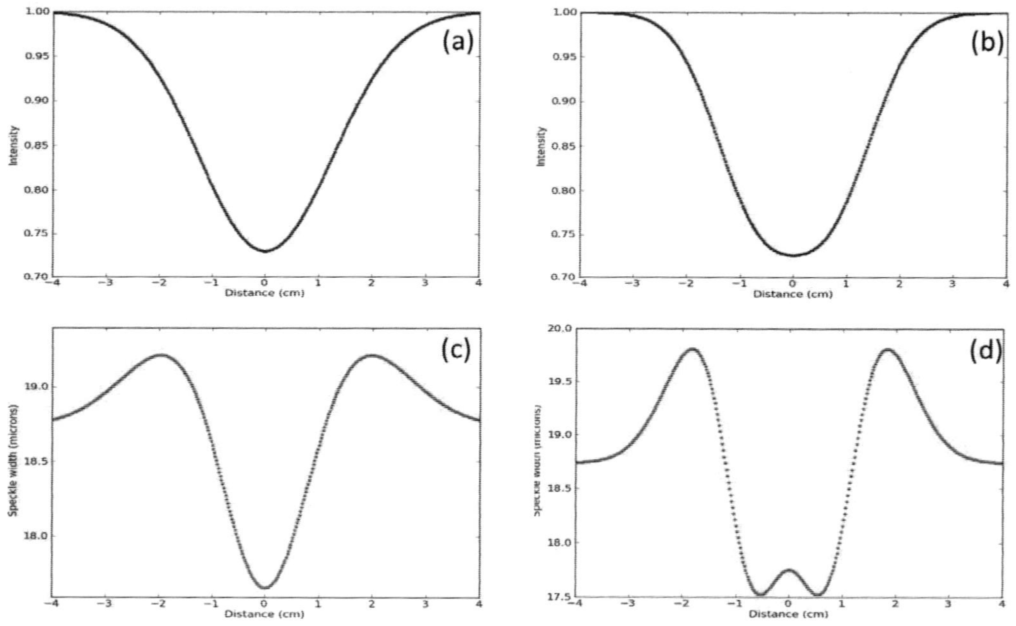

Figure 9. (a): A shadow from a test object (23) with $R_o = 3$ mm and $f = 1$ m inserted between the source and detectors as a function of its distance from the line of sight. (b) A shadow from a Gaussian absorber with $R_o = 5$ mm and $T_0 = 0.62$. (c) and (d): The speckle width for (a) and (b), respectively. In this simulation $L_s = L_1 = L_2 = 50$ cm, $R_s = 1$ cm.

4. Unbalanced arms configuration

In this section we consider a more general case of $L_1 \neq L_2$ which may be important for an asymmetric configuration, e.g. when the correlation measurement is performed by a ground-based detector jointly with a distant space-based detector. We will continue to assume perfect time synchronization between the two detectors, or equivalently, a monochromatic light source. However we will not be able to carry out the analytical calculations without making some reasonable approximations. From (13) we see that the *large aperture approximation* (9) holds if

$$\alpha_s \equiv \frac{2Z_a Z_b}{kR_s^2|Z_b - Z_a|} \ll 1. \tag{24}$$

This approximation is appropriate for evaluation of the first-order terms in (10) where $Z_b - Z_a = L_1$ is large. Indeed, for the optics lab geometry $\lambda = 1$ μm and $R_s = L_s = 1$ cm, so we get $\alpha_s = 1.6 \times 10^{-5}$. For the Solar system geometry with $R_s = 7 \times 10^5$ km (Sun radius), $L_s = 1.5 \times 10^8$ km (the distance from Earth to Sun), and $\lambda = 1$ μm, we get $\alpha_s = 5 \times 10^{-14}$. This parameter becomes even smaller for inter-stellar distances. Therefore when we calculate G_{12} (10) for a Gaussian absorber described by (18), the first-order terms can be approximated as

$$G_{12}^{(1)}(\vec{\rho_1}, \vec{\rho_2}) \equiv G_{12}^{(1a)}(L_1, \vec{\rho_1}; L_2, \vec{\rho_2}) + G_{12}^{(1b)*}(L_2, \vec{\rho_2}; L_1, \vec{\rho_1}) \approx -2\int d^2\rho\, e^{-\frac{\rho^2}{2R_o^2}} h_{L_1}^*(\vec{\rho}-\vec{\rho_1}) h_{L_2}(\vec{\rho}-\vec{\rho_2}). \tag{25}$$

The opposite case of (24) occurs when $Z_a = Z_b$. Furthermore, if q^2 is much greater than the sum of all coefficients for the ρ^2 terms in the real and imaginary exponents in (10), then (14) can be proven to approach a δ-function normalized to unity area. This is the *small aperture approximation*, applicable for the second-order terms of (10). Let us point out that within this approximation, the object cannot create coherence between the transmitted and scattered light unless the speckle size in the object plane approaches or exceeds the size of the object itself.

It is easy to see that for the optics-lab geometry as described above, q^2 exceeds all relevant parameters by a factor of at least 3×10^4. The excess factors are much greater in all reasonable astronomical geometries. Therefore we derive

$$G_{12}^{(2)}(\vec{\rho_1}, \vec{\rho_2}) \approx \int d^2\rho\, e^{-\frac{\rho^2}{R_o^2}} h_{L_1}^*(\vec{\rho} - \vec{\rho_1}) h_{L_2}(\vec{\rho} - \vec{\rho_2}). \tag{26}$$

For a Gaussian absorber case with $L_1 \neq L_2$ we obtain the following approximate expressions:

$$\begin{aligned} G_{12}^{(0)}(\vec{\rho_1}, \vec{\rho_2}) &= G_{12}^{(R_s)}(L_1 + L_s, \vec{\rho_1}; L_2 + L_s, \vec{\rho_2}), \\ G_{12}^{(1)}(\vec{\rho_1}, \vec{\rho_2}) &= -2G_{12}^{(\sqrt{2}R_o)}(L_1, \vec{\rho_1}; L_2, \vec{\rho_2}), \\ G_{12}^{(2)}(\vec{\rho_1}, \vec{\rho_2}) &= G_{12}^{(R_o)}(L_1, \vec{\rho_1}; L_2, \vec{\rho_2}). \end{aligned} \tag{27}$$

Let us first evaluate the correlation function $g^{(2)}$ found by substituting (27) into (5) in the absence of the object, by setting $R_o = 0$. In Fig. 10(a) we show this function for a typical set of optical lab parameters ($L_1 = 55$ cm, $L_2 = L_1 + \Delta L$, $L_s = 55$ cm, $\lambda = 1$ μm), where we assumed that the detectors are coplanar with the line of sight and are placed symmetrically on its opposite sides: $\vec{\rho_1} = -\vec{\rho_2}$. This allows us to introduce a single scalar parameter $\Delta\rho = |\vec{\rho_1} - \vec{\rho_2}|$, in the same way it was done in Fig. 6.

The correlation loss due to $\Delta L \neq 0$ is clearly visible. We would like to emphasize that this is not due to a limited longitudinal coherence of the source, but because of its transverse coherence properties. We can interpret this result as following. By placing the first detector in the plane L_1 we define the speckle pattern in this plane as the transverse mode structure. These speckles may be further considered as mutually incoherent light sources. As light from these sources propagates further, the coherence areas expand as well as overlap. The expansion causes the widening of the correlation function while the overlap causes the contrast reduction due to multimode detection. Using the expression $g^{(2)}(0) = 1 + 1/m$ relating the correlation peak for thermal light to the number of detected modes m we can determine that in our example the longitudinal displacement of the detector by $\Delta L = 6$ mm has lead to the number of detected modes $m \approx 3$. Note that this interpretation differs from the speckle pattern behavior that one might observe, e.g. on a screen, in which case the speckles do not overlap and do not appreciably change in size for small longitudinal translations.

Now let us "turn on" the object and investigate its effect on the correlation function. If a small ($R_o = 1$ mm) object is placed half way between the source and the detectors, the correlation function becomes narrower, which means smaller speckles, see Fig. 10(b).

Let us point out that the $\Delta L = 0$ case in Fig. 10 is consistent with the $R_o = 1$ mm case from Fig. 6, which validates the large-aperture and small-aperture approximations for an absorbing object. On the other hand, we notice that when the small aperture approximation is utilized and (14) is treated as a δ-function, a phase object such as a thin large lens is "erased" by taking the absolute-square of $T(\rho_o)$.

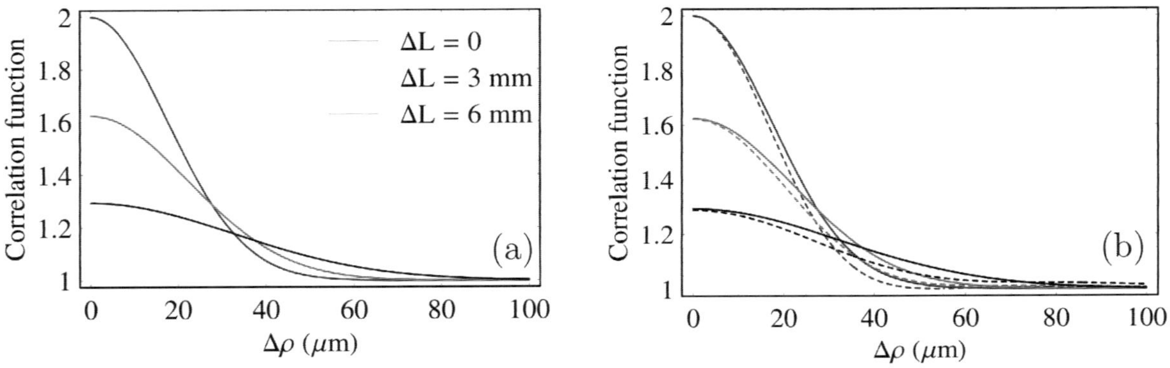

Figure 10. (a): the correlation functions $g^{(2)}(\Delta\rho)$ in the absence of an object for $R_s = 1$ cm $L_1 = 55$ cm and $L_2 = L_1 + \Delta L$. (b): the same correlation functions (solid lines) become narrower (dashed lines) when a small ($R_o = 1$ mm) Gaussian absorbing object is inserted in the line of sight at the distance $L_s = 55$ cm from the source.

5. Signal-to-noise ratio analysis

The sensitivity of the correlation-based measurements can be ascertained by a careful study of the signal-to-noise ratio, which strongly depends on the object of interest and the imaging geometry. As a result, the value added by the correlation-based measurement can range from significant to negligible. The quantitative analysis of the signal-to-noise ratio (SNR) for both types of measurements can leverage SNR calculations for Hanbury Brown-Twiss intensity interferometry, because performing ghost imaging at interstellar distances often yields geometries wherein the detectors are point-like (rather than bucket-like, which is more common in laboratory demonstrations of ghost imaging).

The ghost imaging configuration shown in Fig. 4 consists of two point-like detectors, each measuring the incident far-field irradiance resulting from the thermal source and object combination. The correlation of the two photocurrent outputs isolates the common irradiance fluctuations at their respective transverse locations, yielding information on the mutual coherence function of the effective source created by the true source illuminating the object.

Suppose the baseband envelope of the field incident on detector m, for $m = 1, 2$, is denoted by $E_m(t)$ for $0 < t < T$, having units $\sqrt{\text{photons/s}}$. Because we are considering thermal fields, we assume that the field have zero mean (i.e., $\langle E_m(t)\rangle = 0$) and a complex degree of coherence

$$\gamma_{1,2}(t-u) \equiv \frac{\langle E_1^*(t)E_2(u)\rangle}{\sqrt{\langle |E_1(t)|^2\rangle\langle |E_2(u)|^2\rangle}} \tag{28}$$

that is nonzero. We assume that the the coherence time of the two fields (i.e., the time delay τ for which $\langle E_m^*(t+\tau)E_m(t)\rangle$ is appreciable) is equal and given by T_c.

We will carry the following analysis for the photocurrents $i_1(t)$ and $i_2(t)$ produced by the detectors in response to the incident optical field. This analysis can be easily generalized for the photon-counting detectors. We have

$$\langle i_m(t)|E_m(t)\rangle = \int \mathrm{d}\tau \left(\eta|E_m(\tau)|^2 + \lambda_d\right) h_B(t-\tau)\,, \tag{29}$$

and

$$\langle \Delta i_m(t)\Delta i_m(u)|E_m(t,u)\rangle = \int \mathrm{d}\tau \left(\eta|E_m(\tau)|^2 + \lambda_d\right) h_B(t-\tau)h_B(u-\tau)\,. \tag{30}$$

Here $\Delta i(t) \equiv i(t) - \langle i(t)\rangle$, $h_B(t)$ denotes the real-valued baseband impulse response of the photodetectors with time constant (i.e., inverse bandwidth) equal to T_B, and λ_d is the mean dark photoelectron counts, which are assumed to obey homogeneous Poisson point process statistics. With little loss in generality, we assume that $\int \mathrm{d}t\, h_B(t) = 0$, i.e., that the photodetectors are DC blocking, and that $T_B \ll T_c$, i.e., that the coherence time of the source is much longer than the response time constant of the photodetector. Note that the DC term is useful for direct intensity observations, thus it could be tapped off prior to the $h_B(t)$ filter.

The equal-time photocurrent correlation measurement performed at the receiver corresponds to

$$C \equiv \int_0^T \mathrm{d}t\, i_1(t)i_2(t)\,, \tag{31}$$

where C is repeated for every position of the two detectors. Thus, strictly speaking E_m, i_m and C should have two vector arguments indicating the positions of the two detectors in 2D or 3D space. However, here we are concerned with the SNR at a particular position of these detectors, and therefore we are omitting these variables to avoid clutter.

The SNR of the C measurement is defined then as

$$\mathrm{SNR} \equiv \sqrt{\frac{\langle C\rangle^2}{\langle (C-\langle C\rangle)^2\rangle}}\,. \tag{32}$$

Calculating the first and second moments is tedious, but straightforward [22, 13]. The SNR can be evaluated, in the $T_c \gg T_B$ limit, as [22]

$$\mathrm{SNR} = \sqrt{\frac{TT_B}{T_c^2}\frac{|\gamma_{1,2}|^4N^2}{(1+N_d/N)^2 + 2(1+N_d/N)N(T_B/T_c) + N^2(T_B/T_c)^2[|\gamma_{1,2}|^4 + (1+|\gamma_{1,2}|^2)^2]}}\,, \tag{33}$$

where $\gamma_{1,2} \equiv \gamma_{1,2}(0)$, $N \equiv \eta(T_c/T)\int_0^T \mathrm{d}t\,\langle |E_m(t)|^2\rangle$, for $m = 1, 2$, is the mean photoelectron number per temporal mode (or equivalently per coherence time) of the fields incident on the detectors, and $N_d = \lambda_d T_c$ is the mean dark photoelectron number per coherence time. The terms in the denominator have intuitive interpretations. The first term, which is independent of N, is due to the shot noise of the two detectors. The third term, with the N^2 dependence, is excess noise resulting from the statistical fluctuations of the incident power on the detectors. This term is sometimes referred to as relative intensity noise. The middle term, with the N dependence, is a result of the beating between the intensity fluctuations and shot noise. Asymptotic SNR expressions can be derived for different limiting cases, depending on the particular imaging configuration and photodetector properties. In the following discussions, we assume that the dark-counts from the detector are negligibly small relative to the incident photon counts. In the shot-noise limited regime $N_d \ll N \ll 1$ holds, i.e., the mean number of photoelectrons *per mode* is very small, and the SNR can be approximated as

$$\mathrm{SNR} = \frac{\sqrt{TT_B}}{T_c}|\gamma_{1,2}|^2 N. \tag{34}$$

Therefore the SNR has a linear dependence on the incident photon flux per mode. In the opposite regime with many photoelectrons per mode, thus $N \gg 1$, the SNR saturates to its maximum value

$$\mathrm{SNR} = \sqrt{\frac{T}{T_B}}\frac{|\gamma_{1,2}|^2}{\sqrt{|\gamma_{1,2}|^4 + (1+|\gamma_{1,2}|^2)^2}}. \tag{35}$$

For direct intensity measurements, we assume that a simple photon bucket operation is performed over T-seconds of integration,

$$D = \int_0^T \mathrm{d}t\, i(t), \tag{36}$$

where the statistics in Eqs. (29) and (30) are still true, but we assume that $\int \mathrm{d}t\, h_B(t) = 1$, such that the DC component is no longer filtered out. Again, defining

$$\mathrm{SNR} \equiv \sqrt{\frac{\langle D\rangle^2 - \lambda_d^2}{\langle (D - \langle D\rangle)^2\rangle}}, \tag{37}$$

it is straightforward to derive the SNR as

$$\mathrm{SNR} = \sqrt{\frac{T}{T_c}\frac{N}{1 + N_d/N + N}}, \tag{38}$$

where all parameters are as defined before. Thus, when $N_d \ll N \ll 1$, the system becomes shot-noise limited and we have

$$\mathrm{SNR} = \sqrt{N\frac{T}{T_c}}, \tag{39}$$

whereas when $N \gg 1$ and $N > N_d$ excess noise dominates and the SNR saturates at

$$\mathrm{SNR} = \sqrt{\frac{T}{T_c}}. \tag{40}$$

To separate the SNR dependence on the integration time T, which is common to both the correlation and direct intensity measurements, it is convenient to normalize it to $\sqrt{T/T_c}$. Fig. 11 shows thus normalized SNR of a direct intensity measurement (38) and the asymptotic approximations (34) and (35) of correlation-measurement SNRs, as a function of the mean photoelectron number per mode for the case of $\gamma_{1,2} = 1$. As seen from this figure, the correlation-measurement SNR can approach the direct-intensity measurement SNR for the sources with high spectral brightness, $N \approx 1$. The correlation-based measurements typically have worse SNR than direct-intensity measurements when shot-noise limited due to the dependence on the square of the incident average photon number in correlation measurements rather than just the photon number in direct intensity measurements. However in the excess-noise limited regime, the correlation measurements' SNR may remain shot-noise limited (as also pointed out in [22]) due to the fact that such measurements can distinguish source fluctuations from those caused by an object better than the direct-intensity measurement. Note, however, these plots compare the SNRs when the photoelectrons per mode are equal in both methods. Whereas correlation-based measurements require $T_B < T_c$, thereby limiting the total flux incident on the detector, the direct intensity measurement can integrate over a very wide optical bandwidth, without penalty. We have brielfy discussed this aspect of correlation imaging in Section 2, considering the example of background light derived from the Solar spectrum.

Thus, in general, the SNR in the correlation measurement is worse than in the direct intensity measurements. Nonetheless, in order to have a realistic assessment of the added value of correlation-based measurements for imaging space objects, one must take into account stray light, detector aging, natural variation of source brightness and other practical effects that are usually omitted in the SNR analyses published to date. The detectors' dark noise may be particularly important in cases when the incident photons flux is low. On the other hand, the correlation technique may be particularly beneficial for narrow-band imaging, e.g. imaging using a specific spectral line. A narrow spectral feature will lead to the higher spectral brightness and give the correlation imaging advantage according to Fig. 11.

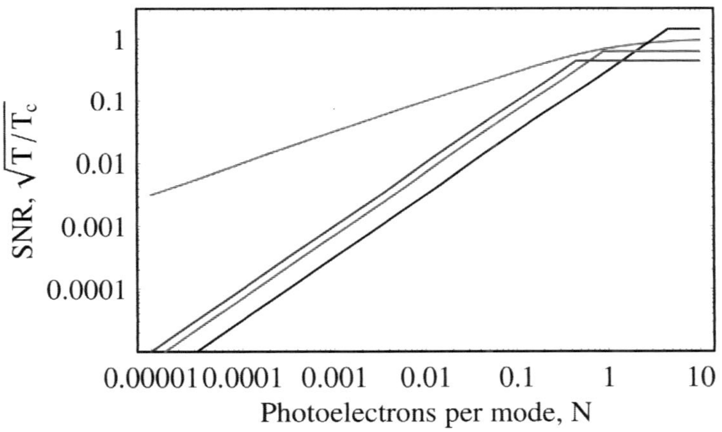

Figure 11. The normalized intensity SNR (red) and correlation measurement SNR for $T_B/T_c = 0.1$ (the lowest), 0.5 and 1 (the highest).

6. Conclusions and summary

In this paper we have analyzed intensity-correlation "ghost imaging" of dark objects in space illuminated by thermal light sources (stars). Our approach hinges on replacing the beam splitter, indispensable for thermal light ghost imaging but infeasible for space imaging, with the object itself. The absorptive and refractive properties of the object are predicted to imprint themselves on the intensity correlation properties of the transmitted and scattered light. Consequently, in principle, they could be extracted from intensity correlation measurements. To provide an insightful exposition of this concept we limited our discussion to a fully analytical model relying on a two-dimensional source and an object with Gaussian distribution of luminosity, absorption, or phase delay (the latter representing a thin lens) in the paraxial approximation. We demonstrated the variation of the far-field speckle size due to the presence of the object. We have shown that the speckle size variation is a non-trivial function of the object's properties and position. In some cases it allows us to distinguish different phase and amplitude objects even when they produce very similar shadows and can hardly be distinguished by a direct intensity measurement. Thus the correlation measurement provides a complementary information to a direct observation.

This understanding has encouraged us to apply our analytical model to a realistic space object imaging scenario, such as the Kepler mission. Our prediction for the flux variation is very close to the actual observation. It also predicts a similar fractional change of the speckle size. We have carried out a preliminary SNR analysis for a correlation measurement, comparing it to a direct flux measurement. The analysis has shown that, for parameters typical of the Kepler mission, the correlation measurement SNR would be significantly worse than the intensity measurement SNR. This analysis however does not include instrument-generated technical noise, that may be detrimental for the intensity measurement more than for the correlation measurement, and could potentially balance or even reverse the SNRs inequality between the two schemes. Examples of such technical noises are dark noise and variation of detector's responsivity (quantum efficiency) due to environmental fluctuations and aging. The ambient background light is another important factor that needs to be considered.

Regardless of the technical noises, the correlation measurement SNR improves relative to the intensity measurement SNR when the spectral brightness of the signal increases. Thus the correlation measurement may be especially advantageous in narrow-band imaging, e.g. imaging based on a selected spectral line. As a final note, comparison of the SNRs of two types of measurement is the decisive criterion when the measurements provide the same type of information. As we have seen, the correlation measurement can provide information additional to the intensity measurement. Moreover, a space object predominantly introducing phase (such as e.g. a gravitational lens or a dilute gas cloud) results in little or no tell-tale shadows, but has a more pronounced affect on the optical coherence. In such scenarios, intensity-correlation ghost imaging may be a preferable option.

7. Acknowledgments

This work was carried out at the Jet Propulsion Laboratory, California Institute of Technology under a contract with the National Aeronautics and Space Administration, sponsored by the NIAC program. D.V.S. acknowledges discussions with Drs. Maria Chekhova and Timur Iskhakov.

References

[1] Hanbury R. Brown and R.Q. Twiss, Proc. of the Royal Society of London, A, **242** 300-324 (1957); Hanbury R. Brown and R.Q. Twiss, Proc. of the Royal Society of London, A, **243** 291-319 (1958).
[2] I. Klein, M. Guelman, and S. G. Lipson, Appl. Opt. 46, 4237-47 (2007).
[3] C. K. Hong, Z. Y. Ou, and L. Mandel, Phys. Rev. Lett. **59**, 2044 (1987).
[4] T.B. Pittman, Y.H. Shih, D.V. Strekalov, and A.V. Sergienko, Phys. Rev. A, **52**, R3429 (1995);
[5] A. Valencia, G. Scarcelli, M. D'Angelo, and Y. Shih, Phys. Rev. Lett. **94**, 063601 (2005).
[6] F. Ferri, D. Magatti, A. Gatti, M. Bache, E. Brambilla, and L. A. Lugiato, Phys. Rev. Lett. **94**, 183602 (2005).
[7] X.-H. Chen, Q. Liu, K.-H. Luo, and L.-A. Wu, Opt. Lett. **34**, 695–697 (2009).
[8] K.W.C. Chan, M.N. OSullivan, and R.W. Boyd, Phys. Rev. A, **79**, 033808 (2009).
[9] A. Gatti, E. Brambilla, M. Bache, and L. A. Lugiato, Phys. Rev. A, **70**, 013802 (2004); A. Gatti, E. Brambilla, M. Bache, and L. A. Lugiato, Phys. Rev. Lett, **93**, 93602 (2004).
[10] R. Meyers, K.S. Deacon and Y. Shih, Phys. Rev. A, **77**, 041801(R) (2008).
[11] N. Hardy and J. H. Shapiro, Phys. Rev. A **84**, 063824 (2011).
[12] L.-G. Wang, S.Qamar,S.-Y. Zhu, and M.S. Zubairy, Phys. Rev. A, **79**, 033835 (2009).
[13] B.I. Erkmen, and J.H. Shapiro, Advances in Optics and Photonics **2**, 405450 (2010).
[14] R. J. Glauber, Phys. Rev. **130**, 2529-2539, (1963).
[15] E.D. Lopaeva and M. Chekhova, JETP Lett. **91**, 447 (2010).
[16] G.Brida, M.V. Chekhova, G.A. Fornaro, M. Genovese, E. Lopaeva, I. RuoBerchera, Phys Rev. A **83**, 063807 (2011).

[17] T. Iskhakov, A. Allevi, D. A. Kalashnikov, V.G. Sala, M. Takeuchi, M. Bondani, and M.V. Chekhova, Eur. Phys. J. Special Topics **199**, 127-138, (2011).
[18] `http://www.nrel.gov/solar_radiation/`
[19] J.G. Cramer, Phys. Rev. D, **22**, 362-376 (1980); J.G. Cramer, Rev. of Mod. Phys., **58**, 647 (1986); D.N. Klyshko, Phys. Lett. A, **128**, 133-137, (1988).
[20] J.J. Lissauer et al., Nature **470**, 53-58 (2011).
[21] F. Fressin et al., Nature **482**, 195-198 (2011).
[22] H. Gamo, JOSA **56**, 441-451 (1966).

Journal of Physics: Conference Series **414** (2013) 012038 doi:10.1088/1742-6596/414/1/012038

Ideal quantum reading of optical memories

Michele Dall'Arno

Graduate School of Information Science, Nagoya University, Chikusa-ku, Nagoya, 464-8601, Japan

ICFO-Institut de Ciencies Fotoniques, Mediterranean Technology Park, E-08860 Castelldefels (Barcelona), Spain

E-mail: michele.dallarno@math.cm.is.nagoya-u.ac.jp

Alessandro Bisio, Giacomo Mauro D'Ariano

Quit group, Dipartimento di Fisica, via Bassi 6, I-27100 Pavia, Italy

Istituto Nazionale di Fisica Nucleare, Gruppo IV, via Bassi 6, I-27100 Pavia, Italy

Abstract. Quantum reading is the art of exploiting the quantum properties of light to retrieve classical information stored in an optical memory with low energy and high accuracy. Focusing on the ideal scenario where noise and loss are negligible, we review previous works on the optimal strategies for minimal-error retrieving of information (ambiguous quantum reading) and perfect but probabilistic retrieving of information (unambiguous quantum reading). The optimal strategies largely overcome the optimal coherent protocols (reminiscent of common CD readers), further allowing for perfect discrimination. Experimental proposals for optical implementations of optimal quantum reading are provided.

1. Introduction

In the engineering of optical memories (such as CDs or DVDs) and readers, a tradeoff among several parameters must be taken into account. High precision in the retrieving of information is surely an indefeasible assumption, but also energy requirements, size and weight can play a very relevant role for applications. Clearly, size and weight of the device increase with the energy, and using a low energetic radiation to read information reduces the heating of the physical bit, thus allowing for smaller implementation of the bit itself. Moreover, many physical media (e.g., superconducting devices) dramatically change their optical properties if the energy flow overcomes a critical threshold.

In the problem of quantum reading [1, 2, 3, 4, 5, 6, 7, 8, 9] of optical devices one's task is to exploit the quantum properties of light in order to retrieve some classical digital information stored in the optical properties of a given media, making use of as few energy as possible. The quantum reading of optical memories was first introduced in Ref. [1]. A realistic model of digital memory was considered, where each cell is composed of a beamsplitter with two possible reflectivities. A single optical port is available to probing the beam splitter, while the other port introduces thermal noise in the reading process, so that the problem considered is the discrimination of two lossy and thermal Gaussian channels. It was shown that, for fixed mean number of photons irradiated over each memory cell, even in the presence of noise and loss, a quantum source of light can retrieve more information than any classical source - in particular in the regime of few photons and high reflectivities. This provided the first evidence that the use of quantum light can provide great improvements in applications in the technology of digital memories such as CDs or DVDs.

In practical implementations noise can sometimes be noticeably reduced [1]. On the other hand, in general loss inherently affects quantum optical setups. Nevertheless, a theoretical analysis of the ideal, i.e. lossless and noiseless, quantum reading provides a theoretical insight of the problem and a meaningful benchmark for any experimental realization. In this hypothesis quantum reading of optical devices can be recasted to a discrimination among optical devices with low energy and high precision.

In the ideal reading of a classical bit of information from an optical memory, namely in the discrimination of a quantum optical device from a set of two, different scenarios can be distinguished. A possibility is the on-the-fly retrieving of information (e.g. multimedia streaming), where the requirement is that the reading operation is performed fast - namely, only once, but a modest amount of errors in the retrieved information is tolerable. This scenario corresponds to the problem of minimum energy ambiguous discrimination of optical devices [13, 14, 15], where one guesses the unknown device and the task is to minimize the probability of making an error.

On the other hand, in a situation of criticality of errors and very reliable technology, the perfect retrieving of information is an issue. Then, unambiguous discrimination of optical devices [16], where one allows for an inconclusive outcome (while, in case of conclusive outcome, the probability of error is zero) becomes interesting.

In Ref. [2] an optimal strategy for the first scenario - namely, the minimum energy ambiguous discrimination of optical devices - has been provided for the ideal case. This strategy, that exploits fundamental properties of the quantum theory such as entanglement, allows for the ambiguous discrimination of beamsplitters with probability of error under any given threshold, while minimizing the energy requirement. The proposed optimal strategy has been compared with a coherent strategy, reminiscent of the one implemented in common CD readers, showing that the former saves orders of magnitude of energy if compared with the latter, and moreover allows for perfect discrimination with finite energy.

In Ref. [6] the results of Ref. [2] were extended to the case of unambiguous ideal quantum reading - namely, the minimum energy unambiguous discrimination of optical devices. The optimal strategy for unambiguous discrimination of beamsplitters with probability of failure under a given threshold, while minimizing the energy requirement, was provided. It was shown that the optimal strategy does not require any ancillary mode - while in the presence of noise and loss ancillary states improve the performance of the quantum reading setup [17, 1]. Both strategies for ambiguous and unambiguous quantum reading reduce to the same optimal strategy for perfect discrimination if the probability of error (in the former case) or the probability of failure (in the latter case) is set to zero. Then, some experimental setups implementing such optimal strategies which are feasible with present day quantum optical technology, in terms of preparations of single-photon input states, linear optics and photodetectors, were provided.

This paper comprehensively reviews the main results that we obtained in Refs. [2, 6] reformulating them in a coherent and homogeneous presentation. The paper is structured as follows. In Section 2 we formally introduce and discuss the ideal quantum reading of optical memories. In Section 3 we consider the particular case where each memory cell is represented by a beamsplitter. In Section 4 we compare the optimal quantum reading strategy with the optimal coherent protocol, making use of coherent input states and homodyne detection. In Section 5 we propose some experimental optical implementations of quantum reading. We conclude summarizing our results in Section 6.

[1] This fact is no longer true for example in the analogous context of quantum illumination [10, 11, 12], where one's task is to perform a low energy detection of the presence (or absence) of a far object in a noisy environment.

2. Ideal Quantum Reading of Optical Memories

A M-modes quantum optical device [18] is described by a unitary operator U relating M input optical modes with annihilation operators a_i on $\mathcal{H}_i$, to M output optical modes with annihilation operators a'_i on $\mathcal{H}_{i'}$, where $\mathcal{H}_i$ denotes the Fock space of the optical mode i. We denote the total Fock space as $\mathcal{H} = \bigotimes_i \mathcal{H}_i$.

An optical device is called *linear* if the operators of the output modes are related to the operators of the input modes by a linear transformation, namely

$$\begin{pmatrix} \mathbf{a}' \\ \mathbf{a}'^{\dagger} \end{pmatrix} = S \begin{pmatrix} \mathbf{a} \\ \mathbf{a}^{\dagger} \end{pmatrix}, \qquad S := \begin{pmatrix} A & B \\ \bar{B} & \bar{A} \end{pmatrix} \tag{1}$$

where S is called scattering matrix, $\bar{X}$ denotes the complex conjugate of X, $\mathbf{a} = (a_1, \ldots a_N)$ is the vector of annihilation operators of the input mode, and analogously $\mathbf{a}'$ for the output modes. If $B = 0$ in Eq. (1) the device is called *passive* and conserves the total number of photons, that is $\langle\psi| N |\psi\rangle = \langle\psi| U^{\dagger}NU |\psi\rangle$ with $N := \sum_i a_i^{\dagger}a_i$ the *number operator* on $\mathcal{H}$. In the following, for any pure state $|\psi\rangle$, we denote with $\psi := |\psi\rangle\langle\psi|$ the corresponding projector. For any Fock space $\mathcal{H}$, we denote with $|n\rangle$ a Fock basis in $\mathcal{H}$ ($|0\rangle$ denotes the state of the vacuum).

Suppose we want to discriminate between two linear optical passive devices U_1 and U_2. If a single use of the unknown device is available, the most general strategy consists of preparing a bipartite input state $\rho \in \mathcal{B}(\mathcal{H} \otimes \mathcal{K})$ ($\mathcal{K}$ is an ancillary Fock space with mode operators b_i), applying locally the unknown device and performing a bipartite POVM Π on the output state $(\mathcal{U}_x \otimes \mathcal{I}_{\mathcal{K}})\rho = (U_x \otimes I_{\mathcal{K}})\rho(U_x^{\dagger} \otimes I_{\mathcal{K}})$ (x can be either 1 or 2).

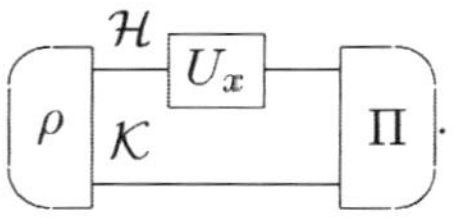

(2)

The choice of Π in Eq. (2) depends on the figure of merit taken into account. For example, for ambiguous discrimination $\Pi = \{\Pi_1, \Pi_2\}$ and one's task is to minimize the probability of error

$$P_E(\rho, U_1, U_2) := \mathrm{Tr}[(\mathcal{U}_1 \otimes \mathcal{I}_{\mathcal{H}})(\rho)\Pi_2 + (\mathcal{U}_2 \otimes \mathcal{I}_{\mathcal{H}})(\rho)\Pi_1],$$

with $0 \leq P_E(\rho, U_1, U_2) \leq 1/2$. When $p_1 = p_2 = 1/2$ the minimal probability of error has been proven to be given by the following function [19] of ρ,

$$P_E(\rho^*, U_1, U_2) = \frac{1}{2}\left(1 - ||\left[(\mathcal{U}_1 - \mathcal{U}_2) \otimes \mathcal{I}_{\mathcal{K}}\right]\rho||_1\right), \tag{3}$$

where $||X||_1 = \mathrm{Tr}[\sqrt{X^{\dagger}X}]$ denotes the trace norm.

For unambiguous discrimination $\Pi = \{\Pi_1, \Pi_2, \Pi_F\}$, $\mathrm{Tr}[(\mathcal{U}_1 \otimes \mathcal{I}_{\mathcal{H}})(\rho)\Pi_2] = \mathrm{Tr}[(\mathcal{U}_2 \otimes \mathcal{I}_{\mathcal{H}})(\rho)\Pi_1] = 0$ and one's task is to minimize the probability of inconclusive outcome (failure)

$$P_F(\rho, U_1, U_2) := \mathrm{Tr}[(\mathcal{U}_1 \otimes \mathcal{I}_{\mathcal{H}} + \mathcal{U}_2 \otimes \mathcal{I}_{\mathcal{H}})(\rho)\Pi_F], \tag{4}$$

with $0 \leq P_F(\rho, U_1, U_2) \leq 1$.

In the following, whenever the results we present hold for $P_E(\rho, U_1, U_2)$ (in an ambiguous discrimination scenario) as well as for $P_F(\rho, U_1, U_2)$ (in an unambiguous discrimination scenario), we will simply write $P(\rho, U_1, U_2)$.

Upon denoting with $E_D(\rho) := \mathrm{Tr}[\rho(N \otimes I_{\mathcal{K}})]$ the energy that flows through the unknown device, the total energy of the input state is $E(\rho) := E_D + \mathrm{Tr}[\rho(I_{\mathcal{H}} \otimes N_{\mathcal{K}})]$.

We can now introduce the ideal quantum reading problem [2, 6]. For any set of two optical devices $\{U_1, U_2\}$ and any threshold q in the probability of error (failure), find the minimum energy input state ρ^* that allows to ambiguously (unambiguously) discriminate between U_1 and U_2 with probability of error (failure) not greater than q, namely

$$\rho^* = \arg \min_{\rho \text{ s.t. } P(\rho, U_1, U_2) \leq q} E(\rho). \tag{5}$$

where $P(\rho, U_1, U_2) = P_E(\rho, U_1, U_2)$ for the ambiguous discrimination problem and $P(\rho, U_1, U_2) = P_F(\rho, U_1, U_2)$ for the unambiguous discrimination problem.

First, notice that for any POVM Π we have $P((\mathcal{U}_1 \otimes \mathcal{I}_\mathcal{K})\rho, I, U_2 U_1^\dagger) = P(\rho, U_1, U_2)$ and $E((\mathcal{U}_1 \otimes I_\mathcal{K})\rho) = E(\rho)$, so we can restrict our analysis to the case in which $U_1 = I$ and $U_2 = U$, and identify $P(\rho, I, U) = P(\rho, U)$.

Then, notice that without loss of generality the constraint in Eq. (5) can be restated as $P(\rho, U) = q$. Indeed, for any POVM Π we have that $P(\rho, U)$ is a continuous function maximized in $|0\rangle\langle 0|$ [indeed $P_E(|0\rangle\langle 0|, U) = 1/2$ and $P_F(|0\rangle\langle 0|, U) = 1$]. So for any ρ with $P(\rho, U) < q$ there exists a $0 < \alpha \leq 1$ such that $P[(1-\alpha)\rho + \alpha|0\rangle\langle 0|, U] = q$. Since $E[(1-\alpha)\rho + \alpha|0\rangle\langle 0|] < E(\rho)$, the constraint in Eq. (5) becomes $P(\rho, U) = q$.

Proposition 1 (Optimal state is pure). *For any optical device U and any threshold q in the probability of error $P_E(\rho, U)$ [probability of failure $P_F(\rho, U)$], there exists a state ρ^* which minimizes Eq. (5) such that ρ^* is pure.*

Proof. Notice that Eq. (5) is equivalent to $C(\rho, U) := pP(\rho, U) + (1-p)E(\rho)$, for any fixed value of p. If ρ^* is the state that minimizes $C(\rho, U)$, for $q := P(\rho^*, U)$ we have that $E(\rho^*)$ gives the minimum possible value for the energy. Since $P(\rho, U)$ and $E(\rho)$ are linear functions of ρ, it follows that $C(\rho, U)$ is a linear function of ρ and its minimum is attained on the boundary of its domain, namely for a pure state $|\psi^*\rangle$. □

As a consequence of Proposition 1, Eq. (5) can be restated as

$$\psi^* = \arg \min_{\psi \text{ s.t. } P(\psi, U) = q} E(\psi). \tag{6}$$

For pure states, the probability of error in the ambiguous discrimination when $p_1 = p_2 = 1/2$ given by Eq. (3) becomes

$$P_E = \frac{1}{2}\left(1 - \sqrt{1 - |\langle\psi| (U \otimes I_\mathcal{K}) |\psi\rangle|^2}\right). \tag{7}$$

For pure states, the probability of failure in the unambiguous discrimination when $p_1 = p_2 = 1/2$ given by Eq. (4) has been proved to be given by [16]

$$P_F(\psi^*, U) = |\langle\psi| U \otimes I_\mathcal{K} |\psi\rangle|. \tag{8}$$

Proposition 2 (No ancillary modes are required). *For any optical device U and any threshold q in the probability of error $P_E(\rho, U)$ [probability of failure $P_F(\rho, U)$], there exists a state ρ^* which minimizes Eq. (5) such that $\rho^* \in \mathcal{H}$.*

Proof. We show that for any pure input state ψ there exists a pure state ψ' that does not resort to ancillary modes and that allows for quantum reading with the same probability of error (failure) but with lower energy. Let us denote with $|\mathbf{n}\rangle = |n_1, \ldots, n_{\dim \mathcal{H}}\rangle$ a Fock basis in $\mathcal{H}$ with respect to which U is diagonal and with $|\mathbf{m}\rangle = |m_1, \ldots, m_{\dim \mathcal{K}}\rangle$ a Fock basis in $\mathcal{K}$. Let us denote with $e^{i\delta_i}$ the eigenvalue of U corresponding to mode i-th, and let $\boldsymbol{\delta} = (\delta_1, \ldots, \delta_{\dim \mathcal{H}})$, namely

$U|\mathbf{n}\rangle = e^{i\boldsymbol{\delta}\cdot\mathbf{n}}|\mathbf{n}\rangle$. Any pure input state can be written as $|\psi\rangle = \sum_{\mathbf{n},\mathbf{m}} c_{\mathbf{n},\mathbf{m}}|\mathbf{n},\mathbf{m}\rangle$ for some $c_{\mathbf{n},\mathbf{m}}$, then one has $\langle\psi|U\otimes I_K|\psi\rangle = \sum_{\mathbf{n},\mathbf{m}}|c_{\mathbf{n},\mathbf{m}}|^2 e^{i\boldsymbol{\delta}\cdot\mathbf{n}}$ and $E(\psi) = \sum_{\mathbf{n},\mathbf{m}}|c_{\mathbf{n},\mathbf{m}}|^2(\sum_i n_i + \sum_j m_j)$. For any $|\psi\rangle$ let us define $|\psi'\rangle := \sum_{\mathbf{n}} c'_{\mathbf{n}}|\mathbf{n},\mathbf{0}\rangle$ with $c'_{\mathbf{n}} := \sqrt{\sum_{\mathbf{m}}|c_{\mathbf{n},\mathbf{m}}|^2}$, then one has $\langle\psi'|U\otimes I_K|\psi'\rangle = \langle\psi|U\otimes I_K|\psi\rangle$ and $E(\psi') = \sum_{\mathbf{n},\mathbf{m}}|c_{\mathbf{n},\mathbf{m}}|^2(\sum_i n_i)$. Since $P(\psi',U) = P(\psi,U)$ and $E(\psi') \le E(\psi)$ - the former immediately following from Equations (7) and (8) - the statement is proved. □

Since no ancillary modes are required, the energy $E_D(\psi)$ that flows through the unknown device is equal to the total energy of the input state $E(\rho)$, so minimizing the former instead than the latter - namely, replacing $E(\psi)$ with $E_D(\psi)$ in Eq. (6) - does not change the optimal state.

3. Ideal Quantum Reading of Beamsplitters

A beamsplitter is a two-mode linear passive quantum optical device such that $A \in SU(2)$ in Eq. (1). In the following we will use the basis $\{|n,m\rangle\}$ with respect to which A is diagonal with eigenvalues $e^{\pm i\delta}$, $0 \le \delta \le \pi$. With this choice, for any $|\psi\rangle = \sum_{n,m=0}^{\infty}\alpha_{n,m}|n,m\rangle$, we have $U|n,m\rangle = e^{i\delta(n-m)}|n,m\rangle$, so that $\langle\psi|U|\psi\rangle = \sum_{n,m=0}^{\infty}|\alpha_{n,m}|^2 e^{i\delta(n-m)}$ and $\langle\psi|N|\psi\rangle = \sum_{n,m=0}^{\infty}|\alpha_{n,m}|^2(n+m)$. We notice that both these expressions only depend on the squared modulus of the coefficients $\alpha_{n,m}$, so we can assume $\alpha_{n,m}$ to be real and positive.

Here $\lfloor x\rfloor$ ($\lceil x\rceil$) denotes the maximum (minimum) integer number smaller (greater) than x.

Proposition 3 (Optimal quantum reading of beamsplitters). *For any beamsplitter U and for any threshold q in the probability of error (probability of failure), there exists a state ψ^* which minimizes Eq. (6) such that*

$$|\psi^*\rangle = \alpha\frac{|0,n^*\rangle + |n^*,0\rangle}{\sqrt{2}} + \sqrt{1-\alpha^2}\,|00\rangle\,, \tag{9}$$

where

$$|\alpha| = \sqrt{\frac{1-K(q)}{1-\cos(\delta n^*)}}, \qquad K(q) = \begin{cases} 2\sqrt{q(1-q)} & \text{for ambiguous reading} \\ q & \text{for unambiguous reading} \end{cases},$$
$$n^* = \arg\min_{\lfloor x^*\rfloor,\lceil x^*\rceil} E(\psi^*), \qquad x^* = \min(x > 0|\delta x = \tan(\delta x/2)).$$

Proof. First we prove that the optimal state in Eq. (6) is a superposition of NOON states. For any state $|\psi\rangle = \sum_{n,m}\alpha_{n,m}|n,m\rangle$, the state $|\psi'\rangle = \sqrt{1/2}\sum_l \alpha'_l(|l,0\rangle + |0,l\rangle)$ with $|\alpha'_l|^2 = \sum_{|n-m|=l}|\alpha_{nm}|^2$ is such that

$$\langle\psi'|N|\psi'\rangle = \sum_{n,m=0}^{\infty}\alpha_{nm}^2|n-m| \le \langle\psi|N|\psi\rangle\,, \tag{10}$$
$$|\langle\psi'|U|\psi'\rangle| = \left|\sum_{n,m=0}^{\infty}\alpha_{nm}^2\cos(\delta|n-m|)\right| \le |\langle\psi|U|\psi\rangle|.$$

So we have $\langle\psi|U|\psi\rangle \in \mathbb{R}$ and the constraint in Eq. (6) becomes $\langle\psi|U|\psi\rangle = K(q)$.

Then we prove that the optimal state is the superposition of two NOON states. Let $|\psi^*\rangle = \sqrt{1/2}\sum_n \alpha_n^*(|n,0\rangle + |0,n\rangle)$ be the optimal state and let the set $\{\alpha_n^*\}$ have $N \ge 3$ not-null elements. Then there exist n_1 and n_2 such that $\alpha_{n_1}, \alpha_{n_2} \ne 0$ and $\cos(\delta n_1) \le K(q) \le$

$\cos(\delta n_2)$. Define $|\chi\rangle := 1/\sqrt{2}\sum_{i=1,2}\beta_{n_i}(|n_i,0\rangle + |0,n_i\rangle)$ such that $\langle\chi|\,U\,|\chi\rangle = K(q)$, and $|\xi\rangle := 1/\sqrt{2}(1-\epsilon)^{-1/2}\sum_n \gamma_n(|n,0\rangle + |0,n\rangle)$, where

$$\gamma_n = \begin{cases} \alpha_n & \text{if } n \neq n_1, n_2 \\ \sqrt{\alpha_n^2 - \epsilon\beta_n^2} & \text{if } n = n_1, n_2 \end{cases},$$

and $\epsilon \leq \min(\alpha_{n_1}/\beta_{n_1}, \alpha_{n_2}/\beta_{n_2})$. Notice that $\langle\xi|\,U\,|\xi\rangle = K(q)$, and $\langle\psi^*|\,N\,|\psi^*\rangle = \epsilon\,\langle\chi|\,N\,|\chi\rangle + (1-\epsilon)\,\langle\xi|\,N\,|\xi\rangle$. If $\langle\chi|\,N\,|\chi\rangle = \langle\psi^*|\,N\,|\psi^*\rangle$ the statement follows with $|\psi\rangle = |\chi\rangle$. If $\langle\chi|\,N\,|\chi\rangle \neq \langle\psi^*|\,N\,|\psi^*\rangle$, either $\langle\chi|\,N\,|\chi\rangle < \langle\psi^*|\,N\,|\psi^*\rangle$ or $\langle\xi|\,N\,|\xi\rangle < \langle\psi^*|\,N\,|\psi^*\rangle$, that contradicts the hypothesis that $|\psi^*\rangle$ is the optimal state.

Finally we prove that the optimal state is the superposition of a NOON state and the vacuum. Let $|\psi^*\rangle = 1/\sqrt{2}\sum_{i=1,2}\alpha_{n_i}(|n_i,0\rangle + |0,n_i\rangle)$. Then

$$\langle\psi^*|\,N\,|\psi^*\rangle = \frac{n_2\cos(\delta n_1) - n_1\cos(\delta n_2) + K(q)(n_1 - n_2)}{\cos(\delta n_1) - \cos(\delta n_2)}.$$

It is lengthy but not difficult to verify (see Ref. [2] for an explicit proof) that it is not restrictive to set $n_2 = 0$, so one has $\langle\psi^*|\,N\,|\psi^*\rangle = [1 - K(q)][1 - \cos(\delta n_1)]^{-1}n_1$. Then one can see that it is not restrictive to choose $\pi/2 \leq \delta n_1 \leq \pi$, where $\langle\psi^*|\,N\,|\psi^*\rangle$ can be proven [20] to be a convex function that attains its minimum for $n_1 = \lfloor x^*\rfloor, \lceil x^*\rceil$, with $x^* = \min(x \in \mathbb{R}^+ | \delta x = \tan(\delta x/2))$. The statement immediately follows. □

Notice that from Proposition 3 it immediately follows that ambiguous (unambiguous) discrimination between beamsplitters U and I can be achieved only if the threshold q in the probability of error (failure) satisfies the inequality $K(q) \geq \cos(\delta n^*)$ with $K(q)$ as in the statement of Proposition 3.

The optimal energy-error tradeoffs in the ambiguous and unambiguous quantum reading are trivial consequences of Proposition 3, given respectively by

$$E(P_E) = \frac{1 - 2\sqrt{P_E(1-P_E)}}{1 - \cos(\delta n^*)}n^*, \qquad E(P_F) = \frac{1 - P_F}{1 - \cos(\delta n^*)}n^*, \tag{11}$$

where n^* is constant (for any fixed δ) and is given in the statement of Proposition 3. Figure 1 shows the optimal energy-error tradeoff for some values of δ.

4. Comparison with Coherent Strategy

Here we consider the minimum energy discrimination that makes use of coherent input states $|\alpha_i\rangle$ and homodyne detections X_{φ_i} to ambiguously[2] discriminate a single use of a n-modes passive linear optical device randomly chosen in the set $\{I, U\}$ with equal probabilities

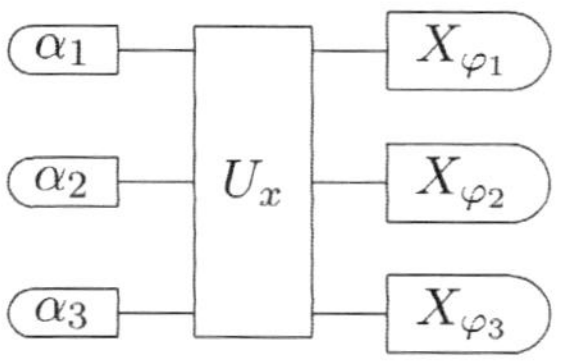

. (12)

[2] Notice that since the conditional probability distribution of the outcome of a homodyne measurement given a coherent state is Gaussian, no outcome has zero probability to occur. For this reason, no coherent strategy exists for the unambiguous discrimination of optical devices.

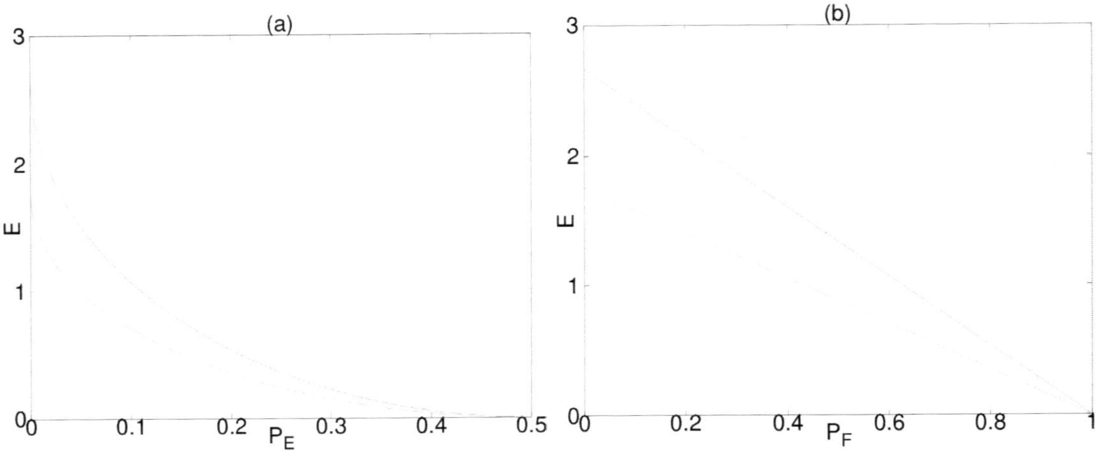

Figure 1. (Color online) Optimal tradeoff between the energy E and the probability of error P_E (probability of failure P_F) in the ambiguous (a) and unambiguous (b) discrimination of I and $U = \exp(i(\delta a_1^\dagger a_1 - \delta a_2^\dagger a_2))$, for $\delta = \pi/6$ (upper line), $\delta = \pi/4$ (middle line), and $\delta = \pi/3$ (lower line), as given by Eq. (11).

If we consider coherent input states $|\alpha_i\rangle$ on mode i the global input state is $|\xi\rangle = \bigotimes_i |\alpha_i\rangle$ which corresponds to an energy value

$$E(\xi) := \langle\xi| N |\xi\rangle = \sum_i |\alpha_i|^2. \tag{13}$$

Since for any passive linear device V we have that $V \bigotimes_i |\alpha_i\rangle = \bigotimes_i |\beta_i\rangle$ where $|\beta_i\rangle$ are coherent states, we can assume U to be diagonal, i.e. $U = \sum_i e^{i\delta_i a_i^\dagger a_i}$. The evolution of $|\xi\rangle$ under the action of U is then given by

$$U|\xi\rangle = \bigotimes_i \left| e^{i\delta_i}\alpha_i \right\rangle. \tag{14}$$

A quantum homodyne detection X_φ is described [21, 22, 23] by the POVM $\{|x,\varphi\rangle\langle x,\varphi|\}$, where $|x,\varphi\rangle$ are the eigenstates of the quadrature $e^{i\varphi}a + e^{-i\varphi}a^\dagger$. The probability of outcome x when the system is prepared in a coherent state $|\alpha\rangle$ with $\alpha = e^{i\phi_\alpha}|\alpha|$ is given by the Gaussian

$$p_\varphi(x|\alpha) = |\langle\alpha|x,\varphi\rangle|^2 = \sqrt{\frac{2}{\pi}} e^{-2(x-|\alpha|\cos(\varphi+\phi_\alpha))^2}. \tag{15}$$

We notice that $p_\varphi(x|\alpha)$ depends on the phases φ and ϕ_α only through the sum $\varphi + \phi_\alpha$. We can then fix $\varphi = 0$ and vary only the α_i. The conditional probabilities of outcome $\mathbf{x} = (x_i)$ given I or U are n-dimensional Gaussians, namely

$$p(\mathbf{x}|I) = (2/\pi)^{n/2} e^{2|\mathbf{x}-\mathbf{v_0}|^2}, \qquad p(\mathbf{x}|U) = (2/\pi)^{n/2} e^{2|\mathbf{x}-\mathbf{v_1}|^2}, \tag{16}$$

with $\mathbf{v_0} = (\mathrm{Re}\,\alpha_i)$ and $\mathbf{v_1} = (\mathrm{Re}\, e^{i\delta_i}\alpha_i)$.

Any classical postprocessing of the outcome $\mathbf{x}$ can be described by a function $q(X|\mathbf{x})$ that evaluates to 1 if one guesses the unitary X from outcome $\mathbf{x}$, and to 0 otherwise, with $X = I, U$.

Journal of Physics: Conference Series **414** (2013) 012038 doi:10.1088/1742-6596/414/1/012038

The probability of error is given by

$$P_E(\xi) = \frac{1}{2}\int d\mathbf{x}\, p(\mathbf{x}|I)q(U|\mathbf{x}) + p(\mathbf{x}|U)q(I|\mathbf{x}), \tag{17}$$

and thus the optimal postprocessing is

$$q(X|\mathbf{x}) = \begin{cases} 1 & \text{if } p(\mathbf{x}|X) \geq p(\mathbf{x}|Y) \\ 0 & \text{if } p(\mathbf{x}|X) < p(\mathbf{x}|Y) \end{cases}. \tag{18}$$

Inserting Eq. (18) and Eq. (16) into the expression (17), the probability of error becomes

$$P_E(\xi) = \frac{1}{2}\left[1 + (2/\pi)^{n/2}\int_A d\mathbf{x}\, \left(e^{-2|\mathbf{x}-\mathbf{v_0}|^2} - e^{-2|\mathbf{x}-\mathbf{v_1}|^2}\right)\right], \tag{19}$$

where we defined the set

$$A = \{\mathbf{x} \text{ s.t. } |\mathbf{x}-\mathbf{v_0}|^2 \geq |\mathbf{x}-\mathbf{v_1}|^2\}. \tag{20}$$

Within this framework it is more convenient to fix the amount of energy, that is the average number of photons η, and find the input state $|\xi^*\rangle$ that minimizes the probability of error in the discrimination, i.e.

$$|\xi^*\rangle = \arg \min_{\langle\xi|N|\xi\rangle=\eta} P_E(\xi). \tag{21}$$

With a little machinery it is possible to prove that $P_E(\xi)$ is a non-increasing function of $|\mathbf{v_0}-\mathbf{v_1}|^2$ and then the minimization of $P_E(\xi)$ can be rephrased as a maximization of $|\mathbf{v_0}-\mathbf{v_1}|^2$. We have then

$$|\mathbf{v_0}-\mathbf{v_1}|^2 = \sum_i [\mathrm{Re}(\alpha_i) - \mathrm{Re}(e^{i\delta}\alpha_i)]^2 \leq \sum_i [2\sin(\delta_i/2)|\alpha_i|]^2 \leq 4\sin^2(\delta/2)\eta, \tag{22}$$

where $\delta^* := \arg\max_{\delta_i} |\delta_i|$, and i^* labels the corresponding mode. The bounds in Eq. (22) are achieved for

$$|\xi^*\rangle = \bigotimes_{i\neq i^*} |0_i\rangle \otimes |\alpha^*_{i^*}\rangle\,, \tag{23}$$

where $\alpha^*_{i^*} = \sqrt{\eta}\exp(i\frac{\pi-\delta^*}{2})$. The corresponding optimal discrimination strategy is

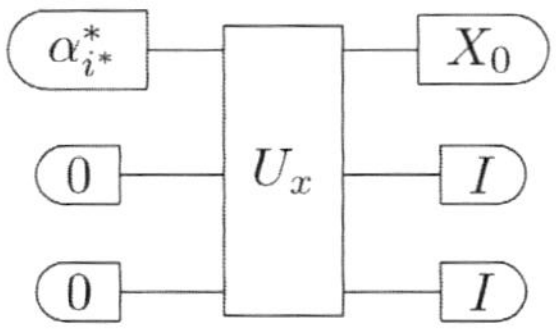

, (24)

where $-\!\boxed{I}$ means that the corresponding mode is discarded. With this choice of the input state the probability of error becomes

$$P_E = \sqrt{\frac{2}{\pi}}\int_{-\infty}^{0} dx\, e^{-2\left(x-\sqrt{\eta}\sin\frac{|\delta^*|}{2}\right)^2} = \frac{1}{2}\left[1 - \mathrm{erf}\left(\sqrt{2\eta}\sin\frac{|\delta^*|}{2}\right)\right]. \tag{25}$$

where $\mathrm{erf}(x) := \frac{2}{\sqrt{\pi}}\int_0^x dt\exp(-t^2)$ denotes the error function.

From Eq. (25) one can obtain the tradeoff between the energy and the probability of error, which is plotted in Fig. 2, for some choices of U_1 and U_2. If we consider the case in which we want to discriminate a 50/50 beamsplitter from the identity, one can notice that, for $P_E = 0.1$, the coherent state - homodyne detection discrimination strategy requires a factor of ~ 4 more photons that the optimal strategy. Moreover, this factor increases as the two devices get closer, i. e. for small values of δ. For example, when $\delta = \pi/12$, the factor is ~ 12. As expected, one notice that this factor increases when the probability of error decreases.

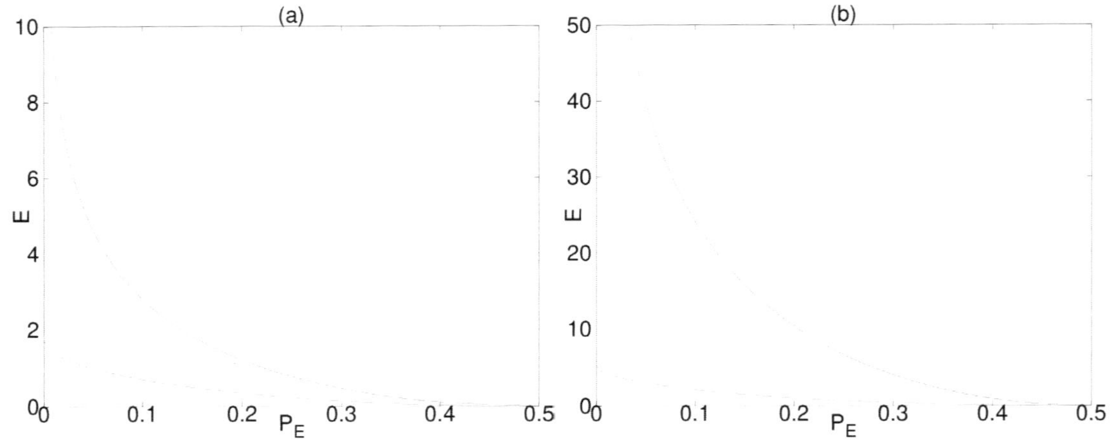

Figure 2. (Color online) Optimal tradeoff between the energy E and the probability of error P_E in the discrimination of I and $U = \exp(i(\delta a_1^\dagger a_1 - \delta a_2^\dagger a_2))$ ($\delta = \pi/4$ in (a) and $\delta = \pi/12$ in (b)). The upper line represents the discrimination with coherent states and homodyne detections, while the lower line represents the optimal discrimination. Comparing (a) and (b), we notice that the improvement provided by the optimal strategy increases as δ decreases.

5. Experimental setup for quantum reading

In this Section we provide experimental setups for ambiguous, unambiguous, and perfect quantum reading, which are feasible with present quantum optical technology. The input is a single-photon state, that can be realized e.g. through spontaneous parametric down conversion or through the attenuation of a laser beam. The evolution is given by a circuit of beamsplitters, one of which is the unknown one, and the final measurement is implemented through photodetectors.

In Proposition 2 we proved that, for the ambiguous (unambiguous) quantum reading of optical devices, no ancillary modes are required. Nevertheless, the proposed setups for quantum reading make use of three-modes input states - namely, an ancillary mode is employed. This choice is due to the requirement to have an input state with fixed number of photons in order to be able to experimentally take into account loss. For this reason, our setup minimizes the energy $E_D(\rho)$ that flows through the unknown device, while the total energy of the input state is fixed.

In the following, for any beamsplitter X we denote with A_X the A matrix of X in Eq. (1), so we write

$$A_X = \begin{pmatrix} r_X & -t_X \\ t_X & r_X \end{pmatrix}, \qquad A_X^\dagger = \begin{pmatrix} r_X & t_X \\ -t_X & r_X \end{pmatrix}.$$

We define the reflectivity R_X and the transmittivity T_X of X as $R_X := |r_X|^2$ and $T_X := |t_X|^2$, respectively, with $R_X + T_X = 1$.

The general setup is given by a Mach-Zender interferometer with beamsplitters B and $B^\dagger$, acting on modes 2 and 3. In one of the harms of the interferometer (corresponding to mode 2), the following beamsplitters are inserted

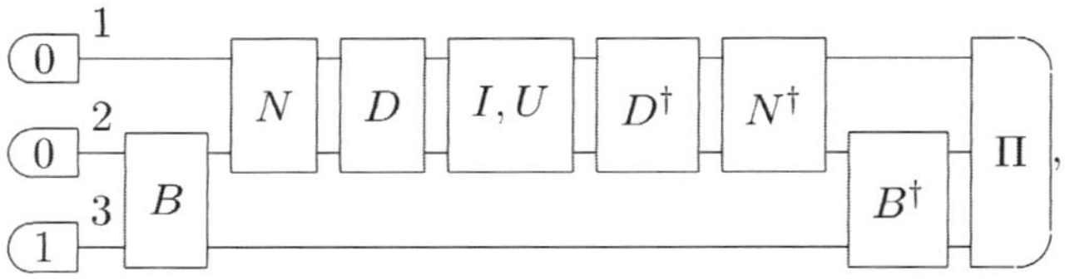

,

where N is a 50/50 beamsplitter, I, U is the unknown beamsplitter, and D is the beamsplitter diagonalizing U. The POVM Π is different for ambiguous and unambiguous quantum reading. It is easy to verify that the composition of beamsplitters DN reduces to a phase shifter on mode 2, namely

$$A_D = \frac{1}{\sqrt{2}} \begin{pmatrix} 1 & 1 \\ i & -i \end{pmatrix}, \qquad A_D A_N = \begin{pmatrix} 1 & 0 \\ 0 & i \end{pmatrix}. \tag{26}$$

It is easy to check that this phase shifter is irrelevant, so in the following we will disregard it.

Here we describe an experimental setup implementing the optimal strategy for ambiguous quantum reading as given by Proposition 3, namely the ambiguous discrimination of a beamsplitter randomly chosen from the set $\{I, U\}$ with equal prior probabilities, with probability of error $P_E(\rho, U)$ under a given threshold q and minimal energy flow through the unknown device. In the following we set $K(q) := 2\sqrt{q(1-q)}$. According to Proposition 3, in order to have $P_E(\rho, U) \leq q$, we must have $K(q) \geq \sqrt{R_U}$.

The experimental setup is then given by

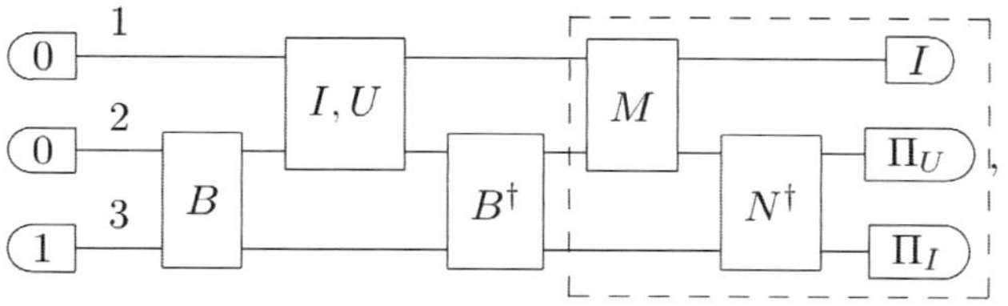

where the reflectivities and transmittivities of beamsplitters B, M and $N^\dagger$ are given by

$$R_B = \frac{K(q) - r_U}{1 - r_U}, \qquad R_M = \frac{[1 - K(q)][K(q) - r_U]}{(1-2q)^2}, \qquad R_N = \sqrt{1-q}.$$

The optimal measurement for ambiguous discrimination [19] is implemented by the two beamsplitters M and $N^\dagger$ and by the two photocounters Π_U and Π_I surrounded by the dashed line (no measurement is performed on output mode 1). The conditional probabilities $p_{X|Y}$ of detecting a photon in photodetector Π_X given that the unknown device is Y are given by

$$p_{U|U} = p_{I|I} = 1 - q, \qquad p_{I|U} = p_{U|I} = q.$$

Detecting a photon in Π_U or Π_I implies that the unknown beamsplitter is U or I, respectively, with probability of error q.

Here we describe an experimental setup implementing the optimal strategy for unambiguous quantum reading as given by Proposition 3, namely the unambiguous discrimination of a beamsplitter randomly chosen from the set $\{I, U\}$ with equal prior probabilities, with probability of failure $P_F(\rho, U)$ under a given threshold q and minimal energy flow through the unknown device. According to Proposition 3, in order to have $P_F(\rho, U) \leq q$, we must have $q \geq \sqrt{R_U}$.

The experimental setup is given by

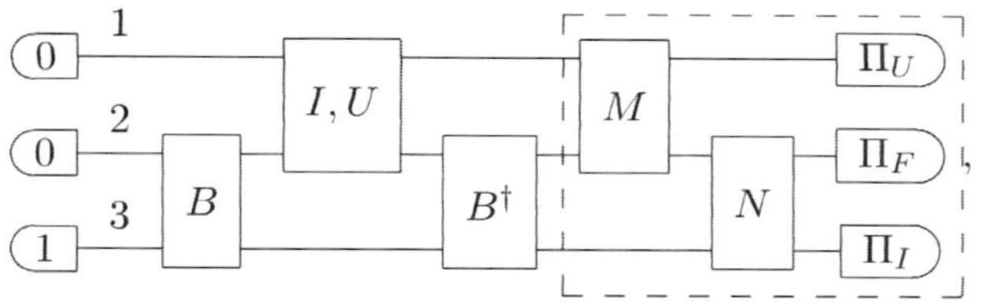

where the reflectivities and transmittivities of beamsplitters B, M and N are given by

$$R_B = \frac{q - r_U}{1 - r_U}, \qquad R_M = \frac{[\sqrt{1 + r_U} - \sqrt{q(q - r_U)}]^2}{(1 + q)^2}, \qquad R_N = \sqrt{1 - q}.$$

The optimal measurement for unambiguous discrimination [16] is implemented by the two beamsplitters M and N and by the three photocounters Π_U, Π_I, and Π_F surrounded by the dashed line. The conditional probabilities $p_{X|Y}$ of detecting a photon in photodetector Π_X given that the unknown device is Y are given by

$$p_{U|U} = p_{I|I} = 1 - q, \qquad p_{I|U} = p_{U|I} = 0, \qquad p_{F|U} = p_{F|I} = q.$$

Detecting a photon in Π_U or Π_I implies that the unknown beamsplitter is certainly U or I, respectively, while detecting a photon in Π_F declares a failure with probability q.

6. Conclusion

In this paper we considered ambiguous and unambiguous quantum reading of optical memories, on the assumption that noise and loss are negligible (Section 2). We provided the optimal strategy for ambiguous and unambiguous quantum reading of beamsplitters (Section 3), showing that the optimal input state is a superposition of a NOON state and the vacuum. In Section 4 we showed that the optimal strategy for ambiguous quantum reading largely overcomes the optimal coherent protocol (reminiscent of common CD readers), further allowing for perfect quantum reading. Finally in Section 5 we proposed some experimental implementations of ambiguous and unambiguous quantum reading, where the input state was fixed to be a single photon state. By making use of an ancillary mode it was possible to tune the amount of energy flowing through the device.

In addition to their relevance in the framework of quantum communication and information theory, the presented results also have obvious connections with experimental quantum optical applications. For these reasons we believe that they will have a relevant impact in the future development of technology for storage and retrieval of digital information.

Acknowledgments

This work was supported by the Japanese Society for the Promotion of Science (JSPS), the Spanish project FIS2010-14830, the Italian Ministry of Education through PRIN 2008 and the European Community through the COQUIT and Q-Essence projects.

[1] S. Pirandola, Phys. Rev. Lett. **106**, 090504 (2011).
[2] A. Bisio, M. Dall'Arno, and G. M. D'Ariano, Phys. Rev. A **84**, 012310 (2011).
[3] R. Nair, Phys. Rev. A **84**, 032312 (2011).
[4] S. Pirandola, C. Lupo, V. Giovannetti, S. Mancini, and S. L. Braunstein, New J. Phys. **13**, 113012 (2011).
[5] O. Hirota, arXiv:quant-ph/1108.4163v2.
[6] M. Dall'Arno, A. Bisio, G. M. DAriano, M. Mikova, M. Jezek, and M. Dusek, Phys. Rev. A **85**, 012308 (2012).
[7] G. Spedalieri, C. Lupo, S. Mancini, S. L. Braunstein, and S. Pirandola, Phys. Rev. A **86**, 012315 (2012).
[8] C. Weedbrook, S. Pirandola, R. García-Patrón, N. J. Cerf, T. C. Ralph, J. H. Shapiro, and S. Lloyd, Rev. Mod. Phys. **84**, 621 (2012).
[9] J. Prabhu Tej, A. R. Usha Devi, and A. K. Rajagopal, arXiv:1210.0791.
[10] S. Lloyd, Science **321**, 1463 (2008)
[11] S. H. Tan, B. I. Erkmen, V. Giovannetti, S. Guha, S. Lloyd, L. Maccone, S. Pirandola, J. H. Shapiro, Phys. Rev. Lett. **101**, 253601 (2008).
[12] J. H. Shapiro and S. Lloyd, New J. Phys. **11** 063045 (2009).
[13] A. Acin, Phys. Rev. Lett. **87**, 177901 (2001).
[14] G. M. D' Ariano, P. Lo Presti, and M. G. A. Paris, J. Opt. B **4**, 273 (2002).
[15] R. Duan, Y. Feng, M. Ying, Phys. Rew. Lett. **98**, 100503 (2007).
[16] M. Sedlák, Act. Phys. Slov. **59**, 653 (2009).

[17] M. F. Sacchi, Phys. Rev. A **72**, 014305 (2005).
[18] U. Leonhardt, Rept. Prog. Phys. **66**, 1207 (2003).
[19] C. W. Helstrom, *Quantum Detection and Estimation Theory* (Academic Press, New York, 1976).
[20] S. Boyd and L. Vandenberghe, *Convex Optimization* (Cambridge, New York, 2004).
[21] W. Vogel and D. G. Welsch, *Quantum Optics* (Wiley-VCH, Weinheim, 2006).
[22] T. Tyc, B. C. Sanders, J. Phys. A **37**, 7341 (2004).
[23] K. Banaszek, K. Wodkiewicz, Phys. Rev. A **55**, 3117 (1997).

Quantum minimax receiver for ternary coherent state signal in the presence of thermal noise

Kentaro Kato

Quantum Communication Research Center, Quantum ICT Research Institute, Tamagawa University, Tokyo, JAPAN

E-mail: kkatop@lab.tamagawa.ac.jp

Abstract. This paper is concerned with the minimax strategy in quantum signal detection theory. First we show a numerical calculation method for finding a solution to the quantum minimax decision problem in the case that the average probability of decision errors is used as the quality function of a quantum communication system. To verify the numerical calculation method, ternary coherent state signal is considered in the absence of thermal noise. After that, the error probability of the quantum minimax receiver for the ternary coherent state signal in the pressure of thermal noise is computed by using this numerical calculation method.

1. Introduction

About three decades ago Hirota and Ikehara discussed the quantum minimax strategy and its application to optical communication systems. Through the discussion, they showed necessary and sufficient conditions of the quantum minimax strategy under the situation that the average probability of decision errors is used as a quality function [1]. Recently their result was extended to the case of the average Bayes cost [2]. The aim of this study is to develop an evaluation method for optical communication systems from the view point of the quantum minimax strategy by showing a concrete example with the help of the necessary and sufficient conditions of the minimax strategy.

In this paper we investigate the error performance of the minimax receiver for ternary coherent state signal, $\{|0\rangle, |\alpha\rangle, |-\alpha\rangle\}$. Section 2 gives a short survey of the minimax problem in quantum signal detection. In Section 3, we show a numerical calculation method for finding a solution to the quantum minimax problem. If there is no thermal noise, one can obtain the closed-form expressions of the minimax decision rule, minimax distribution, and minimax value for the ternary coherent state signal, respectively. The comparison between these analytical results and the corresponding numerical calculation results based on the numerical calculation method is done in Section 4. Finally numerical calculation results for the ternary signal in the presence of thermal noise are shown in Section 5, and we give the conclusion of this paper in Section 6.

2. The minimax problem in quantum signal detection

Suppose that there are M symbols to be transmitted from a sender to a receiver and these symbols are drawn according to a probability distribution $\mathbf{p} = (p_1, p_2, \ldots, p_M)$. The i-th symbol is modulated to a quantum state signal, and the signal is transmitted to the receiver through a communication channel. Let $\hat{\rho}_i$ denote the received signal when the i-th signal was transmitted,

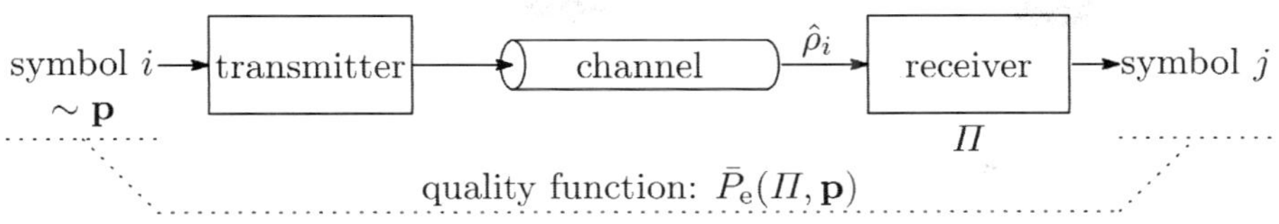

Figure 1. Communication model and its quality function

where $\hat{\rho}_i$ is a density operator on Hilbert space $\mathcal{H}$: $\hat{\rho}_i \geq 0$ and $\mathrm{Tr}\hat{\rho}_i = 1$. Then the receiver performs an appropriate quantum signal detection, which is described by a positive operator-valued measure (POVM) $\Pi = (\hat{\Pi}_1, \hat{\Pi}_2, \ldots, \hat{\Pi}_M)$. At that time, the conditional probability that the receiver chooses symbol j when symbol i was sent is given by $P(j|i) = \mathrm{Tr}\hat{\rho}_i\hat{\Pi}_j$. Similarly, the joint probability $P(i,j)$ is given by $P(i,j) = p_i\mathrm{Tr}\hat{\rho}_i\hat{\Pi}_j$. To evaluate this communication system, we employ the average probability of error,

$$\bar{P}_{\mathrm{e}}(\Pi, \mathbf{p}) \equiv \sum_{i \neq j} p_i \mathrm{Tr}\hat{\rho}_i\hat{\Pi}_j = 1 - \sum_{i=1}^{M} p_i \mathrm{Tr}\hat{\rho}_i\hat{\Pi}_i. \tag{1}$$

In our scenario we assume that the probability distribution $\mathbf{p}$ is unknown, unlike the Bayes decision problem. Under this assumption, we consider the following optimization problem: find $(\Pi^\circ, \mathbf{p}^\circ)$ and $\bar{P}_{\mathrm{e}}^\circ$ such that

$$\min_{\Pi \in \mathcal{D}} \max_{\mathbf{p} \in \mathcal{P}} \bar{P}_{\mathrm{e}}(\Pi, \mathbf{p}) = \bar{P}_{\mathrm{e}}(\Pi^\circ, \mathbf{p}^\circ) = \bar{P}_{\mathrm{e}}^\circ, \tag{2}$$

where $\mathcal{D} = \{\Pi = (\hat{\Pi}_1, \ldots, \hat{\Pi}_M)\}$ and $\mathcal{P} = \{\mathbf{p} = (p_1, \ldots, p_M)\}$. For this problem, Hirota and Ikehara [1] showed that the relation $\min_{\Pi \in \mathcal{D}} \max_{\mathbf{p} \in \mathcal{P}} \bar{P}_{\mathrm{e}}(\Pi, \mathbf{p}) = \bar{P}_{\mathrm{e}}^\circ = \max_{\mathbf{p} \in \mathcal{P}} \min_{\Pi \in \mathcal{D}} \bar{P}_{\mathrm{e}}(\Pi, \mathbf{p})$ holds and there exists a solution to this problem. Hereafter, we call Π° the minimax decision rule, $\mathbf{p}^\circ$ the minimax distribution, and $\bar{P}_{\mathrm{e}}^\circ$ the minimax value. In general, it is not easy to find a closed-form solution to this minimax problem even when the signal states are pure. However, we have obtained necessary and sufficient conditions for the minimax decision rule $\Pi^\circ = (\hat{\Pi}_1^\circ, \hat{\Pi}_2^\circ, \ldots, \hat{\Pi}_M^\circ)$ and minimax distribution $\mathbf{p}^\circ = (p_1^\circ, p_2^\circ, \ldots, p_M^\circ)$ as follows [1, 2]:

$$\begin{aligned} \hat{\Pi}_i^\circ (p_i^\circ \hat{\rho}_i - p_j^\circ \hat{\rho}_j) \hat{\Pi}_j^\circ &= 0 \quad \forall (i,j), & (3) \\ \mathrm{Tr}\hat{\Pi}_i^\circ \hat{\rho}_i &= \mathrm{Tr}\hat{\Gamma}^\circ \quad \forall i \text{ s.t. } p_i^\circ > 0, & (4) \\ \mathrm{Tr}\hat{\Pi}_i^\circ \hat{\rho}_i &\geq \mathrm{Tr}\hat{\Gamma}^\circ \quad \forall i \text{ s.t. } p_i^\circ = 0, & (5) \end{aligned}$$

where $\hat{\Gamma}^\circ = \sum_{k=1}^{M} p_k^\circ \hat{\rho}_k \hat{\Pi}_k^\circ$, and the minimax value is $\bar{P}_{\mathrm{e}}^\circ = 1 - \mathrm{Tr}\hat{\Gamma}^\circ$.

3. Numerical calculation method for finding a minimax solution

As pointed out above it is not easy to find an explicit expression of the minimax solution, $(\Pi^\circ, \mathbf{p}^\circ)$ and $\bar{P}_{\mathrm{e}}^\circ$, in general. This motivates us to consider a numerical calculation method for finding a minimax solution.

To begin with, we choose a probability distribution $\mathbf{p}^{(0} = (p_1^{(0}, p_2^{(0}, \ldots, p_M^{(0})$ of the signals arbitrary. Then we can find a Bayes decision rule $\Pi^{(0} = (\hat{\Pi}_1^{(0}, \hat{\Pi}_2^{(0}, \ldots, \hat{\Pi}_M^{(0})$ for $\mathbf{p}^{(0}$ such that

$$\min_{\Pi \in \mathcal{D}} \bar{P}_{\mathrm{e}}(\Pi, \mathbf{p}^{(0)}) = \bar{P}_{\mathrm{e}}(\Pi^{(0}, \mathbf{p}^{(0)}) = \bar{P}_{\mathrm{e}}^{\mathrm{bayes}}(\mathbf{p}^{(0)}) \equiv \bar{P}_{\mathrm{e}}^{(0}. \tag{6}$$

Our numerical calculation method starts with these initial values — $(\Pi^{(0)}, \mathbf{p}^{(0)})$ and $\bar{P}_{\rm e}^{(0)}$ — and iterates the procedures mentioned below until the change of the average probability of error is small enough.

Suppose that the calculation procedure is in the $(n+1)$-st stage, $(n = 0, 1, \dots)$. At that time, the n-th distribution $\mathbf{p}^{(n)} = (p_1^{(n)}, p_2^{(n)}, \dots, p_M^{(n)})$, Bayes decision rule $\Pi^{(n)} = (\hat{\Pi}_1^{(n)}, \hat{\Pi}_2^{(n)}, \dots, \hat{\Pi}_M^{(n)})$, and error probability $\bar{P}_{\rm e}^{(n)} \equiv \bar{P}_{\rm e}(\Pi^{(n)}, \mathbf{p}^{(n)})$ have been already recorded in the memory of computer. Now we choose a pair of indeces (i, j) such that $i \neq j$. For this (i, j), we consider the set of probability distributions of the form

$$\mathbf{p}^{(\rm temp)} = (p_1^{(n)}, p_2^{(n)}, \dots, \underbrace{s}_{i\text{th}}, \dots, \underbrace{t}_{j\text{th}}, \dots, p_M^{(n)}), \tag{7}$$

where s and t satisfy the conditions $s \geq 0$, $t \geq 0$, and $s+t = 1 - \sum_{k\neq i,\neq j} p_k^{(n)}$. From the concavity of $\min_\Pi \bar{P}_{\rm e}(\Pi, \mathbf{p}) (= \bar{P}_{\rm e}^{\rm bayes}(\mathbf{p}))$ with respect to $\mathbf{p}$, we have the inequality

$$\bar{P}_{\rm e}(\Pi^{(n)}, \mathbf{p}^{(n)}) \leq \max_{s,t} \bar{P}_{\rm e}^{\rm bayes}(\mathbf{p}^{(\rm temp)}) = \max_{s,t} \min_{\Pi} \bar{P}_{\rm e}(\Pi, \mathbf{p}^{(\rm temp)}). \tag{8}$$

Therefore our task at this stage is to find a pair $(\Pi^*, \mathbf{p}^*)$ such that

$$\max_{s,t} \min_{\Pi} \bar{P}_{\rm e}(\Pi, \mathbf{p}^{(\rm temp)}) = \bar{P}_{\rm e}(\Pi^*, \mathbf{p}^*). \tag{9}$$

Since s and t are not independent, the maximization part is a single-parameter maximization problem. Therefore, by changing the value s form $s = 0$ to $t = 1 - \sum_{k\neq i,\neq j} p_k^{(n)}$, we can find a solution to this maximization problem. Recall that the Bayes cost reduction algorithm has been well developed [3, 4]. So, one can find such a pair $(\mathbf{p}^*, \Pi^*)$ without any technical difficulty. Once we could find $(\mathbf{p}^*, \Pi^*)$, we set

$$\Pi^{(n+1)} \leftarrow \Pi^*, \qquad \mathbf{p}^{(n+1)} \leftarrow \mathbf{p}^*, \qquad \bar{P}_{\rm e}^{(n+1)} \leftarrow \bar{P}_{\rm e}(\Pi^{(n+1)}, \mathbf{p}^{(n+1)}), \tag{10}$$

and proceed to the next stage by choosing another (i, j).

Finally we check whether the conditions (3)-(5) are satisfied or not, when the increase of $\bar{P}_{\rm e}^{(\cdot)}$ is small enough. If the conditions are satisfied with sufficiently small numerical error, we halt the calculation; if not, we choose another initial values and repeat the procedures.

4. Ternary pure state signal case

In this section we consider the pure state signal case to verify the numerical calculation method mentioned above. Let $\mathcal{A} = \{0, +, -\}$ be the alphabet of signaling symbols. For each symbol, we set $0 \mapsto |0\rangle \equiv |\psi_0\rangle$, $+ \mapsto |\alpha\rangle \equiv |\psi_+\rangle$, $- \mapsto |-\alpha\rangle \equiv |\psi_-\rangle$, where $|0\rangle$ is the vacuum state, and $|\pm\alpha\rangle$ are the coherent states of light having complex amplitudes α and $-\alpha$, respectively. Now we assume that $\alpha > 0$ for simplicity and define

$$\begin{cases} |\phi_1\rangle \equiv \frac{1}{\sqrt{2(1+\kappa^4)}}|\psi_+\rangle + \frac{1}{\sqrt{2(1+\kappa^4)}}|\psi_-\rangle, \\ |\phi_2\rangle \equiv \frac{1}{\sqrt{2(1-\kappa^4)}}|\psi_+\rangle - \frac{1}{\sqrt{2(1-\kappa^4)}}|\psi_-\rangle, \\ |\phi_3\rangle \equiv \frac{\sqrt{1+\kappa^4}}{1-\kappa^2}|\psi_0\rangle - \frac{\kappa}{(1-\kappa^2)\sqrt{1+\kappa^4}}|\psi_+\rangle - \frac{\kappa}{(1-\kappa^2)\sqrt{1+\kappa^4}}|\psi_-\rangle, \end{cases} \tag{11}$$

where $\kappa = \exp[-|\alpha|^2/2]$. These vectors form an ordered orthonormal basis, $\gamma = \{|\phi_1\rangle, |\phi_2\rangle, |\phi_3\rangle\}$. By using this orthonormal basis γ, the signal states can be represented as

$$|\psi_0\rangle \doteq [|\psi_0\rangle]_\gamma = \begin{pmatrix} w \\ 0 \\ x \end{pmatrix}, \quad |\psi_+\rangle \doteq [|\psi_+\rangle]_\gamma = \begin{pmatrix} y \\ z \\ 0 \end{pmatrix}, \quad |\psi_-\rangle \doteq [|\psi_-\rangle]_\gamma = \begin{pmatrix} y \\ -z \\ 0 \end{pmatrix}, \tag{12}$$

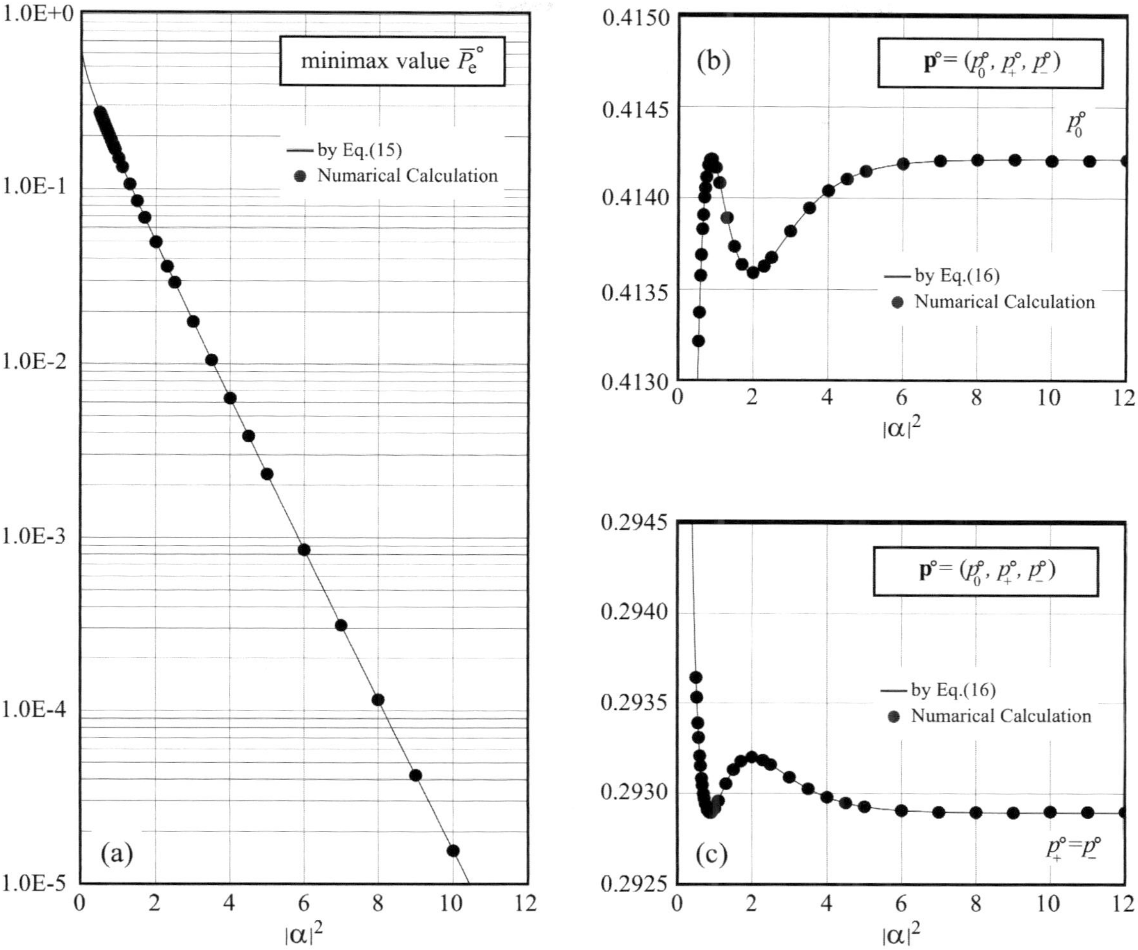

Figure 2. (a) The minimax value $\bar{P}_e^\circ$ of the ternary coherent state signal in the absence of thermal noise versus the average number $|\alpha|^2$ of signal photons. (b)-(c) The corresponding minimax distribution $\mathbf{p}^\circ = (p_0^\circ, p_+^\circ, p_-^\circ)$.

where

$$w = \frac{\sqrt{2}\kappa}{\sqrt{1+\kappa^4}}, \quad x = \frac{1-\kappa^2}{\sqrt{1+\kappa^4}}, \quad y = \frac{\sqrt{1+\kappa^4}}{\sqrt{2}}, \quad z = \frac{\sqrt{1-\kappa^4}}{\sqrt{2}}. \tag{13}$$

Let $\Pi^\circ = (\hat{\Pi}_0^\circ, \hat{\Pi}_+^\circ, \hat{\Pi}_-^\circ)$ denote the minimax decision rule for the ternary coherent state signal. The minimax decision operators are respectively given by $\hat{\Pi}_0 = |d_0\rangle\langle d_0|$, $\hat{\Pi}_+ = |d_+\rangle\langle d_+|$, and $\hat{\Pi}_- = |d_-\rangle\langle d_-|$ with

$$|d_0\rangle \doteq \begin{pmatrix} \sqrt{1-q^2} \\ 0 \\ q \end{pmatrix}, \quad |d_+\rangle \doteq \frac{1}{\sqrt{2}}\begin{pmatrix} q \\ 1 \\ -\sqrt{1-q^2} \end{pmatrix}, \quad |d_+\rangle \doteq \frac{1}{\sqrt{2}}\begin{pmatrix} q \\ -1 \\ -\sqrt{1-q^2} \end{pmatrix}, \tag{14}$$

where $q = (\sqrt{4y\left(y-\sqrt{2}x\right)+2w}+\sqrt{2}xz-yz)/(y^2-2\sqrt{2}xy+2)$. Further, the minimax value $\bar{P}_{\mathrm{e}}^{\circ}$ and minimax distribution $\mathbf{p}^{\circ}=(p_0^{\circ},p_+^{\circ},p_-^{\circ})$ are respectively given by

$$\bar{P}_{\mathrm{e}}^{\circ} = 1-\left(\sqrt{1-q^2}w+qx\right)^2, \tag{15}$$

and

$$p_0^{\circ} = \frac{\sqrt{1-q^2}y}{\sqrt{2}qw+\sqrt{1-q^2}\left(y-\sqrt{2}x\right)}, \quad p_+^{\circ}=p_-^{\circ}=\frac{qw-\sqrt{1-q^2}x}{2qw+\sqrt{1-q^2}\left(\sqrt{2}y-2x\right)}. \tag{16}$$

Using Eqs.(15) and (16) one can simulate their behavior explicitly. Figure 2(a) shows the minimax value $\bar{P}_{\mathrm{e}}^{\circ}$ that is given by Eq.(15), together with the numerical calculation result that is computed by the proposing method. Similarly, Figure 2(b) and 2(c) show the minimax distribution $\mathbf{p}^{\circ}$ that is given by Eq.(16) and the corresponding numerical calculation result, respectively. From these figures we see that the numerical calculation results based on the numerical calculation method capture the analytical results. Thus the numerical calculation method works.

5. For the signals in the presence of thermal noise

Let $\bar{N}_{\mathrm{th}}$ denote the average number of photons of thermal noise. The density operator of the thermal state is given by

$$\hat{\rho}_{\mathrm{th}} = (1-\nu)\sum_{n=0}^{\infty}\nu^n|n\rangle\langle n|, \tag{17}$$

where $|n\rangle$ is the eigenstate of the number operator $\hat{a}^{\dagger}\hat{a}$, and the real parameter ν $(0\le\nu\le 1)$ is defined as $\nu=\bar{N}_{\mathrm{th}}/(1+\bar{N}_{\mathrm{th}})$, *i.e.*, $\bar{N}_{\mathrm{th}}=\nu/(1-\nu)$.

Suppose that the i-th coherent state signal has amplitude α_i. If each signal is in the pressure of thermal noise, then the corresponding quantum state is given as $\hat{\rho}_i=\hat{D}(\alpha_i)\hat{\rho}_{\mathrm{th}}\hat{D}^{\dagger}(\alpha_i)$, where $\hat{D}(\beta)\equiv\exp[\beta\hat{a}^{\dagger}-\beta^*\hat{a}]$ is the displacement operator. In the case of the ternary coherent state signal, the received signal states are respectively given by

$$\hat{\rho}_0=\hat{\rho}_{\mathrm{th}}, \qquad \hat{\rho}_+=\hat{D}(\alpha)\hat{\rho}_{\mathrm{th}}\hat{D}^{\dagger}(\alpha), \qquad \hat{\rho}_-=\hat{D}(-\alpha)\hat{\rho}_{\mathrm{th}}\hat{D}^{\dagger}(-\alpha), \tag{18}$$

and the signal-to-noise ratio is defined as $\mathrm{SNR}=|\alpha|^2/(2\bar{N}_{\mathrm{th}}+1)$ [3].

To perform numerical calculation, we use the following matrix representation of the signals in the basis $\{|n\rangle\}$, according to the literature [5]: $\hat{\rho}_i \doteq \rho_i = [\langle m|\hat{\rho}_i|n\rangle]_{m,n=0,1,2,\ldots}$ with

$$\langle m|\hat{\rho}_i|n\rangle = \sqrt{\frac{m!}{n!}}(1-\nu)\nu^n\left(\frac{\alpha_i^*}{\bar{N}_{\mathrm{th}}}\right)^{n-m}\exp[-(1-\nu)|\alpha_i|^2]L_m^{(n-m)}[-\frac{(1-\nu)^2|\alpha_i|^2}{\nu}] \tag{19}$$

for $m\le n$, and $\langle m|\hat{\rho}_i|n\rangle=(\langle n|\hat{\rho}_i|m\rangle)^*$ for $m>n$, where $L_n^a(x)$ is the the associated Laguerre polynomial. Numerical calculation results are shown in Figure 3 for $\nu=0.01, 0.02, 0.05, 0.1, 0.2$, and 0.5. We observe from Figure 3(b) that every p_i° is positive for each ν. According to *Lemma 3* in the literature [2], the relation $\bar{P}_{\mathrm{e}}(\Pi^{\circ},\mathbf{q})=\bar{P}_{\mathrm{e}}^{\circ}$ holds for any $\mathbf{q}\in\mathcal{P}$. Thus, in each ν, the minimax receiver Π° for the ternary coherent state signal always indicates the same error performance shown in Figure 3(a) even if the true distribution $\mathbf{q}$ of the signals varies.

6. Conclusions

We have shown a numerical calculation method for finding a solution to the quantum minimax decision problem. To verify the numerical calculation method, we have analytically derived the minimax decision operators, minimax distribution, and minimax value for the ternary pure

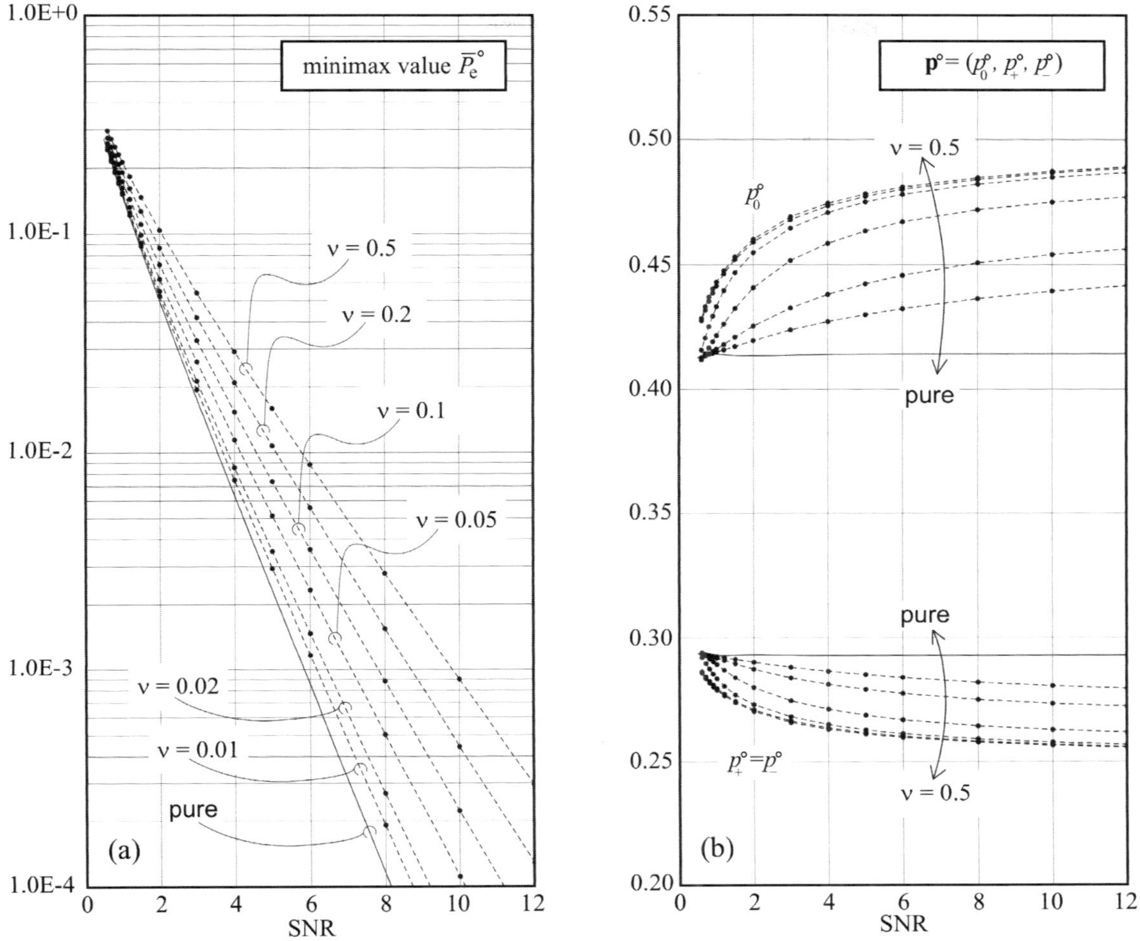

Figure 3. (a) The minimax value $\bar{P}^\circ$ of the ternary coherent state signal in the presence of thermal noise versus the signal-to-noise ratio $|\alpha|^2/(2\bar{N}_{\rm th}+1)$. (b) The corresponding minimax distribution $\mathbf{p}^\circ = (p_0^\circ, p_+^\circ, p_-^\circ)$.

coherent state signal, respectively. Using these closed-form expressions, we have verified that the numerical calculation method works. Finally the error probability of the quantum minimax receiver for the ternary coherent state signal that is in the pressure of thermal noise was computed by using the numerical calculation method. Since every minimax probability obtained in our simulation is positive, we can expect that the minimax receiver for the ternary coherent state signal always indicates the same error performance even if the true distribution of the signals varies, because of *Lemma 3* of the literature [2].

References

[1] O. Hirota and S. Ikehara, *Trans. IECE of Japan* **E65**, 627(1982).
[2] K. Kato, *2012 IEEE Int. Symp. Inform. Theory Proc.*, 1082(2012).
[3] C. W. Helstrom, *IEEE Trans. Inform. Theory* **28**(2), 359(1982).
[4] Y. C. Eldar, A. Megretski, and G. C. Verghese, *IEEE Trans. Inform. Theory* **49**(4), 1007(2003).
[5] B. Mollow and R. J. Glauber, *Phys. Rev.* **160**(5), 1076(1967).

Journal of Physics: Conference Series **414** (2013) 012040 doi:10.1088/1742-6596/414/1/012040

Quantum Tomograms and Their Application in Quantum Information Science

Aleksey K Fedorov[1,2,3] **and Stanislav O Yurchenko**[1]

[1] Center for Photonics and Infrared Engineering, Bauman Moscow State Technical University, 2nd Baumanskaya str., 5, 105005 Moscow, RUSSIA

[2] Russian Quantum Center, Novaya str., 100, 143025, Skolkovo, Moscow region, RUSSIA

[3] Visiting Research Associate at Institute for Quantum Information Science, University of Calgary, 2500 University Dr. NW, T2N 1N4, Calgary, Alberta, CANADA

E-mail: akfedorv@student.bmstu.ru, st.yurchenko@bmstu.ru

Abstract. This note is devoted to quantum tomograms application in quantum information science. Representation for quantum tomograms of continuous variables via Feynman path integrals is considered. Due to this construction quantum tomograms of harmonic oscillator are obtained. Application tomograms in causal analysis of quantum states is presented. Two qubit maximum entangled and "quantum-classical" states have been analyzed by tomographic causal analysis of quantum states.

1. Introduction

Tomographic representation of quantum states is one of the most developing areas of quantum physics [1]-[15]. The main idea of quantum tomography consists in using of nonnegative probability distribution functions (quantum tomograms). Quantum tomography is an equivalent to other approaches to quantum mechanics. In turn, quantum tomogram is directly related to Wigner function, Glauber-Sudarshan function and Husimi function [1]-[12]. This probability distribution functions are called "tomograms", because of their relation with Wigner function by Radon integral transformation, which is used in medicine computed tomography.

There are two main reasons for quantum tomography using. First, the tomograms of quantum states are measurable via method of balanced homodyne detection [13]-[15]. Current developments in continuous-variable quantum-state tomography presented in [15]. Experimental area of quantum tomography is closely connected with quantum state engineering. There are a lot of applications of this framework in quantum information technology [15]-[16].

Second, methods of quantum tomography are widely used for simulation [6], [7]. Most methods of quantum systems numerical simulation are based on solution equations obtained for not positive define functions (wave functions, Wigner function). This gives rise to problems associated with convergence of integrals ("sign problem") [6]. Quantum tomograms are positive defined functions. But numerical method based on quantum tomography requires solution of system of differential equations with local approximation (e.g., exponential). In this note, representation of quantum tomograms by Feynman path integral is used [1]. Feynman path integral representation of quantum tomogram is the basis for Path Integral Monte Carlo (PIMC) numerical methods with positive define quantum tomograms. PIMC method has significant computational advantages, e.g. in condensed matter physics [17].

2. Continuous variables quantum tomography

Quantum tomography is based on a linear canonical transformation of phase space. Transformation from one set of canonical variables (positions and momentums) (q, p, t) to other (Q, P, t) is the canonical one if and only if Jacobi matrix $\mathcal{M}$ of this transformation is symplectic. It's easy to show, that canonical transformation of phase space provides conservation of Poisson bracket in classical mechanics case and commutator (canonical commutation relation) in quantum mechanics case. Thus, Hamiltonian equations and principle of least action are valid for new canonical variables. Canonical transformation with symplectic matrix $\mathcal{M}$ can be presented as follows

$$\begin{pmatrix} Q \\ P \end{pmatrix} = \begin{pmatrix} \mu & \eta \\ \acute{\eta} & \acute{\mu} \end{pmatrix} \begin{pmatrix} q \\ p \end{pmatrix}, \tag{1}$$

where elements $\mu, \eta, \acute{\mu}, \acute{\eta}$ of matrix $\mathcal{M}$ are real numbers and determinant of $\mathcal{M}$ is 1. Note, than canonical transformation is used in canonical quantization (see [20], [2]).

Symplectic tomogram $\mathcal{T}(Q, \mu, \eta)$ of observable $Q = \mu\widehat{q} + \eta\widehat{p}$ can be defined through complex wave function $\psi(q)$ in position representation

$$\mathcal{T}(Q, \mu, \eta) = |\widehat{\mathcal{F}}_{\mu,\eta}[\psi(q)](Q)|^2, \qquad m = \hbar = \omega = 1$$

where $\widehat{\mathcal{F}}_{\mu,\eta}$ is fractional Fourier transformation operator [12]

$$\widehat{\mathcal{F}}_{\mu,\eta}[\psi(q)](Q) = \frac{1}{\sqrt{2\pi|\eta|}} \exp\left[\frac{i\mu q^2}{2\eta}\right] \int \psi(q) \exp\left[i\left(\frac{\mu q^2}{2\eta} - \frac{Qq}{\eta}\right)\right] dq. \tag{2}$$

A definition for symplectic tomogram is obtained from (1) and (2)

$$\mathcal{T}(Q, \mu, \eta) = \frac{1}{2\pi|\eta|} \left| \int \psi(q) \exp\left[i\left(\frac{\mu q^2}{2\eta} - \frac{Qq}{\eta}\right)\right] dq \right|^2. \tag{3}$$

In accordance with definition (3), sympectic tomogram $\mathcal{T}(Q, \mu, \eta)$ is nonnegative and normalize on Q function

$$\mathcal{T}(Q, \mu, \eta) \geq 0, \qquad \int \mathcal{T}(Q, \mu, \eta) dQ = 1, \tag{4}$$

and, therefore, tomogram is a probability distribution function of variable Q with parameters μ and η. Position and momentum probability distribution functions could be calculated based on symplectic tomography

$$\mathcal{T}(Q, 1, 0) = |\psi(q)|^2, \qquad \mathcal{T}(Q, 0, 1) = |\psi(p)|^2. \tag{5}$$

Symplectic tomograms are the most general case of probability distribution function in quantum tomography. Let consider two important examples. Starting form the orthogonal matrix $\mathcal{M}$ case

$$\begin{pmatrix} \cos\theta & \sin\theta \\ -\sin\theta & \cos\theta \end{pmatrix}, \tag{6}$$

where $\theta \in \mathbb{R}/2\pi\mathbb{Z}$ is a phase space rotation angle.

Using (3) expression for *optical tomogram* can be obtained

$$\mathcal{T}(Q, \theta) = \frac{1}{2\pi|\sin\theta|} \left| \int \psi(q) \exp\Big(\frac{i\cot\theta}{2} q^2 - \frac{iQ}{\sin\theta} q\Big) dq \right|^2. \tag{7}$$

Optical tomography is extremely important for quantum states measurements in quantum optics with balanced homodyne detection technique and for quantum state reconstruction process [16]. Also, in quantum optics it is widely used a connection between Wigner function and quantum tomogram $\mathcal{T}(Q,\mu,\eta)$ by Radon transformation

$$\mathcal{W}(q,p)=\frac{1}{2\pi}\int\mathcal{T}(Q,\mu,\eta)\exp(i(Q-\mu q-\eta p))dQd\mu d\eta. \tag{8}$$

Tomogram is a first order homogeneity function. Lets consider inverse Radon transformation

$$\mathcal{T}(Q,\mu,\eta)=\int\mathcal{W}(q,p)\exp(-is(Q-\mu q-\eta p))\frac{dsdqdp}{4\pi^2}$$

and transformation variables $(Q,\mu,\eta)\mapsto(\lambda Q,\lambda\mu,\lambda\eta)$ with arbitrary real λ. Therefore

$$\mathcal{T}(\lambda Q,\lambda\mu,\lambda\eta)=\int\mathcal{W}(q,p)\exp(-is\lambda(Q-\mu q-\eta p))\frac{dsdqdp}{4\pi^2}=|\lambda|^{-1}\mathcal{T}(\varepsilon,\mu,\eta)$$

From Euler lemma for homogenates functions $\mathcal{T}(Q,\mu,\eta)$

$$\left(Q\frac{\partial}{\partial Q}+\mu\frac{\partial}{\partial\mu}+\eta\frac{\partial}{\partial\eta}+1\right)\mathcal{T}(\varepsilon,\mu,\eta)=0. \tag{9}$$

Thus, *Fresnel tomogram* [9] is a solution of equation (9)

$$\mathcal{T}_f(\acute{Q},\acute{\eta})=\frac{1}{|\mu|}f\left(\frac{Q}{\mu},\frac{\eta}{\mu}\right),$$

where f is an arbitrary function.

Quantum tomograms evolution equation is Fokker-Planck type equation. Fokker-Planck equation is a well-known stochastic differential equation of classical probability theory. For quantum system with Hamiltonian

$$\widehat{H}=\frac{\widehat{p}^2}{2}+V(\widehat{q})$$

Fokker-Planck equation is represented as

$$\frac{\partial\mathcal{T}}{\partial t}=\mu\frac{\partial\mathcal{T}}{\partial\eta}+i\Big[-V\Big(\hat{I}_Q\frac{\partial}{\partial\mu}+\frac{i\eta}{2}\frac{\partial}{\partial\mu}\Big)-V\Big(-\hat{I}_Q\frac{\partial}{\partial\mu}+\frac{i\eta}{2}\frac{\partial}{\partial\mu}\Big)\Big]\mathcal{T}, \tag{10}$$

where $\hat{I}_Q: f(Q)\to\int f(Q)dQ$ is an integrity operator on variable Q.

From (10) for free particle

$$\frac{\partial\mathcal{T}}{\partial t}-\mu\frac{\partial\mathcal{T}}{\partial\eta}=0. \tag{11}$$

For harmonic oscillator

$$\frac{\partial\mathcal{T}}{\partial t}-\mu\frac{\partial\mathcal{T}}{\partial\eta}+\eta\frac{\partial\mathcal{T}}{\partial\mu}=0. \tag{12}$$

In case of parametric oscillator with time-dependent frequency $\omega(t)$

$$\frac{\partial\mathcal{T}}{\partial t}-\mu\frac{\partial\mathcal{T}}{\partial\eta}+\omega^2(t)\eta\frac{\partial\mathcal{T}}{\partial\mu}=0. \tag{13}$$

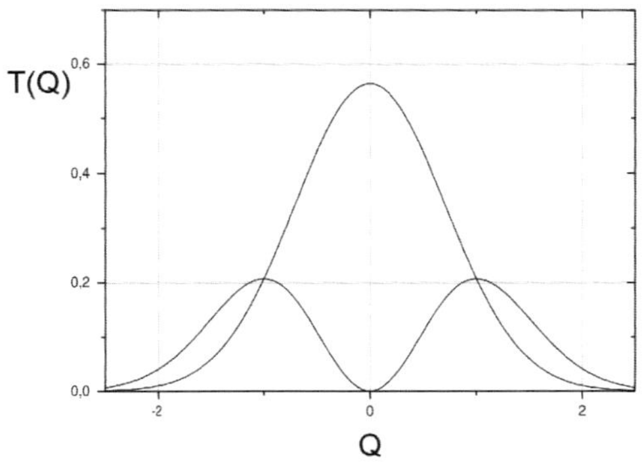

Figure 1. Optical tomograms for ground an first state of harmonic oscillator.

Let's consider symplectic tomograms representation via Feynman path integral

$$\mathcal{T}(Q,\mu,\eta)=\frac{1}{2\pi|\eta|}\left|\int\psi(q_1)\exp\left[i\left(\mathcal{S}(q_1,q_2)+\frac{\mu}{2\eta}q_2^2-\frac{Q}{\eta}q_2\right)\right]\mathcal{D}[q(t)]dq_1dq_2\right|^2, \quad (14)$$

where $\mathcal{S}(q_1,q_2)$ is an action functional, $\psi(q_1)$ is Cauchy problem initial condition for Schrodinger equation

$$-\frac{1}{2}\Delta\psi(q,t)+V(q)\psi(q,t)=i\frac{\partial}{\partial t}\psi(q,t),$$

and $\mathcal{D}[q(t)]$ is "Lebesgue measure" [19]

$$\mathcal{D}[q(t)]=\lim_{N\to\infty}\prod_{j=1}^{N}\Big(-\frac{2\pi i}{\Delta t_j}\Big)^{3/2}d^3q_j.$$

Feynman path integral representation of optical tomograms and Fresnel tomograms can be obtained. Let consider harmonic oscillator as an example. Green function for Schrodinger equation in case of harmonic oscillator

$$G_o(q_2,q_1,t)=\frac{1}{\sqrt{2\pi i\sin(t)}}\exp\left(\frac{i}{2}\cot(q_2^2+q_1^2)-\frac{iq_1q_2}{\sin(t)}\right).$$

Substituting Green function in tomographic Feynman path integral solution for stationary states were obtained

$$\mathcal{T}_n(Q,\mu,\eta)=\frac{2^{-n}}{n!\sqrt{\pi(\mu^2+\eta^2)}}\exp\left(-\frac{Q^2}{\mu^2+\eta^2}\right)\mathcal{H}_n^2\left(\frac{Q}{\sqrt{\mu^2+\eta^2}}\right), \quad (15)$$

where $\mathcal{H}_n(x)$ is n order Hermite polynomials.

Therefore, quantum tomograms of ground state $\mathcal{T}_0(Q,\mu,\eta)$ has the following form

$$\mathcal{T}_0(Q,\mu,\eta)=\frac{1}{\sqrt{\pi(\mu^2+\eta^2)}}\exp\left(-\frac{Q^2}{\mu^2+\eta^2}\right). \quad (16)$$

Tomogram $\mathcal{T}_1(Q,\mu,\eta)$ of the first state

$$\mathcal{T}_1(Q,\mu,\eta) = \frac{2Q^2}{(\mu^2+\eta^2)\sqrt{\pi(\mu^2+\eta^2)}} \exp\left(-\frac{Q^2}{\mu^2+\eta^2}\right). \tag{17}$$

There are ground and the first state optical tomograms presented on fig. 1. Note, that for parameters $\mu = 1$ and $\eta = 0$ quantum tomogram is the ordinary probability distribution in position representation. These results are in agreement with results obtained in [8].

Feynman path integral is one of the main tool in theoretical physics [21]-[23]. Feynman path integral representation of quantum tomograms takes an important part in quantum tomography. First of all, series of perturbation theory can be constructed [3]. Second, this representation is closely connected with Feynman formula (representation of semigroups by limits of n-fold iterated integrals as $n \to \infty$) for Fokker-Planck equation [23] and path integral computation technique in filtering problem [24].

3. Quantum tomography for discrete variables.

Tomography for discrete variables has been proposed in [11]. Let consider density matrix ρ^0 for particle with 1/2 spin, which characterizing spin projectionon Oz axis. For arbitrary direction Oz' which is characterized by identity vector $\vec{n} = \{\sin\theta\cos\varphi, \sin\theta\sin\varphi, \cos\theta\}$, density matrix has the following form

$$\rho(\vec{n}) = U(\vec{n})\rho^0 U^\dagger(\vec{n}), \tag{18}$$

where $U(\vec{n}) = U(\theta,\varphi)$ is an unitary rotation matrx

$$U(\theta,\varphi) = \begin{pmatrix} \cos(\theta/2)\exp(i\varphi/2) & \sin(\theta/2)\exp(i\varphi/2) \\ -\sin(\theta/2)\exp(-i\varphi/2) & \cos(\theta/2)\exp(-i\varphi/2) \end{pmatrix}, \tag{19}$$

with Euler angels φ, θ .

Tomograms of qubit state can be presented as follows

$$\mathcal{T}(m,\vec{n}) = \{\rho(\vec{n})\}_{mm}, \tag{20}$$

with $m = 1$ and $m = 2$ correspond to spin projections $|+1/2,\vec{n}\rangle$ and $|-1/2,\vec{n}\rangle$ respectively. In this way, only diagonal elements of density matrix are considered. They determine the probability to obtain a particular measurement result of spin projection on an arbitrary axis defined by the vector $\vec{n}$.

By definition, quantum tomogram $\mathcal{T}(m,U)$ is nonnegative and normalized

$$\mathcal{T}(m,\vec{n}) \geq 0, \qquad \sum\nolimits_m \mathcal{T}(m,\vec{n}) = 1.$$

Therefore, the formula for von Neumann entropy in tomogaphic representation of quantum mechanics can be obtained in the following way

$$S^t(\vec{n}) = -\sum\nolimits_m \mathcal{T}(m,\vec{n})\log_2 \mathcal{T}(m,\vec{n}). \tag{21}$$

Index t is fixed for tomograhic functions. Tomographic entropy is generalization of von Neumann entropy

$$S = -\operatorname{Tr}[\rho\log(\rho)].$$

Finally, for a two-part system with initial density matrix ρ^0_{AB} tomogram can be presented in the form

$$\mathcal{T}(m_1,m_2,\vec{n_1},\vec{n_2}) = \{U(\vec{n_1},\vec{n_2})\rho^0_{AB}U^\dagger(\vec{n_1},\vec{n_2})\}_{2m_1+m_2\ 2m_1+m_2}, \tag{22}$$

where matrix $U(\vec{n_1}, \vec{n_2})$ can be defined by two unitary matrix tensor product

$$U(\vec{n_1}, \vec{n_2}) = U(\vec{n_1}) \otimes U(\vec{n_2}).$$

Note, than Euler angles presents a rotation of space (19). Problem of representation for space rotation can be solved with quaternions. Construction of quantum tomgraphy via quaternions is presented in [4].

4. Tomographic causal analysis of quantum states

Quantum causal analysis proposes an original approach, based on von Neuman entropy, which helps to understand causality of entangled asymmetric states [25], [26]. Mathematical formalization of causal analysis is bases on a pair of independence functions and on irreversible information flow velocity

$$i_{A|B} = \frac{S_{A|B}}{S_A}, \qquad i_{B|A} = \frac{S_{B|A}}{S_B}, \qquad c_2 = k\frac{(1 - i_{A|B})(1 - i_{B|A})}{i_{A|B} - i_{B|A}}, \qquad k = 1.$$

Tomographic causal analysis of quantum states is based on tomographic von Neumann entropy (21) for independence functions and for the linear velocity of irreversible information flow. In turn, the main idea quantum states tomographic causal analysis is to use tomographic relation (21) for von Neumann entropy [3].

First, let's start from consideration of pure maximum entangled state

$$|\psi\rangle_{AB} = 1/\sqrt{2}(|00\rangle_{AB} + |11\rangle_{AB}),$$

where $|0\rangle$ and $|1\rangle$ correspond to spin positive and negative projections on chosen axis. Thus, density matrix for this system has the following form

$$\rho^0_{AB} = \frac{1}{2}\begin{pmatrix} 1 & 0 & 0 & 1 \\ 0 & 0 & 0 & 0 \\ 0 & 0 & 0 & 0 \\ 1 & 0 & 0 & 1 \end{pmatrix}, \quad \rho^0_A = \rho^0_B = \frac{1}{2}\begin{pmatrix} 1 & 0 \\ 0 & 1 \end{pmatrix}. \tag{23}$$

As the result, $S_{AB} = 0$, $S_A = S_B = 1$, and causality is upset $i_{A|B} = i_{B|A} = -1$, $|c_2| \to \infty$. This state is special, because density matrix of subsystems A and density matrix subsystem B are invariant to rotation matrix action

$$U(\vec{n_1})\rho^0_A U^\dagger(\vec{n_1}) = U(\vec{n_2})\rho^0_B U^\dagger(\vec{n_2}) = \rho^0_A = \rho^0_B.$$

Finally, $S^t_A(\vec{n_1}) = S^t_B(\vec{n_2}) = 1$ and $i^t_{A|B} = i^t_{B|A}$. There are independence functions as the functions of angles θ_1 and θ_2 ($\varphi_1 = \varphi_2 = 0$ are fixed) on fig. 2.

System maximum correlation conditions, obtained after measurements, has been obtained in the following way

$$i^t_{A|B} = i^t_{B|A} = 0.$$

This is a condition of using the same basis for measurements in both subsystem (with $\theta_1 = \theta_2$). These results of tomographic causal analysis are in agreement with the results of ordinary causal analysis of quantum states [27].

Second, let's consider "quantum-classical" state with inequality for entropies

$$S_B < S_{AB} < S_A,$$

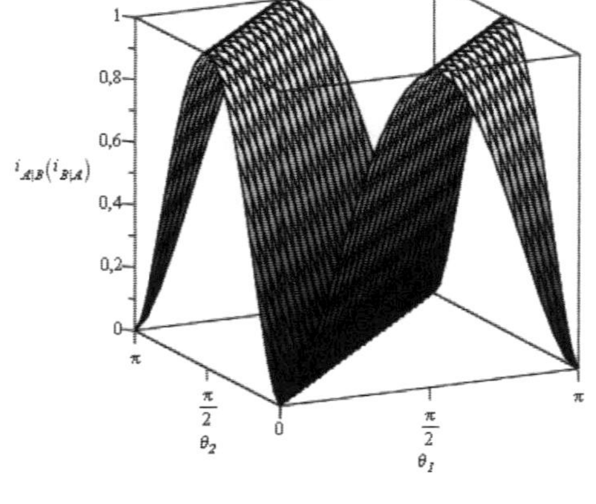

Figure 2. Independence functions $i^t_{A|B}(\theta_1, \theta_2) = i^t_{B|A}(\theta_1, \theta_2)$.

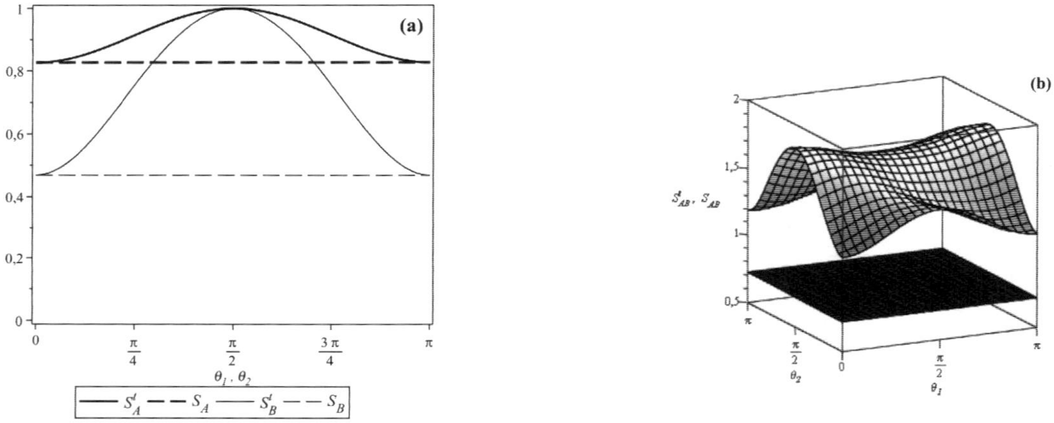

Figure 3. Tomographic entropy and von Neumann entropy for subsystems (a) and system (b) for asymmetric "quantum-classical" state. Angles are fixed.

with density matrix

$$\rho^0_{AB} = q|\psi_1\rangle\langle\psi_1| + (1-q)|\psi_2\rangle\langle\psi_2|,$$

where $|\psi_1\rangle = a|00\rangle_{AB} + \sqrt{1-a^2}|11\rangle_{AB}$ and $|\psi_2\rangle = a|10\rangle_{AB} + \sqrt{1-a^2}|01\rangle_{AB}$ – pure entangled states. In accordance with [4], parameters $q = 0.2$, $a^2 = 0.1$ are chosen.

Von Neumann entropy are $S_{AB} \simeq 0.722$, $S_A \simeq 0.827$, $S_B \simeq 0.469$. Causaulity is $c_2 \simeq 1.604$, in this way, subsystem A is a "cause", subsystem B is "effect". There are tomographic entropy and von Neumann entropy of subsystems A, subsystem B and system for asymmetric "quantum-classical" state with fixed angles on fig. 3.

After tomographic transformation entropy becomes a function of angles $\theta_{1(2)}$: $S^t_A = S^t_A(\theta_1)$, $S^t_B = S^t_B(\theta_2)$. And entropy of system $S^t_{AB} = S^t_{AB}(\vec{n_1}, \vec{n_2}) = S^t_{AB}(\theta_1, \varphi_1, \theta_2, \varphi_2)$. There are ordinary and tomographic velocities of irreversible information flow (c_2 and c^t_2) and information (I and I^t) for asymmetric "quantum-classical" state is shown on fig. 4.

Therefore, tomographic quantum causal analysis is extremely useful for initial asymmetry

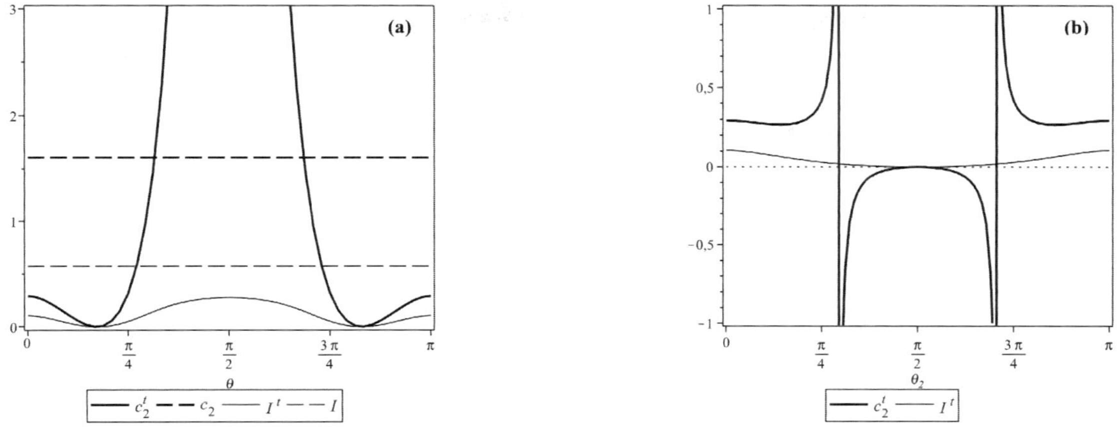

Figure 4. Ordinary causal and tomographic velocities of irreversible information flow (c_2 and c_2^t) and information (I and I^t) for asymmetric "quantum-classical" state: $\varphi_1 = \varphi_2 = 0$, $\theta_1 = \theta_2 = \theta$ (a); $\varphi_1 = \varphi_2 = \theta_1 = 0$ (b)

investigation for system with unknown density matrix.

5. Conclusion

Application of quantum tomograms in quantum information science was discussed. Continuous variables quantum tomograms representation via Feynman path integrals was used. Tomograms construction of harmonic oscillator state was obtained. Application of tomograms in quantum states causal analysis was discussed. Two qubit maximum entangled pure and "quantum-classical" states by tomographic causal analysis of quantum states was considered.

Acknowledgments

Authors thank A.I. Lvovsky, V.I. Man'ko and Yu.E. Lozovik for grateful discussion. Authors thank E.O. Kiktenko and S.M. Korotaev for grateful discussion about causal analysis of quantum states. This work was supported by Russian Foundation for Basic Research (projects 12-08-31104 and 12-08-33112).

References

[1] Fedorov A K, Yurchenko S O 2012 Symplectic Tomograms Presented through Feynman's Path Integrals, J. BMSTU Nat. Science **45** 29-37.

[2] Fedorov A K, Yurchenko S O 2011 Tomographic Methods in the Theory of Secondary Quantization, Sc. Educ.: Electr. Scien. Techn. Period. **13** 62-67.

[3] Fedorov A K, Yurchenko S O 2012 Tomographic Series of Perturbation Theory, J. BMSTU Nat. Sc. Spec. Is. 6-12.

[4] Fedorov A K, Yurchenko S O 2012 Quaternion Representation of Spin Tomograms, J. BMSTU Nat. Sc. Spec. Is. 257-261.

[5] Kiktenko E O, Korotaev S M, Fedorov A K, Yurchenko S O 2012 Causal Analysis of Entangled States in Tomographic Representation of Quantum Mechanics, J. BMSTU Nat. Sc. Spec. Is. 75-85.

[6] Lozovik Yu E, Sharapov V A, Arkhipov A S 2005 Simulation of Tunneling in the Quantum Tomography, Phys. Rev. A **69** 022116.

[7] Arkhipov A S Lozovik Yu E 2003 New Method of Quantum Dynamics Simulation Based on the Quantum Tomography, Physics Letters A **319** 217–224.

[8] Man'ko O V, Man'ko V I 1999 Classical Propagator and Path Integral in the Probability Representation of Quantum Mechanics, J. Russ. Las. Res. **20** 67-76.

[9] De Nicola S, Fedele R, Man'ko M A, Man'ko V I 2005 Fresnel Tomography: A Novel Approach to Wave-Function Reconstruction Based on the Fresnel Representation of Tomograms, Theor. Math. Phys. **144** 1206-1213.

[10] Chernega V N, Man'ko V I 2007 Wave Function of the Harmonic Oscillator in Classical Statistical Mechanics, J. Russ. Las. Res. **28** 535-547.

[11] Andreev V A, Man'ko V I, Man'ko O V, Shchukin E V 2006 Tomography of Spin States, the Entanglement Criterion, and Bell's Inequalities, Theor. Math. Phys. **146** 140-151

[12] Amosov G G, Dnestryan A I 2011 On the Spectrum of a Set of Integral Operators Determining the Symplectic Quantum Tomogram, Proc. MIPT **3** 5-9

[13] Beck M, Smithey D T, Raymer M G 1993 Experimental Determination of Quantum-phase Distributions Using Optical Homodyne Tomography, Phys. Rev. A **48** 890-893.

[14] Smithey D T, Beck M, Raymer M G, Faridani A 1993 Measurement of the Wigner Distribution and the Density Matrix of a Light Mode Using Optical Homodyne Tomography: Application to Squeezed States and the Vacuum, Phys. Rev. A **70** 1244-1247.

[15] Lvovsky A I, Raymer M G 2009 Continuous-variable Optical Quantum-state Tomography, Rev. Mod. Phys. **81** 299-332.

[16] Anis A, Lvovsky A I 2012 Maximum-likelihood Coherent-state Quantum Process Tomography, New J. Phys. **14** 105021.

[17] Pollock E L, Ceperley D M 1987 Path-Integral Computation of Superfluid Densities, Phys. Rev. B **36** 1244-1247.

[18] Feynman R F, Hibbs A R 1965 Quantum Mechanics and Path Integrals, New York, McGraw-Hill.

[19] Garrod C 1966 Hamiltonian Path-Integral Methods, Rev. Mod. Phys. **38** 483-494.

[20] Bogoliubov N N., Shirkov D V 1959 The Theory of Quantized Fields, New York, Interscience.

[21] Berezin F A 1971 Non-Wiener Functional Integrals, Theor. Math. Phys. **6** 141-155.

[22] Bottcher B, Butko Ya A, Schilling R L, Smolyanov O G 2011 Feynman Formulas and Path Integrals for Some Evolution Semigroups Related to τ-quantization, Russ. J. Math. Phys. **18** 387-399.

[23] Drozdov A N, Talkner P 1998 Path Integrals for Fokker-Planck Dynamics With Singular Diffusion: Accurate Factorization for the Time Evolution Operator, J. Chem. Phys. **109** 2080-2091.

[24] Ovseevich A I 2008 Kalman Filter and Quantization, Prob. Inform. Transm. **44** 53-71.

[25] Kiktenko E O, Korotaev S M 2012 Causal Analysis of Asymmetric Entangled States Under Decoherence, Phys. Lett. A. **376** 820-823.

[26] Korotaev S M, Kiktenko E O 2012 Causality and Decoherence in the Asymmetric States, Phys. Scr. **85** 055006.

[27] Zyczkowski K, Horodecki P, Horodecki M, Horodecki R 2002 Dynamics of quantum entanglement, Phys. Rev. A **85** 055006.

The digital randomness of black–body radiation

Sándor Varró
Wigner Research Centre for Physics, Hungarian Academy of Sciences,
Institute for Solid State Physics and Optics, Budapest, Hungary
E-mail: varro.sandor@wigner.mta.hu

Abstract. The statistical properties of the fractional part of the random energy of a spectral component of black–body radiation have been analysed in the frame of classical Kolmogorovian probability theory. Besides the integer part of the energy (which satisfies the well-known Planck–Bose distribution) the realizations of its fractional part (related to 'round-off errors') has been represented by binary sequences, like z = 0.001011000010.... It has been shown that the binary variables realized by the 0-s and 1-s at different positions are *independen*t. From the condition of independence the original distribution of the fractional part z can be recovered. If these binary variables have the *same distribution*, they describe a temperature-independent (random) energy, whose expectation value is the well-known zero–point energy. Thus, the zero–point fluctuations can be considered as a physical representative of an ideal random number generator.

Keywords: black–body radiation, Planck–Bose distribution, infinitely divisible random variables, digital randomness, round–off error, zero–point energy, random number generator

PACS: 01.55.+r, 02.70.Rr, 05.30.-d, 03.65.Ta, 03.67.Hk, 05.20.-y, 05.40.Ca

1. Introduction

In the present paper on 'digital randomness' we mean the following: when we express the measurement data acquired in a physical experiment, then the digits of the representing real numbers are always undergoing some random changes from one measurement act to the other. These uncertainties may come from the imperfect control of the boundary conditions, from the always 'finite resolution' of the measuring apparatus, or from the 'loss of precision' in data acquisition, processing, and in displaying these data. Accurate numbers, like sharp and reproducible sequences of the 1 (yes) and 0 (no) digits within some finite time interval, can be obtained if we can answer with certainty the questions 'what is unity?' and 'what is zero?'. Of course, it is an important matter at what accuracy can we decide (draw the level of some physical quantity for distingtion) between these 'yes'-es and 'no'-s. The uncertainties can also be caused by the 'objective randomness' of the constituents and transitions of the physical system under study, which cannot be avoided by any means. In the present paper we shall deal with this kind of randomness, which manifest itself in the energy distribution of black–body radiation.

Thermal noise, due to the universal presence of black–body radiation (Planck, 1959) is the main basic source of unavoidable fluctuations and uncertainties in physical measurements. It has already been emphasized by Planck (1908) that even the splitting of the total energy of a moving body in two parts; to the 'translation energy' and to the 'internal energy', is conceptionally impossible, because the 'thermal part' of this latter energy necessarily contains heat radiation, whose energy (and mass) depend on the velocity of the body. Taking the simple numerical example due to Planck (1908); if an ideal gas is heated at constant volume, then the ratio, at which the acquired energy is distributed among the black-body radiation and the mechanical energy of the molecules, is ¼ at 0.001 Hgmm pressure and temperature 2063 Kelvin (~ the melting point of platinium). It is also well-known that in order to have a computer reliably working, the proper cooling of it has to be secured. Otherwise – as is expected merely on physical grounds – sooner or later the system becomes to much loaded with thermal noise, and finally a 'physical loss of precision' may occur, even if no parts are mechanically or electronically damaged.

Objective indeterminacy has been widely accepted as an inherent property of quantum systems, and the task of the theory is just to quantify this property. In quantum theory the elementary measurement acts are represented by projectors on Hilbert spaces, which form the non–distributive lattice associated to the physical systems. On the other hand, in real experiments, every piece of information on 'genuine quantum objects' like single photons (Oxborrow and Sinclair, 2005) or biphotons (Bogdanov et al., 2007) is gained through macroscopic interfaces, which are necessarily loaded by additional noise. The 'signal–to–noise ratio' define the reliability of any experiment, and it is calculated from relative frequencies, according to classical probability theory (which is based on the 'common sense' distributive Boole algebra of events). In this context we note that recently, several useful concepts and quantities (like for instance the Rényi entropy) introduced in classical probability have received an increasing importance in quantum optics, too (Man'ko and Man'ko, 2011). In the present paper we base all of our considerations on the classical Kolmogorovian probability theory (see e.g. Feller (1970) or Rényi (1970)).

In our earlier studies concerning black–body radiation, on the basis of classical probability theory we have already proved that from the classical chaotic energy distribution – by subtracting the 'dark', fractional part – one can derive the Planck–Bose distribution (Varró, 2007). The boson gas of photons can further be uniquely decomposed into the assembly of 'binary photons', which follow Fermi–Dirac statistics, and they may serve as a natural physical basis for binary representation of integer numbers (Varró 2006a-b). In the present work we discuss the fractional part of the energy, which may also be considered as a 'round–off error' in the relative count number of photons of energy $h\nu$, or an error in measured correlations of photon count numbers (Varró, 2011). It is assumed that the counter has reached the thermal equilibrium during long enough experimental runs. The (random) fractional part ζ of the energy is represented by binary sequences e.g. z = 0.001011000010.... We analyse the connection between the distribution of ζ and the distribution of zeros and ones, which are being realizations of binary random variables.

In order to have the present paper possibly self–contained, in Section 2 we summarize our results on the fractional part and the integer part of the energy of a mode of black–body radiation. Section 3 is devoted to the proof of the independence of the binary components. In Section 3 we summarize the results of the present paper.

2. The fractional part and the integer part of the energy of a chaotic field component

In the present section we briefly summarize the main steps of the derivation of the Planck–Bose distribution, by subtracting the 'dark', fractional part from the chaotic energy, according to our earlier work (Varró, 2007). In classical physics the black–body radiation, a radiation being in thermal equilibrium in a *Hohlraum* (a cavity with perfectly reflecting walls at absolute temperature T) is considered as a chaotic electromagnetic radiation. The average spatial distribution of such a stationary radiation is homogeneous and isotropic and the electric field strength and the magnetic induction of its spectral components have completely random amplitudes which are built up of infinitely many independent infinitesimal contributions. In this description the electric field strength and the magnetic induction of a mode (characterized by its frequency ν, wave vector and polarization) of the thermal radiation are proportional with the random process

$$a_\nu(t) = a_c \cos(2\pi\nu\cdot t) + a_s \sin(2\pi\nu\cdot t) = \sqrt{a_c^2 + a_s^2}\cos(2\pi\nu\cdot t - \vartheta)\,, \qquad \vartheta = \arg(a_c + ia_s)\,, \tag{1}$$

where a_c and a_s are independent random variables. According to the Central Limit Theorem (see e.g. Feller, 1970), these amplitudes have Gaussian *density functions*,

$$P(q \le a_c < q+dq; p \le a_s < p+dp) = \left\{ \frac{1}{a\sqrt{2\pi}} \exp(-q^2/2a^2)dq \right\} \cdot \left\{ \frac{1}{a\sqrt{2\pi}} \exp(-p^2/2a^2)dp \right\}, \tag{2}$$

where the parameter a is related to the spectral energy density $u_\nu = \left\langle a_\nu^2(t)/8\pi \right\rangle = a^2/8\pi = Z_\nu \bar{\varepsilon}$, with $Z_\nu = 8\pi\nu^2/c^3$ being the well–known spectral mode density in vacuum. The average energy of one mode is denoted by $\bar{\varepsilon}$, and $u_\nu d\nu$ gives the energy density of the chaotic radiation in the spectral range $(\nu, \nu + d\nu)$. By introducing the mode energy as a classical random variable, $E = (a_c^2 + a_s^2)/Z_\nu 16\pi$, and the 'action–angle parameters' $\varepsilon = (q^2 + p^2)/Z_\nu 16\pi$ and $\theta = \arg(q + ip)$, we have from (2)

$$\begin{aligned} &P(q \le a_c < q+dq; p \le a_s < p+dp) \\ &= P(\varepsilon \le E < \varepsilon + d\varepsilon)P(\theta \le \vartheta < \theta + d\theta) = [(1/\bar{\varepsilon})\exp(-\varepsilon/\bar{\varepsilon})d\varepsilon](d\theta/2\pi). \end{aligned} \tag{3}$$

Thus, the mode energy, E satisfies an exponential (Boltzmann type) distribution, which stems from the two-dimensional Gauss distribution (2). Henceforth, we shall not display the uniform phase distribution, represented by the factor $(d\theta/2\pi)$ in (3). In our earlier analysis (Varró, 2007) we have introduced two independent energy parameters, ε_0, $\bar{\varepsilon}$, containing two universal constants, and *derived* their physical meaning, too. $\bar{\varepsilon}$ has turned out to be $\bar{\varepsilon} = kT$, where $k = 1.831 \times 10^{-16} erg/K$ is the Boltzmann constant and T is the absolute temperature of the radiation. $\varepsilon_0 = h\nu$ is proportional woth the Planck's quantum of action, $h = 6.626 \times 10^{-27} erg \cdot \sec$. In terms of the dimensionless energy variable $\eta \equiv E/\varepsilon_0$, equation (3) can be brought to the form

$$P(y \le \eta < y+dy) = f_\eta(y)dy, \quad f_\eta(y) = \beta e^{-\beta y} \quad (0 < y < \infty), \quad \eta \equiv E/\varepsilon_0, \quad \beta \equiv \varepsilon_0/\bar{\varepsilon}. \tag{4}$$

By introducing the Boltzmann entropy $S_\eta = -k\int_0^\infty f_\eta \log f_\eta dy$, and by using the basic relation $\partial S_\eta / \partial E_\eta = 1/T$ of phenomenological thermodynamics, we immediately have $E_\eta \equiv \varepsilon_0 \bar{\eta} = \bar{\varepsilon} = kT$, the law of equipartition of energy. From (4), the distribution function of a mode reads

$$P(\eta < y) \equiv F_\eta(y) = 1 - e^{-\beta y} \quad (0 \le y < \infty), \quad \eta \equiv E/\varepsilon_0, \quad \beta = \varepsilon_0/kT. \tag{5}$$

According to our earlier work (Varró, 2007), the density function and the distribution function of the fractional part $\zeta \equiv \{\eta\} \equiv \eta - [\eta]$ of the mode energy reads

$$P(\zeta < z) = G_\zeta(z) = \frac{1 - e^{-\beta z}}{1 - e^{-\beta}}, \quad f_\zeta(z) = \frac{\beta e^{-\beta z}}{1 - e^{-\beta}} \quad (0 \le z \le 1), \quad \zeta \equiv \{\eta\} \equiv \eta - [\eta], \tag{6}$$

where $[\eta]$ denotes integer part, i.e. the largest integer which is smaller than or equal to η. The expectation of the fractional part of the energy is

$$E_\zeta \equiv \varepsilon_0 \bar{\zeta} = \varepsilon_0 \int_0^1 f_\eta(z) z dz = \varepsilon_0 \left[\frac{1}{(\varepsilon_0/kT)} - \frac{1}{\exp(\varepsilon_0/kT) - 1} \right], \quad \lim_{T \to \infty} E_\zeta = \frac{1}{2}\varepsilon_0, \tag{7}$$

which means that we have derived the Planck factor $[\exp(\varepsilon_0 / kT) - 1]^{-1}$ from the continuous distribution (6). It is remarkable that for very high temperatures the expectation value of the fractional part tends to a temperature-independent constant, as is shown in the second equation of (7). On the basis of (6) and (4), we have proved that the *integer part* $\xi \equiv [\eta]$ must satisfy the *Planck–Bose distribution*, i.e.

$$P(\xi = n) = \frac{\bar{n}^n}{(1+\bar{n})^{1+n}}, \quad \xi \equiv [\eta], \quad \bar{n} \equiv \frac{1}{e^{\varepsilon_0/kT} - 1}, \quad E_\xi \equiv \varepsilon_0 \bar{\xi} = \varepsilon_0 \bar{n},$$

$$S_\xi = k[(1+\bar{n})\log(1+\bar{n}) - \bar{n}\log\bar{n}]. \tag{8}$$

From Wien's displacement law (Planck, 1959), written in the form $S_\xi = \Psi(\nu/T)$, where S_ξ is the entropy and Ψ an universal function, it follows from (8) that the energy parameter ε_0 must be proportional with the frequency, i.e. $\varepsilon_0 = h\nu$. The constant of this proportionality is just Planck's universal quantum of action h, and then the universal function is given in (8). According to $\varepsilon_0 = h\nu$ and the relation $\eta = [\eta] + \{\eta\} \equiv \xi + \zeta$, from (8), (7) and (4), for the spectral energy density u_ν we obtain the well-known *Planck's law of black–body radiation* (Planck, 1959),

$$u_\nu = \frac{8\pi\nu^2}{c^3}\frac{h\nu}{e^{h\nu/kT} - 1}, \quad \bar{U} = \frac{h\nu}{e^{h\nu/kT} - 1} + \frac{h\nu}{2}. \tag{9}$$

In the second equation of (9) we have also displayed the average energy $\bar{U}$ of a Planckian oscillator, containing the *zero–point energy* ½ $h\nu$ (Planck, 1911), which remains finite even at zero absolute temperature. It is interesting to note that, according to (7), the fractional part of the energy of the chaotic field, $\{E_\eta\}$, approaches exactly this value at high temperatures, $E_\zeta \to h\nu/2$. It can also be shown that the integer part $\xi \equiv [\eta]$ can be decomposed into a sum of independent binary random variables $\mathrm{u}_s = \mu_s(\xi) \cdot 2^s$ ($\mu_s = 0$ *or* 1), which satisfy Fermi–Dirac statistics (Varró, 2007), i.e.

$$\xi \equiv [\eta] = \sum_{s=0}^{\infty} \mathrm{u}_s = \sum_{s=0}^{\infty} \mu_s(\xi) \cdot 2^s,$$

$$P(\mu_s(\xi) = 0) = \frac{1}{1+b^{2^s}}, \quad P(\mu_s(\xi) = 1) = \frac{b^{2^s}}{1+b^{2^s}}, \quad b = \exp(-h\nu/kT) \quad (s = 0, 1, 2, \ldots). \tag{10}$$

Equation (8) and (10) shows that there is a natural basis to express the possible photon excitation in terms of excitation of the '*binary photons*', represented by the fermionic variables u_s. In the next section we shall show that a similar decomposition is possible for the fractional part $\zeta \equiv \{\eta\}$, too.

3. The statistics of the binary digits of the fractional part of the energy

In the present section we discuss the interrelation between the statistical properties of the fractional part of the scaled energy, ζ ($0 < \zeta < 1$), and the zeros and ones in its dyadic expansion. We express ζ as

$$\zeta = \sum_{k=1}^{\infty} \mathrm{v}_k = \sum_{k=1}^{\infty} \varepsilon_k(\zeta)/2^k, \quad \mathrm{v}_k \equiv \varepsilon_k(\zeta)/2^k, \quad \varepsilon_k(\zeta) = 0 \text{ or } 1, \tag{11}$$

i.e., a possible realization of ζ may look like $z = 0.011010010\ldots$ (with $\varepsilon_1 = 0$, $\varepsilon_2 = 1$, $\varepsilon_3 = 1, \ldots$). Following Chatterji (1963), the algebraic construction of this problem can be formulated as follows. First, let $[\Omega_k, \mathcal{F}_k, P_k]$ ($k = 1,2,3,\ldots$) be a sequence of measure spaces which defined as

$$\Omega_k = \{0,1\}, \quad \mathscr{E}_k = \{O, \Omega_k, \{0\}, \{1\}\}, \quad P_k(\{1\}) = p_k, \quad P_k(\{0\}) = 1 - p_k = q_k \;\; (0 \le p_k \le 1). \tag{12}$$

In the $\sigma-$algebra $\mathscr{E}_k$ the elementary sets $\{1\}$ and $\{0\}$ represent that in the physical system under discussion there exists an excitation of energy $h\nu/2^k$ (with probability $P(\{1\}) = p_k$) or not (with probability $q_k = P(\{0\})$), respectively. Then we construct the Cartesian product $[\Omega, \mathscr{E}, P]$ of the measure spaces $[\Omega_k, \mathscr{E}_k, P_k]$, i.e.

$$\Omega = \mathop{\times}_{k=1}^{\infty} \Omega_k, \quad \mathscr{E} = \mathop{\times}_{k=1}^{\infty} \mathscr{E}_k, \quad P = \mathop{\times}_{k=1}^{\infty} P_k, \tag{13}$$

and introduce $[I, \mathscr{B}, m]$, a measure space in the closed unit interval, where $I = [0,\ 1]$, the algebra $\mathscr{B}$ is built up from the Borel sets, and $m(A)$ denotes the Lebesgue measure of $A \in \mathscr{B}$. The expansion introduced in equation (11) corresponds to the mapping $\phi(\varepsilon) = \sum_{k=1}^{\infty} \varepsilon_k / 2^k$ of Ω onto I, where $\varepsilon = (\varepsilon_1, \varepsilon_2, ...)$ and the ε_k 's are 0's or 1's. According to this construction, the interrelation between the distributions of ζ and the ε_k 's is the interrelation of the probability measure P on the product space $(\Omega, \mathscr{E})$ and the Lebesgue measure m on the Borel space $(I, \mathscr{B})$. Besides various results, Chatterji (1963) has shown that the ε_k 's are independent, with respect to the absolutely continuous induced measure m, then the *density function of* m have the form

$$f_a(x) = \begin{cases} 1 \;\; if \;\; a = 0 \\ \dfrac{a e^{ax}}{e^a - 1} \;\; if \;\; a \neq 0 \end{cases}, \quad f_{a=-\beta}(z) = f_\zeta(z) = \frac{\beta e^{-\beta \cdot z}}{1 - e^{-\beta}}, \quad a = -\beta = -h\nu / kT, \tag{14}$$

where a is an arbitrary real number. It should be remarked that the measures with exponential densities are the only absolutely continuous measures for which the ε_k 's are independent. In our case, we have to put in (14) $a = -\beta = -h\nu / kT$, and then recover the density function $f_\zeta(z) = \beta e^{-\beta \cdot z} / (1 - e^{-\beta})$, which has been shown already in equation (4), as a result of our earlier work (Varró, 2007).

Before we prove the independence of the ε_k 's, we make some remarks on $\varepsilon_k(x)$'s considered as ordinary functions.

In the interval (0, 1) the functions $\varepsilon_k(x)$ are piece-wise constant, namely

$$\varepsilon_k(x) = 0,\ 1,\ 0,\ 1,\ 0,\ \ldots \text{ on the intervals } \frac{q}{2^k} \le x < \frac{q+1}{2^k} \text{ with } q = 0,\ 1,\ \ldots\ 2^k - 1. \tag{15}$$

Accordingly, the function $\varphi_k(x) \equiv 1 - 2\varepsilon_k(x)$ will take on the values $+1,\ -1,\ +1,\ -1,\ +1,\ \ldots$ on the intervals shown in (15). Thus, the $\varepsilon_k(x)$'s are closely related to the Rademacher (1928) sytem of orthogonal functions $\{\, r_0(x) = 1,\ r_k(x) = sign(\sin 2^k \pi x)\ \ (k = 1,\ 2,\ \ldots)\}$, in fact $\varphi_k(x) = r_k(x)$ for $k \neq 0$.

In order to prove the independence of the $\varepsilon_k(x)$'s in the case $a \neq 0$, by analogy from equation (10) for the distribution of the binary photons, we take

$$P_a(\varepsilon_k(x) = 1) = \frac{e^{a/2^k}}{1 + e^{a/2^k}}, \quad P_a(\varepsilon_k(x) = 0) = \frac{1}{1 + e^{a/2^k}}, \quad P_a(\varepsilon_k(x) = \delta_k) = \frac{e^{a\delta_k/2^k}}{1 + e^{a/2^k}}, \tag{16}$$

where $\delta_k = 1$ or $\delta_k = 0$. The products of such probabilities reads

$$\prod_{k=1}^{n} P_a(\varepsilon_k(x)=\delta_k) = \frac{\exp(a\sum_{k=1}^{n}\delta_k/2^k)}{\prod_{k=1}^{n}(1+e^{a/2^k})}, \qquad \delta_k = 1 \text{ or } \delta_k = 0. \tag{17}$$

Since $(1-x)(1+x)(1+x^2)\cdot\ldots\cdot(1+x^{2^s}) = 1-x^{2^{s+1}}$, by taking $x=(e^{a/2^k})$ we have

$$\frac{1-(e^{a/2^k})^{2^k}}{1-(e^{a/2^k})} = (1+(e^{a/2^k}))\cdot(1+(e^{a/2^k})^2)\cdot(1+(e^{a/2^k})^2)\cdots(1+(e^{a/2^k})^{2^{k-1}}),$$

$$\frac{1}{\prod_{k=1}^{n}(1+e^{a/2^k})} = \frac{e^{a/2^n}-1}{e^a-1}, \tag{18}$$

thus the product of the probabilities can be brought to the closed form

$$\prod_{k=1}^{n} P_a(\varepsilon_k(x)=\delta_k) = \frac{[\exp(a\sum_{k=1}^{n}\delta_k/2^k)](e^{a/2^n}-1)}{e^a-1}. \tag{19}$$

On the other hand, because $x=\sum_{k=1}^{\infty}\varepsilon_k(x)/2^k$, the relation $\sum_{k=1}^{n}\delta_k/2^k \le x < \sum_{k=1}^{n}\delta_k/2^k + 1/2^n$ means the joint event $\{\varepsilon_1(x)=\delta_1\}\cdot\{\varepsilon_2(x)=\delta_2\}\cdots\{\varepsilon_n(x)=\delta_n\}$, and since the density function (14) is in fact the derivative $f_a(x) = \frac{ae^{ax}}{e^a-1} = \frac{1}{e^a-1}\frac{d}{dx}e^{ax}$,

we may write

$$P(\varepsilon_k(x)=\delta_k; 1\le k\le n) = \int_{\sum_{k=1}^{n}\delta_k/2^k}^{\sum_{k=1}^{n}\delta_k/2^k+1/2^n} f_a(x)dx = \frac{1}{e^a-1}\left[\exp(a\sum_{k=1}^{n}\delta_k/2^k)\right](e^{a/2^n}-1). \tag{20}$$

By comparing (20) and (19) we see that the probability (20) of the products of events $\{\varepsilon_k(x)=\delta_k\}$ equals the product (19) of the probabilities of these events, i.e.

$$P(\varepsilon_1(x)=\delta_1, \varepsilon_2(x)=\delta_2, \cdots, \varepsilon_n(x)=\delta_n) = P(\varepsilon_1(x)=\delta_1)P(\varepsilon_2(x)=\delta_2)\cdots P(\varepsilon_n(x)=\delta_n). \tag{21}$$

Equation (20) shows the independence of $\varepsilon_k(x)$-s with respect to the measure whose density is $f_a(x)$, which coincides our $f_\zeta(z)$ in (4), by choosing $a=-\beta=-h\nu/kT$. On the basis of (10) and (16) our results can be summariyed as follows

$$\eta = \xi+\zeta, \quad \xi = \sum_{s=0}^{\infty}\mathrm{u}_s = \sum_{s=0}^{\infty}\mu_s(\xi)\cdot 2^s, \quad \zeta = \sum_{k=1}^{\infty}\mathrm{v}_k = \sum_{k=1}^{\infty}\varepsilon_k(\zeta)\cdot 2^{-k}, \tag{22}$$

or, by introducing a uniform notation λ_r for μ_r and ε_r

$$\eta = \sum_{r=-\infty}^{\infty}\lambda_r\cdot 2^r, \quad \lambda_r = \mu_r(\xi), \ r\ge 0 \text{ and } \lambda_r = \varepsilon_r(\zeta), \ r<0. \tag{23}$$

Equations (22) and (23) show that the energy of the chaotic radiation can be decomposed to an integer part and to a fractional part. The distribution of the integer part is just the Planck – Bose distribution (8), which can be further decomposed to the Fermi – Dirac distributions (10) of binary photons (Varró, 2007). The fractional part of $\eta=[\eta]+\{\eta\}\equiv\xi+\zeta$ has been expanded in the dyadic representation (11), whose factors, $\varepsilon_k(x)$, have been turned out to be independent random variables.

In the first case $(a=0)$ shown in equation (14) the distribution function is unity, $f_{a=0}(x)=1$ and the expectation value of the fractional part of the energy would be $h\nu\int_0^1 f_{a=0}(x)xdx = h\nu/2$, which just the *zero–point energy*. However, this energy ζ_0 cannot stem from the original chaotic distribution, because for the fractional part of that we have already derived the temperature dependent result $f_{a=-\beta}(z)$ in equation (14) and (4). For the study of distribution of the $\varepsilon_k(x)$-s in this case $(a = 0)$ one would need more sofisticated reasonings and more space, thus, we leave this discussion to a separate paper (Varró, 2012).

4. Summary

In Section 2 we summarized our results on the fractional part, $\zeta = \{\eta\}$, and the integer part, $\xi = [\eta]$, of the scaled energy η of a mode of black–body radiation, which has been represented by a Gaussian distribution. In Section 3 is we proved the independence of the binary components of the fractional part. By using a mathematical theorem, we have shown that, if the 'digital randomness' is such that the binary variables at different positions are independent, then we may recover the distribution of the fractional part of the energy of the thermal radiation mode. We have also seen that if, in addition, 'the randoness is maximal', then these digits have the same distribution, and ζ_0 has a uniform distribution, and its expectation value is just the zero–point energy. According to the present description, this cannot be associated to the photon mode, but rather, it can be a property of a resonant absorber , like the Planckian oscillators being in equilibrium with the mode. We think that the above considerations and results may have even technological relevance in the contex of physical round-off errors in data processing.

Acknowledgments
This work has been supported by the Hungarian National Scientific Research Foundation OTKA, Grant No. K 104260. Partial support by the National Development Agency, Grant No. ELI_ 09-1-2010-0010, Helios Project is also acknowledged.

References

Bogdanov Yu I, Galeev R F, Kulik S P and Moreva E V 2007 Mathematical modeling of the accuracy characteristics in problems of precision quantum tomography of biphoton states. *Optics and Spectroscopy* **103** 107-115

Chatterji S D 1963 Certain induced measures on the unit interval *J. London Math. Soc.* **38** 325-331

Feller W 1970 *An Introduction to Probability and Its Applications I.*(Wiley, New York, 1970)

Man'ko M A and Man'ko V I 2011; Probability description and entropy of classical and quantum systems. *Foundations of Physics* **41** 330-344

Oxborrow M and Sinclair A G 2005 Single–photon sources. *Contemporary Physics* **46** 173-206

Planck M 1908 Zur Dynamik bewegter Systeme *Annalen der Physik* (4) **26** 1-34

Planck M 1911 Eine neue Strahlungshypotese *Verh. d. Deutsch. Phys. Ges.* **13** 138-148

Planck M 1959 *The theory of heat radiation* (Dover Publications, Inc., New York, 1959), translation of the second edition of Planck's book by M. Masius from 1914.

Rademacher H 1922 Einige Sätze über Reihen von allgemeinen Orthogonalfunktionen *Math. Ann.* **87** 112-138

Rényi A 1970 *Probability Theory* (Akadémia Kiadó, Budapest, 1970)

Varró S 2006a Einstein's fluctuation formula. A historical overview *Fluctuation and Noise Letters* **6** R11-R46

Varró S 2006b A study on black-body radiation: classical and binary photons *Acta Phys. Hung. B: Quant. El.* **26**, 365-389

Varró S 2007 Irreducible decomposition of Gaussian distributions and the spectrum of black-body radiation *Physica Scripta* **75** 160-169

Varró S 2011 The role of self-coherence in correlations of bosons and fermions in linear counting experiments. Notes on the wave-particle duality *Fortschr. Phys.* **59** 296-324

AUTHOR INDEX

Afanasiev, K 184
Agibalov, A A 212
Ahmed, E 255
Ahrens, S........ 139
Ali, A 255
Alsalhi, M S 199, 255
Anashkina, E A 131
Andriyash, I A........ 105
Ansari, A A 199, 255
Atif, M........ 199, 255, 266
Azarm, A........ 165
Bakhmutov, D 192
Balcou, Ph........ 105
Bandrauk, A D 149
Barzin, J........ 233
Bashinov, A V........ 131
Bauke, H 139
Bienias, P 323
Biryukova, T........ 192
Bisio, A........ 390
Bissell, L J........ 86
Bobkov, P V........ 243
Bolech, C J........ 343
Borisov, G M 277
Buganov, O V 277
Bulanov, S S........ 112
Bulanov, S V 112
Carrasquilla, J........ 292
Ceuppens, S 358
Chin, S L........ 165
Cohen, O........ 100
Correggi, M 351
Dadi, A 122
Dall'Arno, M 390
D'Ariano, G M........ 390
Davuluri, S........ 76
D'Humières, E........ 105
Ding, Y J 19
Dziarmaga, J 301
Erkmen, B I........ 372
Esirkepov, T Z........ 112
Fang, B........ 100
Farisar, S M........ 224
Fedorov, A K 365, 408
Fillion-Gourdeau, F 149
Gajda, M........ 323
Gol'Dort, V G 64
Gonchukov, S 192
Gonoskov, A A........ 131
Heidrich-Meisner, F........ 343
Hernández-García, C........ 159
Hosseini, S........ 165
Ibupoto, Z H 199
Ishchenko, V N 64
Israr, M Q........ 255
Jelvani, S 224, 233
Kain, B 315
Kando, M 112
Karnaukhov, A A 288
Kato, K 402
Khun, K........ 199
Khvorostov, E B 64
Kim, A V 131
King, B 175
Kisel', V E 277
Kochubei, S A 64, 277
Koga, J K 112
Korobtsov, A........ 184
Kotova, S........ 184
Kovalyov, A A........ 277
Kuleshov, N V........ 277
Kurbakov, I L........ 365
Kuril'Chik, S V........ 277
Langer, S........ 343
Li, D........ 19
Ling, H Y........ 315
Lorenz, V O 100
Lorin, E 149
Losevsky, N........ 184
Lozovik, Y E 365
Lukishova, S G 86
Mayorova, A 184
McCulloch, I P........ 343
Moguilnaya, T Y........ 212, 243
Mollabashi, M........ 233
Morozov, V B........ 288
Müller, C........ 122, 139
Müller, T-O........ 139
Nur, O........ 255
Olenin, A N 288
Orso, G 343
Patlan, V........ 184
Pawlowski, K 323
Pazokian, H........ 224, 233
Piazza, A D........ 175
Pinsker, F 351
Pipahl, A........ 131
Plaja, L 159
Popov, A M 165
Preobrazhenskii, V V 277
Putyato, M A........ 277
Reshetov, V A 64
Rigol, M........ 292, 343
Rostovtsev, Y V........ 76
Rougerie, N 351
Rubtsova, N N........ 64, 277
Rzazewski, K........ 323
Sadaf, J R........ 255
Semyagin, B R........ 277
Shamirzaev, T S........ 277

AUTHOR INDEX

Sharifi, S M........165
Skobelev, S A........131
Sridharan, A........165
Strekalov, D V........372
Sukhinina, A........192
Tempere, J........358
Tikhomirov, S A........277
Tikhonchuk, V T........105
Tikhonova, O V........165
Tomilin, V A........243
Tomilin, V I........212
Toncian, M........131
Toncian, T........131
Tylutki, M........301
Varró, S........417
Vermeyen, E........358
Villalba-Chávez, S........139
Volkova, E A........165
Volostnikov, V........184
Wan, J........1
Wang, Q Q........165
Wang, Y........1
Weiss, C........333
Willander, M........199, 255
Willi, O........131
Winkler, J M........86
Yakovlev, D V........288
Yevseyev, I V........64
Yngvason, J........351
Yu, N........372
Yurchenko, S O........408
Zadernovsky, A A........10
Zhang, J........1
Zhao, P........19
Zhu, Y........1
Zou, B........1
Zurek, W H........301